Y

U.S. Department of Defense

Published by

Stiles Publishing, Inc.
Lexington, Kentucky

This publication contains a DoD Publication published on 11 January 2024.

Edited by Chris Stiles

Printed in the United States of America

First Printing: May 2024
Stiles Publishing, Inc.

ISBN-979-8-3249671-8-5

ATP 3-21.8

INFANTRY RIFLE PLATOON AND SQUAD

JANUARY 2024

DISTRIBUTION RESTRICTION: Approved for public release; distribution is unlimited.
This publication supersedes ATP 3-21.8, dated 12 April 2016.

Headquarters, Department of the Army

Army Techniques Publication
No. 3-21.8

Headquarters
Department of the Army
Washington, DC, 11 January 2024

INFANTRY RIFLE PLATOON AND SQUAD

Contents

DISTRIBUTION RESTRICTION: Approved for public release; distribution is unlimited.

*This publication supersedes ATP 3-21.8, dated 12 April 2016.

Figures

Tables

Preface

ATP 3-21.8 provides doctrine for the Infantry rifle platoon and squad of the Infantry rifle company against a peer threat. This publication describes relationships, organizational roles and functions, capabilities and limitations, and responsibilities within the Infantry rifle platoon and squad. Techniques, non-prescriptive ways or methods used to perform missions, functions, or tasks are discussed in this publication and are intended to be used as a guide. They are not prescriptive. This publication supersedes the techniques and employment principles of the Infantry rifle platoon and squad in ATP 3-21.8, dated 12 April 2016.

To comprehend the doctrine contained in this publication, readers must understand the tactics in FM 3-90 and FM 3-96. To comprehend how the Infantry rifle platoon organizes and is doctrinally employed, the reader must understand the techniques in ATP 3-21.10 and ATP 3-21.20.

The principal audience for ATP 3-21.8 is the commanders, staff, officers, noncommissioned officers, and Soldiers within the Infantry battalion. The audience includes the United States Army Training and Doctrine Command institutions and components. This publication serves as an authoritative reference for personnel developing doctrine, materiel and force structure, institutional and unit training, and standard operating procedures for the Infantry rifle platoon and squad.

Commanders, staffs, and subordinates must ensure that their decisions and actions comply with applicable U.S., international, and in some cases host-nation laws and regulations. Commanders at all levels will ensure that their Soldiers operate in accordance with the law of armed conflict and applicable rules of engagement. (See FM 6-27 for additional information.)

ATP 3-21.8 uses joint terms where applicable. Selected joint and Army terms and definitions appear in both the glossary and the text. Terms and acronyms for which ATP 3-21.8 is the proponent publication (the authority) are marked with an asterisk (*) in the glossary. Definitions for which ATP 3-21.8 is the proponent publication are boldfaced in the text and the term is italicized. For other definitions shown in the text, the term is italicized, and the number of the proponent publication follows the definition.

ATP 3-21.8 applies to the Active Army, the Army National Guard/the Army National Guard of the United States, and the United States Army Reserve unless otherwise stated.

The proponent of ATP 3-21.8 is the United States Army Maneuver Center of Excellence. The preparing agency is the Doctrine and Collective Training Division, Directorate of Training and Doctrine, United States Army Maneuver Center of Excellence. Send comments and recommendations on DA Form 2028, (*Recommended Changes to Publications and Blank Forms*) to Commander, Maneuver Center of Excellence, Directorate of Training and Doctrine, Doctrine and Collective Training Division, ATTN: ATZK-TDD, 1 Karker Street, Fort Moore, GA 31905-5410; by email to usarmy.moore.mcoe.mbx.doctrine@army.mil; or submit an electronic DA Form 2028.

This page intentionally left blank.

Introduction

ATP 3-21.8 addresses the tactical application of techniques associated with the offense and defense for the Infantry rifle platoon and squad. ATP 3-21.8 does not discuss defense support of civil authorities (see ADP 3-28 and ATP 3-28.1) and the element of stability (see ATP 3-21.10 chapter 4). Employing the techniques addressed in ATP 3-21.8 requires using and integrating the techniques found in ATP 3-21.10 and ATP 3-21.20 and the tactics and procedures found in FM 3-96, *tactics* are the employment, ordered arrangement, and directed actions of forces in relation to each other (ADP 3-90). Procedures are standard, detailed steps that prescribe how to perform specific tasks.

The techniques addressed in ATP 3-21.8 includes the movement and maneuver of units in relation to each other, the terrain, and the enemy. Techniques vary with terrain and other circumstances; they change frequently as the enemy reacts and friendly forces explore new approaches. Applying techniques usually entails acting under time constraints with incomplete information. Techniques always require judgment in application; they are always descriptive, not prescriptive.

Fictional scenarios, used as discussion vehicles throughout this publication, illustrate different ways an Infantry rifle platoon and squad can accomplish its mission regardless of which inherent element (offense or defense) of conventional warfare currently dominates. *Conventional warfare* is a violent struggle for domination between nation-states or coalitions of nation-states (FM 3-0). Scenarios focus on potential challenges confronting the platoon leader and platoon subordinate leaders in accomplishing their mission but are not intended to be prescriptive of how the Infantry rifle platoon performs any particular operation.

Note. These same scenarios drive the techniques used in ATP 3-21.10 and ATP 3-21.20. The scenarios focus on the techniques used to perform missions, functions, or tasks in support of the Infantry rifle company.

ATP 3-21.8 incorporates the significant changes in Army doctrinal terminology, concepts, constructs, and proven tactics developed during recent operations. It also incorporates changes based on newly published Army capstone doctrine and operational concept.

Note. The Army's operational concept is multidomain operations. *Multidomain operations* are the combined arms employment of joint and Army capabilities to create and exploit relative advantages to achieve objectives, defeat enemy forces, and consolidate gains on behalf of joint force commanders (FM 3-0). Employing Army and joint capabilities makes use of all available combat power from each domain (air, land, maritime, space, and cyberspace) to accomplish missions. Multidomain operations are the Army's contribution to joint campaigns, spanning the competition continuum. (See FM 3-0 for additional information.)

The following is a brief introduction and summary of changes by chapter:

Chapter 1 – Organization

Chapter 1 focuses on the Infantry rifle platoon and squads' role and organization, as well as the platoon's mission, capabilities, and limitations. It provides a discussion on the duties and responsibilities within the platoon and squads.

Chapter 2 – Planning and Preparing for Operations

Chapter 2 provides small-unit leaders with a framework (troop leading procedures) to analyze a mission, develop a plan, and prepare for an operation. In addition, chapter 2:

- Discusses parallel planning.
- Addresses the intelligence preparation of the operational environment.
- Describes mission command as the Army's approach to command and control.
- Provides additional information on the mission variables of METT-TC (I) mission, enemy, terrain and weather, troops and support available, time available, civil considerations, and integrated into the other variables, informational considerations.
- Discusses mission orders and disciplined initiative.

Chapter 3 – Movement and Maneuver

Chapter 3 discusses considerations for tactical movement and maneuver. In addition, chapter 3 discusses route selection, navigation and execution, movement formations, movement techniques, and movement during limited visibility.

Chapter 4 – Offense

Chapter 4 discusses offensive actions to defeat, destroy, or neutralize the enemy. The chapter addresses the characteristics of the offense and describes the four offensive operations (discussion mainly focuses on movement to contact and attack). Chapter 4 also discusses—

- Forms of maneuver (flank attack removed as a form; frontal assault now frontal attack).
- Offensive control measures and the tactical framework of the offense.

- Common offensive planning considerations.
- Operations during limited visibility.

Chapter 5 – Defense

Chapter 5 discusses defensive actions to defeat enemy attacks, gain time, control key terrain, protect critical infrastructure, secure the population, and economize forces. The chapter addresses the characteristics of the defense and describes the three defensive operations (discussion mainly focuses on area defense and retrograde). Chapter 5 also discusses—

- Defensive control measures and tactical framework of the defense.
- Common defensive planning considerations.
- Variations of an area defense and position selection.
- Engagement area development.
- Fighting position construction.

Chapter 6 – Tactical Enabling Operations and Activities

Chapter 6 discusses tactical enabling operations and activities in support of offensive and defensive operations. These operations and activities include breaching, relief in place, passage of lines, linkup, tactical deception, security, security operations, and troop movement.

Chapter 7 – Patrols and Patrolling

Chapter 7 provides an overview of patrolling and the patrols conducted by the Infantry rifle platoon and squad. It discusses in detail combat and reconnaissance patrols. Chapter 7 also discusses—

- Common patrolling considerations, leader's reconnaissance, and patrol base activities.
- Planning and preparation, and coordination measures.
- Essential and supporting tasks to patrolling.
- Pre- and post-departure activities.
- Post patrol activities.

Chapter 8 – Sustainment

Chapter 8 discusses the process that the platoon leader and platoon sergeant use to anticipate the needs of the Infantry rifle platoon. In addition, chapter 8 addresses—

- Sustainment responsibilities and relationships within the platoon.
- Planning, preparation, execution, and assessment for:
  - Supply and field services.
  - Distribution and resupply operations.
  - Maintenance.
- Tactical combat casualty care.

Appendixes

Five appendixes complement the body of this publication. They include:

- Appendix A: Direct Fires.
- Appendix B: Fire Support Planning.
- Appendix C: Machine Gun Employment and Theory.
- Appendix D: Shoulder-Launched Munitions and Close Combat Missile System.
- Appendix E: Battle Drills.

Chapter 1
Organization

The primary mission of the Infantry rifle platoon and squad is to close with the *enemy*—party identified as hostile against which the use of force is authorized (ADP 3-0)—by means of fire and movement to destroy, capture, or repel an assault by fire, close combat, and counterattack. *Fire and movement* is the concept of applying fires from all sources to suppress, neutralize, or destroy the enemy, and the tactical movement of combat forces in relation to the enemy (as components of maneuver, applicable at all echelons). At the squad level, fire and maneuver entails a team placing suppressive fire on the enemy as another team moves against or around the enemy (FM 3-96). To succeed, Infantry rifle platoons and squads are aggressive, physically fit, disciplined, and well trained. The inherent strategic mobility of Infantry units dictates a need to be prepared for rapid deployment in response to situations in different operational environments. This chapter focuses on the Infantry rifle platoon and squads' role as well as the exercise of command and control within the platoon and squad.

SECTION I – ROLE OF THE INFANTRY RIFLE PLATOON AND SQUAD

1-1. The Infantry rifle platoon and squad as part of the Infantry rifle company within the Infantry battalion is organized to conduct *combined arms*—the synchronized and simultaneous application of arms to achieve an effect greater than if each element was used separately or sequentially (ADP 3-0)—operations. An *operation* is a sequence of tactical actions with a common purpose or unifying theme (JP 1, Vol 1). The Infantry rifle platoon and squad can deploy rapidly and execute missions throughout the range of military operations. The rifle platoon and squad conduct effective combat or other operations immediately upon arrival in an operational area. This section addresses the mission, capabilities, and limitations, and the internal organization of the Infantry rifle platoon and squad.

Note. ATP 3-21.8 does not address defense support of civil authorities. (See ADP 3-28 and ATP 3-28.1 for information.) It does not address stability operations tasks. (See ATP 3-21.10 and ATP 3-21.20 for information.)

MISSION, CAPABILITIES, AND LIMITATIONS

1-2. The mission of the Infantry rifle platoon is to close with the enemy using fire and movement to destroy or capture enemy forces, or to repel enemy attacks by fire, close combat, and counterattack to control land areas, including populations and resources. The Infantry rifle platoon leader exercises command and control and directs the operation of the platoon and attached units while conducting combined arms warfare throughout the depth of the platoon's area of operations (AO). Platoon missions, although not inclusive, may include reducing fortified areas, infiltrating and seizing objectives in the enemy's rear, eliminating enemy force remnants in restricted terrain, securing key facilities and activities, and conducting operations in support of stability operations tasks in the wake of maneuvering forces. Reconnaissance and surveillance operations and security operations remain a core competency of the Infantry rifle platoon and squad.

1-3. The following lists capabilities of the Infantry rifle platoon and squad:

- Offensive and defensive operations in all types of environments, day and night.
- Seize, secure, occupy, and retain terrain.
- Destroy, neutralize, suppress, interdict, disrupt, block, canalize, and fix enemy forces.
- Breach enemy obstacles.
- Feint and demonstrate to deceive the enemy.
- Screen and guard friendly units.
- Reconnoiter, deny, bypass, clear, contain, and isolate. These tasks might be oriented on terrain and enemy.
- Small-unit operations.
- Air assault operations.
- Airborne operations (airborne units only).
- Operate with mounted forces.
- Operate with special operations forces.
- Amphibious operations.

1-4. The Infantry rifle platoon and squad has the following limitations:

- Limited vehicle mobility, the foot speed of organic elements may establish the pace of operations.
- Vulnerable to enemy armor, artillery, and air assets when employed in open terrain.
- Vulnerable to enemy chemical, biological, radiological, and nuclear (CBRN) attacks with limited decontamination capability.

ORGANIZATION OF THE INFANTRY RIFLE PLATOON

1-5. The Infantry rifle platoon and its squads (see figure 1-1) can be task organized alone or as a combined arms force based upon the mission variables of METT-TC (I) mission, enemy, terrain and weather, troops and support available, time available, civil considerations, and informational considerations. The platoon's effectiveness increases through the synergy of combined arms including tanks, Bradley fighting vehicles and Stryker Infantry carrier vehicles, engineers, and other support elements. The Infantry rifle platoon and squad as a combined arms force can capitalize on the strengths of the team's elements while minimizing their limitations.

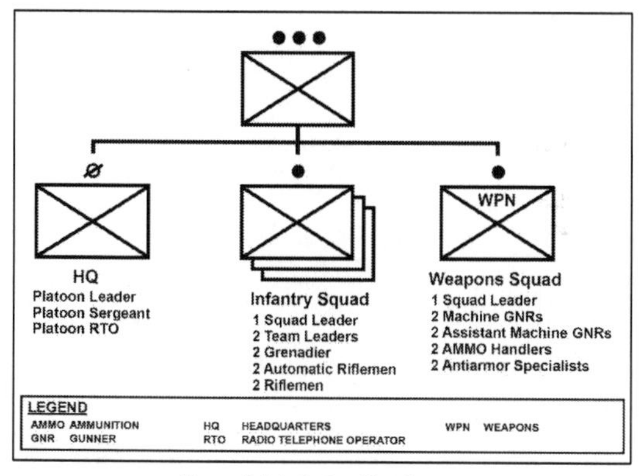

Figure 1-1. Infantry rifle platoon

1-6. Infantry units can operate in all terrain and weather conditions. They might be the dominant force because of rapid strategic deployment. In such cases, they can take and gain the initiative early, seize and retain or control terrain, and mass fires to stop the enemy. Infantry units are particularly effective in urban terrain, where they can infiltrate and move rapidly to the rear of enemy positions. The leader can enhance their mobility by using rotary-wing and fixed-wing aircraft.

1-7. The fundamental considerations for employing Infantry units result from the missions, types, equipment, capabilities, limitations, and organization of units. Other capabilities result from a unit's training program, leadership, discipline, morale, personnel strengths, and many other factors. These other capabilities constantly change based on the current situation.

INFANTRY RIFLE SQUAD

1-8. The Infantry rifle squad's primary role is as a maneuver or base-of-fire element. While the platoon's task organization may change, the Infantry rifle squad's organization generally remains standard. (See figure 1-2 on page 1-4.)

1-9. The squad is comprised of two fire teams and a squad leader. It can establish a base of fire, providing security for another element, or conducting fire and movement with one team providing a base of fire, while the other team moves to the next position of advantage or onto an objective. The squad leader has two subordinate leaders to lead the two teams, freeing the squad leader to control the entire squad.

1-10. Fire teams are the fighting element within the Infantry platoon. Infantry platoons and squads succeed or fail based on the actions of their fire teams.

1-11. Fire teams are designed as a self-contained team. The automatic rifleman provides an internal base of fire with the ability to deliver sustained suppressive small arms fire on area targets. The rifleman provides accurate, lethal direct fire for point targets. The rifleman may be issued a shoulder-launched munition (known as SLM). The grenadier provides high explosives (HEs) indirect fire for both point and area targets. A team leader leads the team by example.

Note. The *combat load*—the minimum mission-essential equipment and supplies as determined by the commander responsible for carrying out the mission, required for Soldiers to fight and survive immediate combat operations (FM 4-40)—for SLMs is two per rifle squad. Either two M72-series light antitank (AT) weapon, M136-series antitank (AT4), M141 bunker defeat munitions (known as BDMs), or a combination of types with each rifleman normally being issued one.

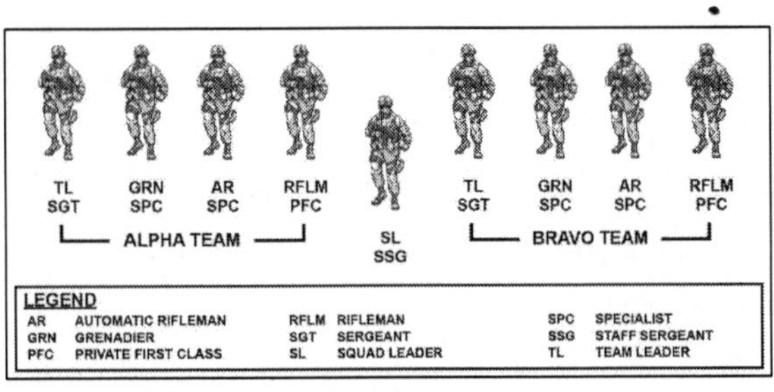

Figure 1-2. Infantry rifle squad

INFANTRY WEAPONS SQUAD

1-12. The Infantry weapons squad provides the primary base of fire for the platoon's maneuver. It is comprised of two medium machine gun teams, two medium close combat missile system (known as CCMS) Javelin teams, and a weapons squad leader. (See figure 1-3.)

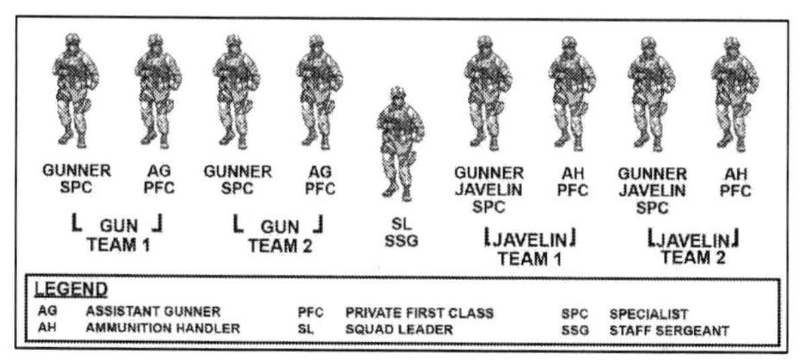

Figure 1-3. Infantry weapons squad

Medium Machine Gun Team

1-13. The two-Soldier medium machine gun team is comprised of a gunner and an assistant gunner. The weapons squad has two medium machine gun teams. These teams provide the platoon with medium-range area suppression at ranges up to 1,100 meters during day, night, and adverse weather conditions. (See appendix C for additional information.)

Close Combat Missile System-Javelin Team

1-14. The two-Soldier CCMS-Javelin team is comprised of a gunner and an ammunition handler. The weapons squad has two Javelin teams. This system provides the platoon with an extremely lethal fire-and-forget, man-portable, direct- and top-attack capability to defeat enemy armored vehicles and destroy fortified positions at ranges up to 2,000 meters. The Javelin has proven effective during day, night, and adverse weather conditions. When the Javelin is replaced with the M3 Multi-role, Antiarmor, Antipersonnel Weapon System (known as MAAWS) the team employs the M3 to destroy enemy personnel, field fortifications, and enemy vehicles. It engages lightly armored targets at ranges up to 700 meters, and soft-skinned vehicles and similar targets at ranges up to 1,300 meters. (See appendix D for additional information.)

INFANTRY RIFLE FORMATIONS

1-15. Infantry rifle formations (specifically the rifle platoon and squad, and rifle company) of the Infantry battalion (of the Infantry brigade combat team [IBCT]) are task-organized alone or as a combined arms force based upon the mission variables of METT-TC (I). Their effectiveness increases through the synergy of combined arms including assault platoon vehicles of the Infantry weapons company, tanks, Bradley fighting vehicles and Stryker Infantry carrier vehicles, engineers, and other enabling elements. Effective application of the Infantry rifle formations as a combined arms force can capitalize on the strengths of the team's elements while minimizing their respective limitations. For example, the Infantry rifle company (see ATP 3-21.10) of the Infantry battalion has a headquarters section, three rifle platoons, a mortar section, and an unmanned aircraft system team. Habitual attachments to the Infantry rifle company include a fire support team (FIST) and combat medics. Attachments from the Infantry battalion may include elements from the Infantry weapons company (specifically, the tube launched, optically tracked, wire guided missiles) as well as other elements (see ATP 3-21.20 appendix D). Figure 1-4 shows the internal organization of the Infantry rifle company within the Infantry battalion.

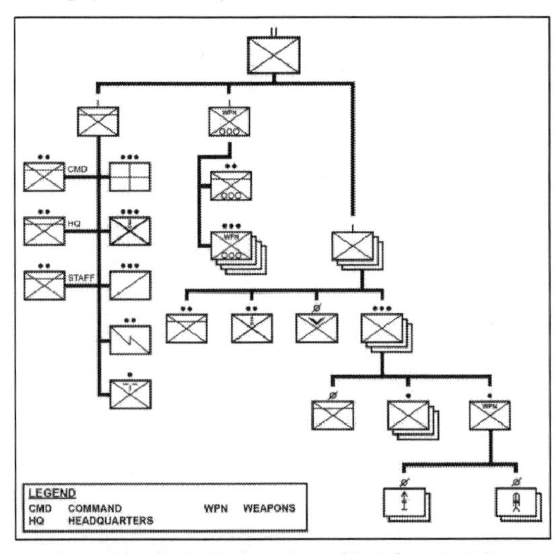

Figure 1-4. Infantry rifle formations of the Infantry battalion

SECTION II – THE EXERCISE OF COMMAND AND CONTROL

1-16. The Infantry rifle platoon and squad within the Infantry rifle company can deploy rapidly and can be sustained by an austere support structure. It conducts operations against conventional and unconventional enemy forces in all types of terrain and climate conditions. The platoon's composition and training uniquely equip it to conduct its mission. In addition to its primary warfighting missions, the rifle platoon and squad may be tasked to perform other types of operations semi-independently or as an integral part of a larger force. This section addresses the exercise of command and control, specifically the duties and responsibilities, and habitual attachments of the Infantry rifle platoon and squad.

DUTIES AND RESPONSIBILITIES

1-17. The duties and responsibilities inherent within the Infantry rifle platoon and squad enable the exercise of command and control throughout the conduct of operations. *Command and control* is the exercise of authority and direction by a properly designated commander over assigned and attached forces in the accomplishment of the mission (JP 1, Vol 2). The *command and control warfighting function* is the related tasks and a system that enable commanders to synchronize and converge all elements of combat power (ADP 3-0). *Combat power* is the total means of destructive and disruptive force that a military unit/formation can apply against an enemy at a given time (JP 3-0).

Note. Combat power is the ability to fight. The complementary and reinforcing effects that result from synchronized operations that yield a powerful blow that overwhelms enemy forces and creates friendly momentum. Army forces deliver that blow through a combination of five dynamics. The dynamics of combat power are leadership, firepower, information, mobility, and survivability. (See FM 3-0 for additional information on the five dynamics.)

1-18. Leadership is the most essential dynamic of combat power. *Leadership* is the activity of influencing people by providing purpose, direction, and motivation to accomplish the mission and improve the organization (ADP 6-22). It is the multiplying and unifying dynamic of combat power, and it represents the qualitative difference between units. Leadership drives command and control but is also dependent upon it. The collaboration and shared understanding inherent in the operations process (see chapter 2) prepare leaders for operations, expand shared understanding, hone leader judgment, and improve the flexibility that leaders apply to the other dynamics of combat power against enemy forces. Leadership is a critical component that allows unit leaders to carry out their duties and responsibilities.

1-19. During operations, no amount of technology or equipment can take the place of competent leadership. The leaders within the platoon and squad lead through a combination of personal example, persuasion, and compulsion. The following paragraphs address the duties and responsibilities of the key personnel common to the Infantry rifle platoon. (See paragraphs 8-2 to 8-5 for sustainment specific responsibilities within the platoon and squad.)

Note. The duties and responsibilities of leaders and platoon members must be executed even in the absence of a particular leader to ensure mission accomplishment in accordance with the commander's intent. A *mission* is the essential task or tasks, together with the purpose, that clearly indicates the action to be taken and the reason for the action (JP 3-0). The *commander's intent* is a clear and concise expression of the purpose of an operation and the desired objectives and military end state (JP 3-0). The *end state* is the set of required conditions that defines achievement of the commander's objectives (JP 3-0). (See ATP 3-21.10 for additional information.)

HEADQUARTERS SECTION

1-20. The headquarters section provides the platoon with command and control, and communications. It consists of the platoon leader, the platoon sergeant, and the platoon radiotelephone operator (known as RTO).

Platoon Leader

1-21. The platoon leader leads Soldiers by personal example and is responsible for accountability of platoon members, assigned equipment, all the platoon does or fails to do, and having complete authority over the subordinates. This centralized authority enables the platoon leader to maintain unit discipline, unity, and to act decisively. The platoon leader must be prepared to exercise initiative within the company commander's intent and without specific guidance for every situation. The platoon leader knows the Soldiers of the platoon, how to employ the platoon, its weapons, and its systems. Relying on the expertise of the platoon sergeant, the platoon leader regularly consults with the platoon sergeant on all platoon matters. During operations, the platoon leader—

- Leads the platoon in supporting the higher headquarters missions. Bases the platoon's actions on the assigned mission and intent and concept of the higher commander(s).
- Conducts troop leading procedures (TLP).
- Maneuvers squads and fighting elements.
- Synchronizes the efforts of squads.
- Looks ahead to the next "move" of the platoon.
- Plans and controls employment of direct and indirect fires.
- Plans the emplacement and employment of key weapon systems and ensures they are integrated to support the platoon's maneuver.
- Knows weapon effects, surface danger zones, and risk estimate distances for the platoon's weapon systems and supporting fires to synchronize the platoon's direct and indirect fires with movement.
- Requests, controls, and synchronizes supporting assets.
- Employs command and control systems available to the squads and platoon.
- Confirms that squad leaders are maintaining coverage of assigned primary and secondary sectors of fire in three dimensions.

- Confirms the emplacement and employment of the platoon's key weapons with the weapons squad leader.
- Issues accurate and timely reports.
- Positions where best to control and accomplish the mission.
- Assigns clear tasks and purposes to the squads.
- Understands the mission and commander's intent two levels up (company and battalion).
- Receives on-hand status reports from the platoon sergeant and squad leaders throughout the operation.
- Coordinates and assists in the development and implementation of the obstacle plan.
- Oversees and is responsible for property management.
- Works to develop and maintain situational awareness, and to the degree possible, situational understanding through the product of four elements:
  - Attempts to know what is happening in present terms of friendly, enemy, neutral, and terrain situations.
  - Knows the end state representing mission accomplishment.
  - Determines the critical actions and events occurring to move the unit from the present to the end state.
  - Assesses the risk throughout the operation.

Platoon Sergeant

1-22. The platoon sergeant is the platoon's most experienced noncommissioned officer and second-in-charge, accountable to the platoon leader for leadership, discipline, training, maintenance and accountability of assigned equipment, and welfare of the platoon's Soldiers. The platoon sergeant sets the example in everything, advises the platoon leader, and assists the platoon leader by upholding standards and platoon discipline. The platoon sergeant's expertise includes tactical maneuver, employment of weapons and systems, sustainment, administration, security, accountability, protection, and Soldier care. As the second-in-charge, the platoon sergeant assumes no formal duties except those prescribed by the platoon leader. However, the platoon sergeant traditionally—

- Ensures the platoon is prepared to accomplish its mission, which includes supervising precombat checks (PCCs), precombat inspections (PCIs), and rehearsals.
- Updates platoon leader on appropriate reports and forwards reports needed by higher headquarters.
- Prepares to assume the role and responsibilities of the platoon leader.
- Takes charge of task-organized elements in the platoon during tactical operations, which may include but is not limited to, quartering parties, special purpose teams, and security patrols.
- Monitors the morale, discipline, and health of the platoon.
- Positions where best needed to help the engagement (either with the assault element or base of fire element).

- Receives squad leaders' administrative, logistical, and maintenance reports.
- Requests logistical support from the higher headquarters, and usually coordinates with the company's first sergeant or executive officer.
- Ensures Soldiers maintain all equipment.
- Ensures ammunition and supplies are properly and evenly distributed after the platoon consolidates on the objective and while the platoon reorganizes.
- Manages the unit's combat load prior to operations and monitors logistical status during operations.
- Establishes and operates the unit's casualty collection point (CCP). This includes directing the platoon combat medic and aid/litter teams in moving casualties, maintaining platoon strength level information, consolidating and forwarding the platoon's casualty reports, and receiving and orienting replacements.
- Employs the available digital command and control systems to the squads and platoon.
- Ensures Soldiers distribute supplies according to the platoon leader's guidance and direction.
- Accounts for Soldiers, equipment, and supplies.
- Coaches, counsels, and mentors squad leaders and team leaders.
- Upholds standards and platoon discipline.
- Understands the mission and commander's intent two levels up (company and battalion).
- Assists with preparing the operation order (OPORD) (for example, determining Soldier load and for example, echeloning loads, load determination, and load tailoring [see ATP 3-21.18]).
- Assists with preparation for rehearsals and priorities of work.

Platoon Radiotelephone Operator

1-23. The RTO primarily is responsible for communication with its controlling headquarters (usually the company). During operations, the RTO—

- Establishes and maintains platoon communications with higher headquarters and subordinate elements; conducts regular radio checks and immediately informs platoon leaders about any change in communications status.
- Conducts radio checks with higher according to unit standard operating procedures (SOPs). If radio contact cannot be made as required, the RTO informs the platoon sergeant or platoon leader.
- Acts as an expert in radio procedures and report formats such as close air support (CAS), call for mortar and artillery fire, medical evacuation (MEDEVAC), or casualty evacuation (CASEVAC).
- Acts as an expert in different types of field expedient antennas.
- Maintains the frequencies and call signs on self in a location known to all Soldiers in the platoon.
- Assists the platoon leader with information management.

- Assists the platoon leader and platoon sergeant employing digital command and control systems with individual squads and platoon.
- Determines combat load prior to operations and manages battery utilization during operations.
- Assists with preparing the OPORD (for example, developing the primary, alternate, contingency, and emergency [known as PACE]) communications plan.
- Assists with preparation for rehearsals (for example, building the platoon sand table).
- Maintains physical security of communications security devices.

RIFLE SQUADS

1-24. The squad leader and two team leaders provide the squad with command and control. Additionally, the squad has two rifleman, two automatic rifleman, and two grenadiers.

Squad Leader

1-25. The squad leader directs team leaders and leads by personal example. The squad leader is responsible for the accountability of squad members, has authority over subordinates, and overall responsibility of those subordinates' actions. Centralized authority enables the squad leader to act decisively while maintaining troop discipline and unity. Under the fluid conditions of close combat, the squad leader accomplishes assigned missions without constant guidance from higher headquarters.

1-26. The squad leader is the senior Infantry Soldier in the squad and is responsible for everything the squad does or fails to do. The squad leader is responsible for the care of the squad's Soldiers, weapons, discipline, and equipment, and leads the squad through two team leaders. During operations, the squad leader—

- Acts as the subject matter expert on all squad-level battle drills and individual Soldier tasks.
- Acts as the subject matter expert for the squad's organic weapons employment and employment of supporting assets.
- Knows weapon effects, surface danger zones, and risk estimate distances for munitions used by and with the squad.
- Uses control measures for direct fire, indirect fire, and tactical movement effectively.
- Controls the movement of the squad and its rate and distribution of direct fires.
- Controls the employment of indirect fires within the squad (including call for and adjust fire).
- Fights the close fight by fire and movement with two fire teams and available supporting weapons.
- Selects the fire team's general location and initial sector of fires.

- Communicates timely and accurate situation reports and status reports including—
  - Size, activity, location, unit, time, and equipment (SALUTE) spot reports.
  - Status to the platoon leader (including squad location and progress, enemy situation, enemy killed in action, and security posture).
  - Status of ammunition, casualties, and equipment to the platoon sergeant.
- Employs digital command and control systems available to the squad and platoon.
- Ensures Soldiers maintain all equipment and account for equipment and supplies.
- Operates in all environments.
- Conducts TLP.
- Coaches, counsels, and mentors team leaders.
- Assumes duties as the platoon sergeant or platoon leader as required.
- Understands the mission and commander's intent two levels up (platoon and company).

Team Leader

1-27. The team leader leads the rifle team members by personal example, is responsible for accountability of team members, and has authority over subordinates and overall responsibility of their actions. Centralized authority enables the team leader to maintain discipline and unity and to act decisively. Under the fluid conditions of close combat, the team leader accomplishes assigned missions using initiative without needing constant guidance from higher headquarters.

1-28. The team leader's position on the battlefield requires immediacy and accuracy in all performed actions and is a fighting leader who leads by example. The team leader is responsible for all the team does or fails to do, and is responsible for caring of the team's Soldiers, discipline, weapons, and equipment. During operations, the team leader—

- Acts as a subject matter expert for all the team's weapons and duty positions and all squad battle drills and individual Soldier tasks.
- Leads the team in fire and movement.
- Controls the movement of the team and its rate and distribution of fire.
- Employs digital command and control systems available to the squad and platoon.
- Ensures the security of the team's AO.
- Assists the squad leader as required.
- Prepares to assume the duties of squad leader and platoon sergeant.
- Enforces field discipline and preventive medicine measures.
- Manages the team's combat load and its available classes of supply as required.
- Understands the mission two levels up (squad and platoon).

Grenadier

1-29. The grenadier is equipped with a 40-millimeter (mm) grenade launcher. The grenadier provides the fire team with a high trajectory and an HE capability. Grenadier fires enable the fire team to achieve complementary effects with high trajectory, HE munitions, and flat trajectory ball ammunition from the team's weapons. The grenade launcher allows the grenadier to produce three effects: suppress and destroy enemy Infantry and lightly armored vehicles with HE or high explosive dual purpose (known as HEDP); provide obscurants to screen and cover the squad's fire and movement; and employ illumination rounds to increase the squad's visibility and mark enemy positions. The grenadier—

- Accomplishes the same tasks of the rifleman.
- Engages targets with appropriate type of rounds both day and night.
- Covers dead space and denies dead space to the enemy.
- Identifies 40-mm rounds by shape and color and knows how to employ each type of round and its minimum safety constraints.
- Knows the maximum ranges for each type of target of the grenade launcher.
- Knows the leaf sight increments without seeing the markings.
- Knows how to make an adjustment from the first round fired so a second-round hit can be attained.
- Loads the grenade launcher quickly in all firing positions and while running.
- Prepares to assume the duties of the automatic rifleman and team leader.
- Understands the mission two levels up (squad and platoon).

Note. In some cases, the team leader will assume the primary weapon system and responsibilities of the grenadier in order to mitigate for a missing team member or create an extra rifleman to assume additional duties within the squad.

Automatic Rifleman

1-30. The automatic rifleman's primary weapon is the light machine gun. The automatic rifleman provides the unit with a high volume of sustained suppressive direct fires of area targets. The automatic rifleman employs the weapon system to suppress enemy Infantry and bunkers, destroy enemy automatic rifle and AT teams, and enable the movement of other teams and squads. The automatic rifleman is normally the senior Soldier of the fire team and must—

- Be able to accomplish all tasks of the rifleman and grenadier.
- Be prepared to assume the duties of team leader and squad leader.
- Be able to engage groups of enemy personnel, thin-skinned vehicles, bunker doors or apertures, and suspected enemy locations with automatic fire.
- Be able to provide suppressive fire on these targets so other team members can close with and destroy the enemy.
- Be familiar with field expedient firing aids to enhance the effectiveness of assigned weapon (an example is aiming stakes).

- Be able to engage targets from the prone, kneeling, and standing positions with and without night observation devices.
- Understand the mission two levels up (squad and platoon).

Rifleman

1-31. The rifleman provides the baseline standard for all Infantry Soldiers and is an integral part of the fire team. The rifleman is an expert in handling and employing the weapon and placing well-aimed fire on the enemy. Additionally, the rifleman must—

- Be an expert on assigned weapon system, its optics, and its laser-aiming device, and is effective with this weapon system day or night.
- Be capable of engaging all targets with well-aimed shots (specifically through correct positioning on the battlefield and marksmanship proficiency).
- Employ all weapons of the squad, as well as common munitions.
- Construct and occupy a hasty firing position and know how to fire from it. Know how to occupy covered and concealed positions in all environments and what protection they provide from direct fire weapons and is competent in the performance of these tasks while using night vision devices.
- Fight as part of the team, which includes proficiency in individual tasks and drills.
- Know the duties of team members and is prepared to fill in with their weapons, if needed (for example, assume the duties of the automatic rifleman and team leader).
- Contribute as a member of special teams, including detainee search, aid/litter, and demolitions and wire/mine breach teams.
- Inform team leader of everything heard and seen when in a tactical situation.
- Perform individual preventive medical measures.
- Administer buddy aid as required.
- Manage supplied food, water, and ammunition during operations.
- Understand the mission two levels up (squad and platoon).

WEAPONS SQUAD

1-32. The weapons squad leader provides the squad with command and control. The squad has two machine gunners, two assistant machine gunners, two Javelin (or M3) gunners, and two Javelin (or M3) ammunition handlers. The weapons squad provides the platoon with responsive crew served weapons and an antiarmor capability under the platoon leader's direct control. The M3 MAAWS, when replacing the Javelin, provides the platoon with an antiarmor and antipersonnel capability. The team employs the M3 to destroy enemy personnel, field fortifications, and enemy vehicles.

Weapons Squad Leader

1-33. The weapons squad leader leads the squad's teams by personal example. The weapons squad leader is responsible for the accountability of squad members and has complete authority over squad subordinates and overall responsibility for those subordinates' actions. This centralized authority enables the weapons squad leader to act decisively while maintaining troop discipline and unity. Under the fluid conditions of modern warfare, the weapons squad leader accomplishes assigned missions using disciplined initiative without needing constant guidance from higher headquarters.

1-34. The weapons squad leader is usually the senior squad leader, second only to the platoon sergeant, and performs all the duties of the rifle squad leader. In addition, the weapons squad leader—

- Acts as the subject matter expert on all squad battle drills, crew drills, and individual Soldier tasks.
- Acts as the subject matter expert for the squad's organic weapons employment and employment of supporting assets.
- Knows weapon effects, surface danger zones, and risk estimate distances for munitions used by and with the squad.
- Controls fires and establishes fire control measures.
- Monitors ammunition expenditure.
- Recommends medium machine gun and Javelin-CCMS (or M3 MAAWS) employment to the platoon leader, for example:
  - Coordinates directly and plans accordingly with the platoon leader in placement of the medium machine gun to best cover (base-of-fire effect) the final protective line (FPL) in the defense, *overwatch*—a task that positions an element to support the movement of another element with immediate fire (ATP 3-21.10)—positions during movement techniques, and support by fire positions in the offense (see appendix C).
  - Coordinates directly and plans accordingly with the platoon leader in placement of the Javelin CCMS (or M3 MAAWS) to best cover armored avenues of approach in the defense, overwatch positions during movements techniques, and *attack by fire*—a tactical mission task using direct and indirect fires to engage an enemy from a distance (FM 3-90)—positions in the offense (see appendix D).
- Employs command and control systems available to the squad and platoon.
- Performs the role of the platoon sergeant as required.
- Conducts TLP.
- Understands the mission and commander's intent two levels up (platoon and company).

Machine Gunner

1-35. The gunner is normally the senior member of the medium machine gun team. During operations, the gunner—

- Is responsible for the assistant gunner and all the gun equipment.
- Is responsible for putting the gun in and out of action.
- Acts as the subject matter expert for information contained in TC 3-22.240, specifically information about the weapon, aiming devices, attachments, and employment/engagement.

- When attached to a rifle squad, is the subject matter expert for employment of the medium machine gun and advises the rifle squad leader of the best way to employ the medium machine gun.
- Enforces field discipline while the gun team is employed tactically.
- Knows the ballistic effects of the weapon on all types of targets.
- Assists the weapons squad leader and is prepared to assume the responsibilities.
- Understands the mission two levels up (squad and platoon).

Assistant Machine Gunner

1-36. The assistant gunner is the second member of the gun team. The assistant gunner is prepared to assume the gunner's role in any situation. During operations, the assistant gunner—

- Provides a supply of ammunition to the gun when employed.
- Spots rounds and reports recommended corrections to the gunner.
- Constantly updates the weapons squad leader on the round count and serviceability of the medium machine gun.
- Watches for Soldiers to the flanks of the target area or between the gun and target.
- Obtains ammunition from other Soldiers who are carrying 7.62-mm machine gun ammunition.
- Immediately assumes the role of gunner if the gunner is unable to continue duties.
- Understands the mission two levels up (squad and platoon).

Javelin Gunner

1-37. The gunner is normally the senior member of the Javelin CCMS (M3 MAAWS when replacing the Javelin) team. During operations, the gunner—

- Is responsible for the ammunition handler and all the missile equipment.
- Is responsible for putting the Javelin (or M3) in and out of action.
- Acts as the subject matter expert for information contained in TC 3-22.37, specifically information about the missile, launch tube assembly, battery coolant unit, command launch unit (known as CLU), and employment/engagement. (See TC 3-22.84 for information on the M3.)
- When attached to a rifle squad, is the subject matter expert for employment of the Javelin (or M3) and advises the rifle squad leader of the best way to employ the missile.
- Enforces field discipline while the Javelin (or M3) team is employed tactically.
- Knows the ballistic effects of the missile on all types of targets.
- Assists the weapons squad leader and is prepared to assume the responsibilities.
- Understands the mission two levels up (squad and platoon).

Javelin Ammunition Handler

1-38. The ammunition handler is the second member of the Javelin CCMS (M3 when replacing the Javelin) team. The ammunition handler is prepared to assume the gunner's role in any situation. During operations, the ammunition handler—

- Provides a supply of missile rounds to the gun when employed.
- Spots rounds and reports recommended corrections to the gunner.
- Constantly updates the weapons squad leader on the round count and serviceability of the Javelin missile system (or M3 MAAWS).
- Watches for Soldiers to the flanks of the target area or between the Javelin (or M3) and target.
- Obtains ammunition (when carried) from other Soldiers within the platoon.
- Immediately assumes the role of gunner if the gunner is unable to continue duties.
- Understands the mission two levels up (squad and platoon).

ADDITIONAL DUTIES

1-39. Additional duties assigned within the platoon normally include the assignment of squad-designated marksmen and CLSs with aid bag within each squad. These additional duties and the numbers required are dependent on the mission variables of METT-TC (I).

Squad-Designated Marksman

1-40. The squad-designated marksman (known as SDM) employs an optically enhanced general-purpose weapon. The SDM receives training available within the unit's resources to improve the squad's precision engagement capabilities at short and medium ranges.

1-41. A rifleman may be assigned as the SDM. The SDM is chosen for the rifleman's demonstrated shooting ability, maturity, reliability, good judgment, and experience. The SDM must be able to execute the entire range of individual and collective rifleman tasks within the squad.

1-42. The SDM is not the squad sniper, though the SDM is a fully integrated member of the rifle squad and provides an improved capability for the rifle squad. The SDM does not operate as a semi-autonomous element on the battlefield as a sniper, nor does the SDM routinely engage targets at the extreme ranges common to snipers. (See TC 3-22.9 for additional information.)

Combat Lifesaver

1-43. The CLS is a nonmedical Soldier trained to provide appropriate tactical combat casualty care (TCCC) skills beyond the level of self-aid or buddy aid. The CLS is not intended to take the place of medical personnel. Using specialized training, the CLS can slow deterioration of a wounded Soldier's condition until treatment by medical personnel is possible. Each certified CLS is issued a CLS aid bag and should have a laminated quick reference GTA 08-01-004 (*MEDEVAC Request Card).* Whenever possible, the platoon leader ensures each fire team includes at least one CLS. (See chapter 8 section II for additional information on CLSs.) The CLS—

- Ensures that the squad CLS bags and litters are properly packed and stored.
- Identifies Class VIII shortages to the platoon medic.
- Applies appropriate TCCC skills for injuries and participates in all manual and litter-carry drills.
- Uses appropriate TCCC skills in the field until casualties can be evacuated or medical personnel take over.
- Assists medical personnel in providing TCCC and preparing patients for evacuation.
- Knows the location of the CCP and the unit's tactical SOP for establishing it.
- Initiates DD Form 1380 (*Tactical Combat Casualty Care [TCCC] Card*), when required (see paragraph 8-52).

Note. Leaders within the platoon should make the CLS program a training priority.

HABITUAL ATTACHMENTS

1-44. Habitual attachments for the Infantry rifle platoon normally include a platoon forward observer (FO) team and a platoon combat medic. These habitual attachments are normally attached whenever the platoon deploys.

FORWARD OBSERVER

1-45. The FO, along with FO's fire support RTO, is the platoon subject matter expert on indirect fires planning and execution. The FO advises the platoon leader and subordinate leader on the employment and execution for all fire support assets (if assigned), including company mortars, battalion mortars, field artillery, and other allocated fire support assets. The FO is responsible for locating targets and calling for and adjusting indirect fires. The platoon FO team also knows the mission and concept of operations, specifically the platoon's scheme of maneuver and concept of fires. The FO also—

- Informs the company FIST of the platoon's situation, location, and indirect fire support requirements.
- Prepares and uses maps, overlays, and terrain sketches.
- Calls for and adjusts indirect fires.
- Operates as a team with the fire support RTO.
- Selects targets to support the platoon's mission.
- Selects observation posts and movement routes to and from selected targets.
- Operates digital message devices and maintains communication with the company and battalion fire support officers.

- Maintains grid coordinates of current location.
- Prepares to employ Army aviation attack and reconnaissance, and joint and multinational CAS assets.

COMBAT MEDIC

1-46. Combat medics are allocated to Infantry rifle companies based on one combat medic per platoon, and one senior combat medic per company. The platoon combat medic or the company senior combat medic goes to the casualty's location, or the casualty is brought to the combat medic at the CCP. The CCP combat medic makes an assessment, administers initial medical care, initiates a DD Form 1380 (see paragraph 8-31), then requests evacuation or returns the individual to duty.

1-47. The Infantry platoon combat medic usually locates with, or near, the platoon sergeant. When the platoon moves on foot in the platoon column formation, the combat medic positions near the platoon sergeant. If the platoon is mounted, the combat medic usually rides in the same vehicle as the platoon sergeant. Emergency medical treatment procedures performed by the combat medic may include opening an airway, starting intravenous fluids, controlling hemorrhage, preventing, or treating for shock, splinting fractures, or suspected fractures, and providing relief for pain.

1-48. The Infantry platoon combat medic is trained under the supervision of the battalion surgeon or physician's assistant and medical platoon leader. The platoon combat medic is under the direct supervision of company senior combat medic when attached to the company. The platoon combat medic—

- Triages injured, wounded, or ill friendly and enemy personnel for priority of treatment.
- Conducts sick call screening.
- Assists in the evacuation of sick, injured, or wounded personnel under the direction of the platoon sergeant.
- Assists in the training of the platoon's CLSs in enhanced first-aid procedures.
- Requisitions Class VIII supplies from the battalion aid station for the platoon (individual first aid kit, CLS, aid bag) according to the tactical SOP.
- Recommends locations for platoon CCP, when established.
- Provides guidance to the platoon's CLSs as required.

This page intentionally left blank.

Chapter 2

Planning and Preparing for Operations

The platoon leader is the primary planner that receives the warning order (WARNORD), operation order (OPORD), and fragmentary order (FRAGORD) from the company and develops the platoon plan to support the company's mission. The platoon sergeant, squad leaders, forward observer (FO), radiotelephone operator (known as RTO) and other personnel may assist the platoon leader. The platoon leader employs troop leading procedures (TLP) to develop the plan and prepare for the mission. (See ATP 3-21.10 for additional information.)

SECTION I – INFANTRY SMALL-UNIT LEADERS

2-1. Infantry small-unit leaders use TLP when working alone or with a small group to solve tactical problems. For example, an Infantry rifle platoon leader may use the platoon sergeant, squad leaders, RTO, and the FO to assist during TLP. The type, amount, and timeliness of information passed from higher to lower directly impact the lower unit leader's TLP.

Operational Framework and Terrain Management

In the context of the *operational framework*—a cognitive tool used to assist commanders and staffs in clearly visualizing and describing the application of combat power in time, space, purpose, and resources in the concept of operations (ADP 1-01)—the platoon will be tasked to conduct combat operations within some portion of a higher command's operational framework. As a visualization tool, the operational framework bridges the gap between a commander's assessment of the operational environment, including all domains and dimensions, and its need to generate detailed orders that direct operations.

Within an operation, commanders will designate appropriate graphics, direct fire control measures, and defensive positions (or offensive actions) to synchronize the employment of combat power. They may create new models to fit the circumstances, but they generally apply a combination of common models according to doctrine. Commanders may use any operational framework model they find useful, but they must remain synchronized with their higher echelon headquarters' operational framework. The three models commonly used to build an operational framework are:

- Assigned areas (types include area of operations [AO], zone [see chapter 4], and sector [see chapter 5]).
- Deep, close, and rear operations (within deep, close, and rear areas respectively).
- Main effort, supporting effort, and reserve.

Assigning areas to subordinates is a key operational framework and *terrain management*—the process of allocating terrain by specifying locations for units and activities to deconflict activities that might interfere with each other (FM 3-90)—consideration for headquarters at every echelon. For example, a platoon is assigned a battle position to provide direct fires into an engagement area or an axis of advance or direction of attack to control its movement to an objective. When appropriate, leaders may further subdivide their assigned area into assigned areas for their subordinate formations. A higher headquarters remains responsible for any area not assigned to a subordinate unit. A unit moving through or delivering effects into another unit's assigned area must coordinate with the assigned unit. (See FM 3-0 and FM 3-90 for additional information.)

OPERATIONS PROCESS

2-2. TLP begin when the small-unit leader receives the first indication (for example, a WARNORD or FRAGORD) of an upcoming mission and continues throughout the *operations process*—the major command and control activities performed during operations: planning, preparing, executing, and continuously assessing the operation (ADP 5-0). During the operations process, TLP are not a hard and fast set of rules. Some actions may be performed simultaneously or in an order different from those shown in the example outline in figure 2-1 on page 2-4. The outline is a guide to apply consistent with the situation and experience of the platoon leader and platoon subordinate leaders. The tasks involved in some actions (such as initiate movement, issue the WARNORD, and conduct reconnaissance) may recur several times during the process. (See ATP 3-21.10 for additional information on the operations process.)

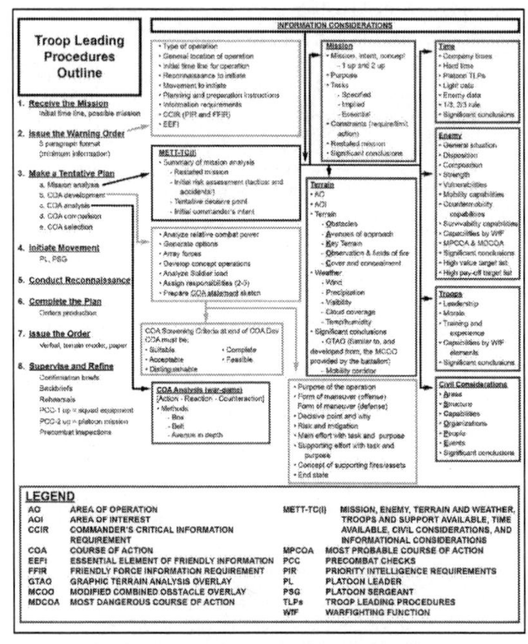

Figure 2-1. Troop leading procedures outline

2-3. In addition to the major activities of the operations process, the small-unit leader uses several integrating processes to synchronize specific functions throughout the operations process. Specifically, they include:

- Intelligence preparation of the operational environment (known as IPOE) process, see paragraphs 2-4 to 2-7.
- Targeting, see appendix B.
- Information collection (specifically reconnaissance and surveillance), see note below and paragraphs 2-102 to 2-106.

- Risk management, see paragraphs 2-26 to 2-33.
- Knowledge management, see paragraph 2-9.

Note. *Reconnaissance* is a mission undertaken to obtain information about the activities and resources of an enemy or adversary, or to secure data concerning the meteorological, hydrographic, geographic, or other characteristics of a particular area, by visual observation or other detection methods (JP 2-0). Reconnaissance is one part of information collection; the other three parts that comprise information collection are security, intelligence operations, and surveillance. *Surveillance* is the systematic observation of aerospace, cyberspace, surface or subsurface areas, places, persons, or things by visual, aural, electronic, photographic, or other means (JP 3-0). Although there are similarities between reconnaissance and surveillance to facilitate information collection, reconnaissance collection uses multiple means, including surveillance, to find information by systematically checking multiple locations in a designated AO.

2-4. IPOE is the systematic process of analyzing the mission variables of enemy, terrain, weather, and civil considerations, and integrated into these variables, informational considerations in an area of interest to determine their effect on operations. An *area of interest* is that area of concern to the commander, including the area of influence, areas adjacent to it, and extending into enemy territory (JP 3-0). Although there are four steps to the IPOE process, it is important to note that IPOE is a continuous process and one required by the small-unit leader for use during TLP. Continuous analysis and assessment (specifically informal assessments by the small-unit leader) are necessary to maintain situational understanding of an *operational environment*—the aggregate of the conditions, circumstances, and influences that affect the employment of capabilities and bear on the decisions of the commander (JP 3-0)—in constant flux. The ranges of a small unit's tactical movement and maneuver capabilities typically define its *area of influence*—an area inclusive of and extending beyond an operational area wherein a commander is capable of direct influence by maneuver, fire support, and information normally under the commander's command or control (JP 3-0); however, leaders consider all forms of contact they can make with enemy forces when visualizing their area of influence. (See ATP 2-01.3 for additional information.) The four steps of the IPOE process are:

- Define the operational environment.
- Describe environmental effects on operations.
- Evaluate the threat.
- Determine threat COAs (most likely and least likely [additionally most dangerous]).

2-5. Although the platoon leader does prepare IPOE products for use in preparation of analysis, and orders, the leader does not generally prepare individual products for subordinates. The platoon leader must be able to use the products of the battalion's IPOE effectively during mission analysis and be able to inform subordinates based on the compilation of platoon and higher IPOE products.

2-6. The intelligence staff at the battalion intelligence cell develops and provides the IPOE products required by the small-unit leader for use during TLP. Small-unit leaders should not need to do any other refinement of these products. The following includes standard IPOE products provided by the battalion to assist the small-unit leader during TLP:

- Threat situation templates (known as SITEMPs) and course of action (COA) statements.
- Terrain and weather products.
- Tactical decision aids (such as modified combined obstacle overlays [MCOOs] and terrain effects evaluations, weather forecast charts, weather effects matrices, and light and illumination data tables).
- Civil considerations tools and products.

Note. Company commanders coordinate (and provide to their platoon leaders as required) with the battalion intelligence cell for any IPOE products or tools they may need.

2-7. Due to the lack of a staff and resources, as well as time constraints, the small-unit leader depends on the timely delivery of IPOE products developed by higher headquarters tailored to support small-unit planning. Specifically, the components of IPOE inform TLPs steps 2 through 6 and actions (for example, information collection) within the TLP.

2-8. The last action (activities associated with supervising and refining the plan) occurs continuously throughout TLP and execution of the mission. The information in section II concerning TLP assumes the small-unit leader will plan in a time-constrained environment. All steps should be done, even if in abbreviated fashion. As such, the suggested techniques are oriented to help a leader quickly develop and issue a mission order.

Note. Mission command is the Army's approach to command and control that empowers subordinate decision making and decentralized execution appropriate to the situation (ADP 6-0). Leaders use mission command, with its emphasis on seizing, retaining, and exploiting operational initiative, through mission orders. *Mission orders* are directives that emphasize to subordinates the results to be attained, not how they are to achieve them (ADP 6-0). Disciplined initiative—as it relates to mission command describes individual initiative. Mission command requires the company commander to convey a clear commander's intent and concept of operations to subordinates leaders. *Concept of operations* is a statement that directs the manner in which subordinate units cooperate to accomplish the mission and establishes the sequence of actions the force will use to achieve the end state (ADP 5-0). Often, subordinates acting on the higher commander's intent develop the situation in ways that exploit unforeseen opportunities. (See ATP 3-21.10 for additional information.)

2-9. *Knowledge management* is the process of enabling knowledge flow to enhance shared understanding, learning, and decision making (ADP 6-0). The purpose of knowledge management is to align people, processes, and tools within the organizational structure and culture to achieve shared understanding. This alignment improves collaboration and interaction between leaders and subordinates and information sharing with subordinate units and higher echelon headquarters. (See ATP 3-21.20 for information on knowledge management.)

PROCESS COMPARISONS AND INITIAL CONSIDERATIONS

2-10. Leaders start the operations process by identifying their units' mission—the essential task or tasks, together with the purpose, that clearly indicates the action to be taken and the reason for the action (see JP 3-0). Throughout the process leaders remain flexible. They adapt TLP to fit the situation rather than try to alter the situation to fit a preconceived idea of how events should flow. At all levels, developing and describing the vision of leaders requires time, explanation, and ongoing clarification.

PARALLEL PLANNING

2-11. All leaders understand that their next higher commander's concept of operations continues to mature, and continue parallel planning as it does so, up until execution. Parallel planning occurs when two or more echelons plan the same operation at about the same time. Parallel planning is easiest when the higher headquarters continuously shares information on future operations with subordinate units. For example, rather than waiting until the platoon leader finishes planning, the squad leader starts to develop the squad's mission as information is received and fleshes out the mission as more information becomes available. Figure 2-2 on page 2-8 illustrates the parallel sequences of the TLP of the company with the TLP of its platoons and the TLP of the platoon with the TLP of its squads.

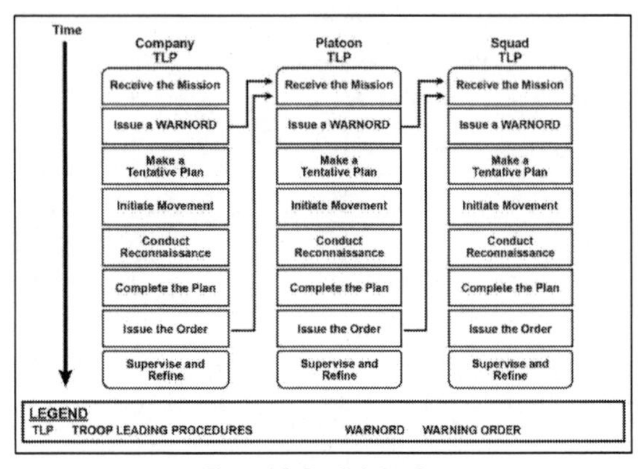

Figure 2-2. Parallel planning

SEQUENCING OF STEPS

2-12. Normally, the first three steps (receive the mission, issue a WARNORD, and make a tentative plan) of TLP (see paragraph 2-16) occur in order. However, the sequence of subsequent steps is based on the situation. The tasks involved in some steps for example, initiate movement and conduct reconnaissance may occur several times. The last step, supervise and refine, occurs throughout.

2-13. A tension exists between executing current operations and planning for future operations. Small-unit leaders must balance both. If engaged in a current operation, there is less time for TLP. If in a lull, transition, or an *assembly area*—an area a unit occupies to prepare for an operation (FM 3-90), leaders have more time to use TLP thoroughly. In some situations, time constraints or other factors may prevent leaders from performing each step of TLP as thoroughly as they would like. For example, during Step 3 – Make a Tentative Plan, small-unit leaders often develop only one acceptable COA vice multiple COAs. If time permits, leaders may develop, compare, and analyze several COAs before arriving at a decision on which one to execute.

INITIAL WARNING ORDER OR NEW MISSION

2-14. Small-unit leaders begin TLP once the initial WARNORD or a new mission is received. As each subsequent order arrives, the leader modifies assessments, updates tentative plans, and continues to supervise and assess preparations. In most situations, leaders will not receive or issue more than one WARNORD; security considerations and/or tempo make it impractical to do so. Leaders carefully consider decisions to eliminate WARNORDs. Subordinates always need to have enough information to plan and prepare for their mission. In other cases, TLP are started before receiving a WARNORD based on existing plans and orders (contingency plans or be-prepared missions) and on subordinate leaders' understanding of the situation.

Note. Leaders filter relevant information into the categories of the mission variables of METT-TC (I) mission, enemy, terrain and weather, troops and support available, time available, civil considerations, and informational considerations. METT-TC (I) represents the variables leaders use to analyze and understand a situation in relationship to the unit's mission. The first six variables are not new. However, the increased use of information (both military and civilian) to generate cognitive effects requires leaders to assess the informational aspects and impacts on operations continuously. Because of this, informational considerations, represented with (I) has been added to the familiar METT-TC mnemonic. Informational considerations is expressed as a parenthetical variable in that it is not an independent variable, but an important component of each variable of METT-TC that leaders pay particular attention to when developing understanding of a situation. (See paragraph 2-89 for additional information.)

2-15. Parallel planning hinges on distributing information as it is received or developed. Subordinate leaders cannot complete their plans until they receive their unit mission. If each successive WARNORD contains enough information, the higher command's final order should confirm what subordinate leaders have already analyzed and put into their tentative plans. In other cases, the higher command's order may cause subordinates to conduct additional planning.

SECTION II – SMALL-UNIT TROOP LEADING PROCEDURES

2-16. *Troop leading procedures* are a dynamic process used by small-unit leaders to analyze a mission, develop a plan, and prepare for an operation (ADP 5-0). TLP comprise a sequence of actions helping the leader use available time effectively and efficiently to issue orders and execute an assigned mission. (See GTA 07-10-003 [*Infantry Reference Card for Small Unit Leaders {Troop Leading Procedures}*] for additional information.) TLP eight steps are as follows:

- Step 1. Receive the mission.
- Step 2. Issue a WARNORD.
- Step 3. Make a tentative plan.
- Step 4. Initiate movement.
- Step 5. Conduct reconnaissance.
- Step 6. Complete the plan.
- Step 7. Issue the order.
- Step 8. Supervise and refine.

STEP 1 – RECEIVE THE MISSION

2-17. Leaders may receive one or more WARNORDs prior to receipt of the company OPORD. In a time-constrained environment, they may not receive their mission until receipt of the company OPORD. In the worst case, units may receive a change of mission while in execution, or even realize themselves the requirement for a new mission to achieve a higher commander's intent. When units conduct parallel planning, subordinate leaders will have a greater understanding of future missions, whether an order is issued or not.

2-18. Upon receipt of the order, leaders conduct an initial analysis of the mission variables of METT-TC (I) to understand their likely mission, the threat, and other details necessary to start preparations for combat. Critical to success is the platoon leader's time analysis. The leader's primary objective is to maximize the time available to the unit to prepare the upcoming mission through effective time management.

Note. The HOPE-LW format (higher, operational, planning, enemy, light/weather) is an effective tool leaders use to analyze time variables impacting their operation.

2-19. Using the variables mentioned above, the leader develops a timeline to manage the platoon's planning and preparation for operations. While this timeline will be updated with the plan, it is essential to time management that the leader identify necessary tasks and starts movement on them in accordance with the timeline. The leader should—

- Analyze the time the unit has available, accounting for hours of darkness.
- Apply one third of available time to planning, leaving two thirds for preparation.
- Identify externally imposed times (for example, rehearsals, orders, and so forth).
- Identify required movements (for example, reposition, reconnaissance, and so forth).
- Prepare an initial timeline for planning, preparation, and execution of the mission.

Note. The WARNORD is the leader's best tool to manage time. Even when the company does not issue a WARNORD, the platoon leader can and should issue a WARNORD providing any known information.

STEP 2 – ISSUE WARNING ORDER

2-20. A WARNORD is a preliminary notice of an order or action to follow. Though less detailed than a complete OPORD, a WARNORD aids in parallel planning. After the leaders receive new missions and assess the time available for planning, preparing, and executing the mission, they immediately issue a WARNORD to their subordinates. By issuing the initial WARNORDs as quickly as possible, they enable subordinates to begin their own planning and preparation (parallel planning) while they begin to develop the OPORDs. When they obtain more information, they issue updated WARNORDs, giving subordinates as much as they know.

2-21. Leaders can issue WARNORDs to their subordinates right after they receive higher command's initial WARNORDs. In their own initial WARNORDs, they include the same elements given in their higher headquarters' initial WARNORDs. If practical, leaders brief their subordinate leaders face-to-face, on the ground. Otherwise, they use a terrain model, sketch, or map. (See figure 2-3 for an example of a WARNORD format.)

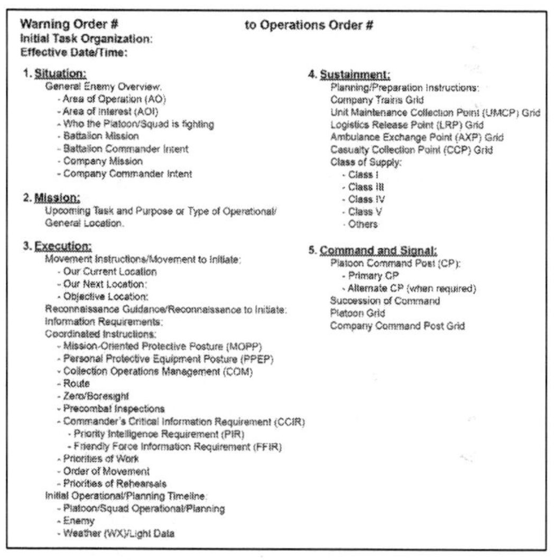

Warning Order # **to Operations Order #**
Initial Task Organization:
Effective Date/Time:

1. **Situation:**
General Enemy Overview.
- Area of Operation (AO)
- Area of Interest (AOI)
- Who the Platoon/Squad is fighting
- Battalion Mission
- Battalion Commander Intent
- Company Mission
- Company Commander Intent

2. **Mission:**
Upcoming Task and Purpose or Type of Operational/General Location.

3. **Execution:**
Movement Instructions/Movement to Initiate:
- Our Current Location
- Our Next Location:
- Objective Location:

Reconnaissance Guidance/Reconnaissance to Initiate:
Information Requirements:
Coordinated Instructions:
- Mission-Oriented Protective Posture (MOPP)
- Personal Protective Equipment Posture (PPEP)
- Collection Operations Management (COM)
- Route
- Zero/Boresight
- Precombat Inspections
- Commander's Critical Information Requirement (CCIR)
  - Priority Intelligence Requirement (PIR)
  - Friendly Force Information Requirement (FFIR)
- Priorities of Work
- Order of Movement
- Priorities of Rehearsals

Initial Operational/Planning Timeline:
- Platoon/Squad Operational/Planning
- Enemy
- Weather (WX)/Light Data

4. **Sustainment:**
Planning/Preparation Instructions:
Company Trains Grid
Unit Maintenance Collection Point (UMCP) Grid
Logistics Release Point (LRP) Grid
Ambulance Exchange Point (AXP) Grid
Casualty Collection Point (CCP) Grid
Class of Supply:
- Class I
- Class III
- Class IV
- Class V
- Others

5. **Command and Signal:**
Platoon Command Post (CP):
- Primary CP
- Alternate CP (when required)

Succession of Command
Platoon Grid
Company Command Post Grid

Figure 2-3. Warning order format

2-22. The WARNORD follows the five-paragraph OPORD format and normally contains minimal information. The WARNORD generally alerts the leader that a new mission is pending. (See FM 5-0 for information on when the higher headquarters issues the full sequence of WARNORDs, specifically WARNORD two and three and the information contained in each.) The WARNORD to the platoon generally contains the following information:

- Type of operation.
- General location of the operation.
- Initial operational timeline (planning, preparation, and execution [assessing throughout the timeline]).
- Any movements necessary to initiate.
- Any collaborative planning sessions directed by the commander.
- Planning and preparation instructions (precombat checks [PCCs], precombat inspections [PCIs] and rehearsal guidance).
- Initial information requirements or commander's critical information requirements (CCIRs) (priority intelligence requirement [PIR] and friendly force information requirement).
- Initial information collection tasks (reconnaissance and surveillance to initiate).

STEP 3 – MAKE A TENTATIVE PLAN

2-23. The platoon leader begins TLP step 3 after having issued the WARNORD, or at any point after having enough information to proceed. The platoon leader need not wait for a complete OPORD before developing a tentative plan. Key steps include mission analysis, COA development, COA analysis, COA comparison, and COA selection.

MISSION ANALYSIS

2-24. Mission analysis is the process by which a platoon leader gains a full understanding of what the platoon has been tasked to accomplish, and an understanding of the commander's intent, both of which inform mission planning. Mission analysis is the most important part of making a tentative plan. There is no specific timeline allocated to mission analysis, but leaders should always adhere to the one-third/two-thirds rule, allocating roughly one-third of available planning and preparation time to themselves, and allowing their subordinates the remaining two-thirds.

2-25. The platoon leader uses the mission variables of METT-TC (I) to conduct mission analysis. This form of mission analysis is a continuation of the initial assessment performed in TLP step 1. The potential informational considerations are embedded within each of the other variables. At the conclusion, the platoon leader will understand—

- Commander's intent at company and battalion.
- Factors of the mission variables as they relate to what the platoon must achieve.
- Tentative decisive point(s) essential to mission success.
- Tentative communications and electromagnetic warfare (EW) activities.
- Type, nature, and probable location of enemy contact throughout the depth of the operation.

- Threat-based and accident-based risks hazards to platoon operations.
- The platoon's mission statement, nested with higher mission and intent.

Note. Typically, leaders below company level do not issue a commander's intent. They reiterate the intent of their higher and next higher commanders.

Risk Management

2-26. *Risk management* is the process to identify, assess, and mitigate risks and make decisions that balance risk cost with mission benefits (JP 3-0). It is an ongoing and iterative process that is integral to planning and preparation during TLPs. WARNORDs are important to risk management (RM) since it is the leader's first opportunity to identify and address the risks that may be associated with the operation. RM continues from the receipt of an initial WARNORD to the receipt of the OPORD. For the small-unit leader, information and assessments are continually revised and updated to reflect the current situation. The process of receiving, distributing, revising, and updating the hazards, risks, and controls for threat-based and accident-based hazards continue simultaneously until the order is issued. Leaders at the platoon and squad level must ensure they identify the RM information they receive to enable their planning and preparation for the mission. To be effective, small-unit leaders must remain especially alert for those controls that require coordination with adjacent and supporting units (for example, fire control measures).

2-27. The RM principles and five-step process help leaders identify hazards, assess risk, and make sound risk decisions-the determination to accept or not accept the risk(s) associated with an action the leader will take or will direct others to take.

2-28. The principles of RM are—

- Integrate RM into all phases of missions and operations.
- Make risk decisions at the appropriate level.
- Accept no unnecessary risk.
- Apply RM cyclically and continuously.

2-29. The five steps of RM are—

- Step 1-Identify the hazards.
- Step 2-Assess the hazards.
- Step 3-Develop controls and make risk decisions.
- Step 4-Implement controls.
- Step 5-Supervise and evaluate.

2-30. Steps 1 and 2 of RM are assessment steps-risk assessment is the identification and assessment of hazards. Steps 3 through 5 of RM are management steps that allow a leader to implement measures to control hazards. The mission variables of METT-TC (I) serve as a standard format for identifying hazards (see ATP 5-19). Mission variables used primarily as part of TLPs for planning and preparing for operations provide a pattern for addressing threat-based and accident-based hazards and associated risk for any activity. Leaders assess risk to protect the force and aid in mission accomplishment.

2-31. During mission analysis, the platoon leader uses informational and intelligence tools (for example, the IPOE and other intelligence products [see paragraphs 2-4 to 2-10]) to help develop a picture of enemy capabilities and vulnerabilities. During analysis, the leader looks for enemy presence or capabilities that pose hazards and risks to an operation and evaluates what the enemy or outside influences could do to defeat or spoil an operation. Informational and intelligence tools support threat-based risk assessments by identifying opportunities and any constraints the operational environment offers to both enemy and friendly forces.

2-32. Threat-based (tactical) risk is associated with hazards existing due to the enemy's presence. The consequences of tactical risk take two major forms:

- Enemy action, where the leader has accepted risk such as an enemy attack where the friendly leader has chosen to conduct an economy of force.
- Lost opportunity, such as a decision to traverse across severely restricted terrain that would restrict the unit's ability to quickly mass the effects of combat power once contact with the enemy is made.

2-33. Accident-based risks include all operational risks other than tactical risk and can include hazards concerning friendly personnel, equipment readiness, and environment. Once the platoon leader identifies an accidental risk based on the results of mission analysis, the risk is reduced through controls. For example, fratricide is a hazard categorized as an accident risk; surface danger zones and risk estimate distance are used to identify the controls, such as target reference points (TRPs), restricted fire lines, and phase lines, to reduce this accidental risk. When the leader decides what risk to accept, the leader also must decide in the COA how to reduce risk to an acceptable level. (See ATP 3-21.10 and ATP 5-19 for additional information.)

Note. Each commander must ensure a thorough understanding of the mission, including the senior commander's intent and the risk tolerance, while not overly restricting subordinate initiative or ability to turn potential risk into opportunity. An unnecessary risk is any risk that, if taken, will not contribute meaningfully to mission accomplishment or will needlessly endanger lives or resources. Army leaders accept only a level of risk in which the potential benefit outweighs the potential loss. The process of weighing risks against opportunities and benefits helps to maximize unit capability, save lives, and preserve resources. The appropriate level of command makes risk acceptance decisions after applying RM and weighing potential gain against potential loss. Commanders need not be risk averse. Forces may undertake even high-risk endeavors when commanders determine that the sum of the benefits exceeds the sum of the costs. Commanders establish the basis for risk acceptance decisions through RM. (See ATP 5-19 for additional information.)

Communications

2-34. The ability to communicate during operations is essential. Initially the platoon establishes radio nets to link the platoon leader with higher headquarters and subordinate leaders. During operations, subordinate elements may have their own control net with the platoon leader or primary command post. Other communications activities such as signals are established and rehearsed. The platoon must be prepared to operate with digital networks or during periods of degraded communications. Messengers and visual signals are excellent means of communication during movement or when in static positions. The platoon may move under radio silence, or may use radios only in emergencies, or when it can use no other means of communication.

2-35. In an environment where every transmission is potentially targeted, the primary, alternate, contingency, and emergency (known as PACE) communications plan enables both mission success and survivability. To be effective, the PACE plan must be understood and rehearsed by every participant. The PACE plan is a communication plan that exists for a specific mission or task, not a specific unit, as the plan considers both intra- and inter-unit sharing of information. The platoon leader's ability to exercise command and control during an operation can suffer due to communication systems not working properly or otherwise unavailable.

2-36. When the platoon does not have four viable methods of communications, it is appropriate to issue a PACE plan that may only have two or three systems listed. If the platoon cannot execute the full PACE plan to the company, the platoon leader must inform the company commander with an assessment of shortfalls, gaps, and possible mitigations as part of the mission analysis process during the TLP. During COA development, the platoon leader nest the platoon's plan with the company's plan (in turn the company's plan nest with the battalion's plan) whenever practical. This aids in maintaining continuity of effort. (See ATP 3-21.10 for additional information.)

2-37. During TLP units must prepare for denied, degraded, and disrupted command and control systems and reduced access to cyberspace and space operations capabilities. Key indicators that command and control systems are being degraded include:

- Degraded voice communications.
- Increased latency for data transmissions.
- Frequent and accurate targeting by threat lethal and nonlethal effects.
- Increased pings/network intrusions.
- Inconsistent digital common operational picture, for example, spoofing.
- Inaccurate Global Positioning System (GPS) data/no satellite lock and inconsistency between inertial navigation aids and GPS-enabled systems.
- Uncharacteristically few voice or digital transmissions.

2-38. Efforts to increase survivability and prevent degraded command and control systems include—

- Reduce radio power settings.
- Use messengers and hand and arm signals.
- Minimize length of frequency modulation transmissions.

- Use terrain to mask transmission signatures.
- Employment of directional antennas.
- Use remote antennas/carry spare antennas.
- Require physical presence of leaders at briefings, for example, distribute information via analog means in person.
- Use of camouflage and deception in all environments.
- Dig in command post.
- Use of communications windows to reduce transmissions.
- Employment of encryption/cypher techniques.

2-39. Efforts that counter the effects of degraded command and control systems include—

- Train to recognize indicators.
- Develop and rehearse contingency plans during the planning process and preparations.
- Maintain analog common operational pictures at all echelons.
- Train to operate from the commander's intent, and analog graphics and synchronization matrixes.
- Keep plans as simple as possible that are less susceptible to friction.

Electromagnetic Warfare

2-40. *Electromagnetic warfare* is military action involving the use of electromagnetic and directed energy to control the electromagnetic spectrum or to attack the enemy (JP 3-85). Commanders and subordinate leaders at each echelon integrate EW activities into operations through cyberspace electromagnetic activities. EW capabilities are applied from the air, land, sea, space, and cyberspace by manned, unmanned, attended, or unattended systems. Cyberspace electromagnetic activities are the process of planning, integrating, and synchronizing cyberspace and EW operations. *Cyberspace operations* are the employment of cyberspace capabilities where the primary purpose is to achieve objectives in or through cyberspace (JP 3-0). (See ATP 3-21.10 for additional information.)

Electromagnetic Warfare Capabilities

2-41. EW capabilities assist in shaping the operational environment to gain an advantage. For example, EW may be used to set favorable conditions for cyberspace operations by stimulating networked sensors, denying wireless networks, or other related actions. Operations in cyberspace and the electromagnetic spectrum depend on EW activities maintaining freedom of action in both. EW consists of three functions, electromagnetic attack, electromagnetic protection (EP), and electromagnetic support. In any environment, the primary focus at the platoon and company level is on EP considerations as it relates to communications (see paragraphs 2-34 to 2-39) within small-unit operations.

Electromagnetic Protection

2-42. *Electromagnetic protection* is a division of electromagnetic warfare involving actions taken to protect personnel, facilities, and equipment from any effects of friendly or enemy use of the electromagnetic spectrum that degrade, neutralize, or destroy friendly combat capability (JP 3-85). For example, EP includes actions taken by the commander and subordinate leaders to ensure friendly use of the electromagnetic spectrum, such as frequency agility in a radio or variable pulse repetition frequency in radar. They avoid confusing EP with self-protection. Both defensive electromagnetic attack and EP protect personnel, facilities, capabilities, and equipment. However, EP protects from the effects of electromagnetic attack (friendly and enemy) and electromagnetic interference, while defensive electromagnetic attack primarily protects against lethal attacks by denying enemy use of the electromagnetic spectrum to guide or trigger weapons. (See ATP 3-21.20 and ATP 3-12.3 for additional information on EW functions and activities.)

Emission Control

2-43. Leaders are responsible for emission control (known as EMCON) as an element of EP that inhibits enemy electromagnetic warfare (EW) capabilities from detecting, intercepting, finding, fixing, or engaging emitters. *Emission control* is the selective and controlled use of electromagnetic, acoustic, or other emitters to optimize command and control capabilities while minimizing, for operations security: a. detection by enemy sensors; b. mutual interference among friendly systems; and/or c. enemy interference with the ability to execute a military deception plan (JP 3-85).

2-44. In large-scale combat operations against peer threats, the enemy is expected to use EW capabilities to detect, intercept, deny, degrade, disrupt, destroy, or manipulate friendly communications, command and control, and intelligence capabilities. EMCON is a planning aid designed to help leaders develop standard procedures and battle drills for their unit's unique suite of emitters using an appropriate mix of the EMCON considerations listed in table 2-1 on page 2-18.

Table 2-1. Emission control considerations

Tactics, Techniques, and Procedures	
Minimize length and frequency of radio transmissions.*	Use satellite communications (SATCOM) information on these practices.
Use appropriate power settings. *	Use high frequency (HF) transmissions.
Plan radio messages. *	Use electronic counter-countermeasures (ECCM).
Use electronic terrain masking. *	Train while employing radio silence.
Establish and enforce a primary, alternate, contingency, and emergency (PACE) communication plan (see paragraph 2-33). *	Ensure electronic equipment is properly grounded and has shield cables.
Use remote antennas.	Train on land navigation (without GPS).
Use brevity codes and proword execution matrixes.	Set radar cueing cycles.
Use secure landlines.	Execute survivability moves.
Use directional antennas.	Ensure electronic equipment is properly grounded and has shield cables.
Use line-of-sight communications parallel to the forward line of own troops (FLOT).	Understand the impact of terrain composition on emissions.
Use alternate means of communications for planning /preparation; use primary for execution.	Recognize communications jamming (reporting criteria).
Use data-burst transmissions.	Recognize GPS jamming (reporting criteria).
Mask with camouflage netting.	Recognize radar jamming (reporting criteria).
Use encrypted Global Positioning System (GPS).	Recognize satellite jamming (reporting criteria).
* These emission control considerations should always be practiced, but leaders emphasize them more as threats involving the electromagnetic spectrum (EMS) elevate.	

2-45. EMCON prevents the threat discovering and attacking the locations of friendly forces with EW. When establishing EMCON best practices, it is important to understand the general categories and status criteria for EMCON levels. Based on the tactical situation, the commander can dictate the appropriate EMCON level to the platoon. During operations, commanders consider EMCON level 3 (amber) as the baseline condition. Figure 2-4 captures the five EMCON levels and the general descriptive criteria associated with each level. (See ATP 3-12.3 and ATP 6-02.53 for additional information.)

EMCON Status	Description
EMCON 5 Green	Describes a situation where there is no apparent hostile activity against friendly emitter operations. Operational performance of all EMS-dependent systems is monitored, and password-encryption-enabled systems are used as a layer of protection.
EMCON 4 Yellow	Describes an increased risk of attack after detection. Increased monitoring of all EMS activities is mandated, and all end users must make sure their systems are secure, encrypted, power levels monitored, and transmissions limited. EMS usage may be restricted to certain emitters, and rehearsals for elevated EMCON is ideal.
EMCON 3 Amber	Describes when a risk has been identified. Counter ECM (encryption, FH, directional antennas) on important systems is a priority, and the CEWO's alertness is increased. All unencrypted systems are disconnected.
EMCON 2 Red	Describes when an attack has taken place but the EMCON system is not at its highest alertness. Non-essential emitters may be taken offline, alternate methods of communication may be implemented and modifications are made to standard lower EMCON configurations (for example, power levels and antenna types).
EMCON 1 Black	Describes when attacks are taking place based on the use of the EMS. The most restrictive methods of EP are enforced. Any compromised systems are isolated from the rest of the network.

LEGEND

CEWO	CYBER ELECTROMAGNETIC WARFARE OFFICE	EMS	ELECTROMAGNETIC SPECTRUM
ECM	ELECTROMAGNETIC COUNTERMEASURES	EP	ELECTROMAGNETIC PROTECTION
EMCON	EMISSION CONTROL	FH	FREQUENCY HOP

Figure 2-4. Emission control status

Decisive Point

2-46. A *decisive point* is a key terrain, key event, critical factor, or function that, when acted upon, enables commanders to gain a marked advantage over an enemy or contribute materially to achieving success (JP 5-0). Identifying a tentative decisive point(s) during mission analysis and verifying it during COA development is the most important aspect of the TLP. Visualizing a valid decisive point is how leaders determine how to achieve success and accomplish their purpose. Leaders develop their entire COA from a decisive point. Without determining a valid decisive point, leaders cannot begin to develop a valid or tactically sound COA. Leaders, based on their initial analysis of METT-TC (I), situational awareness, vision, and insight into how such factors can affect the unit's mission, should visualize where, when, and how their unit's ability to apply combat power (for example, firepower, protection, maneuver, leadership, and information) achieves their assigned task(s). A decisive point might orient on terrain, enemy, time, or a combination of these. A decisive point might be where or how, or from where, the unit will combine the effects of combat power against the enemy. A decisive point might be the event or action (with respect to terrain, enemy, or time, and generation of combat power) which will ultimately and irreversibly lead to the unit achieving its purpose.

2-47. A decisive point(s) does not simply restate the unit's essential task or purpose; it defines how, where, or when the unit will accomplish its purpose. The unit's designated main effort always focuses on a decisive point(s), and always accomplishes the unit's purpose. Designating a decisive point(s) is critical to the leader's vision of how to use combat power to achieve the purpose, how to task-organize units, and how to shape operations to support the main effort, and how the main effort will accomplish the unit's purpose. This tentative decisive point(s) forms the basis of the leader's planning and COA development. Leaders should clearly explain what the decisive point(s) is to subordinate leaders and why it is decisive; this summary, in conjunction with the commander's intent, facilitates subordinate initiative. A valid decisive point enables the leader to clearly and logically link how the application of combat power with respect to terrain, enemy, and time allows the unit to accomplish its purpose. If the leader determines the tentative decisive point(s) is not valid during COA development or analysis, then the leader must determine another decisive point(s) and restart COA development.

Actions on Contact

2-48. *Actions on contact*—is a process to help leaders understand what is happening and to take action (FM 3-90). Leaders analyze the enemy throughout TLP to identify all likely contact situations that may occur during an operation. This process should not be confused with battle drills such as Battle Drill-React to Contact see appendix E. Battle drills are the actions of individual Soldiers and small units when they meet the enemy. Thorough planning and rehearsals conducted during TLP, enable leaders and Soldiers to develop and refine a COA to deal with probable enemy actions. The COA becomes the foundation for a scheme of maneuver.

2-49. Infantry platoons and squads execute actions on contact using a logical, well-organized process of decision making and action entailing these four steps:

- React.
- Develop the situation.
- Choose an action.
- Execute and report.

2-50. This four-step process is not intended to generate a rigid, lockstep response to the enemy. Rather, the goal is to provide an orderly framework enabling the Infantry rifle platoon and its squads to survive the initial contact and apply sound decision making and timely actions to complete the operation. Ideally, the unit sees the enemy (visual contact) before being seen by the enemy; it then can initiate direct contact on its own terms by executing the designated COA.

2-51. During movement once the lead element of a force (for example, during the conduct of a movement to contact) encounters an enemy, they conduct actions on contact. Obstacles encountered during movement are treated like enemy contact, in that, the unit assumes the obstacles are covered by fire. Once contact is made the unit's security force will seek to gain a tactical advantage over an enemy by using tempo and initiative to conduct these actions, allowing it to gain and maintain contact without becoming *decisively engaged,* a fully committed force or unit that cannot maneuver or extricate itself (FM 3-90). How quickly the unit develops the situation is directly related to its security, and the tempo is directly related to the unit's initial planning and use of well-rehearsed tactical standard operating procedure (SOP) and battle drills.

2-52. Leaders understand properly executed actions on contact require time at the squad and platoon levels. To develop the situation, a platoon may have to attack an assailable flank, conduct *reconnaissance by fire*—a technique in which a unit fires on a suspected enemy position (FM 3-90) see paragraph A-71, or call for and adjust indirect fires.

Mission Variables of METT-TC (I)

2-53. Analyzing the mission variables of METT-TC (I) is a continuous process. Leaders constantly receive information from the time they begin planning through execution. During execution, their continuous analysis enables them to issue well-developed *fragmentary order*—an abbreviated operation order issued as needed to change or modify an order or to execute a branch or sequel (JP 5-0). They must assess if the new information affects their missions and plans. If so, then they must decide how to adjust their plans to meet these new situations. They need not analyze METT-TC (I) in a particular order. How and when they do so depend on when they receive information as well as on their experience and preferences. One technique is to parallel the TLP based on the products received from higher. Using this technique, they would, but need not, analyze mission first; followed by terrain and weather; enemy; troops and support available; time available; and civil considerations and integrated into these variables, informational considerations.

Analysis of Mission

2-54. A mission is a task and purpose clearly indicating the action to be taken and reason for the action. In common usage, especially when applied to lower military units, a mission is a duty or task assigned to an individual or unit. The mission is always the first factor leaders consider and most basic information question: "*What have I been told to do, and why?*"

2-55. Leaders at every echelon must understand the mission, intent, task, purpose, location, and concept of operation one and two levels higher. This understanding makes it possible to exercise disciplined initiative. Leaders capture their understanding of what their units are to accomplish in their revised mission statement. A *mission statement* is a short sentence or paragraph that describes the organization's essential task(s), purpose, and action containing the elements of who, what, when, where, and why (JP 5-0). Leaders consider the following elements when analyzing the mission—

- Higher headquarters' (two levels up) mission, commander's intent, and concept of operations.
- Immediate higher headquarters' (one level up) mission, commander's intent, concept of operations, available assets, and timeline.
- Unit's assigned area of operation (AO), task, and purpose.

- The mission of adjacent, supporting, and supported units and their relationships to the higher headquarters' plan.
- Assumptions and constraints.
- Specified, implied, and essential tasks.
- Restated mission.

2-56. Specific to platoon and squad mission, leaders must identify and understand tasks required to accomplish a given mission. The three types of tasks are specified, implied, and essential.

2-57. Specified tasks. Specified tasks are assigned to a unit by a higher headquarters and are found throughout the OPORD. Specified tasks also may be found in annexes and overlays; for example, "*Seize Objective Fox;*" "*Reconnoiter route Blue;*" "*Assist the forward passage of 1st platoon, B Company.*" "*Send two Soldiers to assist in the loading of ammunition.*"

2-58. Implied tasks. Implied tasks are those being performed to accomplish a specified task, but that are not stated in a higher headquarters' order. Implied tasks derive from a detailed analysis of higher up orders, from the enemy situation and COA, from the terrain, and from knowledge of doctrine and history. Analyzing the unit's current location in relation to future AO as well as the doctrinal requirements for each specified task might reveal the implied tasks. For example, if the specified task is "*Seize Objective Fox,*" and an updated intelligence assessment determines that Objective Fox is surrounded by reinforcing obstacles, this intelligence assessment would drive the implied task of "*Breach reinforcing obstacles vicinity Objective Fox.*"

2-59. Essential task. The essential task is the mission task—it accomplishes the assigned purpose. It, along with the platoon's purpose, is usually assigned by the higher headquarters' OPORD in concept of operations or tasks to maneuver units. For main efforts, since the purposes are the same (nested concept), the essential task also accomplishes the higher headquarters' purpose. For supporting efforts, it accomplishes the assigned purpose, which supports the main effort.

2-60. Leaders conclude their analysis of the mission by restating their mission in terms of the platoon's (or squad's) mission statement. To do this, they answer the five Ws:

- Who (the unit).
- What (the unit's essential task and type of operation).
- When (this is the time given in the company OPORD).
- Where (the objective or location stated in company OPORD).
- Why (the unit's purpose, taken from the company's concept of operations).

2-61. Example platoon mission statement in the defense, "1st Platoon/B Company/1-31 Infantry defends no later than 281700(Z) December 2022 from GL 375652 to GL 394660 to prevent the envelopment of A Company, the battalion main effort."

Analysis of Terrain and Weather

2-62. Leaders consider the cumulative effects of the natural terrain, man-made features (including obstacles), climate, and weather on both friendly and enemy operations. In general, terrain, climate, and weather do not favor one side over the other unless one is better prepared to operate in the environment or is more familiar with it. The terrain, however, may favor defending or attacking. Analysis of terrain answers the question: "*What is the terrain's effect on the operation?*" Leaders analyze terrain using the categories of observation and fields of fire, avenues of approach, key terrain, obstacles, and cover and concealment (OAKOC).

> ***Note.*** See ATP 3-21.10 for additional information on analyzing the terrain using the categories of OAKOC.

2-63. The MCOO shows the critical military aspects of terrain. It not only facilitates planning, but also aids in briefing subordinates. Specific questions on the aspects of the MCOO include but are not limited to avenues of approach, key terrain, mobility corridors, natural and man-made obstacles, and terrain mobility classifications. The information within the MCOO depicts the terrain according to the mobility classification. These classifications are severely restricted, restricted, and unrestricted.

2-64. Severely restricted terrain severely hinders or slows movement in movement formations unless some effort is made to enhance mobility, such as committing engineer assets to improving mobility or deviating to different movement formations. Severely restricted terrain for armored and mechanized forces is typically characterized by steep slopes and large or dense obstacle compositions with few bypasses.

2-65. Restricted terrain hinders movement to some degree. Little effort is needed to enhance mobility, but units may have difficulty maintaining preferred speeds, moving in movement formations, or transitioning from one formation to another. Restricted terrain slows movement by requiring zigzagging or frequent detours. Restricted terrain for armored or mechanized forces typically consists of moderate-to-steep slopes or moderate-to-dense obstacle compositions, such as restrictive slopes or curves. Swamps or rugged terrain are examples of restricted terrain for dismounted Infantry forces.

2-66. Unrestricted terrain is free from any restriction to movement. Nothing is required to enhance mobility. Unrestricted terrain for armored or mechanized forces is typically flat to moderately sloping terrain with few obstacles such as limiting slopes or curves.

> ***Note.*** At company level and below, the commander and subordinate leaders develop a graphic terrain analysis overlay (known as GTAO). This product is similar to the MCOO in it shows the critical military aspects of terrain. Not only does it facilitate planning, but it also aids in briefing subordinates.

2-67. From the MCOO developed by higher headquarters (generally the battalion intelligence staff officer [S-2]), leaders can understand the general nature of the ground and effects of weather. However, at company level and below, they must conduct their own detailed analyses to determine how terrain and weather uniquely affects their units' missions and the enemy. They must go beyond merely passing along the MCOO to their subordinate leaders and making general observations of the terrain such as "*This is high ground,*" or "*This is a stream.*" During the development of the GTAO, they determine how the terrain and weather will affect the enemy and their units. Additionally, they apply these conclusions when they develop COA for both enemy forces and their units. (See ATP 3-21.10 and ATP 2-01.3 for additional information.)

2-68. Terrain analysis should produce several specific conclusions for example:

- Battle, support by fire, and/or *attack by fire position*—the general position from which a unit performs the tactical task of attack by fire (ADP 3-90).
- Engagement areas and ambush sites.
- Immediate and intermediate objectives.
- Asset locations such as enemy command posts or ammunition caches.
- Assembly areas (AAs), attack positions, assault positions, objective rally points (known as ORPs), and/or casualty collection points (CCPs).
- Observation posts.
- Artillery firing positions.
- Air defense artillery system positions.
- Reconnaissance, surveillance, and target-acquisition positions.
- Forward area arming and refueling points.
- Ford site (wet/dry) (*gap*—is a ravine, mountain pass, river, or other terrain feature that presents an obstacle that may be bridged [ATP 3-90.4]—locations).
- Landing and drop zones.
- Breach locations.
- Infiltration lanes.

2-69. The five military aspects of weather are visibility; winds; precipitation; cloud cover; temperature, humidity, and atmospheric pressure (see ATP 2-01.3). Consideration of the weather's effects is an essential part of the leader's mission analysis. Leaders go past observing to application. They determine how the weather will affect the visibility, mobility, and survivability of the unit and that of the enemy. Leaders review their commander's conclusions and identifies their own. They apply the results to the friendly and enemy COA they develop. (See ATP 3-21.10 for additional information.)

Analysis of Enemy

2-70. Leaders analyze the enemy's general situation, dispositions, compositions, strengths, doctrine, equipment, and capabilities by warfighting function, vulnerabilities, and probable COA. Analyzing the enemy answers, the question, "*What is the enemy doing and why?*" Leaders also answer—

- What is the composition and strength of the enemy force?
- What are the capabilities of the weapons? Other systems?

- What is the location of current and probable enemy positions?
- What is the enemy's most likely COA (additionally most dangerous)? (Memory aid defend, reinforce, attack, withdraw, or delay [known as DRAW-D].)

2-71. Platoon leaders must understand the assumptions their company commander, in coordination with the battalion S-2, used to portray the enemy's COA. Their own assumptions about the enemy should be consistent with those of their commander. When not, their assumptions must be known by the commander.

2-72. Leaders continually improve their situational understanding of the enemy and update their enemy SITEMP as new information or trends become available. Deviations or significant conclusions reached during their enemy analysis that could positively or negatively affect the battalion's and company's plan should be shared immediately with the battalion, company commander, and S-2.

2-73. In analyzing the enemy, platoon leaders must understand the IPOE (see paragraphs 2-4 to 2-7). Although platoon leaders usually do not prepare IPOE products for their subordinates, platoon leaders must be able to use the products of the higher headquarters' IPOE.

2-74. Leaders must know more than just the number and types of vehicles, Soldiers, and weapons the enemy has. The leaders must thoroughly understand when, where, and how the enemy prefers or tends to use their assets. A SITEMP is a visual illustration of how the enemy force might look and act without the effects of weather and terrain. Leaders look at specific enemy actions during a given operation and use the appropriate SITEMP to gain insights into how the enemy may fight. Likewise, leaders must understand enemy doctrinal objectives. In doctrinal terms, leaders ask "Is the enemy oriented on terrain objectives, friendly force objectives, civilian objective, or some combination? What effect will this have on the way the enemy fights?"

2-75. Depending on the situation, some threat forces may lack structured doctrine. In such a situation, a leader must rely on information provided by battalion or higher echelon reconnaissance and surveillance assets and, most importantly, the leader and the leader's higher headquarters' pattern analysis and deductions about the enemy in their AO. The leader also may make sound assumptions about the enemy, human nature, and local culture. Questions to determine the enemy's general situation include:

- Who are they?
- What are they doing?
- When did they get there?
- Where are they generally located?
- Why are they there?
- What is the nature of the threat they pose?

2-76. Gaining complete understanding of the enemy's intentions can be difficult when the enemy SITEMPs, composition, and disposition are unclear. In all cases, the enemy's recent activities must be understood, because they can provide insight into the enemy's future activities and intentions. If time permits, a leader might be able to conduct a pattern analysis of the enemy's actions to predict future actions. In the operational environment, this might be the most important analysis the leader conducts and is likely to yield the most useful information to the leader.

2-77. Leaders develop their enemy SITEMP after terrain and weather analysis so they better understand how the enemy will fight in a specific scenario. The leader includes in this SITEMP the likely sectors of fire of the enemy weapons and tactical and protective obstacles, either identified or merely template, which support defensive operations. Table 2-2 shows recommended SITEMP items. (See ATP 3-21.10 and ATP 2-01.3 for additional information.)

Table 2-2. Recommended enemy situation template items

Defensive and Offensive	
Primary, alternate, subsequent positions	Attack formations and axes of advance
Engagement area	Battle positions, trenches, and area of operations
Individual vehicles	Coordinated fire line and restrictive firing lines
Crew-served weapons	Objective(s), phase lines, and support (attack) by fire position(s)
Tactical (directed, situational, and reserved) obstacles (for example, persistent and nonpersistent minefields)	Maximum engagement line
	Planned indirect-fire targets
Planned indirect-fire targets	Electromagnetic warfare capabilities
Observation posts	Reconnaissance/security objectives
Command and control positions	Reconnaissance force routes
Final protective fires and final protective line	Fires (indirect fires and manned and unmanned aircraft systems)
Locations of reserves	Fires (indirect fires and manned and unmanned)
Routes for reserve commitment	Chemical, biological, radiological, and nuclear employment
Travel time for reserve commitment	Protective obstacle locations
Reserve force commitment triggers	Enemy scheme of maneuver
Sectors of fire	Unmanned aircraft system capabilities

Analysis of Troops and Support Available

2-78. Leaders study their task organization to determine the number, type, capabilities, and condition of available friendly troops and other support. Analysis of troops follows the same logic as analyzing the enemy by identifying capabilities, vulnerabilities, and strengths. Leaders should know the disposition, composition, strength, and capabilities of their forces one and two levels down. This information can be maintained in a checkbook-style matrix for use during COA development (specifically array forces). They maintain understanding of subordinates' readiness, including maintenance, training, strengths and weaknesses, leaders, and logistics status (known as LOGSTAT). Analysis of troops and support answers the question: What assets are available to accomplish the mission? Leaders also answer these questions:

- What are the strengths and weaknesses of subordinate leaders?

- What is the supply status of ammunition, water, fuel (if required), and other necessary items?
- What is the present physical condition of Soldiers (morale, sleep)?
- What is the condition of equipment?
- What is the unit's training status and experience relative to the mission?
- What additional Soldiers or units will accompany?
- What additional assets are required to accomplish the mission?
- What are the command and support relationships for all supporting or attached forces?

2-79. Perhaps the most critical aspect of mission analysis is determining the combat potential of one's own force. Leaders must realistically and unemotionally determine all available resources and new limitations based on level of training or recent fighting. This includes troops who are attached either to or in direct support of the unit. It also includes understanding the full array of assets in support of the unit. Leaders must know what fires are available (and when) by type, quantity, and priority of fires.

2-80. Throughout planning and preparation for an operation, leaders continually assess their unit's combat effectiveness—the ability of a unit to perform its mission. Information factors such as ammunition, personnel, fuel status, and weapon systems are evaluated and rated. (See FM 1-02.2.) The ratings used are—

- Fully operational – green (85 percent or greater).
- Substantially operational – amber (70 to 84 percent).
- Marginally operational – red (50 to 69 percent).
- Not operational – black (less than 50 percent).
- Unknown.

2-81. Because of the uncertainty always present in operations at the small-unit level, leaders cannot be expected to think of everything during their analysis. This fact forces leaders to determine how to get assistance when the situation exceeds their capabilities. Therefore, a secondary product of analysis of troops and support available should be an answer to the question, how do I get help?

Analysis of Time Available

2-82. The fifth mission variable of METT-TC (I) is time available. Time refers to many factors during the operations process (plan, prepare, execute, and assess). The five categories leaders consider include the memory aide HOPE-LW:

- Next higher echelon's timeline.
- Operational.
- Planning and preparation.
- Enemy timeline.
- Light and weather.

2-83. During all phases, leaders consider critical times, unusable time, the time it takes to accomplish activities, the time it takes to move, priorities of work, and tempo of operations. Other critical conditions to consider include visibility and weather data, and events such as higher headquarters tasks and required rehearsals. Implied in the analysis of time is leader prioritization of events and sequencing of activities.

2-84. As addressed in step 1 of the TLP, time analysis is a critical aspect to planning, preparation, and execution. Time analysis is often the first thing a leader does. The leader must not only appreciate how much time is available, but also must be able to appreciate the time/space aspects of preparing, moving, fighting, and sustaining. The leader must be able to see the leader's own tasks and enemy actions in relation to time. Most importantly, as events occur and more information is available, the leader must adjust the time available and assess its impact on what the leader wants to accomplish. Finally, the leader must update previous timelines for subordinates, listing all events affecting the platoon and its subordinate elements.

2-85. As an example, leaders estimate the time available to plan and prepare for the mission. Leaders begin by identifying the times they must complete major planning and preparation events, including rehearsals. Reverse planning assists in this process. Leaders identify critical times specified by higher headquarters and work backwards, estimating how much time each event will consume. Critical times might include times to load aircraft, cross the line of departure (LD), or the reach the start point for movement.

Analysis of Civil Considerations

2-86. Civil considerations include the influences of man-made infrastructure, civilian institutions, and attitudes, activities of civilian leaders, populations, and organizations within an AO, with regard to the conduct of military operations. Civil considerations generally focus on the immediate impact of civilians on operations in progress. Civil considerations of the environment can either help or hinder friendly or enemy forces; the difference lies in which leader has the information needed and taken the time to learn the situation and its possible effects on the operation. Analysis of civil considerations answers critical questions:

- How do civilian considerations affect the operation?
- How does the operation affect the civilians?
- What operations security measure can we implement to protect our critical information?
- For support we need from headquarters, when is it available?

2-87. The battalion, through the company commander, provides the platoon leader with civil considerations affecting the mission within the platoon's AO. The memory aid: areas, structures, capabilities, organizations, people, and events (ASCOPE) is used to analyze and describe these civil considerations.

2-88. The line between enemy combatants and civilian noncombatants is sometimes unclear. This requires the leader to understand the laws of war, the rules of engagement (ROE), and local situation. (See ATP 3-21.10 and FM 3-24.2 for additional information.)

Informational Considerations

2-89. *Informational considerations* are those aspects of the human, information, and physical dimensions that affect how humans and automated systems derive meaning from, use, act upon, and are impacted by information (FM 3-0). Understanding the physical, information, and human dimensions of each *domain*—a physically defined portion of an operational environment requiring a unique set of warfighting capabilities and skills (FM 3-0), helps the leader assess and anticipate the impacts of their operations. Their definitions follow:

- *Physical dimension* is the material characteristics and capabilities, both natural and manufactured, within an operational environment (FM 3-0).
- *Information dimension* is the content and data that individuals, groups, and information systems communicate and exchange, as well as the analytics and technical processes used to exchange information within an operational environment (FM 3-0).
- *Human dimension* encompasses people and the interaction between individuals and groups, how they understand information and events, make decisions, generate will, and act within an operational environment (FM 3-0).

Note. Within the context of an operational environment, a domain (land, maritime, air, space, and cyberspace) is a physically defined portion of an operational environment requiring a unique set of warfighting capabilities and skills. Each military Service and branch trains and educates its leaders to be experts about operations in a primary domain, although each Service has some capability in each of the domains, and each develops shared understanding of how to integrate capabilities from different domains. (See FM 3-0 for additional information.)

2-90. Operations reflect the reality that war is an act of force (in the physical dimension) to compel (in the information dimension) the decision making and behavior of enemy forces (in the human dimension). Actions in one dimension influence factors in the other dimensions. Understanding the interrelationship enables decision making about how to create and exploit advantages in one dimension and achieve objectives in the others without causing undesirable consequences.

2-91. The platoon leader integrates information into all operations and activities to create favorable support and circumstances for friendly action, limit enemy or adversary action, and minimize unintended consequences. Informational considerations are the relevant friendly, threat, and neutral (both military and civilian) individuals, organizations, and systems capable of generating cognitive effects and influencing behavior. Informational considerations and example information related questions within each mission variable are addressed in paragraphs 2-54 to 2-88. These considerations and example questions are not intended to be an all-inclusive, but they can help guide the platoon leader to integrate informational considerations into the analysis of the mission variables. (See FM 5-0 and FM 3-13 for additional information.)

Specific Operational Environments

2-92. Specific operational environments include urban, mountain (includes cold weather regions), jungle, and desert. Subsurface areas are conditions found in all operational environments. Offensive, defensive, and stability operations in these environments follow the same planning process as operations in any other environment, but they do impose specific techniques and methods for success. The uniqueness of each environment may affect more than their physical aspects but also their informational systems, flow of information, and decision-making. As such, mission analysis must account for the information environment, space operations, and cyberspace within each specific operational environment. Each specific operational environment has a specific publication because of their individual characteristics. They include:

- Urban terrain—see ATTP 3-06.11 and ATP 3-06 for additional information.
- Mountainous terrain and cold weather environments—see ATP 3-90.97, ATP 3-21.50, and ATP 3-21.18 for additional information.
- Jungle terrain—see ATP 3-90.98 for additional information.
- Desert terrain—see ATP 3-90.99 for additional information.
- Subsurface areas:
  - See ATP 3-21.51 for information on threat and hazardous subterranean structures existing or operating in concealment or hidden or when utilized in secret by an enemy or adversary.
  - See ATTP 3-06.11, ATP 3-06.1, ATP 3-21.50, ATP 3-06, and ATP 3-34.81 for additional information on subsurface areas.

Note. Subsurface areas are below the surface level (ground level) that may consist of underground facilities, passages, subway lines, utility corridors or tunnels, sewers and storm drains, caves, or other subterranean spaces. This dimension includes areas both below the ground and below water. In older cities, subsurface areas can include preindustrial age hand-dug tunnels and catacombs.

COURSE OF ACTION DEVELOPMENT

2-93. The purpose of COA development is to determine one or more ways to accomplish the mission consistent with the immediate higher commander's intent. A COA describes how the unit might generate the effects of overwhelming combat power against the enemy at the decisive point(s) with the least friendly casualties. Each COA the leader develops must be detailed enough to describe clearly how the leader envisions using all of the assets and combat multipliers to achieve the unit's mission-essential task and purpose.

Note. Decisive points are not centers of gravity; they are keys to attacking or protecting them. A center of gravity is the source of power that provides moral or physical strength, freedom of action, or will to act. (See FM 5-0 for additional information.)

2-94. To develop a COA, leaders focus on the actions the unit must take at the decisive point(s) and works backward to the start point. For example, leaders ensure sufficient combat power is available to enable the required actions against anticipated enemy contact(s) during movement. Leaders focus their efforts to develop at least one well-synchronized COA; if time permits, leaders should develop several. The result of the COA development process is paragraph 3 of the OPORD. A COA should position the unit for future operations and provide flexibility to meet unforeseen events during execution. It also should give subordinates the maximum latitude for initiative.

Screening Criteria

2-95. Leaders use screening criteria to ensure solutions they consider can solve the tactical problem. Screening criteria defines the limits of an acceptable solution. They are tools to establish the baseline products for analysis. Leaders may reject a solution based solely on the application of screening criteria. (See ATP 3-21.10 for additional information.) Leaders apply five categories of screening criteria to test a possible solution:

- Feasible—fits within available resources.
- Acceptable—worth the cost or risk.
- Suitable—solves the problem and is legal and ethical.
- Distinguishable—differs significantly from other solutions.
- Complete—contains the critical aspects of solving the problem from start to finish.

Note. Leaders assess risk continuously throughout COA development.

Actions (Generate Possible Solutions)

2-96. After gathering information relevant to the tactical problem and developing criteria, the platoon leader formulates possible solutions to accomplish the mission. The leader carefully considers the guidance provided by the commander or their superiors and develop several alternatives to solve the tactical problem. Too many possible solutions may result in time wasted on similar options. Experience and time available determine how many solutions the leader considers. The leader should consider at least two solutions. Limiting solutions enables the problem solver to use both analysis and comparison as problem-solving tools. Developing only one solution to "save time" may produce a faster solution but risks creating more problems from factors not considered. The leader analyzes relative combat power, generates options, arrays forces (generally two levels below their echelon), develops a concept of operations, analyzes Soldier load, assigns responsibility, and prepares a COA statement and sketch for each COA. (See ATP 3-21.10 appendix B for additional information.)

COURSE OF ACTION ANALYSIS

2-97. COA analysis begins with both friendly and enemy COA and, using a method of action-reaction-counteraction war game, results in a synchronized friendly plan, identified strengths and weaknesses, updated risk assessment, and a gaining of insights into possible branches and sequels to the basic plan. After developing the COA, leaders analyze it to determine its strengths and weaknesses, visualize the flow of the battle, identify the conditions or requirements necessary to enhance synchronization, and gain insights into actions at the decisive point(s) of the mission. If leaders developed more than one COA, they apply this same analysis to each COA developed.

2-98. Leaders do this analysis through war gaming or "fighting" the COA against at least one enemy COA. For each COA, leaders think through the operation from start to finish. Leaders compare their COA with the enemy's most probable COA. During the war game, leaders visualize a set of enemy and friendly actions and reactions. War gaming is the process of determining "what if" factors of the overall operations. The object is to determine what can go wrong and what decision leaders will likely have to make as a result. COA analysis allows leaders to synchronize their assets, identify potential hazards, and develop a better understanding of the upcoming mission.

2-99. The best way for the platoon leader to war game is to start at the platoon's current location and go through the mission from start to finish or start at a critical point such as the objective or engagement area. Using the action-reaction method, the platoon leader can think through the engagement beforehand. As war gaming proceeds, the leader can either record observations into a matrix or keep notes in a notebook. The most important aspect of this process is not the method but the output, meaning a more in-depth understanding of the operation. Depending on the time available and the platoon leader's personal preference, the leader may use the box, belt, and avenue-in-depth war gaming techniques. (See ATP 3-21.10 appendix B for additional information on each technique.)

COURSE OF ACTION COMPARISON AND SELECTION

2-100. If leaders developed more than one COA, they must compare them by weighing the specific advantages, disadvantages, strengths, and weaknesses of each as noted during the war game. These attributes may pertain to the accomplishment of the unit purpose, the use of terrain, the destruction of the enemy or other aspect of the operation they believe is important. Leaders make the final selection of a COA based on their own judgment, the start time of the operation, the AO, the scheme of maneuver, and subordinate unit tasks and purposes. (See ATP 3-21.10 for additional information.)

STEP 4 – INITIATE MOVEMENT

2-101. Leaders initiate movements necessary to continue mission preparation or to posture the unit for starting the mission. This step can be executed anytime throughout the sequence of the TLP. It can include movement to a tactical assembly area (TAA), battle position, attack position, or new AO, or the movement of guides or quartering parties. (See chapter 6 section VIII for additional information on troop movement.)

STEP 5 – CONDUCT RECONNAISSANCE

2-102. To exploit the principles of speed and surprise, leaders should weigh the advantages of reconnoitering personally against just using the supplied information from higher echelon information systems. They realistically consider the dangers of reconnoitering personally, and time required to conduct them. Leaders might be able to plan their operations using combat information provided by higher echelon information collection assets. Combat information is the unevaluated data, gathered by or provided directly to the tactical commander which, due to its highly perishable nature or the criticality of the situation, cannot be processed into tactical intelligence in time to satisfy the user's tactical intelligence requirements. It can be extremely important in a time-constrained environment. However, if time permits, leaders should verify higher headquarters' intelligence requirement by reconnoitering visually. They should seek to confirm the PIR supporting their tentative plans. These PIRs usually consists of assumptions or critical facts about the enemy. This can include strength and location, especially at templated positions. It also can include information about the terrain. For example, verification that a tentative *support by fire position*—the general position from which a unit performs the tactical mission task of support by fire (ADP 3-90)—can *suppress*—a tactical mission task in which a unit temporarily degrades a force or weapon system from accomplishing its mission (FM 3-90)—the enemy, or an avenue of approach is useable.

2-103. If possible, leaders should include their subordinate leaders in their reconnaissance efforts. This allows the subordinates to see as much of the terrain and enemy as possible. The reconnaissance also helps subordinate leaders gain insight into the leaders' visions of the operation.

2-104. The leaders' reconnaissance (see paragraphs 7-15 to 7-16) might include moving to or beyond the LD, reconnaissance and surveillance of an AO, or walking from the forward edge of the battle area back to and through the platoon AO or battle position along likely enemy avenues of approach. If possible, leaders should select vantage points with the best possible view of the decisive point(s). In addition to the leaders' reconnaissance efforts, units can conduct additional reconnaissance operations. Examples include surveillance of an area by subordinate elements, patrols (see chapter 7) to determine enemy locations, and establishment of observation posts/combat outpost (see chapter 6) to gain additional information. Leaders also can incorporate Javelin command launch unit (known as CLU) and small unmanned aircraft systems as surveillance tools (day or night), based on an analysis of METT-TC (I).

2-105. The nature of the reconnaissance and surveillance, including what it covers and how long it lasts, depends on the tactical situation and time available. Leaders should use the results of the COA development process to identify information and security requirements of the unit's reconnaissance and surveillance mission.

2-106. Leaders must include disseminating results and conclusions arrived from reconnaissance and surveillance into their time analysis. They also must consider how to communicate changes in the COA to their subordinates and how these changes affect their plans, actions of the subordinates, and other supporting elements. (See ATP 3-21.10 for additional information.)

STEP 6 – COMPLETE THE PLAN

2-107. During this step, leaders expand their selected (or refined) COA into a complete OPORD. The COA is finalized, maturing into a mission statement and intent, describing the decisive point, the concept of operations, and the detailed scheme of maneuver. They prepare overlays, refine the indirect fire list, complete sustainment and command and control requirements and, of course, update the tentative plan based on the latest reconnaissance or information. They prepare briefing sites and other briefing materials they might need to present the OPORD directly to their subordinates.

2-108. Using the five-paragraph OPORD format helps them to explain all aspects of the operation: terrain, enemy, higher and adjacent friendly units, unit mission, execution, support, and command and control. The format also serves as a checklist to ensure they cover all relevant details of the operation. It also gives subordinates a smooth flow of information from beginning to end.

STEP 7 – ISSUE OPERATION ORDER

Note. The platoon leader focuses the OPORD on the purpose of the operation through mission orders (see note on page 2-6). Mission orders allow the leader's subordinates the greatest possible flexibility to accomplish assigned tasks.

2-109. The OPORD precisely and concisely explains both the leader's intent and concept of how the leader envisions the unit accomplishing the mission. The order does not contain unnecessary information. The OPORD is delivered quickly and in a manner allowing subordinates to concentrate on understanding the leader's vision and not just copying what the leader says verbatim. Leaders must prepare adequately and deliver the OPORD confidently and quickly to build and sustain confidence in their subordinates.

2-110. When issuing the OPORD, leaders must ensure their subordinates understand and share their vision of what must be done and when and how it must be done. Subordinates must understand how all platoon elements work together to accomplish the mission. They also must understand how the platoon mission supports the intentions of the immediate higher commander and how the platoon interacts with adjacent units. When a leader has finished issuing the order, subordinate leaders should leave with a clear understanding of what the leader expects their elements to do. Leaders are responsible for ensuring their subordinates understand.

2-111. Leaders must issue the order in a manner instilling subordinates with confidence in the plan and a commitment to do their best to achieve the plan. Whenever possible, leaders must issue the order in person either on a terrain model or the actual terrain on which the operation will occur. Leaders look into the eyes of their subordinate leaders to ensure each one understands the mission and what the element must achieve.

2-112. Complete the order with a *confirmation brief*—a brief subordinate leaders give to the higher commander immediately after the operation order is given to confirm understanding (ADP 5-0). At a minimum, subordinate leaders should be able to brief the unit mission and intent, the immediate higher commander's intent, their own tasks and purpose, and time they will issue their unit's OPORD. Subordinate leaders should confirm they understand the commander's vision and how the mission is accomplished with respect to the decisive point(s). This confirmation brief provides an opportunity to highlight issues or concerns.

2-113. The five-paragraph OPORD format (see figure 2-5), help leaders paint a picture of all aspects of the operation, from the terrain to the enemy, and finally to the unit's own actions from higher to lower. The format helps them decide what relevant details they must include and in providing subordinates with a smooth flow of information from beginning to end. At the same time, leaders must ensure the order is not only clear and complete but also as brief as possible. If leaders have already addressed an item adequately in a previous WARNORD, they can simply state "no change," or provide necessary updates. Leaders are free to brief the OPORD in the most effective manner to convey information to their subordinates.

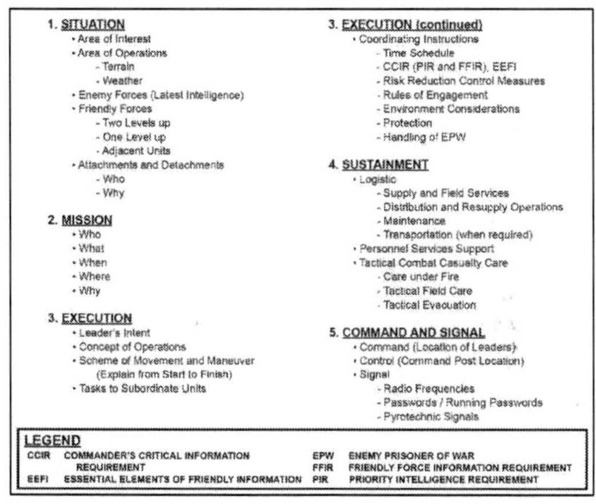

Figure 2-5. Operation order format

STEP 8 – SUPERVISE AND REFINE

2-114. This final step of the TLP is crucial. After issuing the OPORD, leaders and subordinate leaders must ensure the required activities and tasks are completed in a timely manner prior to mission execution. Supervision is the primary responsibility of all leaders. Both officers and noncommissioned officers must check everything important for mission accomplishment. These checks include, but are not limited to—

- Conducting numerous backbriefs on all aspects of the platoon and subordinate unit operations.
- Ensuring the second in command in each element is prepared to execute in the leaders' absence.
- Listening to subordinates' OPORD.
- Observing rehearsals of subordinate units.
- Checking load plans to ensure units are carrying only what is necessary for the mission or what the OPORD specified.
- Checking the status and serviceability of weapons.
- Checking on maintenance activities of subordinates.
- Ensuring local security is maintained.

REHEARSALS

2-115. A *rehearsal* is a session in which the commander and staff or unit practices expected actions to improve performance during execution (ADP 5-0). They are essential in ensuring thorough preparation, coordination, and understanding of the commander's plan and intent. Leaders should never underestimate the value of rehearsals.

2-116. Rehearsals require leaders and, when time permits, other platoon Soldiers to perform required tasks, ideally under conditions as close as possible to those expected for the actual operation. At their best, rehearsals are interactive, while verbalizing their elements' actions. During every rehearsal, the focus is on the how element, allowing subordinates to practice the actions called for in their individual scheme of maneuver.

Note. Rehearsals are different from a discussion of what is supposed to happen during the actual event. Leaders can test subordinates understanding of the plan by ensuring they push the rehearsal forward rather than waiting to dictate each step of the operation.

2-117. Leaders use well-planned, efficiently run rehearsals to accomplish the following:

- Reinforce training and increase proficiency in critical tasks.
- Reveal weaknesses or problems in the plan, leading to more refinement of the plan or development of additional branch plans.
- Integrate the actions of subordinate elements, to include actions on contact during movement and maneuver.

- Confirm coordination requirements within the platoon and between the platoon and adjacent units.
- Improve each Soldier's understanding of the concept of operations, the direct fire plan, anticipated contingencies, and possible actions and reactions for various situations may arise during the operation.
- Ensure seconds-in-command are prepared to execute in their leaders' absence.

Types of Rehearsals

2-118. Each rehearsal type achieves different results and has a specific place in the timeline. (See ATP 3-21.10 and FM 6-0 for additional information.) The four types of rehearsals are—

- Backbrief.
- Combined arms rehearsal.
- Support rehearsal.
- Battle drill or SOP rehearsal.

Methods of Rehearsals

2-119. Rehearsals should follow the crawl-walk-run training methodology whenever possible. This prepares the platoons and subordinate elements for increasingly difficult conditions. (See ATP 3-21.10 and FM 6-0 for additional information.) Units may conduct these methods of rehearsals dependent on the situation and as time permits—

- Full-dress rehearsal.
- Reduced-force rehearsal.
- Terrain-model rehearsal.
- Digital terrain-model rehearsal.
- Sketch-map rehearsal.
- Map rehearsal.
- Network rehearsal.

Note. Because time is scarce during TLP, leaders should direct certain rehearsals (for example, special teams, breach, battle drills) to occur by way of the WARNORD.

PRECOMBAT CHECKS AND INSPECTIONS

2-120. PCCs and PCIs are critical to the success of missions. These checks and inspections are leader tasks and cannot be delegated below the team-leader level. For example, at the platoon echelon, the platoon sergeant spot checks throughout the unit's preparation for combat. The platoon leader and platoon sergeant make a final inspection. For the squad echelon, team leaders spot check throughout preparation, and the squad leader makes a final inspection. They ensure the Soldier is prepared to execute the required individual and collective tasks supporting the mission. Checks and inspections are part of the TLP protecting against shortfalls endangering Soldiers' lives and jeopardize the execution of a mission.

2-121. PCCs and PCIs must be tailored to the specific unit and mission requirements. Each mission and each patrol may require a separate set of checklists. A rifle squad will have its own established set of PCCs and PCIs, but each squad within its rifle platoon should have identical checklists. Weapons squads will have a different checklist than rifle squads, but each weapons squad within a company should be the same.

2-122. One of the best ways to ensure PCCs and PCIs are complete and thorough is with full-dress rehearsals. These rehearsals, run at combat speed with communication and full-battle equipment, allow the leader to envision minute details, as they will occur in the AO. If the operation is to be conducted at night, Soldiers should conduct full-dress rehearsals at night as well. PCCs and PCIs should include backbriefs on the mission, the task and purpose of the mission, and how the Soldiers' role fits into the scheme of maneuver. The Soldiers should be familiar with the latest intelligence assessments, ROE, and be versed casualty evacuation (CASEVAC) and in medical evacuation (MEDEVAC) procedures and sustainment requirements.

2-123. Table 2-3 lists sensitive items, high dollar value items, pieces of equipment, and supplies that a unit may be equipped with during a particular mission. A unit's SOP generally addresses PCC and PCI specific to an individual unit. Table 2-3 is only provided for discussion purposes and is not all inclusive for every mission.

Table 2-3. Precombat checks and precombat inspection checklist, example

Sensitive Items, High-Dollar Value Items, Pieces of Equipment, and Supplies, Example		
- Identification card/identification tags - Canteens/hydration bladder - Protective mask -All clothing items in packing list - Poncho - Improved rain suit top and bottom - Earplugs - Camouflage stick - Ammunition/magazines - Individual weapons - Bayonet and scabbard - Weapons cleaning kit - Weapon tie downs - Pocket mirror - Water purification tablets - Watch - Lip balm and sunscreen - Advanced combat helmet - Combat boots - Entrenching tool - Flexi cuffs - Litter - Protractor/map - Lensatic compass - Notebook/pen/pencil	- Traversing and elevating mechanisms - Spare barrels - Spare barrel bags - Extraction tools - Asbestos gloves - Barrel changing handles - Tripods - Pintles - Counterradio/electromagnetic warfare device - Vehicle tools (if applicable) - Binoculars - Meals and food - Visual/language translator card - Concertina wire gloves - Equipment packed according to standard operating procedure - Call for fire procedures - 9-line medical evacuation procedures	- Grappling hook - Sling sets - Flashlights - VS-17 panel - Pick-up zone marking kit - Radios and backup communications - Global Positioning System/laser range finder - Handheld microphones - Night vision devices - Night vision goggles mounting plate - Batteries and spare batteries - Improved outer tactical vest - Ruck/assault pack systems - Knee and elbow pads - Ballistic spectacles - Insect repellent - Infrared strobe - Chemical lights - Combat lifesaver bags

This page intentionally left blank.

Chapter 3
Movement and Maneuver

Movement is the positioning of combat power to establish the conditions for maneuver (ADP 3-90). *Maneuver* is movement in conjunction with fires (ADP 3-90). Maneuver happens once a unit has made contact with the enemy and combines movement with fires to gain a position of advantage over the enemy. Leaders use their understanding of the enemy situation to identify where and when to deliberately transition from movement to maneuver before making contact with the enemy. This chapter discusses route movement selection, navigation, and execution and situations of tactical movement and maneuver within the Infantry rifle platoon and squad.

SECTION I – ROUTE SELECTION, NAVIGATION, AND EXECUTION

3-1. Leaders analyze the terrain from two perspectives. First, to see how it can provide tactical advantage to friendly and enemy forces. Second, they look at the terrain to determine how it can aid navigation. Leaders identify areas or terrain features dominating their movement *route*—the prescribed course to be traveled from a point of origin to a destination (FM 3-90). These areas can become possible intermediate (commonly called march) and final objectives.

ROUTE SELECTION AND NAVIGATION

3-2. Leaders identify good ground along the route that facilitates navigation and the destruction of enemy forces in the event contact occurs. If the leader wants to avoid contact, the leader chooses terrain that conceals the unit's movement. If the leader wants to make contact, the leader chooses terrain from where the leader can easily scan and observe the enemy. On other occasions, the leader may require terrain allowing stealth or speed. Regardless of the requirement, the leader must ensure most of the terrain along the route provides some tactical advantage.

AID OF TECHNOLOGY

3-3. Route selection and navigation are made easier with the aid of technology. The latest command and control systems enhance the Infantry platoon's and squad's ability to ensure they are in the right place at the right time, and to determine the location of adjacent units.

Note. Soldiers must be proficient in land navigation standard map, terrain association, and compass skills unassisted by technology.

NAVIGATION AIDS

3-4. There are two categories of navigational aids: linear and point. Linear navigational aids are terrain features such as trails, streams, ridgelines, wood lines, power lines, streets, and contour lines. Point terrain features include hilltops and prominent buildings. Navigation aids usually are assigned control measures to facilitate communication during the movement. Typically, linear features are labeled as phase lines while point features are labeled as checkpoints (or rally points). Three navigational aid techniques include catching features, handrails, and navigational attack points.

Catching Features

3-5. Catching features are obvious terrain features which go beyond a waypoint or control measure and can be either linear or point. The general idea is if the unit moves past the objective, limit of advance (LOA), or checkpoint, the catching feature will alert it that it has traveled too far. Two catching features methods include the offset-compass and boxing-in the route.

The Offset-Compass Method

3-6. If there is the possibility of missing a particular point along the route (such as the endpoint or a navigational attack point), it is sometimes preferable to aim deliberately at the leg to the left or right of the end point toward a prominent catching feature. Once reached, the unit simply turns the appropriate direction and moves to the desired endpoint. This method is especially helpful when the catching feature is linear.

Boxing-In the Route Method

3-7. Another method leaders use to prevent themselves from making navigational errors is to "box in" the leg or the entire route. This method uses catching features, handrails, and navigational attack points to form boundaries. Creating a box around the leg or route assists in easily recognizing and correcting deviation from the planned leg or route.

Handrails

3-8. Handrails are linear features parallel to the proposed route. The general idea is to use the handrail to keep the unit oriented in the right direction. Guiding off a handrail can increase the unit's speed while also acting as a catching feature.

Navigational Attack Points

3-9. Navigational attack points are an obvious landmark near the objective, LOA, or checkpoint that can be found easily. Upon arriving at the navigational attack point, the unit transitions from rough navigation (terrain association or general azimuth navigation) to point navigation (dead reckoning). Navigational attack points typically are labeled as checkpoints.

ROUTE PLANNING

3-10. Route planning must consider enabling tasks specific to tactical movement. These tasks facilitate the overall operation. Tactical movement normally contains some or all the following enabling tasks:

- Planning movement with Global Positioning System (GPS) waypoints or checkpoints utilizing navigation skills.
- Movement to and passage of friendly lines.
- Movement to an objective rally point (known as ORP).
- Movement to a probable line of deployment (PLD).
- Movement to an LOA.
- Linkup with another unit.
- Movement to a patrol base or tactical assembly area (TAA).
- Movement back to and reentry of friendly lines.

3-11. Leaders first identify where they want to end up (the objective or LOA). Then, working back to their current location, they identify all the critical information and actions required as they relate to the route. For example, navigational aids, tactical positions, known and suspected enemy positions, and friendly control measures. Using this information, they break up their route in manageable parts called legs. Finally, they capture their information and draw a sketch on a route chart. There are three decisions leaders make during route planning:

- The type of (or combination of) navigation to use.
- The type of route during each leg.
- The start point and end point of each leg.

3-12. All routes have a designated start point and release point that are easily recognizable on the map and on the ground. The *start point* is a designated place on a route where elements fall under the control of a designated march commander (FM 3-90). It is far enough from the TAA to allow units to organize and move at the prescribed speed and interval when they reach the start point. A *release point* is a designated place on a route where elements are released from centralized control (FM 3-90). Once released from centralized control they are released back to the authority of their respective commanders. Each start point must have a corresponding release point, which must also be easy to recognize on the ground. Marching units do not stop at the release point; instead, as they move through the release point and continue toward their own appropriate destination.

3-13. The leader assesses the terrain in the proposed AO. In addition to the standard Army map, the leader may have aerial photographs and terrain analysis overlays from the parent unit, or information is provided by someone familiar with the area.

3-14. To control movement, a leader may use an axis of advance, direction of attack, infiltration lanes, phase lines, PLD, checkpoints (waypoints), final coordination line, rally points, TAA, and routes. Assigning an AO both restricts and facilitates the movement of units and use of fires.

TYPES OF NAVIGATION

3-15. The general types of navigation are terrain association, general azimuth, and point navigation. Leaders use whichever type or combination best suits the situation. The principles of land navigation while mounted are basically the same as while dismounted.

Terrain Association

3-16. Terrain association is the ability to identify terrain features on the ground by the contour intervals depicted on the map. The leader analyzes the terrain using the factors of observation and field of fire, avenues of approach, key terrain, obstacles, and cover and concealment (OAKOC), and identifies major terrain features, contour changes, and man-made structures along the axis of advance. The leader uses these features to orient the unit and to associate ground positions with map locations. The use of terrain association is that it forces the leader to assess the terrain continually. This leads to identifying tactically advantageous terrain and using terrain to the unit's advantage.

General Azimuth

3-17. For this type of navigation, the leader selects linear terrain features; then while maintaining map orientation and a general azimuth, the leader guides on the terrain feature. An advantage the general azimuth navigation has is it speeds movement, avoids fatigue, and often simplifies navigation because the unit follows the terrain feature. The disadvantage is it usually puts the unit on a natural line of drift. This type of navigation ends like terrain association, with the unit reaching a catching feature or a navigational attack point, then switching to point navigation.

Point Navigation

3-18. Point navigation, also called dead reckoning, is done by starting from a known point and strictly following a predetermined azimuth and distance. This type of navigation requires a high level of leader control because even a slight deviation over the course of a movement can cause navigation errors. One method for this type of navigation uses a handheld compass and a distance from the pace Soldier to follow a prescribed route. Another method can involve a Soldier with a GPS following the steering arrow from waypoint to waypoint. Point navigation requires the leader to follow these steps:

- Use the compass to maintain direction.
- Use the pace Soldier's pace (or when mounted use a vehicle odometer) to measure the distance traveled for each leg or part.
- Review the written description of the route plan to help prevent navigational errors.

Note. Do not take compass reading from inside vehicles. Move away from vehicles when using a lensatic compass.

3-19. When performed correctly, point navigation is very reliable, but time-consuming. It is best used when the need for navigational accuracy outweighs the importance of using terrain. Point navigation is particularly useful when recognizable terrain features do not exist or are too far away to be helpful. One of the problems with point navigation is negotiating severely restrictive terrain or danger areas. For example, deserts, swamps, and thick forest make terrain association difficult. Using point navigation early on in a long movement can stress the compass Soldier and it may be advisable to switch the compass Soldier. One of the problems with point navigation is negotiating severely restrictive terrain or danger areas.

Combination

3-20. Leaders can benefit from combining these types of navigation. Terrain association and general azimuth method enable leaders to set a rough compass bearing and move as quickly as the situation allows toward a catching feature or a navigational attack point. Once reached, leaders switch to point navigation by paying close attention to detail, taking as much time as necessary to analyze the situation and find their point. Terrain association and general azimuth method allow for some flexibility in the movement, and do not require the same level of control as point navigation. Point navigation, on the other hand, enables leaders to locate their objective or point precisely.

Mounted Land Navigation

3-21. The principles of land navigation while mounted are basically the same as while dismounted. The major differences are the speed and distances associated with mounted movement. To be effective at mounted land navigation, the travel speed must be considered. When preparing to move, the effects of terrain on navigating mounted vehicles must be determined. Soldiers will cover great distances very quickly, and they must develop the ability to estimate the distance they have traveled. Using the odometer on the vehicle can assists with distance traveled but can be misleading on a map due to turns and going up and down hills for instance. Having a mobility advantage helps while navigating. Mobility makes it much easier if Soldiers get disoriented to move to a point where they can reorient themselves. When determining a route to be used when mounted, consider the capabilities of the vehicles to be used. Most military vehicles are limited in the degree of slope they can climb and the type of terrain they can negotiate. Swamps, thickly wooded areas, or deep streams may present no problems to dismounted Soldiers, but the same terrain may completely stop mounted Soldiers.

ROUTE TYPES

3-22. There are three types of routes leaders can choose from: those which follow linear terrain features; those which follow a designated contour interval; and those which go cross compartment. Terrain association can be used with all three route types. The general azimuth type of navigation is used with the contour and terrain feature route types. Point navigation is used primarily with cross compartment.

Terrain Feature

3-23. Following a terrain feature is nothing more than moving along linear features such as ridges, valleys, and streets. The advantage of this method is the unit is moving with the terrain. This is normally the least physically taxing of the methods. The disadvantage of following terrain features also means following natural lines of drift, which leads to a higher probability of chance contact with the enemy.

Contouring

3-24. Contouring (remaining at the same height the entire leg) follows the imaginary contour line around a hill or along a ridgeline. Contouring has two advantages. First, it prevents undue climbing or descending. Second, following the contour acts as handrail or catching feature. The disadvantage of contouring is it can be physically taxing.

Cross Compartment

3-25. Cross compartment means following a predetermined azimuth and usually means moving against the terrain. The advantage of this method is it provides the most direct route from the start point to the end point of the leg or route. There are two primary disadvantages to this type of route. First, this method can be physically taxing. Second, the unit might expose itself to enemy observation.

Develop a Leg

3-26. The best way to manage a route is to divide it into segments called "legs." By breaking the overall route into several smaller segments, the leader is able to plan in detail. Legs typically have only one distance and direction. A change in direction usually ends the leg and begins a new one.

3-27. A leg must have a definite beginning and ending, marked with a control measure such as a checkpoint or phase line. (When using GPS, these are captured as waypoints.) When possible, the start point and end point should correspond to a navigational aid (catching feature or navigational attack point).

3-28. To develop a leg, leaders first determine the type of navigation and route best suiting the situation. Once these two decisions are made, the leader determines the distance and direction from the start point to the end point. The leader then identifies critical METT-TC (I) mission, enemy, terrain and weather, troops and support available, time available, civil considerations, and informational considerations as it relates to the specific leg. Finally, leaders capture this information and draw a sketch on a route chart. (See figure 3-1.)

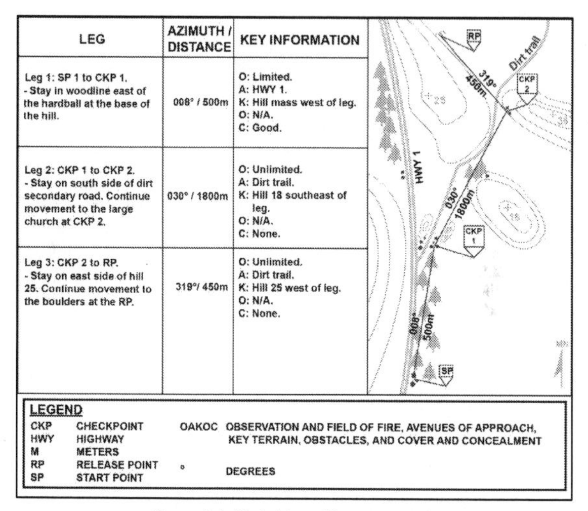

LEG	AZIMUTH / DISTANCE	KEY INFORMATION
Leg 1: SP 1 to CKP 1. - Stay in woodline east of the hardball at the base of the hill.	008° / 500m	O: Limited. A: HWY 1. K: Hill mass west of leg. O: N/A. C: Good.
Leg 2: CKP 1 to CKP 2. - Stay on south side of dirt secondary road. Continue movement to the large church at CKP 2.	030° / 1800m	O: Unlimited. A: Dirt trail. K: Hill 18 southeast of leg. O: N/A. C: None.
Leg 3: CKP 2 to RP. - Stay on east side of hill 25. Continue movement to the boulders at the RP.	318°/ 450m	O: Unlimited. A: Dirt trail. K: Hill 25 west of leg. O: N/A. C: None.

LEGEND

CKP	CHECKPOINT	OAKOC	OBSERVATION AND FIELD OF FIRE, AVENUES OF APPROACH, KEY TERRAIN, OBSTACLES, AND COVER AND CONCEALMENT
HWY	HIGHWAY		
M	METERS		
RP	RELEASE POINT	°	DEGREES
SP	START POINT		

Figure 3-1. Sketching of legs, example

ROUTE EXECUTION

3-29. Using decisions about the route and navigation made during planning and preparation, leaders execute their route and direct their subordinates. In addition to executing the plan, leaders—

- Determine and maintain accurate location.
- Designate rally points.

DETERMINE AND MAINTAIN ACCURATE LOCATION

3-30. A leader must always know the unit's location during movement. Without an accurate location, the unit cannot rapidly receive help from supporting arms, integrate reserve forces, or accomplish their mission. To ensure an accurate location, a leader uses many techniques, including:

- Executing common skills.
- Designating a compass Soldier and pace Soldier.
- Using command and control systems (for example, GPS).

Common Skills

3-31. All Infantry Soldiers, particularly leaders, must be experts in land navigation. Important navigation tasks common to all include:

- Locating a point using grid coordinates. Using a compass (day/night).
- Determining location using resection, intersection, or modified resection.
- Interpreting terrain features.
- Measuring distance and elevation.
- Employing command and control systems.

Compass Soldier

3-32. The compass Soldier assists in navigation by ensuring the lead fire team leader remains on course at all times. The compass Soldier should be thoroughly briefed. Instructions to the compass Soldier must include an initial azimuth with subsequent azimuths provided as necessary. The platoon leader or squad leader should also designate an alternate compass Soldier and validate the patrol's navigation with GPS devices.

Pace Solder

3-33. The pace Soldier maintains an accurate pace at all times. The platoon leader or squad leader should designate how often the pace Soldier reports the pace. The pace Soldier also should report the pace at the end of each leg. The platoon leader or squad leader should designate an alternate pace Soldier.

Global Positioning Systems

3-34. During planning, leaders enter their waypoints into the GPS. Once entered, the GPS can display information such as distance and direction from waypoint to waypoint. During execution, leaders use the GPS to establish their exact location.

3-35. GPS signals are subject to denial by threat forces. Also, GPS signals may be inaccurate in dense foliage. Both mean that GPS should always be backed with map, compass, and pace count.

Note. Leaders need to remember GPS and digital displays are not the only navigational tools they can use. The best use of GPS or digital displays is for confirming the unit's location during movement. Terrain association and map- reading skills still are necessary, especially for point navigation. Over reliance of GPS and digital displays can cause leaders to ignore the effects of terrain, travel faster than conditions allow, miss opportunities, or fail to modify routes when necessary.

DESIGNATE RALLY POINTS

3-36. A rally point is a place designated by the leader where the unit moves to reassemble and reorganize if it becomes dispersed. It also can be a place for a temporary halt to reorganize and prepare for actions at the objective, to depart from friendly lines, or to reenter friendly lines. Planned rally points are common control measures used during tactical movement. (See chapter 7 for additional information.)

SECTION II – TACTICAL MOVEMENT (APPROACH MARCH)

3-37. This section discusses the tactical movement of Infantry small units (specifically the fire team, squad, and platoon) during the conduct of an approach march. Platoon leaders use tactical movement for several purposes: to conduct movement, to relate one squad to another on the ground; to position firepower to support the direct-fire plan; to establish responsibilities for security among squads; and/or to aid in the execution of battle drills. Just as leaders do with maneuver, leaders plan tactical movements based on where they expect enemy contact, and how they are to react to contact (see chapter 2). Platoon leaders evaluate the situation and decide what type of tactical movement best suits the situation and mission. During approach marches, units use movement formations and movement techniques to balance security and speed throughout the operation. (See chapter 6 section VIII for additional information on tactical and nontactical movement.)

Note. Tactical movement involves movement of a unit assigned a mission under combat conditions when not in direct ground contact with the enemy. Tactical movement is based on the anticipation of early ground contact with the enemy, either en route or shortly after arrival at the destination. Tactical movement ends when ground contact is made, or the unit reaches its destination. Tactical movement is not maneuver.

MOVEMENT FORMATIONS

3-38. A *movement formation* is an ordered arrangement of forces for a specific purpose and describes the general configuration of a unit on the ground (ADP 3-90). They are composed of two variables: lateral frontage, represented by the line formation; and depth, represented by the column formation. The advantages attributed to one of these variables are disadvantages to the other. Leaders combine the elements of lateral frontage and depth to determine the best formation for their situation. In addition to the line and column/file, the other five types of formations—box, vee, wedge, diamond, and echelon—combine these elements to varying degrees. Each does so with different degrees of emphasis resulting in unique advantages and disadvantages.

MOVEMENT CONSIDERATIONS

3-39. Movement formations do not demand parade-ground precision. Platoons and squads must retain the flexibility needed to vary their formations to the situation. Using formations allows Soldiers to execute battle drills quickly and gives them the assurance their leaders and buddy team members are in the expected positions and performing the right tasks. Formations also provide 360-degree security and allow units to give most of their firepower to the flanks or front in anticipation of enemy contact.

3-40. Sometimes platoon and company formations differ due to the mission variables of METT-TC (I). For example, the platoons could move in wedge formations within a company vee. It is not necessary for platoon formations to be the same as the company formation unless directed by the company commander. However, platoon leaders coordinate their formation with other elements moving in the main body's formation.

Note. Formation illustrations shown in this chapter are examples only. They might not depict actual situation or circumstances. Leaders must be prepared to adapt their choice of formation to the specific situation. Leaders always should position themselves where they can best control their formations.

3-41. The seven movement formations can be grouped into two categories: formations with one lead element, and formations with more than one lead element. The formations with more than one lead element, as a general rule, are better for achieving *fire superiority*—the dominating fires of one force over another force that permits that force to maneuver at a given time and place without prohibitive interference by the other (FM 3-90)—to the front but are more difficult to control. Conversely, the formations with only one lead element are easier to control but are not as useful for achieving fire superiority to the front.

3-42. Leaders attempt to maintain flexibility in their formations. Doing so enables them to react when unexpected enemy actions occur. The line, echelon, and column formations are the least flexible of the seven formations. The line formation masses fires to the front but has vulnerable flanks. The echelon is optimized for a flank threat, something units want to avoid. The column has difficulty reinforcing an element in contact. Leaders using these formations should consider ways to reduce the risks associated with their general lack of flexibility. For example, by expanding or contracting the formation depending on the terrain.

3-43. During movement, every squad and Soldier has a standard position within a tactical movement. During movement Soldiers can see their fire team leaders. Fire team leaders can see their squad leaders. Leaders control their units using dismounted hand and arm signals and intra-squad and team communications (see TC 3-21.60 and ATP 6-02.53). Movement formation considerations and hand and arm (dismounted) signals are addressed and illustrated, respectively in table 3-1.

Table 3-1. Movement formation considerations with hand and arm signals

NAME/FORMATION/ SIGNAL (APPLICABLE)	CHARACTERISTICS	ADVANTAGES	DISADVANTAGES
Line Formation	- All elements arranged in a row - Majority of observation and direct fires oriented forward; minimal to the flanks - Each subordinate unit on the line must clear its own path forward - One subordinate designated as base on which the other subordinates cue their movement	Ability to: - Generate fire superiority to the front - Clear a large area - Disperse - Transition to bounding overwatch, base of fire, or assault	- Control difficulty increases during limited visibility and in restrictive or close terrain - Difficult to designate a maneuver element - Vulnerable assailable flanks - Potentially slow - Large signature
Column/File Formation	- One lead element - Majority of observation and direct fires oriented to the flanks; minimal to the front - One routed means unit only influenced by obstacles on that one route	- Easiest formation to control (as long as leader can communicate with lead element) - Ability to generate a maneuver element - Secure flanks - Speed	- Reduced ability to achieve fire Superiority to the front - Clears limited area and concentrates the unit - Transitions poorly to bounding overwatch, base of fire, and assault - Column's depth makes it a good target for close air attacks and machine gun beaten zone
Vee Formation	- Two lead elements - Trail elements move between the two lead elements - Used when contact to the front is expected - "Reverse wedge" - Unit required to two lanes/routes forward	The ability to: - Generate fire superiority to the front - Generate a manuever element - Secure flanks - Disperse - Transition to bounding overwatch, base of fire, or assault	- Control difficulty increases during limited visibility and in restrictive or close terrain - Potentially slow
Box Formation	- Two lead elements - Trail elements follow lead elements - All-around security	Same as vee formation advantages	Same as vee formation disadvantages
Wedge Formation	- One lead element - Trail elements paired off abreast of each other on the flanks - Used when the situation is uncertain	The ability to: - Control, even during limited visibility, in restrictive terrain, or in close terrain - Transition trail elements to base of fire or assault - Secure the front and flanks - Easy transition to line and column	- Trail elements are required to clear their path forward - Frequent need to transition to column in restrictive, close terrain
Diamond Formation	- Similar to the wedge formation - Fourth element follows lead element	Same as wedge formation advantages	Same as wedge formation disadvantages
Echelon Formation (Right)	- Elements deployed diagonally left and right - Observation and fire to both the front and one flank - Each subordinate unit on the line clears its own path forward	- Ability to assign sectors that encompass both the front and flank	- Difficult to maintain proper relationship between subordinates - Vulnerable to the opposite flanks

FIRE TEAM MOVEMENT FORMATIONS

3-44. The term fire team formation refers to the Soldiers' relative positions within the fire team. The most common fire team formations among the movement options addressed in table 3-1 on page 3-11 include the fire team wedge and fire team column/file. (See table 3-2.) Both formations have advantages and disadvantages. Regardless of which formation the team employs, Soldiers must know their location in the formation relative to the other fire team members and team leader. Each Soldier covers a set area of responsibility for observation and direct fire as the team is moving. To provide the unit with all-around protection, these areas interlock. Team leaders are constantly aware of their teams' sectors of fire and correct them as required.

Table 3-2. Comparison of fire team formations

Movement Formation	*When Most Often Used*	*Movement Characteristics*			
		Control	*Flexibility*	*Fire Capabilities and Restrictions*	*Security*
Fire team wedge	Basic fire team formation	Easy	Good	Allows immediate fires in all direction	All-round
Fire team column/file	Close terrain, limited visibility, dense vegetation	Easiest	Less flexible than the wedge	Allows immediate fires to the flanks, masks most fires to the rear	Least

3-45. The team leader adjusts the team's formation as necessary while the team is moving. The distance between Soldiers will be determined by the mission, the nature of the threat, the closeness of the terrain (such as swamps, woods, hilly and mountainous areas, and urban areas), and by the visibility. As a rule, the unit should be dispersed up to the limit of control. This allows for a wide area to be covered, makes the team's movement difficult to detect, and makes it less vulnerable to enemy ground and air attack. Fire teams rarely act independently. However, in the event they do, when halted, they use a perimeter defense to ensure all-around security.

Fire Team Wedge

3-46. The wedge (see figure 3-2) is the basic formation of the fire team. The interval between Soldiers in the wedge formation is normally 10 meters. The wedge expands and contracts depending on the terrain. Fire teams modify the wedge when encountering rough terrain, poor visibility, or other factors make control of the wedge difficult. The normal interval is reduced so all team members still can see their team leader and all team leaders still can see their squad leader. The sides of the wedge can contract to the point where the wedge resembles a single file. Soldiers expand or resume their original positions when moving in less rugged terrain where control is easier.

3-47. In this formation, fire team leaders are in the lead position with their Soldiers echeloned to the right and left behind them. The positions for all but the leader may vary. This simple formation permits the fire team leader to lead by example. The leader's standing order to the Soldiers is, "Follow me and do as I do." When the leader moves to the right, the Soldiers should move to the right. When the leader fires, the Soldiers fire. When using the lead-by-example technique, it is essential for all Soldiers to maintain visual contact with their leader.

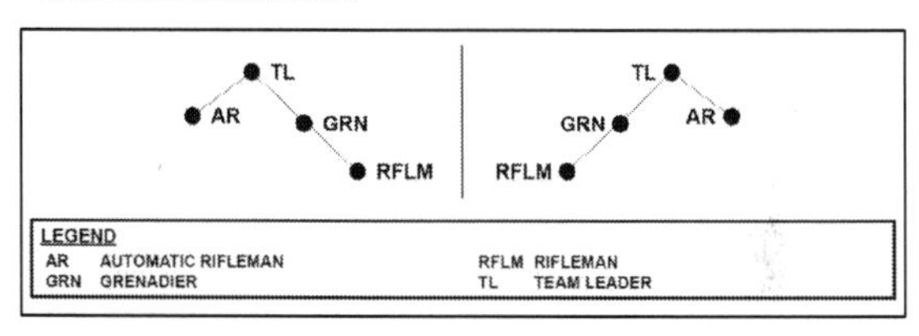

Figure 3-2. Fire team wedge

Fire Team Column/File

3-48. Team leaders use the column/file when employing the wedge is impractical. This formation most often is used in severely restrictive terrain, when inside a building, in dense vegetation, experiencing limited visibility, and so forth. The distance between Soldiers in the column/file changes due to constraints of the situation, particularly when in urban operations. (See figure 3-3.)

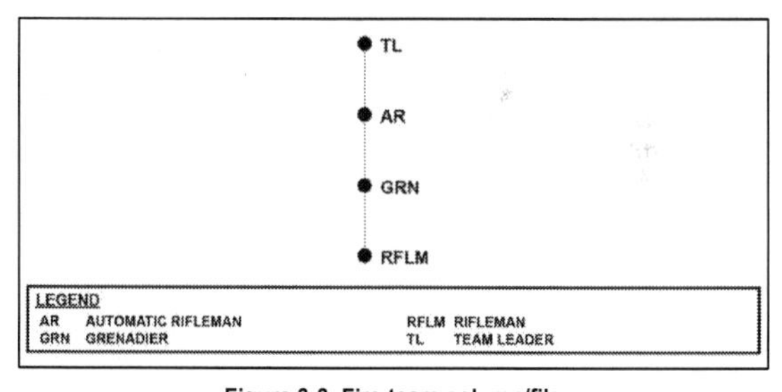

Figure 3-3. Fire team column/file

SQUAD MOVEMENT FORMATIONS

3-49. The term squad movement formation refers to the relative locations of the fire teams. Squad formations include the squad column, the squad line, and squad file. Table 3-3 compares squad formations.

Table 3-3. Comparison of squad formations

Movement Formation	*When Most Often Used*	*Movement Characteristics*			
		Control	*Flexibility*	*Fire Capabilities and Restrictions*	*Security*
Squad column	The main squad formation	Good	Aids maneuver, good dispersion laterally and in-depth	Allows large volume of fire to the flanks but only limited volume to the front	All-around
Squad line	For maximum firepower to the front	Not as good as the column	Limited maneuver capability (both fire teams committed)	Allows maximum immediate fire to the front	Good to the front, little to the flank and rear
Squad file	Close terrain, dense vegetation, limited visibility conditions	Easiest	Most difficult formation to maneuver from	Allows immediate fire to the flanks, masks most fire to the front and rear	Least

3-50. Squad leaders adjust their squad's formation as necessary while moving, primarily through the three movement techniques (see paragraphs 3-75 to 3-96). Squad leaders exercise command and control primarily through the two team leaders and move in the formation where they can best achieve this. Squad leaders are responsible for 360-degree security, for ensuring the team's sectors of fire are mutually supporting, and for being able to rapidly transition the squad upon contact.

3-51. Squad leaders designate one of the fire teams as the base fire team. Squad leaders control the squad's speed and direction of movement through the base fire team while the other team and attachments cue their movement off the base fire team. This concept applies when not in contact and when in contact with the enemy.

3-52. Weapons from the weapons squad (a medium machine gun or a Javelin) (or M3 Multi-role, Antiarmor, Antipersonnel Weapon System [known as MAAWS]) may be attached to the squad for movement or throughout the operation. These high value assets need to be positioned so they are protected and can be quickly brought into the engagement when required. Ideally, these weapons should be positioned so they are between the two fire teams.

Squad Column

3-53. The squad column is the squad's main formation for movement unless preparing for an *assault*—a short and violent well-ordered attack against a local objective (FM 3-90). (See figure 3-4.) It provides good dispersion both laterally and in-depth without sacrificing control. It also facilitates maneuver. The lead fire team is the base fire team. Squads can move in either a column wedge (squad column-team wedge) or a modified column wedge. A modified column wedge occurs when rough terrain, poor visibility, and other factors require the squad to modify the wedge into a file for control purposes. As the terrain becomes less rugged and control becomes easier, the Soldiers resume their original positions.

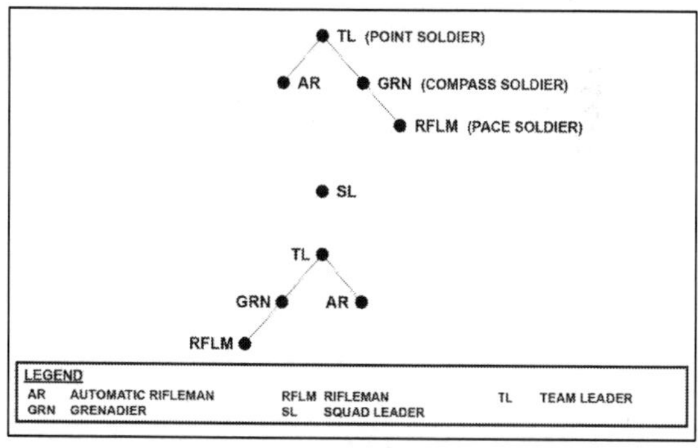

Figure 3-4. Squad column, fire teams in wedge

Squad Line

3-54. The squad line provides maximum firepower to the front and is used to assault or as a pre-assault formation. (See figure 3-5 on page 3-16.) To execute the squad line, the squad leader designates one of the teams as the base team. The other team cues its movement off the base team. This applies when the squad is in close combat as well. From this formation, the squad leader can employ any of the three movement techniques or conduct fire and movement.

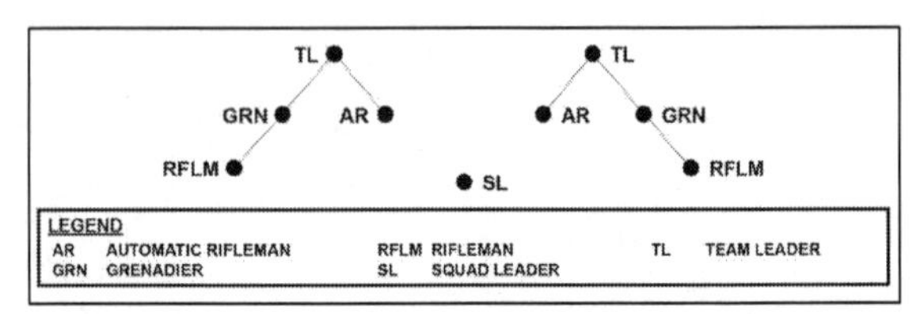

Figure 3-5. Squad line

Squad File

3-55. The squad file has the same characteristics as the fire team file. (See figure 3-6.) In the event the terrain is severely restrictive or extremely close, teams within the squad file also may be in file. This disposition is not optimal for enemy contact but provides squad leaders with maximum control. They increase control over the formation by moving forward to the second or third position. Moving forward enables them to exert greater morale presence by leading from the front, and to be immediately available to make vital decisions. Moving a team leader to the last position can provide additional control over the rear of the formation.

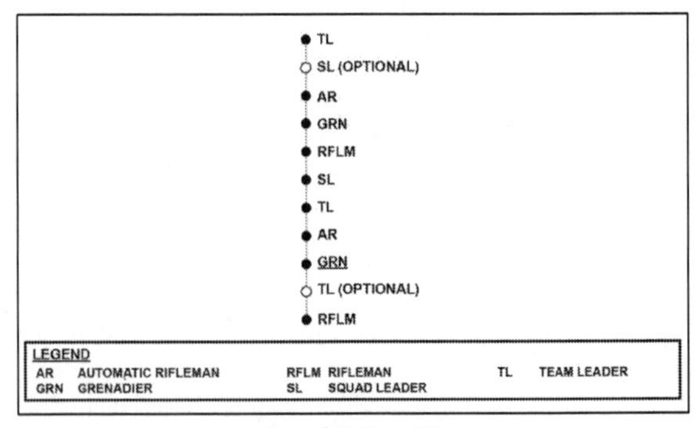

Figure 3-6. Squad file

WEAPONS SQUAD MOVEMENT FORMATION

3-56. The weapons squad is not a rifle squad and should not be treated as such. During tactical movement, platoon leaders have one of two options when it comes to positioning the weapons squad. The weapons squad can either travel as a separate entity or can be broken up and distributed throughout the formation. The advantage to keeping the weapons squad together is the ability to generate a *support by fire*—a tactical mission task in which a unit engages the enemy by direct fire in support of another maneuvering force (FM 3-90)—quickly and gain fire superiority under the direction of the weapons squad leader. The disadvantage to this approach is the lack of redundancy throughout the formation. The advantage to distributing the weapons squad throughout the rifle squads is the coverage afforded to the entire formation. The disadvantage is losing the weapons squad leader as a single command and control element and time required reassembling the weapons squad, if needed.

3-57. When the weapons squad travels dispersed, it can either be attached to squads or attached to the essential leaders like the platoon leader, platoon sergeant, and weapons squad leader. There is no standard method for its employment. Rather, the platoon leader places the weapons using three criteria: ability to generate fire superiority quickly, ability to protect high value assets, and the ability to control or deny enemy avenues of approach.

3-58. Like the rifle squad, the weapons squad, when traveling as a squad, uses either a column or line formation. Within these formations, the two sections can be in column or line formation.

PLATOON MOVEMENT FORMATIONS

3-59. The actual number of useful combinations of squad and fire team movement formations within the platoon movement formations is numerous, creating a significant training requirement for the unit. Add to the requirement to modify formations with movement techniques, immediate action drills, and other techniques, and it is readily apparent that what the platoon leader needs a few simple methods. These methods should be detailed in the unit tactical standard operating procedure (SOP).

Platoon Leader Responsibilities

3-60. Like squad leaders, platoon leaders exercise command and control primarily through their subordinates and move in the formation where they can best achieve this. The squad leader and team leader execute the movement formations and movement techniques within their capabilities based on the platoon leader's guidance.

3-61. Platoon leaders are responsible for 360-degree security, for ensuring each subordinate unit's sectors of fire are mutually supporting, and for being able to transition the platoon rapidly upon contact. They adjust their platoon's formation as necessary while moving, primarily through the three movement techniques. Like the squad and team, this determination is a result of the task, the nature of the threat, the closeness of terrain, and visibility.

3-62. Platoon leaders are responsible for ensuring their squads can perform their required actions. They do this through training before combat and rehearsals during combat. Well-trained squads can employ movement formations, movement techniques, actions on contact (see chapter 2), and stationary formations (for example, security halts, see paragraphs 6-70 to 6-75).

Platoon Headquarters

3-63. Platoon leaders must decide how to disperse the platoon headquarters elements (themselves, their radiotelephone operator [known as RTO], forward observer [FO], platoon sergeant, and combat medic). These elements do not have fixed positions in the formations. Rather, they should be positioned where they can best accomplish their tasks. Platoon leaders position their element where best to conduct actions on contact, supervise navigation, and communicate with higher. The FO team should be where they can best see the battlefield and where they can communicate with the platoon leader and company fire support officer. This is normally near the platoon leader. The platoon sergeant's element should be wherever the platoon leader is not. Typically, this means the platoon leader is toward the front of the formation, while the platoon sergeant is toward the rear of the formation. Because of the platoon sergeant's experience, the platoon sergeant should be given the freedom to assess the situation and advise the platoon leader accordingly.

Base Squad

3-64. Platoon leaders designate one of the squads as the base squad. They control the platoon's speed and direction of movement through the base squad, while the other squads and attachments cue their movement off the base squad.

Moving as Part of a Larger Unit's Movement

3-65. Infantry platoons and squads often move as part of a larger unit's movement. The next higher commander assigns the platoon a position within the formation. Platoon leaders assign their subordinates an appropriate formation based on the situation and use the appropriate movement technique. Regardless of the platoon's position within the formation, it must be ready to make contact or to support the other elements by movement, by fire, or by both.

3-66. When moving in a company formation, the company commander normally designates a base platoon to facilitate control. The other platoons cue their speed and direction on the base platoon. This permits quick changes and lets the commander control the movement of the entire company by controlling only the base platoon. The company commander normally locates within the formation where best to see and direct the movement of the base platoon. The base platoon's center squad is usually its base squad. When the platoon is not acting as the base platoon, its base squad is its flank squad nearest the base platoon.

Platoon Movement Formation Comparisons and Examples

3-67. Platoon formations include the column, the line (squads online or in column), the vee, the wedge, and the file. Leaders should weigh these carefully to select the best formation based on the unit's mission and on their METT-TC (I) analysis. Comparisons of the different formations are in table 3-4. Figures 3-7 to 3-12 on pages 3-21 through 3-26 are examples and do not dictate the location of the platoon leader or platoon sergeant.

Table 3-4. Comparison of platoon movement formations

MOVEMENT FORMATION	Platoon line, squads in column	Platoon line, squads on line	Platoon column
WHEN MOST OFTEN USED	May be used when the leaders do not want everyone on line; but want to be prepared for contact: when crossing a line of departure near an objective	When the leaders want all Soldiers forward for maximum firepower to the front and the enemy situation is known	Platoon primary movement formation
MOVEMENT CHARACTERISTICS — *Control*	Easier than platoon column, squads on line, but less than platoon line, squads on line	Difficult	Good for maneuver (fire and movement)
MOVEMENT CHARACTERISTICS — *Flexibility*	Greater than platoon column, squads on line, but less than platoon line, squads on one	Minimal	Provides good dispersion laterally and in depth
MOVEMENT CHARACTERISTICS — *Fire or Fires Capabilities and Restrictions*	Good firepower to the front and rear, minimum fires to the flanks; not as good as platoon column, better than platoon line	Allows maximum firepower to the front, little to flanks and rear	Allows limited firepower to the front and rear, but high volume to the flanks
MOVEMENT CHARACTERISTICS — *Security*	Good security all around	Less secure than other formations because of the lack of depth, but provides excellent security for the higher formations in the direction of the echelon	Extremely limited overall security
MOVEMENT CHARACTERISTICS — *Movement*	Slower than platoon column, faster than platoon line, squads on line	Slow	Good

Table 3-4. Comparison of platoon movement formations (continued)

MOVEMENT FORMATION		Platoon file	Platoon wedge	Platoon vee
WHEN MOST OFTEN USED		When visibility is poor due to terrain, vegetation, or light	When the enemy situation is vague, but contact is not expected	When the enemy situation is vague, but contact is expected from the front
MOVEMENT CHARACTERISTICS	*Control*	Easiest	Difficult but better than platoon vee and platoon line, squads in line	Difficult
	Flexibility	Most difficult formation from which to maneuver	Enables leader to make a small element and still have two squads to maneuver	Provides two squads up front for immediate firepower and one squad to the rear for movement (fire and movement) upon contact from the flank
	Fire Capabilities And Restrictions	Allows immediate fires to the flanks, masks most fires to front and rear	Provides heavy volume of firepower to the front or flanks	Immediate heavy volume of firepower to the front or flanks, but minimum fires to the rear
	Security	Extremely limited overall security	Good security to the flanks	Extremely limited overall security
	Movement	Fastest for dismounted movement	Slow, but faster than platoon vee	Slow

Platoon Column

3-68. In the platoon column formation, the lead squad is the base squad. (See figure 3-7.)

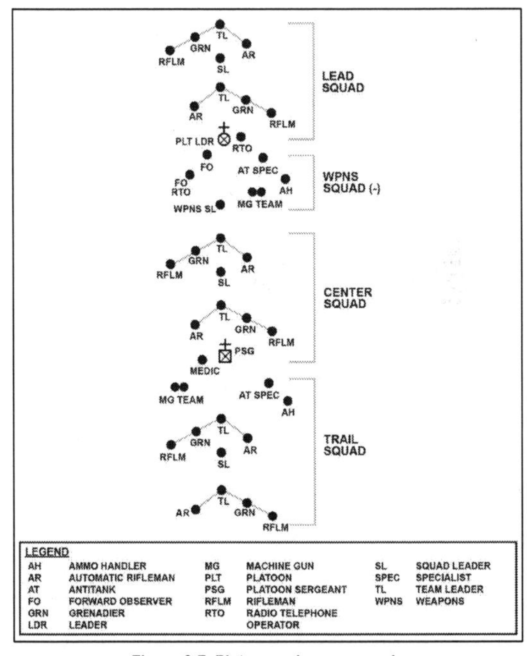

Figure 3-7. Platoon column, example

Note. METT-TC (I) considerations determine where the weapons squad or medium machine gun teams locate in the platoon formation.

Platoon Line, Squads Online

3-69. In the platoon line, squad's online formation, or when two or more platoons are attacking, the company commander chooses one of them as the base platoon. The base platoon's center squad is its base squad. When the platoon is not acting as the base platoon, its base squad is its flank squad nearest the base platoon. The weapons squad may move with the platoon, or it can provide the support by fire position. This is the basic platoon assault formation. (See figure 3-8.)

3-70. The platoon line with squads online is the most difficult formation from which to make the transition to other formations. It may be used in the assault to maximize the firepower and shock effect of the platoon. This is normally done when there is no intervening terrain between the unit and the enemy when antitank (AT) systems are suppressed, or when the unit is exposed to artillery fire and must move rapidly.

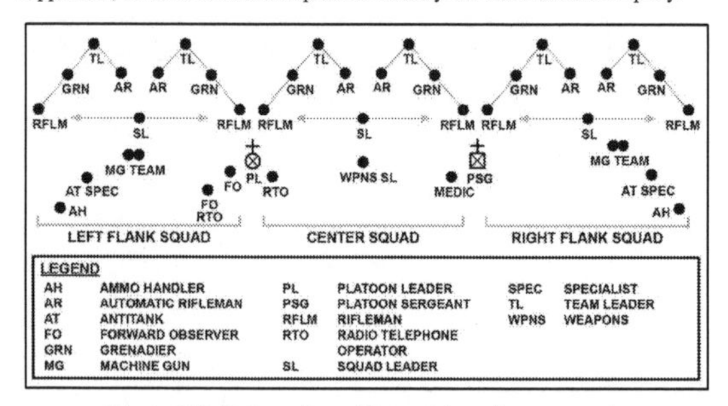

Figure 3-8. Platoon line with squads online, example

Platoon Line, Squads in Column

3-71. When two or more platoons are moving, the company commander chooses one of them as the base platoon. The base platoon's center squad is its base squad. When the platoon is not the base platoon, its base squad is its flank squad nearest the base platoon. (See figure 3-9.) The platoon line with squads in column formation is difficult to transition to other formations.

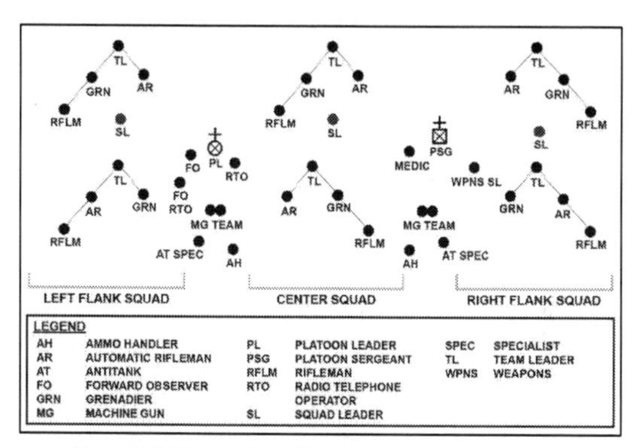

Figure 3-9. Platoon line with squads in column, example

Platoon Vee

3-72. This formation has two squads upfront to provide a heavy volume of fire on contact. (See figure 3-10 on page 3-24.) It also has one squad in the rear either overwatching or trailing the other squads. The platoon leader designates one of the front squads as the platoon's base squad.

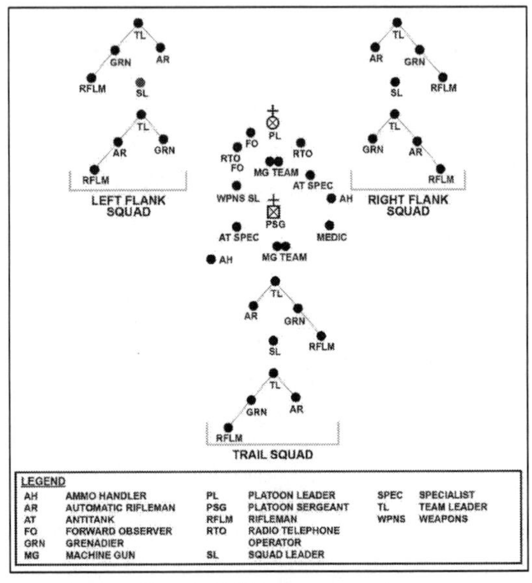

Figure 3-10. Platoon vee, example

Platoon Wedge

3-73. This formation has two squads in the rear overwatching or trailing the lead squad. (See figure 3-11.) The lead squad is the base squad. The wedge formation—

- Can be used with the traveling and traveling overwatch techniques.
- Allows rapid transition to bounding overwatch.

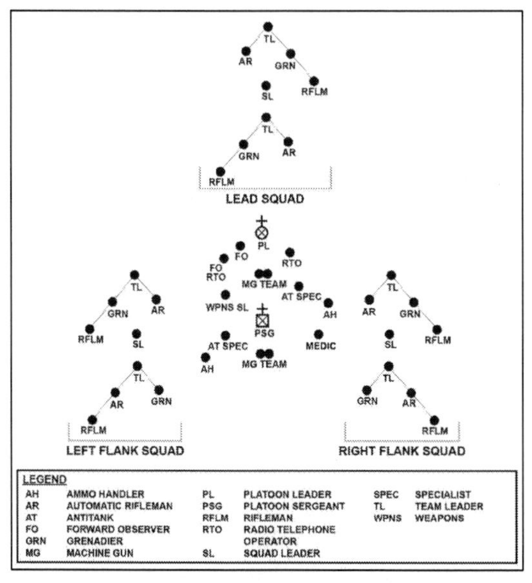

Figure 3-11. Platoon wedge, example

Platoon File

3-74. This formation may be set up in several methods. (See figure 3-12 on page 3-26.) One method is to have three-squad files follow one another using one of the movement techniques. Another method is to have a single platoon file with a front security element (point) and flank security elements. The distance between Soldiers is less than normal to allow communication by passing messages up and down the file. The platoon file has the same characteristics as the fire team and squad files. It normally is used for traveling only.

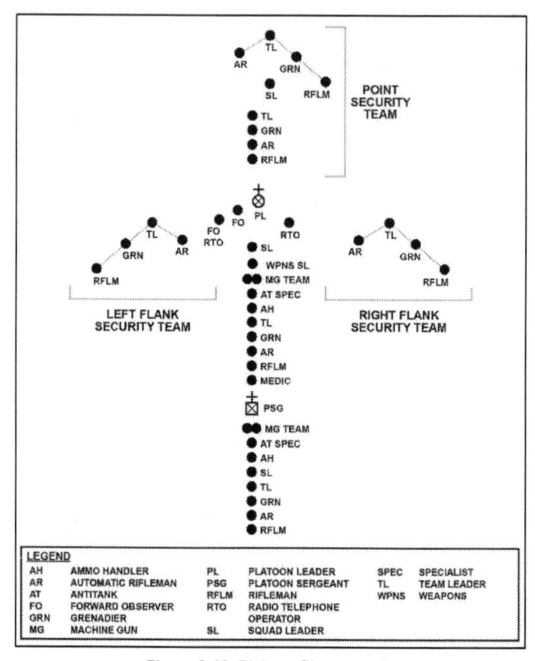

Figure 3-12. Platoon file, example

MOVEMENT TECHNIQUES

3-75. Movement techniques are not fixed formations. They refer to the distances between Soldiers, teams, and squads, and vary based on mission, enemy, terrain, visibility, and other factors affecting control. There are three movement techniques: traveling, traveling overwatch, and bounding overwatch.

- *Traveling* is a movement technique used when speed is necessary and contact with enemy forces is not likely (FM 3-90).

- *Traveling overwatch* is a movement technique used when contact with enemy forces is possible (FM 3-90).
- *Bounding overwatch* is a movement technique used when contact with enemy forces is expected (FM 3-90).

3-76. The selection of a movement technique is based on the likelihood of enemy contact and need for speed. Factors to consider for each technique are control, dispersion, speed, and security. (See table 3-5.)

Table 3-5. Movement techniques and characteristics

Movement Formation	*When Normally Used*	*Characteristics*			
		Control	*Dispersion*	*Speed*	*Security*
Traveling	Contact not likely	More	Less	Fastest	Least
Traveling overwatch	Contact possible	Less	More	Slower	More
Bounding overwatch	Contact expected	Most	Most	Slowest	Most

Note. In addition to walking Soldiers may move, dependent on the situation, in one of three other methods known as individual movement techniques. The three methods include the low crawl, high crawl, or rush. The low crawl method provides the Soldier with the lowest silhouette. Use the low crawl to cross places where the cover and/or concealment are very low and enemy fire or observation prevents a Soldier from getting up. The high crawl lets the Soldier move faster than the low crawl while still providing the Soldier with a low silhouette. Use the high crawl when there is good cover and concealment, but enemy fire prevents you from getting up. The rush is the fastest way to move from one covered position to another. Each rush should last from 3 to 5 seconds. Rushes are kept short to prevent enemy machine gunners or riflemen from tracking you. However, do not stop and hit the ground in the open just, because 5 seconds have passed. Always try to hit the ground behind some cover. Before moving, pick out your next covered and concealed position and the best route to it. (See TC 3-21.75 for additional information on each method.)

SQUAD MOVEMENT TECHNIQUES

3-77. The platoon leader determines and directs which movement technique the squad will use. Squad movement techniques include traveling, traveling overwatch, and bounding overwatch.

Squad Traveling

3-78. Traveling is a movement technique used when speed is necessary and contact with enemy forces is not likely. (See figure 3-13.)

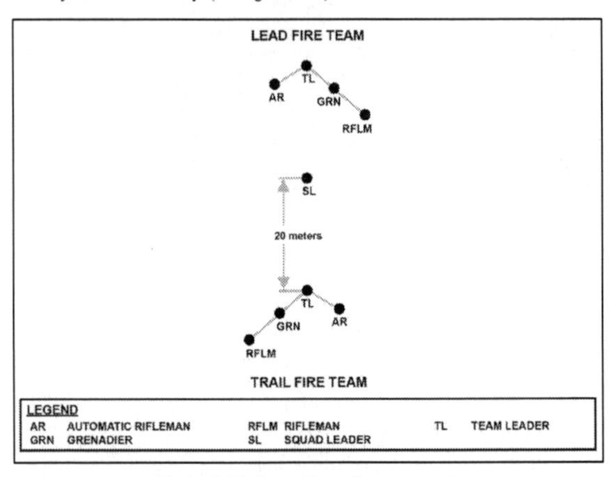

Figure 3-13. Squad traveling, example

Squad Traveling Overwatch

3-79. Traveling overwatch is used when contact with enemy forces is possible. Attached weapons move near and under the control of the squad leader so they can employ quickly. Rifle squads normally move in column or wedge formation. (See figure 3-14.) Ideally, the lead team moves at least 50 meters in front of the rest of the element.

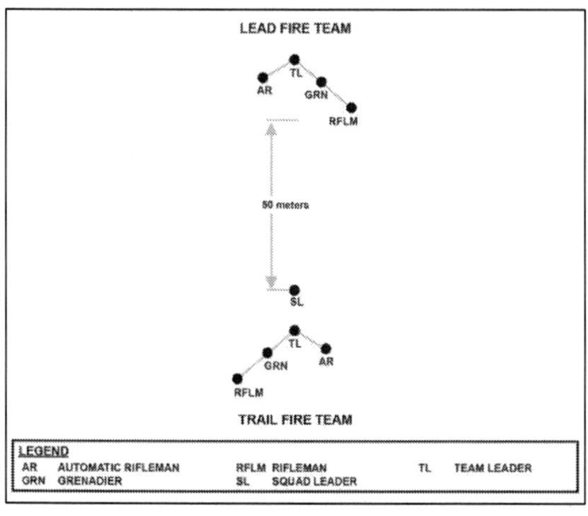

Figure 3-14. Squad traveling overwatch, example

Squad Bounding Overwatch

3-80. Bounding overwatch is used when contact with enemy forces is expected, the squad leader feels the enemy is near (based upon movement, noise, reflection, trash, fresh tracks, or even a hunch), or a large open danger area must be crossed. The lead fire team overwatches first. Soldiers in the overwatch team scan for enemy positions. The squad leader usually stays with the overwatch team. The trail fire team bounds and signals the squad leader when the team completes its bound and is prepared to overwatch the movement of the other team.

3-81. Both team leaders must know with which team the squad leader will be. The overwatching team leader must know the route and destination of the bounding team. The bounding team leader must know the team's destination and route, possible enemy locations, and actions to take when they arrive there. The bounding team leader must know where the overwatching team will be and how the bounding team will receive their instructions. (See figure 3-15 on page 3-30.) The cover and concealment on the bounding team's route dictates how its Soldiers move.

Alpha Team Leader: "We saw movement in the tree-line 100 meters to the right front, but we can't see it now."

Squad Leader: "Alpha Team overwatch from your current position. Ensure all weapon systems have clear fields of fire into that area."

Squad Leader: "Bravo Team, bound around to the right through the brush along the side of the stream until you get to that big boulder about 75 meters away."
"Setup an overwatch and be prepared to engage into that suspicious area Alpha Team identified." "I'll stay here with Alpha Team overwatching your move until you signal. Then I'll move up with Alpha Team."

Figure 3-15. Squad bounding overwatch, example

Alternate Bounds or Successive Bounds

3-82. During bounding overwatch, the squad can employ either of two methods of this technique: alternate bounds or successive bounds (see figure 3-16). In the alternate and successive bound techniques, the overwatching elements cover the bounding elements from covered, concealed positions with good observation and fields of fire against possible enemy positions. Overwatching elements can immediately support the bounding elements with fires if the bounding elements make contact. Unless they make contact en route, the bounding elements move via covered and concealed routes into the next set of support by fire positions. The length of the bound is based on the terrain and the range of overwatching weapons.

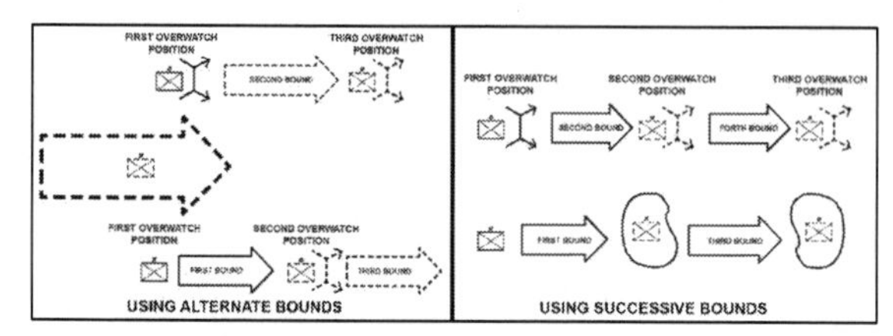

Figure 3-16. Squad successive and alternate bounds, example

Alternate Bounds

3-83. Covered by the rear element, the lead element moves forward, halts, and assumes overwatch positions. The rear element advances past the lead element and takes up overwatch positions. This sequence continues as necessary with only one element moving at a time. This method is usually more rapid than successive bounds.

Successive Bounds

3-84. In the successive bounding method, the lead element, covered by the rear element, advances and takes up overwatch positions. The rear element then advances to an overwatch position roughly abreast of the lead element, halts, and takes up overwatch. The lead element then moves to the next position, and so on. Only one element moves at a time, and the rear element avoids advancing beyond the lead element. This method is easier to control and more secure than the alternate bounding method, but it is slower.

Platoon Movement Techniques

3-85. A movement technique is the manner a platoon uses to traverse terrain. The platoon leader determines and directs which movement technique the platoon uses. Movement at the company level and above often entails use of a security force and main body; however, Infantry platoons and squads are not large enough to separate their forces into separate security forces and main body forces. Platoons can accomplish these security functions by employing movement techniques.

3-86. As the probability of enemy contact increases, the platoon leader adjusts the movement technique to provide greater security. The essential factor to consider is that the overwatching element must be able to ensure observation and supporting fires throughout, and beyond, the depth of the bounding element's move. Additionally, Soldiers must be able to see their fire team leader. Squad leaders must be able to see their fire team leaders. The platoon leader should be able to see the lead squad leader.

Traveling

3-87. Platoons often use the traveling technique when contact is not likely, and speed is needed. (See figure 3-17.) When using the traveling technique, all unit elements move continuously. In continuous movement, all Soldiers travel at a moderate rate of speed, with all personnel alert. During traveling, formations are essentially not altered except for effects of terrain.

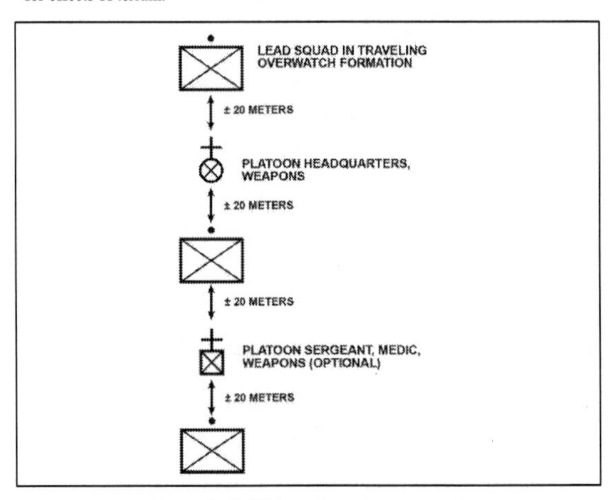

Figure 3-17. Platoon traveling, example

Traveling Overwatch

3-88. Traveling overwatch is an extended form of traveling in which the lead element moves continuously but trailing elements open the distance between themselves and the lead element by dropping back, while varying their speed and sometimes pausing to provide overwatch of the lead element. (See figure 3-18 on page 3-34.) Traveling overwatch is used when enemy contact is possible but not expected. Caution is justified but speed is desirable.

3-89. The trail element maintains dispersion based on its ability to provide immediate suppressive fires in support of the lead element. The intent is to maintain depth, provide flexibility, and continue movement in case the lead element is engaged. The trailing elements cue their movement to the terrain, overwatching from a position where they can support the lead element if needed. Trailing elements overwatch from positions and at distances that do not prevent them from firing or moving to support the lead element. The idea is to put enough distance between the lead units and trail units so that if the lead unit comes into contact, the trail units will be out of contact but have the ability to maneuver on the enemy.

3-90. Traveling overwatch require leaders to control their subordinates' spacing to ensure mutual support. This involves a constant process of concentrating (close it up) and dispersion (spread it out). The primary factor is mutual support, with its two critical variables being weapon ranges and terrain. Infantry platoons' and squads' weapon range and terrain limitations dictate the distance between two units that their direct fires can cover effectively. In compartmentalized terrain, this distance is closer, but in open terrain this distance is greater.

Note. Mutual support has two aspects—supporting range and supporting distance. Supporting range is the distance one unit may be geographically separated from a second unit yet remain within the maximum range of the second unit's weapons systems. For small units (such as squads, sections, and platoons), it is the distance between two units that their direct fires can cover effectively. Visibility may limit the supporting range. If one unit cannot effectively or safely fire to support another, the first may not be in supporting range, even though its weapons have the required range. Supporting distance is the distance between two units that can be traveled in time for one to come to the aid of the other and prevent its defeat by an enemy. (See ATP 3-21.10 for additional information.)

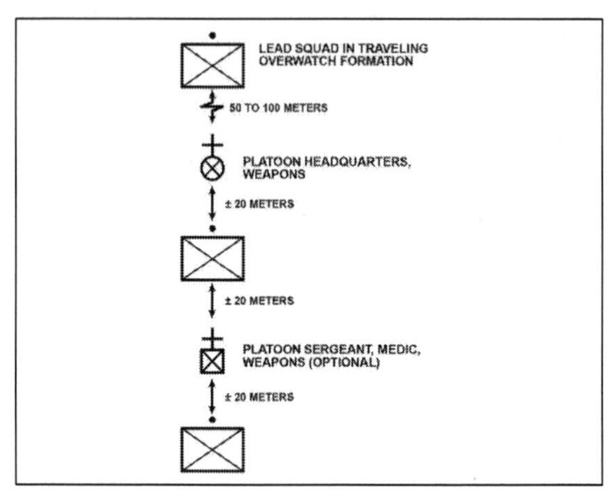

Figure 3-18. Platoon traveling overwatch, example

Bounding Overwatch

3-91. Bounding overwatch is like fire and movement in which one unit overwatches the movement of another. (See figure 3-19.) The difference is there is no actual enemy contact. Bounding overwatch is used when the leader expects contact. The key to this technique is the proper use of terrain.

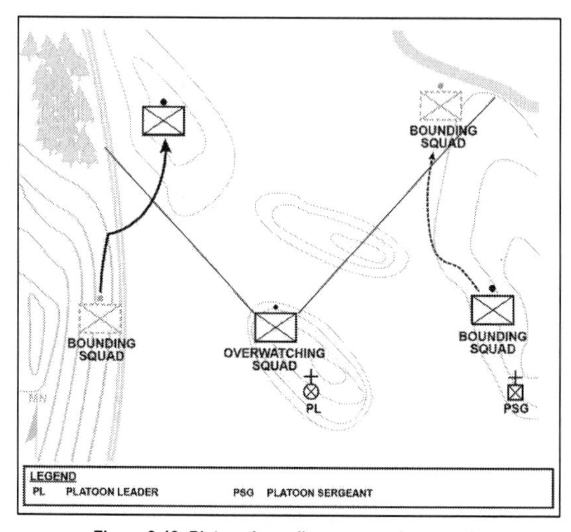

Figure 3-19. Platoon bounding overwatch, example

One Squad Bounding

3-92. One squad bounds forward to a chosen position; it then becomes the overwatching element unless contact is made en route. The bounding squad can use traveling overwatch, bounding overwatch, or individual movement techniques (low and high crawl, and 3- to 5-second rushes by the fire team or buddy teams [see note on page 3-27]).

3-93. The mission variables of METT-TC (I) dictate the length of the bounds. However, the bounding squads never should move beyond the range at which the base-of-fire squads can suppress known, likely, or suspected enemy positions. In severely restrictive terrain, the bounding squads make shorter bounds than they would in more open areas. The destination of the bounding element is based on the suitability of the next location as an overwatch position. When deciding where to send the bounding squad, a platoon leader considers—

- The requirements of the mission.
- Where the enemy is likely to be.

- The routes to the next overwatch position.
- The ability of an overwatching element's weapons to cover the bound.
- The responsiveness of the rest of the platoon.

One Squad Overwatching

3-94. One squad overwatches the bounding squad from covered positions and from where it can see and suppress likely enemy positions. The platoon leader remains with the overwatching squad. Normally, at least one of the platoon's medium machine guns are located with the initial overwatching squad.

One Squad Awaiting Orders

3-95. Based on the situation, one squad is uncommitted and ready for employment as directed by the platoon leader. The platoon sergeant and leader of the squad awaiting orders position themselves close to the platoon leader. On contact, this unit should be prepared to support the overwatching element, move to assist the bounding squad, or move to another location based on the platoon leader's assessment.

Weapons Squad

3-96. Medium machine guns normally are employed in one of two ways—

- Attached to the overwatch squad or the weapons squad supporting the overwatching element.
- Awaiting orders to move (with the platoon sergeant) or as part of a bounding element.

Command and Control of the Bounding Element

3-97. Ideally, the overwatch element maintains visual contact with the bounding element. However, the leader of the overwatch element may have the ability to track the location of the bounding element digitally without maintaining visual contact. This provides the bounding element further freedom in selecting covered and concealed routes to its next location. Before a bound, platoon leaders give an order to their squad leaders from the overwatch position. (See figure 3-20.) They tell and show them the following:

- The direction or location of the enemy (if known).
- The positions of the overwatching squad.
- The next overwatch position.
- The route of the bounding squad.
- What to do after the bounding squad reaches the next position.
- What signal the bounding squad will use to announce it is prepared to overwatch.
- How the squad will receive its next orders.

Figure 3-20. Platoon leader order for bounding overwatch, example

MOVEMENT DURING LIMITED VISIBILITY

3-98. At night or when visibility is poor, a platoon must be able to function in the same way as during daylight. It must be able to control, navigate, and maintain security, move, and stalk at night or during limited visibility. (See ATP 3-21.10 for additional information.) Additional considerations include—

- When visibility is poor, the following methods aid in control:
  - Use of night vision devices.
  - Infrared chemical lights.
  - Leaders move closer to the front.
  - The platoon reduces speed.
  - Soldiers use two small strips of luminous tape on the rear of their helmet, allowing Soldiers behind them to see them from the rear.

  - Leaders reduce the interval between Soldiers and units to make sure they can see each other.
  - Leaders conduct headcounts at regular intervals and after each halt to ensure personnel accountability.
- To assist in navigation during limited visibility, leaders use—
  - Terrain association (general direction of travel coupled with recognition of prominent map and ground features).
  - Dead reckoning, compass direction and specific distances or legs. (At the end of each leg, leaders should verify their location.)
  - Movement routes that parallel identifiable terrain features.
  - Guides or marked routes.
  - Command and control systems.
- For stealth and security in night moves, squads, and platoons—
  - Designate a point Soldier to maintain alertness, the lead team leader to navigate, and a pace Soldier to count the distance traveled.
  - Alternate compass and pace Soldiers are designated.
  - Ensure good noise and light discipline.
  - Use radio-listening silence.
  - Camouflage Soldiers and equipment.
  - Use terrain to avoid detection by enemy surveillance or night vision devices.
  - Make frequent listening halts.
  - Mask the sounds of movement with artillery fires.

SECTION III – MANEUVER

3-99. Maneuver begins once a unit has made contact with the enemy. Direct fire is inherent in maneuver, as is close combat. At the platoon level, maneuver forms the heart of every tactical operation and task. Platoon leaders maneuver their squads to close with, gain positional advantage over, and ultimately destroy the enemy.

BASE-OF-FIRE ELEMENT

3-100. Combining fire and movement requires a base of fire. Some platoon elements (usually a squad, for example, the weapons squad) remain stationary to provide protection for bounding elements by suppressing or destroying enemy elements. For example, the rifle squads of the platoon can maneuver while protected by the weapons squad in a base-of-fire position and then establish another base of fire with the weapons or a rifle squad.

3-101. Because maneuver is decentralized in nature, platoon leaders determine from their terrain analysis where and when they want to establish a base of fire. During actions on contact, they adjust maneuver plans as needed. Making maneuver decisions normally falls to the leader on the ground, who knows what enemy elements can engage the maneuvering element and what friendly forces can provide the base of fire.

3-102. The base-of-fire element occupies positions that afford the best possible cover and concealment, a clear view, and clear fields of fire. Platoon leaders normally designate a general location for the base of fire, and the element leader selects the exact location. Once in position, the base-of-fire element suppresses known, likely, or suspected enemy elements while aggressively scanning its assigned sector. It identifies previously unknown elements and then suppresses them with direct and indirect (when available) fires. The base-of-fire element allows the bounding unit to keep maneuvering so it can retain the initiative even when the enemy can see and fire on it. While maneuvering to or in position, the base-of-fire element leader is constantly looking for other locations that may provide better support for the maneuvering element. If a more suitable position is identified, the base of fire element leader will direct buddy team maneuver, while the rest of the element continues supporting the maneuvering element until the base of fire element has reached its new position.

BOUNDING ELEMENT

3-103. Maneuver is inherently dangerous. Enemy weapons, unknown terrain, and other operational factors all increase the danger. When maneuvering, the platoon leader considers the following:

- The bounding element must take full advantage of whatever cover and concealment the terrain offers.
- Squad members must always maintain all-around security and continuously scan their assigned sector of fire.
- The mission variables of METT-TC (I) dictate the length of the bounds. However, the bounding element should never move beyond the range at which the base-of-fire element can effectively suppress known, likely, or suspected enemy positions. General practice is to limit movement to no more than two-thirds the effective range of the supporting weapon system.
- In severely restricted terrain, the bounding element makes shorter bounds than it would in more open areas.
- The bounding element must focus on its ultimate goal—gaining a positional advantage. Once achieved, the element uses this advantage to destroy the enemy with direct fires and dismounted Infantry assault.

Note. Skillful integration of supporting direct fires improves the survivability and lethality of the Infantry rifle platoon. For example, the supporting heavy weapon systems of the Infantry weapons company of the Infantry battalion (see ATP 3-21.20 appendix D) may focus on destroying armored enemy vehicles and dismounted Soldiers at long ranges or become the base of fire while an Infantry maneuver element bounds.

PLATOON AS THE RESERVE

3-104. The designation of a reserve allows the commander to retain flexibility during an attack or the defense. The commander should be prepared to commit the reserve to exploit success and to continue the attack or to reinforce the defense. The reserve may repulse counterattacks during consolidate and reorganization. The reserve is normally under a commander's control and positioned where it can best exploit the success of the attack or defense. The reserve should not be so close that it loses flexibility during the assault or defense.

3-105. During the attack or defense, the Infantry rifle platoon may be designated the company or battalion reserve. A reserve is an uncommitted unit, though it may have a be-prepared mission. The company or battalion commander commits the reserve to reinforce the main effort and to maintain the momentum. To exploit the success of the other attacking units, the reserve should attack the enemy from a new or unexpected direction. Because of the many missions the platoon may be assigned, the platoon leader has to maintain situational awareness, know the missions and tactical plans of the other units, and be familiar with the terrain and enemy situation in the whole AO. It must react quickly and decisively when committed.

3-106. The reserve platoon may be assigned one or more of the following be-prepared missions:

- Protect the flank and rear of the unit.
- Conduct a counterattack or establish a blocking position.
- Maintain contact with adjacent units.
- *Clear*—a tactical mission task in which a unit eliminates all enemy forces within an assigned area (FM 3-90)—a position that has been overrun or bypassed by another unit.
- *Reduce*—a tactical mission task in which a unit destroys an encircled or bypassed enemy force (FM 3-90)—a position that has been encircled or bypassed enemy.
- Establish a support by fire position.
- Assume the mission of an attacking or defending unit.
- Attack from a new or unexpected direction.
- Protect or assist in the consolidate and reorganization on the objective.

Chapter 4
Offense

Platoon leaders and squad leaders must understand the principles and tactics, techniques, and procedures associated with the offense. They must comprehend their role when operating within a larger organization's operations, and when operating independently. Leaders must recognize the complementary and reinforcing effects of other maneuver elements and supporting elements with their own capabilities and understand the impact of open or restrictive terrain on their operations. The primary purpose of the offense for the Infantry rifle platoon is to decisively defeat, destroy, or neutralize the enemy force, or to seize key terrain. This chapter covers the basic components during the conduct of the offense, common offensive planning considerations, movement to contact, and attack.

SECTION I – CONDUCT OF THE OFFENSE

4-1. Offensive action is the critical part of any engagement. Successful offensive operations require units to seize, retain, and exploit the initiative to *defeat*—to render a force incapable of achieving its objectives (ADP 3-0)—the enemy. Even when conducting primarily defensive operations, taking the initiative from the enemy requires offensive operations that are force- or terrain-oriented. Force-oriented tasks focus on the enemy while terrain-oriented tasks focus on seizing, retaining, and controlling terrain and facilities. The tactical mission tasks, seize, retain, and control are defined as follows:

- *Seize* is a tactical mission task in which a unit takes possession of a designated area by using overwhelming force (FM 3-90).
- *Retain* is a tactical mission task in which a unit prevents enemy occupation or use of terrain (FM 3-90).
- *Control* is a tactical mission task in which a unit maintains physical influence over an assigned area (FM 3-90).

CHARACTERISTICS OF THE OFFENSE

4-2. The offense is characterized by audacity, concentration, surprise, and tempo. The ability to gain and maintain the initiative to defeat an enemy greatly depends upon the proper application of the characteristics of the offense. (See FM 3-96 for additional information.)

Audacity

4-3. Audacity is a willingness to take bold risks. The bold execution of plans increases the chance for surprise and enables leaders to control the tempo. This depends upon the leader's ability to see opportunities for action and accept prudent risks commensurate with the value of their objectives to take advantage of those opportunities. Leaders dispel uncertainty by acting decisively and inspire Soldiers to overcome adversity and danger. They compensate for a lack of information by developing the situation aggressively and seizing the initiative. The Infantry platoon and squad demonstrate audacity through—

- Building flexible plans that allow leaders to adapt to changing circumstances.
- Maintaining situational awareness to identify and take advantage of opportunities.
- Understanding the true value of an objective to the larger operation and the risks involved in achieving it.
- Maintain continuous communications to ensure others are aware of changes to the plan.
- Requesting additional support as required to mitigate unnecessary risks.

Concentration

4-4. Concentration is massing the effects of combat power in time and space at the decisive point(s) to achieve a single purpose. This requires coordination and an awareness of the environment to position forces in ways that enable concentration without losing surprise, sacrificing tempo, or creating lucrative targets for the enemy. The Infantry platoon and squad achieve concentration through—

- Careful planning and coordination to synchronize efforts based on a thorough terrain and enemy analysis and accurate reconnaissance and surveillance.
- Designating a main effort and allocating resources to support it.
- Integrating effects from higher-level assets, including indirect fires, Army and joint and multinational aviation assets, and electromagnetic warfare (EW) with maneuver.
- Maintaining continuous information flow to keep operations synchronized.

Surprise

4-5. Surprise is an effect achieved by attacking the enemy at a time or place they do not expect or in a manner for which they are unprepared. Surprise delays enemy reactions, overloads and confuses enemy command and control systems, induces psychological shock in enemy Soldiers and leaders, and reduces the coherence of defensive missions. By diminishing enemy combat power, surprise enables the attackers to exploit enemy paralysis and hesitancy. The Infantry platoon and squad achieve surprise by—

- Conducting thorough reconnaissance and surveillance to confirm enemy disposition, determine what actions they are prepared for, and identify opportunities to exploit in unexpected ways.

- Changing the tempo of operations or accepting risk in a way the enemy was unable to anticipate.
- Conducting counterreconnaissance, employing stealth, and disrupting communications to deny enemy situational understanding.
- Striking the enemy from an unexpected direction, at an unexpected time, and with a combination of capabilities that they are unable to defeat.

TEMPO

4-6. *Tempo* is the relative speed and rhythm of military operations over time with respect to the enemy (ADP 3-0). Tempo is not simply moving fast, but the idea that the cumulative effects of offense operations place enemy forces under continuous pressure. Even if portions of friendly forces are in a tactical pause, someone is always applying pressure, resulting in an enemy force that is reactive. Controlling or altering tempo is necessary to retain the initiative. A faster tempo allows attackers to quickly penetrate barriers and defenses and destroy enemy forces in-depth before they can react. Leaders can also purposely slow the tempo to lull the enemy into a false sense of security, but this is rarely done at the platoon or squad level. Leaders adjust tempo as tactical situations, sustainment necessity, or operational opportunities allow. They ensure synchronization and proper coordination, but not at the expense of losing opportunities to defeat the enemy. Rapid tempo denies the enemy the chance to rest while continually creating offensive opportunities but demands quick decisions and places ever increasing strain on Soldiers, equipment, and systems. The Infantry platoon and squad control the tempo by—

- Streamlining planning, preparation, and recovery activities to start movement more quickly than expected from a previous action.
- Moving more quickly than expected, especially across restrictive terrain.
- Engaging the enemy at multiple locations within a short time or simultaneously.
- Maintaining situational awareness and audaciously exploiting opportunities as they appear.

OFFENSIVE OPERATIONS

4-7. An *offensive operation* is an operation to defeat or destroy enemy forces and gain control of terrain, resources, and population centers (ADP 3-0). Offensive operations impose the leader's will on an enemy. The offense is the most direct means of seizing, retaining, and exploiting the initiative to gain a physical and psychological advantage. In the offense, the platoon's main effort is a sudden action directed toward enemy weaknesses and capitalizing on speed, surprise, and shock. If that effort fails to destroy an enemy, operations continue until enemy forces are defeated. The offense compels an enemy to react, creating new or larger weaknesses the attacking force can exploit. The four types of offensive operations are movement to contact, attack, exploitation, and pursuit.

Movement to Contact

4-8. *Movement to contact* is a type of offensive operation designed to establish or regain contact to develop the situation (FM 3-90). The goal of a movement to contact is to make initial contact with the smallest friendly force possible, consistent with protecting the force while retaining enough combat power to develop the situation and mitigate the associated risk. The unit conducts a movement to contact when the enemy situation is vague or not specific enough to conduct an attack. A movement to contact creates favorable conditions for subsequent tactical actions. Once the platoon makes contact with the enemy force, the leader has five options: attack (platoon assault), defend, bypass, delay, or withdraw. Subordinate variations of a movement to contact include search and attack, and cordon and search operations. (See section III of this chapter for additional information.)

Attack

4-9. An *attack* is a type of offensive operation that defeats enemy forces, seizes terrain, or secures terrain (FM 3-90). An attack differs from a movement to contact because enemy main body dispositions are at least partially known, allowing the leader to achieve greater synchronization. This enables more effective massing of an attacking forces' combat power than in a movement to contact. The attack can be hasty or deliberate depending upon the time available for assessing the situation, planning, and preparing. (See section IV of this chapter for additional information.)

4-10. Variations of the attack are ambush, counterattack, raid, and spoiling attack (see section IV of this chapter). The platoon leader's intent and the mission variables guide which of these variations of attack to employ. Units conduct each of these variations, except for a raid, as either a hasty or a deliberate operation. (See chapter 7 for additional information on the ambush and raid.)

Exploitation

4-11. An *exploitation* is a type of offensive operation following a successful attack to disorganize the enemy in depth (FM 3-90). While Infantry platoons and squads might be part of a higher echelon exploitation as a fixing or striking force, they will not plan or execute one on their own. (See ATP 3-21.20 for additional information.)

Pursuit

4-12. A *pursuit* is a type of offensive operation to catch or cut off a disorganized hostile force attempting to escape, with the aim of destroying it (FM 3-90). There are two variations of the pursuit: frontal and combination. A pursuit normally follows a successful exploitation but can follow any type of offensive operation if the enemy resistance breaks down and they begin fleeing the battlefield. Infantry platoons and squads do not plan pursuits at their levels and are unlikely to participate in a higher echelon pursuit due to a lack of mobility. (See ATP 3-21.20 for additional information.)

FORMS OF MANEUVER

4-13. *Forms of maneuver* are distinct tactical combinations of fire and movement with a unique set of doctrinal characteristics that differ primarily in the relationship between the maneuvering force and the enemy (ADP 3-90). Leaders select the form of maneuver based on their analysis of the mission variables of METT-TC (I) mission, enemy, terrain and weather, troops and support available, time available, civil considerations, and informational considerations. Leaders then synchronize the contributions of all warfighting functions to the selected form of maneuver. A platoon operation may contain several forms of offensive maneuver, such as a frontal attack to clear enemy security forces prior to establishing a support by fire position, followed by an envelopment focused on attacking an *assailable flank*—a flank exposed to attack or envelopment (ADP 3-90)—to avoid the enemy's strength to the front of an objective. A form of maneuver can also be used in the defense (see chapter 5). For example, in an area defense or mobile defense the counterattacking force or striking force, respectively, conducts an envelopment to destroy the enemy. While Infantry platoons and squads do not have the combat power to conduct all forms of maneuver on their own, they will participate as part of a larger organization. The five forms of maneuver are—

- Frontal attack.
- Envelopment.
- Infiltration.
- Turning movement.
- Penetration.

FRONTAL ATTACK

4-14. A *frontal attack* is a form of maneuver in which an attacking force seeks to destroy a weaker enemy force or fix a larger enemy force in place over a broad front (FM 3-90). The frontal attack is usually the least desirable form of maneuver because it exposes the majority of the offensive force to the concentrated fires of the defender. The Infantry rifle platoon normally conducts a frontal attack as part of a larger operation against a stationary or moving enemy force. (See figure 4-1 on page 4-6.) Unless frontal attacks are executed with overwhelming and well-synchronized speed and strength against a weaker enemy, they are seldom decisive. The frontal attack may be appropriate in an attack (see section VI) or meeting engagement (see section III) where speed and simplicity are paramount to maintain tempo and, ultimately, the initiative or in a supporting effort to fix an enemy force.

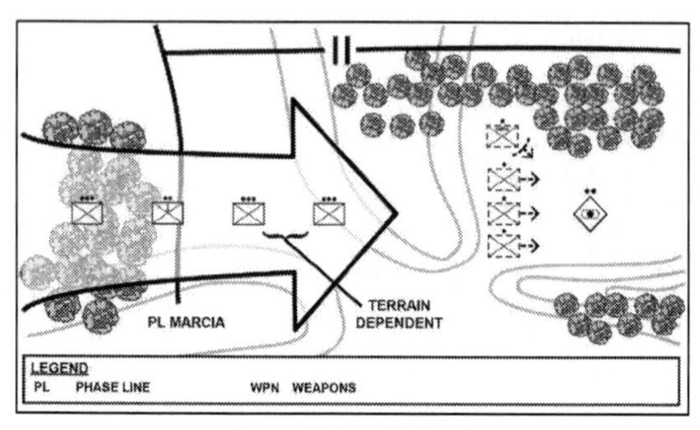

Figure 4-1. Frontal attack

ENVELOPMENT

4-15. *Envelopment* is a form of maneuver in which an attacking force avoids an enemy's principal defense by attacking along an assailable flank (FM 3-90). The attacking element's main effort focuses on attacking an assailable flank. It avoids the enemy's strength at the front where the effects of fires and obstacles are greatest. Generally, an envelopment is the preferred form of maneuver instead of a penetration or frontal attack because the attacking force tends to suffer fewer casualties while having the most opportunities to destroy the enemy. If no assailable flank is available, the attacking force creates one. At the tactical level (specifically at platoon and company echelons), the enveloping force's main effort is focused on the enemy's flank while the supporting effort of the envelopment usually engages the enemy's front by fire. The supporting effort diverts the enemy's attention from the threatened flank and prevents maneuver against the main effort. This form of maneuver is often used for a hasty operation (specifically an attack) or meeting engagement where speed and simplicity are paramount to maintaining battle tempo and, ultimately, the initiative. The three variations of the envelopment are—

- *Single envelopment*—a variation of envelopment where a force attacks along one flank of an enemy force (FM 3-90). (See figure 4-2.)
- *Double envelopment*—a variation of envelopment where forces simultaneously attack along both flanks of an enemy force (FM 3-90). (See ATP 3-21.20.)
- *Vertical envelopment*—a variation of envelopment where air-dropped or airlanded troops attack an enemy forces rear, flank, or both (FM 3-90). (See ATP 3-21.20 and FM 3-99.)

Note. When the Infantry rifle company and platoon are involved in an envelopment as part of a larger attacking force (generally conducted on a deeper axis), the envelopment will generally focus on seizing terrain, destroying specific enemy forces, and interdicting enemy withdrawal routes. *Interdict* is a tactical mission task in which a unit prevents, disrupts, or delays the enemy's use of an area or route in any domain (FM 3-90). (See ATP 3-21.20 for additional information on the double and vertical envelopments.)

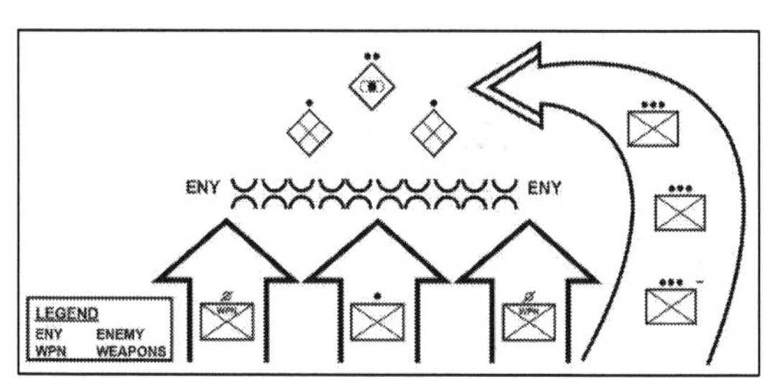

Figure 4-2. Single envelopment

INFILTRATION

4-16. An *infiltration* is a form of maneuver in which an attacking force conducts undetected movement through or into an area occupied by enemy forces (FM 3-90). The platoon leader uses infiltration to—

- Attack lightly defended positions or stronger positions from the flank and rear.
- Secure key terrain in support of the main effort.
- Disrupt or harass enemy defensive preparations/operations.
- Relocate the platoon by moving to battle positions around an engagement area.
- Reposition to attack vital facilities or enemy forces from the flank or rear.

4-17. The platoon leader can impose measures to control the infiltration including checkpoints, linkup points, phase lines, and *assault position*—a covered and concealed position short of the objective from which final preparations are made to assault the objective (ADP 3-90)—on the flank or rear of enemy positions. If it is not necessary for the entire infiltrating unit to reassemble to accomplish its mission, the *objective area*—a geographical area, defined by competent authority, within which is located an objective to be captured or reached by the military forces (JP 3-06)—may be broken into smaller objectives or positions. Each infiltrating element would move through a release point or move directly to its objective or position to conduct operations. (See figure 4-3.) A *checkpoint*—is a predetermined point on the ground used to control movement, tactical maneuver, and orientation (FM 3-90). The commander designates checkpoints along the route to assist marching units in complying with the timetable. A *linkup point*—is a designated place where two forces are scheduled to meet (FM 3-90). A linkup point (see paragraphs 6-44 to 6-52) should be an easily identifiable point on the ground, large enough for all infiltrating elements to assemble, and offer cover and concealment. Additional control measures for an infiltration may include—

- An area of operations (AO) for the infiltrating unit.
- One or more infiltration lanes.
- A line of departure (LD) or point of departure (PD).
- Movement routes with their associated start points and release points, or a direction or axis of attack.
- Linkup or rally points, including objective rally points (known as ORPs).
- Support by fire, attack by fire, or assault positions.
- A limit of advance (LOA).

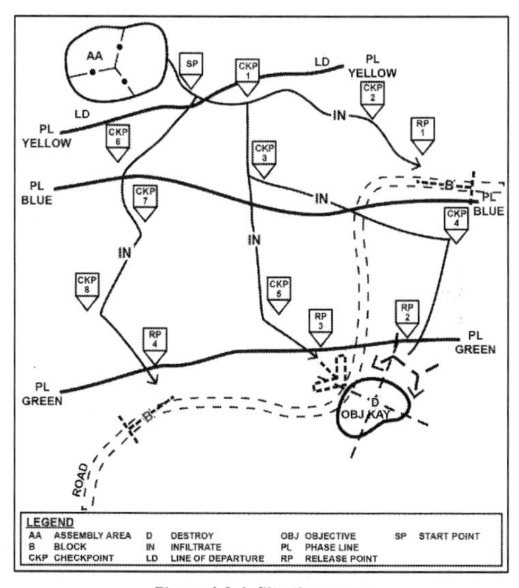

Figure 4-3. Infiltration route

4-18. Mission analysis will dictate the location of the infiltration route(s), the number of infiltration lanes, the size of the infiltration lane(s), the anticipated speed of movement, and the time of departure. An infiltration should be planned during limited visibility through areas the enemy does not occupy or cover by surveillance and fire. Planning should incorporate infiltration lanes, rally points along the route or axis, and contact points. Platoons and companies usually conduct infiltrations, but it is possible to execute at a battalion or squad level also. Planning considerations for the integration of attachments must be identified to increase effectiveness and to identify the potential for detection. Additional considerations during mission analysis can include:

- Level of cover and concealment.
- Movement distance and timeframe.
- Size of infiltrating force.
- Time required to linkup/reassemble.
- Movement of units simultaneously vice sequentially.

4-19. An *infiltration lane* is a control measure that coordinates forward and lateral movement of infiltrating units and fixes fire planning responsibilities (FM 3-90). Single or multiple infiltration lanes can be planned. Using a single infiltration lane facilitates navigation, control, and reassembly, reduces susceptibility to detection, reduces the area requiring detailed intelligence, and increases the time required to move the force through enemy positions. Using multiple infiltration lanes reduces the possibility of compromise and allows more rapid movement but makes control more challenging. (See figure 4-4.)

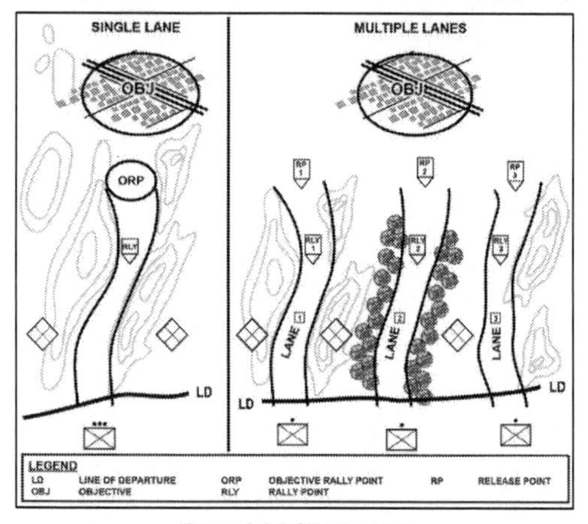

Figure 4-4. Infiltration lane

4-20. Infantry forces conducting an infiltration, or a patrol (see chapter 7) commonly use a rally point as a control measure. A *rally point* is an easily identifiable point on the ground at which units can reassemble and reorganize if they become dispersed (FM 3-90). An *objective rally point* is an easily identifiable point where all elements of the infiltrating unit assemble and prepare to attack the objective (ADP 3-90). This rally point is typically near the infiltrating unit's objective; however, there is no standard distance from the objective to the ORP. It should be far enough away from the objective so that the enemy will not detect the infiltrating unit's attack preparations.

TURNING MOVEMENT

4-21. A *turning movement* is a form of maneuver in which the attacking force seeks to avoid the enemy's principal defensive positions by attacking to the rear of their current positions forcing them to move or divert forces to meet the threat (FM 3-90). A turning movement seeks to make the enemy force displace from their current locations, whereas an enveloping force seeks to engage the enemy in their current locations from an unexpected direction. A turning movement is particularly suited for division-sized or larger forces possessing a high degree of tactical mobility. The Infantry rifle platoon participates in a turning movement as part of that larger force, most likely as a supporting effort or fixing force as opposed to the main effort involving the turning force. (See ATP 3-21.20 for additional information.)

PENETRATION

4-22. A *penetration* is a form of maneuver in which a force attacks on a narrow front (FM 3-90). Destroying the continuity of defense allows the enemy's subsequent isolation and defeat in detail by exploiting friendly forces. The penetration extends from the enemy's security area through main defensive positions into the enemy support area. The leader employs a penetration when there is no assailable flank, enemy defenses are overextended and weak spots are detected in the enemy's positions, or time pressures do not permit envelopment. Penetrations are normally only conducted at battalion echelon and above. When essential to the accomplishment of the mission, intermediate objectives should be planned for the attack. Usually, when the penetration is successfully completed, Infantry units within the battalion transition to another form of maneuver. (See ATP 3-21.20 for additional information.)

COMMON OFFENSIVE CONTROL MEASURES

4-23. The higher commander defines the leader's intent and establishes control measures allowing for decentralized execution and leader initiative to the greatest extent. Common control measures for the offense are the—

- Assault position, assault time, and tactical assembly area (TAA).
- Attack by fire position, attack position, and axis of advance.
- Battle handover line (BHL).
- Direction of attack.
- Final coordination line and forward line of own troops (FLOT).
- LOA and LD.
- Objective and ORP.
- Phase line.
- Passage line, passage point, and PD
- Probable line of deployment (PLD).
- Rally point and release point.
- Support by fire position.
- Time of attack.
- Zone.

4-24. An AO defines the location where subordinate units conduct their offensive actions. One technique breaks the battalion and company AO into many-named smaller AO. Units remain in designated AO as they conduct their missions. Battalion and higher reconnaissance assets might be used to observe an AO not assigned to a subordinate unit, while platoons or companies provide their own reconnaissance in their AO. This technique, along with target reference points (TRPs), help avoid fratricide in noncontiguous operations. Using a TRP facilitates the responsiveness of fixing and finishing elements once a reconnaissance element detects the enemy. Objectives and checkpoints guide the movement of subordinates and help leaders control their organizations. Contact points help coordination among the unit operating in adjacent areas.

4-25. When looking for terrain features to use as control measures, leaders consider three types: contiguous; point; and area. Contiguous features follow major natural and man-made features such as ridgelines, valleys, trails, streams, power lines, and streets. Point features can be identified by a specific feature or a grid coordinate including, hilltops and prominent buildings. Area features are significantly larger than point features and require a combination of grid coordinates and terrain orientation.

4-26. A *zone* is an operational area assigned to a unit in the offense that only has rear and lateral boundaries (FM 3-0). The non-bounded side of a zone is open towards enemy forces. A higher echelon headquarters uses fire support coordination and maneuver control measures such as an LOA and a coordinated fire line to synchronize its deep operations with those of a subordinate unit. Zones allow higher headquarters to adjust deep operations without having to change unit boundaries. This gives greater flexibility to the higher headquarters for controlling deep operations, allowing subordinate units to focus on close and rear operations. A zone is best for front-line units executing high-tempo offensive operations characterized by direct fire contact with the enemy and a fluid FLOT. Units treat everything behind the FLOT as an AO. Zone can be further subdivided as needed. (See FM 3-90 appendix A for additional information on zones.) (See ATP 3-21.10 and FM 3-96 for additional information.)

TACTICAL FRAMEWORK OF THE OFFENSE

4-27. Offensive operations are typically executed in a four-step tactical framework. (See paragraphs 4-81 to 4-87 and 4-137 to 4-154.) The framework is for discussion purposes only and is not the only way of executing offensive operations. These steps overlap during the conduct of the offense. Normally the first two steps are supporting efforts, while the finish step is the main effort. Follow through is usually a sequel or branch to the plan based upon the situation. The four-step tactical framework of the offense during execution is—

- Find the enemy. Intel drives fires and maneuver.
- Fix the enemy. Prevent repositioning or reinforcement making them easier to destroy.
- Finish the enemy. Mass available combat power to accomplish the mission.
- Follow through. Defeat in detail, consolidate, reorganize, and transition.

SECTION II – COMMON OFFENSIVE PLANNING CONSIDERATIONS

4-28. Platoon leaders begin with a designated AO, identified mission, and assigned forces. They develop and issue planning guidance based on visualization relating to the physical means to accomplish the mission. Within the planning process, a warfighting function is a group of tasks and systems united by a common purpose that the platoon leader uses to accomplish missions. Warfighting functions provide an intellectual organization for common critical functions. The following paragraphs discuss activities, functions, and specific operational environments as the framework for discussing offensive planning considerations.

> ***Note.*** A *warfighting function* is a group of tasks and systems united by a common purpose that commanders use to accomplish missions and training objectives (ADP 3-0). The six warfighting functions are command and control, movement and maneuver, intelligence, fires, sustainment, and protection. (See FM 3-96 for additional information.)

COMMAND AND CONTROL

4-29. In the Infantry rifle platoon, the platoon leader is the central figure in command and control and is essential to integrating the capabilities of the warfighting functions. Command and control provides the greatest possible freedom of action to subordinates, facilitating their abilities to develop the situation, adapt, and act decisively through disciplined initiative within the platoon leader's intent. It focuses on empowering subordinate leaders and sharing information to facilitate decentralized execution.

> ***Note.*** The art of command comprises the creative and skillful exercise of authority through timely decision making and leadership. Commanders constantly use judgment gained from experience and training to delegate authority, make decisions, determine the appropriate degree of control, and allocate resources. Control, as contrasted with command, is more science than art. As such, it relies on objectivity, facts, empirical methods, and analysis. Control requires a realistic appreciation of time and distance factors, including the time required to initiate, complete, and assess directed actions. Commanders exercise control to direct and adjust operations as conditions dictate. The science of control supports the art of command. (See FM 3-96 for additional information.)

4-30. Command and control conveys the leader's intent and an appreciation of METT-TC (I), with special emphasis on—

- Enemy positions, strengths, and capabilities.
- Missions and objectives, including task and purpose, for each subordinate element's shared understanding.

- Higher commander's intent and scheme of maneuver.
- AO for use of each subordinate element with associated control graphics.
- Time the operation is to begin.
- Scheme of maneuver and leaders positioning that allows for timely decisions and the application of judgement.
- Special tasks required to accomplish the mission.
- Risk acceptance.
- Options for accomplishing the mission.

MOVEMENT AND MANEUVER

4-31. The platoon conducts tactical movement when contact with the enemy is possible or anticipated. The platoon maintains integrity throughout the movement and plans for enemy interference en route to or shortly after arrival at its destination. During tactical movements, the platoon leader must be prepared to maneuver against an enemy force. Once deployed in its assigned AO, the platoon moves using proper techniques for assigned missions. When contact is made, fire and movement is executed.

4-32. The platoon leader's mission analysis helps to facilitate the decision on how to move most effectively. When planning platoon movements, the platoon leader ensures the unit is moving in a way that supports a rapid transition to maneuver. Platoon movement should be as rapid as the terrain, mobility of the force, and enemy situation permit while ensuring Soldiers arrive in a condition to accomplish their mission. The ability to gain and maintain the initiative often depends on movement being undetected by the enemy. The platoon depends heavily upon the terrain for protection from enemy fire. Once contact with the enemy is made, squads and platoons execute the appropriate actions on contact (see chapter 2), and leaders begin to maneuver their units.

4-33. The platoon leader conducts movement and maneuver to avoid enemy strengths and create opportunities that increase the effects of combat power. Surprise is achieved by making unexpected maneuvers, rapidly changing the tempo of ongoing operations, avoiding observation, and using deceptive techniques and procedures. The platoon leader seeks to overwhelm the enemy with one or more unexpected actions before it has time to react in an organized fashion. This occurs when the attacking force can engage the defending enemy force from positions of advantage with respect to the enemy, such as engaging from an assailable flank.

INTELLIGENCE

4-34. The platoon leader uses threat event templates, the situation template (known as SITEMP), the likely threat course of action (COA), the most dangerous threat COA, civil consideration products, terrain analysis products, and other intelligence products and assessments. During offensive operations, the evaluation of avenues of approach leads to a recommendation of the best avenues of approach to a command's objective and to the identification of avenues of approach available to the threat for counterattack, withdrawal, or the movement of reinforcements or reserves. By studying the terrain, the leader tries to determine the principal enemy heavy and light avenues of approach to the objective. The leader also tries to determine the most advantageous area from which enemy may defend, routes the enemy may use to conduct counterattacks, and other factors such as observation and field of fire, avenues of approach, key terrain, obstacles, and cover and concealment (OAKOC) (see paragraph 2-62). The attacking platoon continuously conducts information collection during the battle because it is unlikely the platoon leader has complete knowledge of the enemy's intentions and actual actions. (See chapter 2 section II for additional information.)

FIRES

4-35. Prior to movement, the platoon leader plans for fires throughout the depth of the operation, specifically along the route to cover anticipated places of enemy contact, during actions on the objective, and during consolidate and reorganization activities. These targets (for example, TRPs and final protective fires [FPFs]) are a product of the platoon leader's analysis of the mission variables of METT-TC (I) and must be incorporated into the company's fire support plan.

4-36. The platoon leader conducts fire support planning concurrently with the Infantry rifle company and Infantry battalion. The Infantry brigade combat team (IBCT) and Infantry battalion typically use top-down fire support planning, with bottom-up refinement of plans. As part of the top-down fire planning system, the company commander refines the fire support plan from higher headquarters to meet mission requirements, ensuring these refinements are incorporated into the higher headquarters' plan.

4-37. A clearly defined concept of fires enables the platoon leader and forward observer (FO) to articulate precisely how they want indirect fires to affect the enemy during the different phases of the operation. In turn, this allows the company fire support officer to facilitate the development of fires supporting accomplishment of the company's mission down to the squad level. (See appendix B for additional information.)

SUSTAINMENT

4-38. The objective of sustainment in the offense is to assist the platoon in maintaining the momentum. The platoon leader wants to take advantage of windows of opportunity and launch offensive operations with minimum advance warning time. The platoon sergeant and squad leaders must anticipate these events and maintain flexibility to support the offensive plan accordingly.

4-39. A key to an offense is the ability to carry 72 hours of supplies specific to the platoon's assigned mission, and to anticipate the requirement for resupply, specifically regarding ammunition, fuel, and water. This anticipation helps maintain the momentum of attack by resupplying as far forward as possible. Leaders use throughput distribution, and preplanned contingency resupply, preconfigured packages of essential items to help maintain offensive momentum and tempo. (See chapter 8 for additional information and ATP 3-21.10.)

PROTECTION

4-40. The rapid tempo and changing nature of the offense presents challenges to the protection of friendly assets. The forward movement of subordinate units is critical if the leader is to maintain the initiative necessary for offensive operations. Denying the enemy, a chance to plan, prepare, and execute a response to the friendly offense by maintaining a high operational tempo is a vital means the leader employs to ensure the survivability of the platoon. Using multiple routes, dispersion, piecemeal destruction of isolated enemy forces, scheduled rotation, and relief of forces before they culminate, and wise use of terrain are techniques for maintaining a high tempo of offense. During the attack, the leader integrates and synchronizes the use of obscuration to support critical actions such as breaching or assaults. The mission variables of METT-TC (I) determine the exact techniques employed in a specific situation.

4-41. Depending on the threat, primary protection concerns of the platoon leader may be enemy air threats, to include hostile small, unmanned aircraft systems. (See Battle Drill-React to Aircraft While Dismounted, appendix E.) If these threats exist, the leader prepares the unit and adjusts the scheme of maneuver accordingly. In the face of an enemy air threat, the platoon usually has minimal active and passive (with its organic weapons) air defenses. Air defense units are usually not assigned below IBCT level. However, air defense assets may be located near the platoon and may provide coverage. If air defense elements are assigned, the platoon leader with the advisement of the air defense element leader determines likely enemy air avenues of approach and plans movement routes accordingly. The platoon leader establishes priorities for protection during movement and during actions on the objective. (See ATP 3-01.81 for additional information.)

4-42. Planning for air defense and implementing all forms of active and passive air defense measures are imperative to minimize the vulnerability to enemy air attack. The platoon leader integrates the fire support plan with any attached or supporting air defense assets. Should hostile aircraft attack during movement, the subordinate unit under attack moves off the road into a defensive posture and immediately engages the aircraft with all available automatic weapons. The rest of the platoon moves to covered and concealed areas until the engagement ends. Air defense units, when available, may move with the platoon. In this case, the air defense unit leader ensures both the current and projected weapons control status (WCS) and air defense warning are known. (See ATP 3-01.8 for additional information.)

4-43. During the conduct of a passage of lines, units participating in the operation present a lucrative target for air attack. The passing unit leader coordinates air defense (when available) protection with the stationary force during the passage of lines. This method allows the passing force supporting air defense assets to conduct a move at the same time. If the passing force requires static air defense, then it coordinates the terrain with the stationary commander.

4-44. When attached in a direct support role during the offense, air defense assets can increase protection for command posts. In addition, air defense assets may provide security to maneuver and sustainment units and positioned to overwatch key air and ground routes or avenues of approach. During the attack, active and passive air defense measures are integrated and synchronized with the use of obscuration to support critical actions such as breaching or assaults. (See paragraphs 5-78 to 5-79 for additional information on active and passive air defense.)

> ***Note.*** Survivability encompasses capabilities of military forces both while on the move and when stationary, see chapter 5 for additional information on survivability focus more on stationary capabilities—constructing fighting and protective positions and hardening facilities.

4-45. The platoon leader integrates chemical, biological, radiological, and nuclear (CBRN) operations considerations (see paragraphs 5-80 to 5-101) into offensive planning depending on the CBRN threat. This includes CBRN passive defense principles, such as contamination avoidance, individual and collective protection, and decontamination. CBRN protective measures may slow the tempo, degrade combat power, and increase sustainment requirements. Soldiers wearing individual protective equipment find it difficult to work or fight for an extended period. CBRN considerations include the following:

- Understand threat capabilities and likelihood to use CBRN.
- Employ organic CBRN capabilities to maximize protection across the platoon.
- Disseminate information regarding observed CBRN threats or hazards to the platoon immediately, send report to higher.
- Understand duration of protection gear effectiveness (see paragraph 5-92 to 5-101).
- Ensure Soldiers have necessary medical countermeasures (such as antidote treatment nerve agent auto-injector) and decontamination kits.
- Development of decontamination plans based on the platoon leader's priorities and vulnerability assessment.
- Dissemination of information regarding planned and active decontamination sites.
- Preparate to conduct CBRN reconnaissance and surveillance (normally conducted by the scout platoon) to assist in CBRN reconnaissance and surveillance efforts.

4-46. Like the protection requirements that are reflected in the mission-oriented protective posture (MOPP) level (see chapter 5 section II), personal protective equipment posture (PPEP) provides a threat-based framework for commanders and subordinate leaders to reference regarding the level of body armor and personal protective equipment Soldiers need to wear while conducting an operation. The PPEP levels range from PPEP 0 to PPEP 3 and act as a common language to standardize protection levels and suggested protection level based on the possible threats present on the battlefield. (See ATP 3-21.18 for additional information on PPEP.)

4-47. The most basic protection from enemy EW is radio discipline. This can take the form of limiting transmissions, antenna masking, and use of low power. Limiting the number of overall transmissions and transmission time when a radio microphone is keyed makes it more difficult for enemy EW operators to target friendly forces for geolocation and jamming. (See paragraphs 2-34 to 2-39 for additional information on radio discipline and paragraph 2-42 for additional information on EP.)

4-48. Detainees and captured enemy equipment or materiel often provide excellent combat information (see paragraph 2-102). This information is of tactical value only if the platoon processes and evacuates detainees and materiel to the rear quickly. In all tactical situations, the platoon will have specific procedures and guidelines for processing detainees and captured materiel.

4-49. As a matter of policy, all detainees are treated as enemy prisoners of war until their appropriate legal status is determined by a competent authority according to the criteria enumerated in the Geneva Convention Relatives to the Treatment of Prisoners of War. However, as a practical matter, when a platoon captures enemy personnel, they must provide the initial processing and holding for detainees. Detainee processing is a resource-intensive and politically sensitive operation requiring detailed training, guidance, and supervision.

4-50. All detainees immediately shall be given humanitarian care and treatment. U.S. Armed Forces never will torture, maltreat, or purposely place detained persons in positions of danger. There is never a military necessity exception to violate these principles.

4-51. Soldiers must process detainees using the "search, silence, segregate, speed, safeguard, and tag" (5 Ss and T) technique. The steps are as follows:

- Search. Neutralize a detainee and confiscate weapons, personal items, and items of potential intelligence or evidentiary value.
- Silence. Prevent detainees from communicating with one another or making audible clamor. Silence uncooperative detainees by muffling them with a soft, clean cloth tied around their mouths and fastened at the backs of their heads. Do not use duct tape or other adhesives, place a cloth or objects inside the mouth, or apply physical force to silence detainees.
- Segregate. Segregate detainees according to policy and SOPs. (Segregation requirements differ from operation to operation.) The ability to segregate detainees may be limited by the availability of manpower and resources at the point of contact. At a minimum, try to segregate detainees by grade, gender, age (keeping adults from juveniles and small children with mothers), and security risk. Military intelligence and military police personnel can provide additional guidance and support in determining the appropriate segregation criteria.
- Safeguard. Protect detainees and ensure the custody and integrity of all confiscated items. Soldiers must safeguard detainees from combat risk, harm caused by other detainees, and improper treatment or care. Report all injuries. Correct and report violations of U.S. military policy that occur while safeguarding detainees. Acts, omissions, or both that constitute inhumane

treatment are violations of the law of war and, as such, must be corrected immediately. Simply reporting violations is insufficient. If a violation is ongoing, a Soldier has an obligation to stop the violation and report it.

- Speed to a safe area/rear. Quickly move detainees from the continuing risks associated with other combatants or sympathizers who still may be in the area of capture. If there are more detainees than the Soldiers can control, call for additional support, search the detainees, and hold them in place until reinforcements arrive. Evacuate detainees from the battlefield to a holding area or facility as soon as possible. Transfer captured documents and other property to the forces assuming responsibility of the detainees.
- Tag. Ensure that each detainee is tagged using DD Form 2745 (*Enemy Prisoner of War [EPW] Capture Tag*). Confiscated equipment, personal items, and evidence will be linked to the detainee using the DD Form 2745 number. When a DA Form 4137 (*Evidence/Property Custody Document*) is used to document confiscated items, it will be linked to the detainee by annotating the DD Form 2745 control number on the form or by field expedient means. Field expedient means should include tagging with date and time of capture, location of capture, capturing unit, and circumstances of capture. There are three parts to this form. DD Form 2745, Unit Record Card, Part B, is the unit record copy. DD Form 2745, Document/Special Equipment Weapons Card, Part C, is for detainee confiscated property. Tagging is critical. If it does not happen, the ability of higher headquarters to obtain pertinent tactical information quickly is reduced greatly.

4-52. Detainees should be evacuated as soon as is practical to the higher echelon detainee collection point. Military police teams, when attached, can assist in the rapid evacuation of detainees to these collection points. Tactical questioning of detainees is allowed relative to collection of commander's critical information requirements (CCIRs). However, detainees must always be treated in accordance with Department of Defense (DOD) programs and policies that ensure compliance with the laws of the United States; the law of war, including the Geneva Conventions, and humane treatment during all intelligence interrogations, detainee debriefings, or tactical questioning. (See DODD 2310.01E, DODD 3115.09, and DODD 2311.01.)

4-53. Soldiers capturing equipment, documents, and detainees should tag them (using the DD Form 2745, Part A), take digital pictures, and report the capture immediately. Detainees are allowed to keep individual protective equipment such as a protective mask. (See FM 3-63 for additional information.)

4-54. The platoon leader protects subordinate squads to deny the enemy the capability to interfere with their ongoing operations. Protection also meets the leader's legal and moral obligations to the organization's Soldiers. Additional protection tasks (see ATP 3-21.10 for additional information) conducted by the platoon in the offense may include:

- Employ safety techniques (including *fratricide*—the unintentional killing or wounding of friendly or neutral personnel by friendly firepower [ADP 3-37]—avoidance).
- Implement *operations security*—a capability that identifies and controls critical information, indicators of friendly force actions attendant to military operations, and incorporates countermeasures to reduce the risk of an adversary exploiting vulnerabilities (JP 3-13.3).
- Implement *physical security*—that part of security concerned with physical measures designed to safeguard personnel; to prevent unauthorized access to equipment, installations, material, and documents; and to safeguard them against espionage, sabotage, damage, and theft (JP 3-0)—procedures.

SECTION III – MOVEMENT TO CONTACT

4-55. Movement to contact is a type of offensive operation designed to develop the situation and establish or regain contact. It ends when enemy contact is made. The platoon usually conducts movement to contact as part of an Infantry rifle company or larger element. Based upon the mission variables of METT-TC (I), the platoon may conduct the operation independently. Search and attack, and cordon and search are variations of a movement to contact (see note below).

Note. In an environment focused on stability operations, the Infantry platoon typically operates under more restrictive rules of engagement (ROE) and will typically conduct either a cordon and search or search and attack when conducting a movement to contact. *Cordon and search* is a variation of movement to contact where a friendly force isolates and searches a target area (FM 3-90). *Search and attack* is variation of a movement to contact where a friendly force conducts coordinated attacks to defeat a distributed enemy force (FM 3-90). (See ATP 3-21.10 and ATP 3-21.20 for additional information on each variation.)

CONDUCT OF A MOVEMENT TO CONTACT

4-56. Purposeful and aggressive movement, decentralized control, and hasty deployment of formations from the march to conduct offensive operations characterize the movement to contact. The fundamentals of movement to contact—

- Focus all efforts on finding the enemy.
- Make initial contact with the smallest force possible, consistent with protecting the force.
- Make initial contact with small, mobile, self-contained forces to avoid decisive engagement of the main body on ground chosen by the enemy. This allows the leader maximum flexibility to develop the situation.
- Task-organizes the force and uses movement formations to deploy and attack rapidly in all directions.

- Keep subordinate forces within supporting distances to facilitate a flexible response.
- Maintains contact regardless of the COA adopted once contact is made.

ORGANIZATION OF FORCES

4-57. The Infantry rifle platoon normally conducts movement to contact as part of a company within the battalion; however, based on the mission variables of METT-TC (I) it can conduct the operation independently. As an example, the platoon may conduct a movement to contact prior to occupation of a screen line (see FM 3-98) or to expand an airhead line, assault objective, or security area boundary (see FM 3-99). Because the enemy situation is not clear, the platoon moves in a way that provides security and supports a rapid buildup of combat power against enemy units once they are identified. If no contact occurs, the platoon might be directed to conduct consolidate on the line, objective, or boundary. The platoon leader analyzes the situation and selects the proper movement formation and movement technique (see chapter 3 section II) to conduct the mission. The leader reports all information rapidly and accurately and strives to gain and maintain contact with the enemy. The platoon leader retains freedom of maneuver by moving the platoon in a manner that—

- Employs a movement formation and technique that keeps the bulk of the platoon out of contact and able to maneuver when making contact.
- Rapidly develops the situation to gain understanding of the disposition of threat forces.
- Ensures continuous, effective overwatch of moving elements.
- Ensures the platoon's elements are not fixed in position by enemy fires.
- Gains fire superiority when in contact to suppress enemy forces.

Note. In both the offense and defense, forms of contact occur when a unit encounters any situation that requires an active or passive response to a threat or potential threat. The nine forms of contact are direct; indirect; nonhostile; obstacle; CBRN; aerial; visual; electromagnetic; and influence. The conduct of tactical offensive and defensive operations most often involves contact using the visual, direct, and indirect forms.

4-58. When the Infantry rifle platoon conducts a movement to contact within the Infantry rifle company as part of the Infantry battalion's independent movement to contact, the battalion normally organizes (as a minimum) with its companies and platoons allocated within the forward security force and a main body. (See figure 4-5 on page 4-22.) The forward security force generally comprises a reconnaissance and surveillance force (can be detailed from both the maneuver companies and the scout platoon), an advance guard, and flank and rear security. The main body consists of forces not detailed to security duties. Combat elements of the main body prepare to respond to enemy contact with security forces. Fire support teams (FISTs) may displace forward so they can respond to calls for fire immediately. The main body follows the advance guard and keeps enough distance between itself and the advance guard to maintain flexibility.

The Infantry battalion commander may designate a portion of the main body as the reserve. (See ATP 3-21.10 for additional information.)

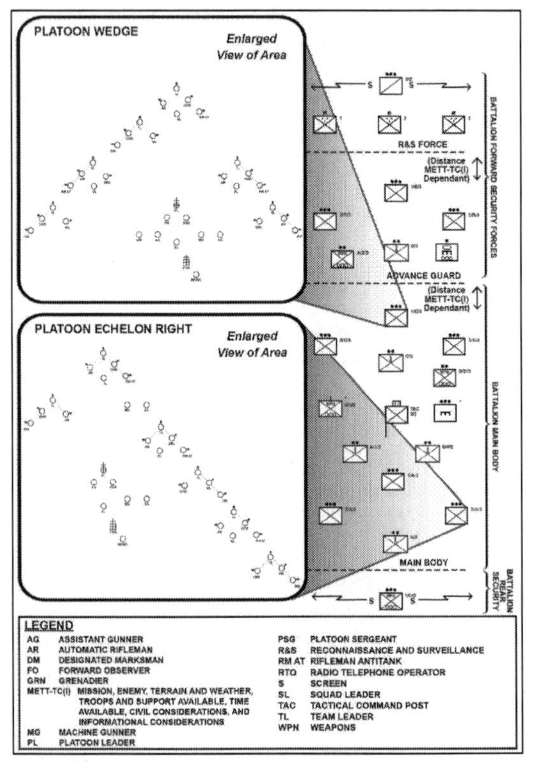

Figure 4-5. Notional organization of forces for a movement to contact

CONTROL MEASURES

4-59. Execution of a movement to contact usually starts from an LD at the time specified in the operation order (OPORD). The unit's movement is controlled by phase lines, contact points, and checkpoints as required. The depth of the movement to contact is usually controlled by a *limit of advance*—a phase line used to control forward progress of the attack (ADP 3-90)—or a *forward boundary*—a boundary that delineates the forward edge of a unit's area of operation (FM 3-90). A *lateral boundary* is a boundary defining the left or right limit of a unit's assigned area (FM 3-90). One or more objectives is designated to limit the extent of the movement to contact and to orient the force. However, these are often terrain-oriented (commonly called march objectives) and used only to guide movement. Although movement to contact may result in taking a terrain objective, the primary focus should be on the enemy force. The leader can designate a series of phase lines successively becoming the new *rear boundary*—a boundary that delineates the rearward limits of a unit's assigned area (FM 3-90)—of forward security elements as the force advances. Each rear boundary becomes the forward boundary of the main body and shifts as the security force moves forward. The rear boundary of the main body designates the limit of responsibility of the rear security element. This line also shifts as the main body moves forward. When enough information is available to locate a significant enemy force(s), then another type of offensive operation (for example, an attack) should be conducted. (See FM 3-90 for additional information.)

SEQUENCE OF EVENTS, EXAMPLE

4-60. Most platoon movement to contacts will follow a sequence of events as part of a larger movement to contact. This sequence of events is used for discussion purposes and is not the only way to sequence a movement to contact. With any sequence, leaders understand events will vary depending on the mission variables of METT-TC (I) and to some degree, events will overlap. The sequence of events includes:

- TAA.
- Reconnaissance and surveillance.
- Movement to the LD.
- Movement after the LD.
- Initial engagement (transition to a maneuver plan).

TACTICAL ASSEMBLY AREA

4-61. A *tactical assembly area* is an area that is generally out of the reach of light artillery and the location where units make final preparations (precombat checks and inspections) and rest, prior to moving to the line of departure (JP 3-35). Prior to establishing the TAA, a *quartering party*—a group dispatched to a new assigned area in advance of the main body (FM 3-90)—secures, reconnoiters, and organizes the TAA before the main body's arrival and occupation. A quartering party is a group of unit representatives dispatched to a probable new site of operations in advance of the main body. Guides within the quartering party guide the different elements of the main body into assigned locations within the TAA.

4-62. Once the TAA is established, the platoon prepares for upcoming operations and leaders plan, direct, and supervise mission preparations. This time allows the platoon and squads to conduct precombat checks (PCCs), precombat inspections (PCIs), rehearsals, and sustainment activities. The platoon typically conducts these preparations within a company TAA, as it rarely occupies its own TAA. (See ATP 3-21.10 for additional information.)

Reconnaissance and Surveillance

4-63. All leaders should aggressively seek information about the terrain and enemy. Because the enemy situation and available planning time may limit a unit's reconnaissance and surveillance activities, the platoon usually conducts reconnaissance to answer the CCIRs. An example is reconnoitering and timing routes from the TAA to the LD. The platoon may augment the efforts of the battalion scout platoon to answer the CCIRs. For example, the platoon may conduct an area or zone reconnaissance to reconnoiter an identified named area of interest (NAI). A *named area of interest* is a geospatial area or systems node or link against which information that will satisfy a specific information requirement can be collected, usually to capture indications of enemy and adversary courses of action (JP 2-0). Other forms of reconnaissance and surveillance provided to the platoon may include information from maps and terrain software/databases and unmanned aircraft systems. Updates from reconnaissance and surveillance activities can occur at any time while the platoon and squad are planning for, preparing for, or executing the mission. As a result, leaders must be prepared to adjust their plans.

Movement to the Line of Departure

4-64. The platoon and squad typically move from the TAA to the LD as part of the company movement plan. This plan may direct the platoon or squad to move to an *attack position*—the last position an attacking force occupies or passes through before crossing the line of departure (ADP 3-90)—and await orders to cross the LD. If so, the platoon leader reconnoiters, times, and rehearses the route to the attack position. Squad leaders know where they are to locate within the assigned attack position. The company commander may order all platoons to move within a company formation from the TAA directly to the *point of departure*—the point where the unit crosses the line of departure and begins moving along a direction of attack (ADP 3-90). If one PD is used, it is important the lead platoon and trail platoons reconnoiter, time, and rehearse the route to the point. This allows the company commander to maintain synchronization. To maintain flexibility and to maintain synchronization, a PD along the LD may be designated for each platoon. Movement to and/or across the LD may entail a forward passage of line (see chapter 6 section III for additional information).

Movement after the Line of Departure

4-65. A movement to contact usually starts from a *line of departure*—in land warfare, a line designated to coordinate the departure of attack elements (JP 3-31)—at the time specified in the OPORD or fragmentary order (FRAGORD). The platoon leader controls the movement to contact of the platoon by using phase lines, contact points, and checkpoints as required and controls the depth of the movement to contact by using an LOA or a forward boundary. March objectives (one or more) may be used to limit the extent of the movement to contact and orient the force. This movement is often terrain-oriented and used only to guide the force. Although a movement to contact may result in taking a terrain objective, the primary focus should be on the enemy force. When the platoon leader has enough information to locate significant enemy forces, the leader should plan another type of offensive action.

4-66. Platoon leaders plan the approach for the movement to contact, ensuring synchronization, security, speed, and flexibility by selecting the platoon's routes, movement formation, and movement technique. Leaders must be prepared to make contact with the enemy. They must plan accordingly to reinforce the company commander's needs for synchronization, security, speed, and flexibility. During the movement to contact, the platoon leader may exercise disciplined initiative and alter the platoon's movement formation and movement technique or speed to maintain synchronization with the other platoons and squads. This retains flexibility for the company commander.

Initial Engagement

4-67. Usually within the platoon's movement technique (see paragraphs 3-85 to 3-97) during a movement to contact, the lead element, (usually a fire team or rifle squad) will make the initial contact with the enemy. Upon making contact, the lead element deploys and report, and evaluated and develops the situation. The lead element determines the size and activity of the enemy force and avoids being fixed or destroyed. If possible, the lead element avoids detection. When the enemy is moving, the lead element determines the direction of movement and the size and composition of the force. After evaluating and developing the situation, the element leader proposes a COA to use against the enemy. The element leader or attached FO can disrupt lead enemy forces by placing indirect fires on the enemy forces. Speed of selecting a COA and execution is critical when the enemy is moving. When the enemy is stationary, the lead element determines if the enemy is occupying prepared positions and is reinforced by obstacles and minefields. The lead element tries to identify any crew-served weapon or antitank (AT) weapon positions, the enemy's flanks, and gaps in positions. The lead element passes this combat information and proposed COA to the platoon leader to assess the information and further develop the situation, if required. Developing the situation and achieving an understanding of the enemy force is critical to deciding how to continue the engagement.

4-68. When the rifle platoon (or the platoon is the lead element of a company advance guard) is committed as the advance guard, the squad (see figure 4-6 on page 4-27), or platoon (see figure 4-7 on page 4-40) maneuvers to overpower and destroy a small security team or squad-size element, respectively. Commitment against a larger force or an enemy strong point normally requires the deployment of the movement to contact's main body (see ATP 3-21.10). The advance guard protects the main body by fixing enemy forces, which allows the movement to contact's main body to retain its freedom to maneuver. In developing the situation, the advance guard platoon leader/commander maintains pressure on the enemy by fire and movement. The advance guard probes and conducts a vigorous reconnaissance of the enemy's flanks to determine the enemy's exact location, composition, and disposition. Once contact is made with an enemy force, there are one of five planned options—attack, defend, bypass, delay, or withdraw.

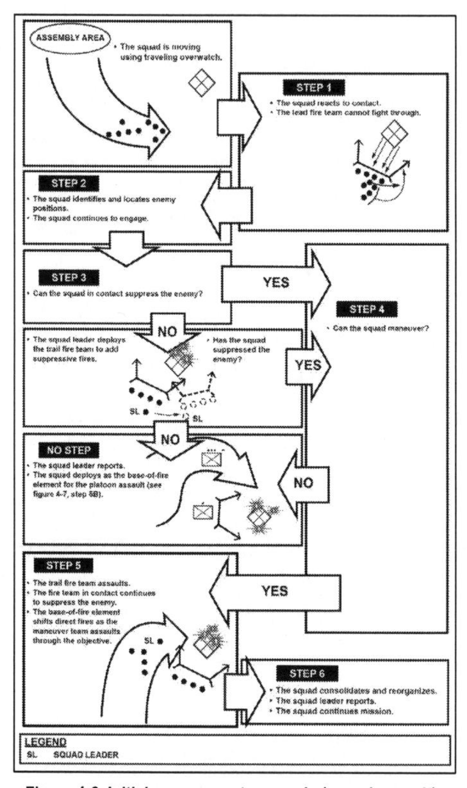

Figure 4-6. Initial engagement, example (squad assault)

4-69. During planning, the five options are war-gamed, generating branch plans and decision points based on the current situation. The lead element within the platoon/company, most likely the lead team/squad of the lead squad/platoon respectively, conducting the advanced guard develops the situation and reports, generating a decision by the higher commander. The decision points determined during war gaming define the conditions in terms of the enemy and friendly strengths and dispositions that are likely to trigger the execution of each option. Likely locations of enemy engagements are based on known or suspected enemy locations. The higher commander states the bypass criteria for the advance guard recognizing the loss of tempo that is created by the lead element fighting every small enemy force it encounters. *Bypass criteria* are measures established by higher echelon headquarters that specify the conditions and size under which enemy units and contact may be avoided (FM 3-90). Normally, the higher commander makes the final decision for execution of an option based on the progress of the initial engagement of the advance guard. The movement to contact generally ends with the commitment of the main body. (See ATP 3-21.10 for additional information.)

OPERATIONS PROCESS

4-70. In addition to the fundamentals of the operations process described in ADP 5-0, when planning and preparing for a movement to contact the platoon leader and subordinate leaders within the platoon recognize complexity; balance resources, capabilities, and activities; recognize planning horizons; and avoid pitfalls throughout the operations process.

Note. The Army's framework for exercising command and control is the operations process (plan, prepare, execute, and assess). See ADP 5-0 and FM 5-0 for additional information.

PLAN

4-71. As in any type of operation, the platoon leader plans to focus operations on finding the enemy and then delaying, disrupting, and destroying each enemy force as much as possible before direct-fire range. During troop leading procedures (TLP), the platoon leader analyzes the terrain to include enemy air avenues of approach and the enemy's most dangerous COA as determined in the war-gaming portion of the TLP. (See chapter 2 for additional information on TLP.) Due to the uncertainty of the enemy's precise location, the analysis during TLP may consist of multiple contingencies and branch plans. Because of the platoon's vulnerability, by the nature of a movement to contact, the enemy must not be underestimated.

4-72. The plan for the movement to contact addresses not only actions anticipated by the platoon leader based on available intelligence products and assessments, but also the conduct of *meeting engagements*—combat action that occurs when a moving force engages an enemy at an unexpected time and place (FM 3-90)—at anticipated times and locations where they might occur. The platoon leader enhances platoon security through a thorough analysis of the enemy as part of step 3 of TLP (make a tentative plan). The interpretation of the intelligence preparation of the operational environment (known as IPOE) products provided by the battalion through the company, and on-going understanding of the company commander's visualization of the enemy and indicating danger areas where the platoon is most likely to make contact shapes the outcome of the enemy analysis. In analyzing the enemy, the platoon leader must understand the IPOE. Although the platoon leader does prepare IPOE products for use in preparation of analysis, and orders, the leader does not prepare individual products for subordinates. The platoon leader must be able to use the products of the battalion's IPOE effectively during mission analysis and be able to inform subordinates based on the compilation of platoon and higher IPOE products.

4-73. During the mission analysis conducted during step 3 of TLP, consideration of terrain and weather help to identify potential danger areas that are likely enemy defensive locations, engagement areas, observation posts, and obstacles. Fire support plans target potential danger areas, and they become on-order targets placed into effect and cancelled as the lead element can confirm or deny enemy presence. Air-ground operations enable fire superiority when organized correctly to fire immediate suppression missions to help maneuver forces get within direct-fire range of the enemy. The platoon movement to contact order must address coverage of these danger areas not only based on actions anticipated but also on available intelligence products and assessments.

4-74. Lead elements of the movement to contact should have sufficient uncommitted forces to develop the situation without requiring the deployment of the main body in most cases. The platoon leader can rely on fire support assets to weight lead element combat power, but still provides the lead element with the combat multipliers (for example, internal to the platoon—a machine gun team, external to the platoon—company mortars, and/or external to the company—combat engineers) they need to accomplish their mission. The frontage assigned to the platoon and delegated to lead elements in a movement to contact must allow them to apply sufficient combat power to maintain the momentum of the operation. The frontage should also provide for efficient movement of the force.

4-75. Unless there is a specific need to designate a main effort, the platoon leader may choose to retain control of any attached or organic assets. In this case, the platoon leader would designate a main effort and weight it accordingly upon contact with the enemy. Bypass criteria, which are measures during the conduct of a movement to contact, may be established by higher headquarters that specify the conditions and size under which enemy units and contact may be avoided. Bypass criteria are clearly stated by the higher commander, and dependent on the mission variables of METT-TC (I). Criteria may also include maneuver around an obstacle or position to maintain the momentum of the operation.

4-76. A platoon leader uses the minimal number and type of control measures possible in a movement to contact because of the uncertain enemy situation. Control measures from higher may include designation of a platoon AO with left, right, front, and rear boundaries, or a separate AO bounded by a continuous boundary (noncontiguous operations). Boundaries are used to separate elements and clearly establish responsibilities between different elements. The platoon leader may further divide the AO into subordinate unit AO to facilitate subordinate unit actions.

4-77. The movement to contact usually starts from an LD at the time specified in the OPORD. The platoon leader controls the operation by using phase lines, contact points, and checkpoints as required. The platoon leader may control the depth of the movement to contact by using an LOA or a forward boundary. One or more objectives to limit the extent of the movement to contact and orient the force may be established. These control measures along with mission orders, coupled with battle drills and formation discipline, synchronize the movement to contact. When an axis of advance is used, for example, in limited visibility, the risk of enemy forces outside the axis not being detected, and thus being inadvertently bypassed must be understood. The platoon leader plans for the array and disposition of forces (for example, a hasty defense) when contact with the enemy is not made, or at the end of the movement to contact. (See ATP 3-21.10 for additional information.)

PREPARE

4-78. During preparation, the platoon leader and subordinate leaders must receive the most current information from organic and higher echelon information collection assets. The company commander must ensure that FRAGORDs are published and that plans are updated to reflect any changes to the platoon's mission. The platoon leader must ensure subordinates understand the company commander's intent and concept of operations and the platoon's concept of operations and platoon's intent, in addition to their individual missions even as new information becomes available. The platoon leader uses confirmation briefs, backbriefs, and rehearsals to ensure missions are understood and all actions are integrated and synchronized. Simple plans that are flexible and rehearsed repetitively against various enemy conditions and that rely on established tactical standard operating procedures (SOPs) are essential to success.

4-79. Subordinate unit preparations are reviewed to ensure they are consistent with the platoon leader's intent and concept of operations. Subordinate rehearsals should emphasize movement through danger areas, actions on contact, passage of lines, and transitions. The platoon leader and subordinate leaders ensure subordinate units (to include attachments) understand assigned missions during movement and maneuver options during execution. Plans are war-gamed and rehearsed against enemy COAs that would cause the platoon and/or squads to execute various maneuver options at different times and locations. The goal is to rehearse situations that may arise during execution to promote flexibility while reinforcing the platoon leader's intent.

4-80. The platoon leader seeks to rehearse the operation from initiation to occupation of the final march objective or LOA. Rehearsals include decision points and actions taken upon each decision. Often, the platoon leader prioritizes maneuver options and enemy COAs to be rehearsed based on the time available. The rehearsal focuses on locating the enemy, developing the situation, executing transition to a maneuver option, exercising direct and indirect fire control measures, and exploiting success. The rehearsal must consider the potential of encountering stationary or moving enemy forces. (See chapter 2 for additional information on preparation activities, specifically inspections.) Other actions to consider during rehearsals include—

- Lead element actions on contact.

- Actions to cross danger areas.
- Main body (participating as or in advance guard), making contact with a small enemy force.
- Main body (participating as or in advance guard), making contact with a large force beyond its capabilities to defeat.
- Main body (participating as or in advance guard), making contact with an obstacle reconnaissance forces and surveillance assets did not identify and report.
- Flank/rear security force (participating in higher echelon movement to contact), making contact with a small force.
- Flank/rear security force (participating in higher echelon movement to contact), making contact with a large force beyond its capability to defeat.
- Actions to report and *bypass*—a tactical mission task in which a unit deliberately avoids contact with an obstacle or enemy force (FM 3-90)—of an enemy force (based on bypass criteria).
- Maneuver options and transitions.

EXECUTE

4-81. During execution, the platoon moves rapidly to maintain the advantage of a rapid tempo. However, the platoon leader must balance the need for speed with the requirement for security. The platoon leader bases the decision on the effectiveness of higher echelon reconnaissance and surveillance efforts, effects of terrain, and the enemy's capabilities. The platoon leader closely tracks the movement and location of subordinate squads, attachments, and adjacent units. This ensures that battalion and/or company security forces provide adequate security for the main body and that they remain within supporting range of the main body, mortars, and field artillery. The platoon synchronizes its actions with adjacent and supporting units, maintaining contact and coordination as prescribed in orders and unit SOP. The following paragraphs discuss executing a movement to contact (the platoon is subordinate to the company for this discussion) using the four-step tactical framework of the offense discussed earlier in this chapter.

Find the Enemy

4-82. Reconnaissance forces and surveillance assets focus on determining the enemy's dispositions, providing the platoon leader through the company with relevant combat information. This relevant information helps to ensure subordinate squads and attachments are committed under optimal conditions. The platoon leader uses all available sources of intelligence and combat information to find the enemy's location and dispositions. With this information, the platoon leader and squad leaders work to ensure that the platoon, as the lead element of the company, makes enemy contact with the smallest element possible to preserve combat power and conceal the size and capabilities of the company main body following the lead element.

Fix the Enemy

4-83. Once contact is made, security forces and main body forces brings overwhelming fires onto the enemy to prevent the enemy, for example, from conducting a withdrawal, spoiling attack, or organizing a defense. Lead elements maneuver as quickly as possible to find *gaps*—areas free of obstacles that enable forces to maneuver in a tactical formation (FM 3-90)—in the enemy's defense or where to position an attack on an assailable flank of the enemy. While assessing the situation the platoon leader continues to gather as much information as possible about the enemy's dispositions, strengths, capabilities, and intentions. As more information becomes available, lead elements and the main body attack to destroy or disrupt enemy command and control centers, fire control nodes, and communication nets.

4-84. The security force leader, (for example, the platoon leader), initiates the platoon's maneuver at a tempo the enemy cannot match, since success in a meeting engagement (see paragraph 4-72) depends upon actions on contact. The security force, in this case the platoon, does not allow the enemy to maneuver against the protected main body of the company. The organization, size, and combat power of a security force are major factors determining the size of the enemy force it can defeat without deploying the main body. The techniques the security force leader employs to *fix*—a tactical mission task in which a unit prevents the enemy from moving from a specific location for a specific period (FM 3-90)—the enemy when both forces are moving are different from those employed when the enemy force is stationary during the meeting engagement. In both situations, when the security force cannot overrun the enemy by conducting a hasty operation, (for example, a frontal attack [see paragraph 4-14]), a portion of the main body is deployed. When this occurs, the unit is no longer conducting movement to contact but an attack. (See section IV for additional information on hasty versus deliberate operations and the attack.)

Finish the Enemy

4-85. If the security force, (for example, the platoon), cannot overrun the enemy with a frontal attack or by attacking an assailable flank of the enemy, the security force leader then fixes the enemy, if possible. The main body commander then quickly maneuvers units of the main body to conduct an attack (for example, an attack on an assailable flank) that overwhelms the enemy force before it can react or be reinforced. The main body commander attempts to defeat the enemy in detail while still maintaining the momentum of the advance. *Defeat in detail* is concentrating overwhelming combat power against separate parts of a force rather than defeating the entire force at once (ADP 3-90). Defeat in detail can occur sequentially or separately. It may finish during follow through. After an attack, the main body resumes the movement to contact and establishes another lead element (security force) moving ahead of the main body along the direction of attack or to the next march objective. If the enemy is not defeated in detail, the commander has two main options: bypass or transition to a more deliberate operation (for example, to attack or defend).

4-86. Once the forward security element, (for example, the platoon), commits to fixing the enemy main body units deploy rapidly to the vicinity of the contact. The main body commander coordinates forward passage through security forces in contact as required. The intent is to deliver the assault before the enemy can deploy or reinforce engaged forces. The main body commander or a higher commander may order an attack from a march column, for example, platoon(s) or company, from one of the main body's columns, while the rest of the main body deploys. The commander also can wait to attack until bringing the bulk of the main body forward. This avoids piecemeal commitment except when rapidity of action is essential, combat superiority at the vital point is present and can be maintained throughout the attack, or when compartmentalized terrain forces a different COA. When trying to conduct an envelopment (see paragraph 4-15), the commander focuses on attacking the enemy's flanks and rear before the enemy prepares to counter these actions. The commander uses the security force to fix the enemy while the main body maneuvers to look for an assailable flank. Elements of the main body also can be used to fix the enemy while the security force finds the assailable flank. (See ATP 3-21.10 for additional information.)

Follow Through

4-87. If the enemy is defeated, the unit transitions back into movement to contact and continue to advance. The movement to contact terminates when the unit reaches the final march objective or LOA, or transitions to a more deliberate operation, attack, area defense, or retrograde.

ASSESS

4-88. By the nature of a movement to contact, when the enemy situation is vague or not specific enough to conduct another operation, the platoon leader assesses and reassesses the mission prior to and during the movement to contact. Assessment is the continuous monitoring and evaluation of a current situation, and the progress of an operation. It involves deliberately comparing forecasted outcomes to actual events to determine the overall effectiveness of force employment. Assessment allows the platoon leader to maintain accurate situational awareness and to a lesser degree situational understanding. Continued monitoring and evaluation of the current situation helps the platoon leader make timely and accurate decisions. Assessment of effects is determining how the platoon actions have succeeded against the enemy. Effects typically are assessed by measure of performance and measure of effectiveness. (See ATP 3-21.10 for additional information.)

SECTION IV – ATTACK

4-89. An attack destroys or defeats enemy forces, seizes and secures terrain, or both. The attack can be a hasty or deliberate operation depending upon the time available for assessing the situation, planning, and preparing. An attack incorporates coordinated movement supported by direct fires and fire support. An attack masses the effects of overwhelming combat power against selected portions of the enemy force with a tempo and intensity that cannot be matched by the enemy. The resulting combat should not be a contest between near equals. Attackers seek decisions on the ground of their choosing through the deliberate synchronization and employment of the combined arms team. Variations of an attack are ambush, counterattack, raid, and spoiling attack.

HASTY VERSUS DELIBERATE OPERATIONS

4-90. The primary difference between a deliberate operation and hasty operation is the extent of planning and preparation the attacking force conducts. A *deliberate operation* is an operation in which the tactical situation allows the development and coordination of detailed plans, including multiple branches and sequels (ADP 3-90). A *hasty operation* is an operation in which a commander directs immediately available forces, using fragmentary orders, to perform tasks with minimal preparation, trading planning and preparation time for speed of execution (ADP 3-90). At one end of the continuum, an Infantry unit launches a hasty operation as a continuation of an engagement that exploits a combat power advantage and preempts enemy actions. At the other end, an Infantry unit conducts a deliberate operation from a reserve position or TAA with detailed knowledge of the enemy, a task organization designed specifically for attacking, and a fully rehearsed plan. Most attacks fall somewhere between the two extremes.

4-91. A deliberate operation normally is conducted when enemy positions are too strong to be overcome by a hasty operation. It is a fully synchronized operation employing every available asset against the enemy defense and is characterized by a high volume of planned fires, use of major supporting attacks, forward positioning of the resources needed to maintain momentum, and operations throughout the depth of enemy positions. Deliberate operations follow a preparatory period that includes planning, reconnaissance and surveillance, coordination, positioning of follow-on forces and reserve, preparation of troops and equipment, rehearsals, and operational refinement.

4-92. A hasty operation is conducted during movement to contact, as part of a defense, or when the enemy is in a vulnerable position and can be defeated quickly with available resources. This type of operation may cause the attacking force to lose a degree of synchronization. To minimize this risk, the leader maximizes use of standard formations and well-rehearsed, thoroughly understood battle drills and SOPs. A hasty operation is often the preferred option during continuous operations, enabling the leader to maintain momentum while denying the enemy time for defense preparations.

ORGANIZATION OF FORCES

4-93. The Infantry rifle platoon normally conducts an attack as part of a company in an independent operation or as part of the company within a battalion operation. Once the scheme of maneuver is determined, the company commander (when the platoon operates within company) task-organizes the force to ensure platoons have enough combat power to accomplish their assign missions. The company commander normally organizes a main body and reserve (a security force, main body, and reserve is normally established during a battalion or higher echelon attack) for the attack, which are all supported by some type of sustainment organization (for example, company or field trains [see paragraphs 8-22 to 8-25]). The commander and subordinate platoon leaders should complete all changes in task organization on time to allow units to conduct rehearsals with their attached and supporting elements. Based upon METT-TC (I), the platoon may conduct the attack (platoon assault) independently. As an example, the platoon may conduct an attack to seize an assault objective, key terrain, or other enemy position dependent on the size of the enemy force and/or terrain. (See ATP 3-21.10 and ATP 3-21.20 for additional information on an attack at the company and battalion echelons.)

Security Forces

4-94. The Infantry rifle platoon leader, as well as the Infantry rifle company commander, do not generally designate dedicated security forces for independent operations conducting an attack. During a platoon's movement along its direction of attack, platoon leaders are responsible for 360-degree security, for ensuring each subordinate unit's sectors of fire are mutually supporting, and for being able to transition the platoon rapidly upon contact with the enemy. Platoon leaders exercise command and control primarily through their subordinate leaders and move in the formation where they can best achieve this.

4-95. Along the platoon's direction of attack, squad leaders and team leaders execute movement formations and movement techniques based on the platoon leader's guidance. They adjust their formations and techniques as necessary while moving (see chapter 3). This determination is a result of the platoon's mission, the nature of the threat, the closeness of terrain, and visibility. Subordinate elements perform their required actions, through training before combat and rehearsals during combat.

4-96. When the platoon is designated as a security force (flank and/or rear) within the Infantry battalion's organization of forces for the attack, under normal circumstances, the commander resources dedicated security forces during an attack only if the attack uncovers one or more flanks, or the rear of the attacking force as it advances. In this case, the commander designates a flank or rear security force and assigns it a guard or screen mission, depending on METT-TC (I).

Note. Normally an attacking unit does not need extensive forward security forces as most attacks are launched from positions in contact with the enemy, which reduces the usefulness of a separate forward security force. The exception occurs when the attacking unit is transitioning from defense to offense and had previously established a security area as part of the defense.

Main Body

4-97. During an attack (platoon assault, see figure 4-7 on page 4-40), conducted independently of the company, the platoon leader establishes a main effort and supporting effort within the platoon. The *main effort* is a designated subordinate unit whose mission at a given point in time is most critical to overall mission success (ADP 3-0). The platoon leader usually weights the main effort with the preponderance of platoon's combat power, or the operation is designed where the effort is singularly focused. Designating a main effort temporarily prioritizes resource allocation. The platoon leader may shift the main effort during execution. A *supporting effort* is a designated subordinate unit with a mission that supports the success of the main effort (ADP 3-0). The platoon leader resources the supporting effort(s) with the minimum assets necessary to accomplish the mission. The platoon often realizes success of the main effort through the success of the supporting effort(s).

4-98. During a deliberate operation, the platoon leader's main effort accomplishes the destruction of the enemy force, seizure of a terrain objective, or the defeat of the enemy's plan. The platoon's scheme of maneuver identifies the focus of the main effort. Because the potential for obstacles always exists, the platoon leader should designate an assault, breach, and support element within the platoon to conduct a breach during the attack.

4-99. During a hasty operation, the platoon leader retains flexibility by arranging forces in-depth and maintains more of a centralized control of subordinate elements, direct and indirect fires, and fire support systems. As soon as the tactical situation develops enough to allow the platoon leader to identify the decisive point(s), the platoon leader focuses available resources to support the designated main effort's achievement of its objective.

RESERVE

4-100. The reserve is used to exploit success, defeat enemy counterattacks, or restore momentum to a stalled attack. The Infantry rifle platoon leader does not generally designate a dedicated reserve for platoon missions, whether conducted independent of the company or not. For a company mission, the dedicated reserve (when designated) is usually a squad-size element. For a battalion mission, the dedicated reserve (when designated) is usually a platoon-size element. The reserve is not a committed force and is not used as a follow and support force, or a follow and assume force. *Follow and assume* is a tactical mission task in which a committed force follows and supports a lead force conducting an offensive operation and continues mission if lead force cannot continue (FM 3-90). *Follow and support* is a tactical mission task in which a committed force follows and supports a lead force conducting an offensive operation (FM 3-90). Once committed, the reserve's actions normally become or reinforce the echelon's main effort. Once the reserve is committed, every effort is made to reconstitute another reserve from units made available by the revised situation. Often a commander's most difficult and important decision concerns the time, place, and circumstances for committing the reserve.

CONTROL MEASURES

4-101. Units conducting an attack operate within an assigned AO. Regardless of whether the attack takes place in a contiguous or noncontiguous assigned area, the owner of the AO normally designates control measures.

4-102. The owner uses only the control measures necessary to control the attack. Short of the LD or line of contact (LC), the leader may designate TAA and attack positions where the unit prepares for the mission or waits for the establishment of required conditions to initiate the attack. Beyond the LD or LC, leaders may designate checkpoints, phase lines, PLD, lateral boundaries, assault positions, and direct and indirect fire support coordination measures. Between the PLD and objective, a final coordination line, assault positions, support by fire and attack by fire positions, and time of assault to better control the final stage of attack can be used. Beyond the objective, the Infantry leader can impose an LOA or forward boundary. (See ATP 3-21.10 for additional information.)

SEQUENCE OF EVENTS, EXAMPLE

4-103. An attack at the platoon level is normally conducted as part of a larger force and is an offensive action characterized by close combat. *Close combat* is warfare carried out on land in a direct-fire fight, supported by direct and indirect fires and other assets (ADP 3-0). The following sequence of events is typical of an attack, regardless of echelon; however, specifics will vary due to the mission variables and the higher scheme of maneuver. Additionally, some elements may occur simultaneously. (See ATP 3-21.10 for additional information.) Events include:

- TAA.
- Reconnaissance and surveillance.
- Movement to the LD.
- Movement to the PLD.
- Actions on the objective (platoon assault).
- Follow through (consolidate, reorganization, and transition).

TACTICAL ASSEMBLY AREA

4-104. The platoon leader and subordinate leaders within the TAA direct and supervise mission planning and preparation for the attack. Additional activities include coordination, PCCs and PCIs, rehearsals, and sustainment preparations for the mission. (See ATP 3-21.10 for additional information.)

RECONNAISSANCE AND SURVEILLANCE

4-105. Higher echelon reconnaissance and surveillance collection efforts provide combat information on the enemy situation and terrain in the platoon's AO. The collection effort may begin prior to occupying the TAA and continues throughout the operation. The platoon may still receive current reports as it occupies the assault position. Reconnaissance and surveillance activities provide the platoon leader with the information needed to plan, prepare, execute, and assess the mission. The situation and available time may limit the ability to conduct reconnaissance and surveillance activities prior to crossing the LD. Leaders balance the benefits of reconnaissance forces and surveillance assets providing combat information with the level of risk involved with them compromising the operation.

MOVEMENT TO THE LINE OF DEPARTURE

4-106. When attacking from positions not in contact with the enemy, the platoon often stages in a rear TAA or moves to an attack position behind friendly units in contact with the enemy, then conducts passage of lines to begin the attack. When attacking from a position in direct contact, the LD may also be the *line of contact*, a general trace delineating the locations where friendly and enemy forces are engaged (ADP 3-90). In certain circumstances, there may not be an LD, for example, an operation in a noncontiguous assigned area. Additionally, commanders may designate PDs for their attacking units instead of an LD for units attacking on foot using infiltration and stealth.

4-107. The tactical situation and order in which the platoon leader wants subordinate elements to arrive at their attack position(s) (if necessary) and/or LD/PD govern the march formation. The platoon leader typically executes a planned movement through a start point(s), along a route(s) and through a release point(s), into an attack position(s) (if necessary) short of the LD/PD after which the leader will deploy into the appropriate movement formation and movement technique. The platoon leader develops an order of movement and timeline (by squad) to the attack position and/or LD/PD. Lead elements of the attacking force cross the LD/PD at the time specified in the OPORD or FRAGORD. Before movement, a small patrol from the platoon (see chapter 7) can be tasked to reconnoiter and mark the route and check the time it takes to move to the LD/PD. (See chapter 6 section III for additional information on the passage of lines prior to crossing the LD/PD.)

Movement to the Probable Line of Deployment

4-108. The platoon moves rapidly through their attack position(s) and across the LD/PD controlled by friendly forces. The platoon uses its designated attack position only by exception, for example, to adjust the march timing of the unit's movement or when the conditions required to ensure the success of the planned maneuver is not yet established. The platoon should minimize the time it remains stationary in the attack position. Generally, the platoon should initiate actions to protect itself and increase survivability every 10 to 15 minutes. This includes such actions as deploying local security, improving cover and concealment, and starting the construction of fighting and survivability positions. The platoon leader considers the mission variables of METT-TC (I) when choosing the actions within an attack position which best balances firepower, tempo, security, and control.

4-109. When crossing the LD controlled by a stationary unit, battle handover occurs along a general trace line designated as the BHL (see paragraph 5-18). A *battle handover* is a coordinated mission between two units that transfers responsibility for fighting an enemy force from one unit to another (FM 3-90). The stationary unit commander determines the location of the line, generally a phase line forward of the stationary unit. The BHL is the FLOT in the offense or forward edge of the battle area in the defense. The line is drawn where the direct fires of forward combat elements of the stationary unit can effectively support the passing unit until completion of the passage of lines (see chapter 6 section III). The area between the BHL and the stationary force belongs to the stationary unit commander. The BHL is within direct fire range and observed indirect fire range of the stationary force. During the passage of lines, the stationary force commander may employ security forces, obstacles, and fires in the area.

4-110. *Probable line of deployment* is a phase line that designates the location where the commander intends to deploy the unit into assault formation before beginning the assault (ADP 3-90). The attacking platoon splits into one or more assault force (generally the main effort) and support forces (generally the supporting effort) as it reaches the PLD, if not already accomplished. (See figure 4-7 on page 4-40 for an example platoon assault [conducted as the lead platoon for an independent company operation].) All forces supporting the assault should be set in their support by fire position, or other position, before the initiation of the attack by the assault force. The assault force maneuvers against or around the enemy to take advantage of the support force's efforts to suppress targeted enemy positions.

Note. As necessary, the platoon conducts a breach. The preferred method of fighting through a defended obstacle is to conduct an in-stride breach. However, the leader must be prepared to conduct a deliberate breach. (See chapter 6 section I for additional information.)

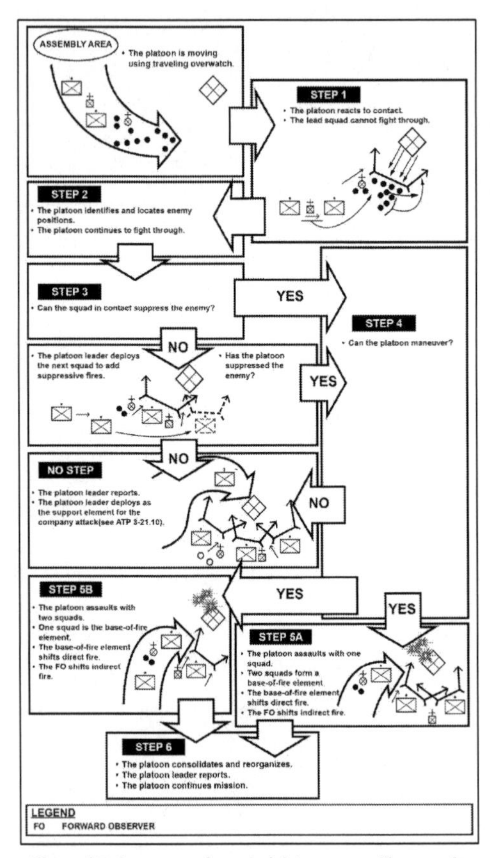

Figure 4-7. Sequence of events (platoon assault), example

ACTIONS ON THE OBJECTIVE (PLATOON ASSAULT)

4-111. The platoon maintains the pace of its advance and deploys into one or more assault forces before or upon reaching the PLD. The platoon leader synchronizes the occupation of support by fire and/or attack by fire positions with the maneuver of the assault force to limit the vulnerability of the forces occupying these positions. The platoon leader maximizes of direct fires and fire support to destroy and to suppress the enemy, and to sustain the momentum of attack.

Support by Fire

4-112. Platoon support forces (supporting effort) occupy support by fire positions that, ideally, afford unobstructed observation, clear fields of fire, and cover and concealment. Selection of support by fire positions are based on a study of the terrain, knowledge of enemy locations, or likely enemy locations, and refined in execution through leader reconnaissance when possible. If the enemy situation is vague or unknown, positions are selected to place effective fire on terrain that dominates the area the assault force will traverse and seize. Once in position, support forces are responsible for both suppressing known enemy forces and for scanning assigned sectors of observation to identify previously unknown enemy elements to suppress them. The protection provided by support forces allow the assault force to continue movement and to retain the initiative even when under enemy observation or within range of enemy weapons.

Assault

4-113. The platoon leader retains the assault force in covered and concealed positions until the support force has set necessary conditions. The assault force must be prepared to deal with unanticipated protective obstacles in/around the objective area, particularly in buildings or fortified areas. Once committed, the platoon's assault is short, violent, and well ordered. The assault force seizes or secures a geographic objective, or destroys, defeats, or disrupts a designated enemy force. From the PLD, the assault force maneuvers against or around the enemy to take advantage of support fires. The assault force must closely follow these supporting fires to gain ground that offers a positional advantage. Depending on the size and preparation of enemy forces, it may be necessary to isolate and destroy portions of the enemy in sequence. It may also be necessary to reposition support forces to continue to engage the depth of, or beyond, the objective area.

4-114. Between the PLD and the objective, an assault position(s) and a final coordination line may be established to control the final stages of the assault. Assaulting forces usually plan to pass through the assault position and deploy to their final assault formation at final coordination line. Stopping at the assault position may allow the enemy to react to the assault and may make it difficult for the assault force to regain the momentum of the assault once the force halts in a covered position. When used, the assault force pauses at the assault position only to—

- Make final equipment preparations, such as to make final preparations for demolitions.
- To ensure all assault forces are in their planned order.

- Wait for preparatory fires to finish.
- Ensure conditions are set for the assault (for example, sufficient suppression, obscuration, or a necessary breach).

4-115. The objective for the assault force may vary from operation to operation. In every case, the assault force's actions on the objective are critical. Assault forces maneuver through fire and movement. Supporting fires must immediately gain fire superiority on the objective in coordination with fires around and beyond the objective to disrupt enemy reinforcement. As the assault closes on the objective, fires shift just forward of the assault force as it moves across the objective or until ordered to stop or shift to other target beyond the assault force or LOA.

Note. The key is to minimize the time between the shifting of fires and the maneuver of the assault force on the objective. The enemy must be suppressed during this time when the assault force is most vulnerable.

Follow Through (Consolidate, Reorganize, and Transition)

4-116. The platoon and squad consolidate and reorganize as required by the situation and mission. During consolidate and reorganization, the platoon may use protective obstacles (see paragraphs 5-56 to 5-58) to aid in flank security and to protect against enemy counterattacks. Additionally, echelons above the platoon may use tactical obstacles (see paragraphs 5-52 to 5-55) to shape enemy routes of withdrawal and to cut off enemy reinforcements.

4-117. Once the platoon takes an enemy position, it consolidates on that position if doing so is tactically necessary or advantageous. *Consolidate* is to organize and strengthen a captured position to use it against the enemy (FM 3-90). Normally, the attacking platoon tries to exploit its success; however, in some situations the platoon may have to pause to consolidate its gains. Consolidated may vary from a rapid repositioning of forces and security elements on the objective, to a reorganization of the attacking force, or to the organization and detailed improvement of the position for defense. Additional actions taken to consolidate include—

- Seeking out and eliminating enemy pockets of resistance (on and off the objective)/secure enemy detainees.
- Searching and marking positions to indicate to other friendly forces that they have been cleared.
- Establishing local security (see paragraph 6-79) and hasty defensive fighting positions (see paragraphs 5-229 to 5-232) while utilizing existing obstacles (see paragraph 5-45) if possible.
- Using security patrols, observation posts/combat outposts, and early warning devices to provide protection and situational awareness.
- Establishing indirect fire targets, FPF, and final protective line (FPL).
- Establishing mutual support between individual fighting positions, adjacent squads/adjacent platoons, to include:

    - Automatic weapons (man position and assign principal direction of fire [known as PDF] to Soldiers manning automatic weapons).
    - Fields of fire (establish sectors of fire and other direct fire control measures for each subunit/Soldier).
    - Protection requirements such as defining type fighting position and building fighting positions.

4-118. *Reorganization* is all measures taken by the commander to maintain unit combat effectiveness or return it to a specified level of combat capability (ATP 3-94.4). While combat effectiveness is the ability of a unit to perform its mission, factors such as ammunition, personnel, status of fuel, and weapon systems are assessed and rated (see paragraph 2-80). Reorganization actions can include cross-leveling ammunition and ensuring key weapon systems are manned and vital leadership positions are filled if the operator or crew became casualties. Additional actions include—

- Reestablishing the chain of command.
- Manning key weapon systems.
- Maintaining communications and reports, to include:
    - Restoring communication with any unit temporarily out of communication.
    - Sending situation reports (at a minimum, subordinates report status of mission accomplishment).
    - Identifying and requesting resupply of critical shortages.
- Resupplying and redistributing ammunition and other critical supplies.
- Performing special team actions such as—
    - Processing, and evacuating causalities, enemy prisoners of wars, enemy weapons, noncombatants/refugees, and damaged equipment (not necessarily in the same location).
    - Evacuating friendly killed in action.

4-119. The platoon executes follow-on orders/missions as directed by the company commander. For example, to reposition the force on the objective and establish security observation posts forward of the objective to prepare for a counterattack, or to organize the force and conduct detailed improvement of fighting positions for defensive missions. A likely mission may be to continue the attack against the enemy within the AO. Regardless of the situation, the platoon and subordinate squads' posture and prepare for continued offensive missions.

OPERATIONS PROCESS

4-120. During an attack, the platoon leader seeks to place the enemy in a position where the enemy can be defeated or destroyed easily. The platoon leader seeks to keep the enemy off-balance while continually reducing the enemy's options. The platoon leader employs movement and maneuver supported by other warfighting capabilities to prevent the enemy from accomplishing its mission and objectives. The amount of time available for planning and preparation will determine whether the attack is conducted as a hasty or deliberate operation. Though a platoon may be tasked to conduct a hasty attack as a continuation of an ongoing engagement, it is always best to plan and prepare for a deliberate attack from a TAA or reserve position.

4-121. In addition to accomplishing the mission, the platoon leader's concept of operations must contain provisions to exploit successes and all advantages that arise during the operation. The platoon leader exploits success and maintains the advantage by continually assessing the process throughout planning and preparation, and by aggressively executing the plan, promoting subordinate leader initiative, and putting subordinate units in a position of advantage. (See ATP 3-21.10 for additional information.)

Note. The operations process (plan, prepare, execute, and assess) is the major command and control activities performed during combat operations. The operations process (specific to TLP) is a platoon leader-led activity. (See chapter 2 for additional information on TLP.)

PLAN

4-122. During the planning process, the platoon leader integrates the scheme of fires with the scheme of maneuver. The *scheme of fires* is the detailed, logical sequence of targets and fire support events to find and engage targets to support the commander's objectives (JP 3-09). The leader identifies the desired effect of fires on enemy weapon systems, such as suppression or destruction, as part of developing the scheme of fires. The scheme of fires identifies critical times and places where the platoon leader needs the maximum effects from fire-support assets. Leaders combine movement with fires to mass effects, achieve surprise, destroy enemy forces, and obtain decisive results.

4-123. The platoon leader assigns subordinate squad leaders and attached element leaders their missions and imposes only those control measures necessary to synchronize and maintain control over the operation. Using the enemy situational and weapons templates, previously developed, the leader determines the probable LC and enemy trigger lines (see paragraph 5-208). As the leader arrays subordinate elements to shape the battlefield, friendly weapon systems are matched against the enemy to determine the PLD. Once the leader determines the PLD, the leader establishes how long it takes subordinates to move from the LD to the PLD and establish all necessary support by fire positions. The leader establishes when and where the force must maneuver into enemy direct-fire range.

4-124. The platoon leader seeks to surprise the enemy by choosing an unexpected direction, time, type, or strength for attacking. Surprise delays enemy reactions, overloads and confuses enemy command and control, induces psychological shock in the enemy, and reduces the coherence of the enemy's defensive operations. The leader achieves tactical surprise by attacking in bad weather and over seemingly impassible terrain, maintaining a high tempo, and employing sound operations security. The leader may plan different attack times for main and supporting efforts to mislead the enemy and allow the shifting of supporting fires. However, simultaneous attacks provide a means to maximize the effects of mass in the initial assault. They also prevent the enemy from concentrating defensive fires against successive attacks.

4-125. In planning, the platoon leader and subordinate leaders focus on the routes, formations, and navigational aids they will use to traverse the ground from the LD or PLD to the objective. Some terrain locations may require the attacking platoon to change its movement formation, direction of movement, or movement technique when it reaches those locations. The platoon can post guides at these critical locations to ensure maintaining control over the movement.

4-126. To employ the proper capabilities and tactics, the platoon leader and subordinate leaders must have detailed knowledge of the enemy's organization, equipment, and tactics. They must understand enemy strengths and weaknesses. The platoon leader may need to request information to answer platoon information requirements.

4-127. Generally, if the platoon leader does not have timely, relevant, and accurate intelligence products and assessments, and does not know where the overwhelming majority of the enemy's units and systems are located, the leader cannot conduct a deliberate operation. The attacking unit must conduct a movement to contact, then conduct a hasty operation once contact is made, or collect more combat information.

4-128. Leaders within the platoon may find themselves as the observer (and executor) of company- and battalion-level fires or exploiting the success of higher echelon military deception operations (for example, feints and demonstrations [see chapter 6 section V]). Understanding the concept of echelon fires is critical for indirect fire planning to be synchronized with the maneuver plan. The purpose of echeloning fires (see appendix B) is to maintain constant fires on a target while using the optimum delivery system up to the point of its risk-estimate distance in combat operations or minimum safe distance in training. Echeloning fires provides protection for friendly forces as they move to and assault an objective, allowing them to close with minimal casualties. It prevents the enemy from observing and engaging the assault by forcing the enemy to take cover, allowing the friendly force to continue the advance unimpeded.

4-129. The goal of the platoon leader's attack criteria is to focus fires on seizing the initiative. The leader emphasizes simple and rapidly integrated direct fires and fire support plans. This is done using quick-fire planning techniques (see ATP 3-21.10 for additional information) and good SOPs. Leaders integrate fire assets as far forward as possible in the movement formation to facilitate early emplacement. Leaders concentrate combat powers at decisive points and times on forward enemy elements to enable maneuver efforts to close with the enemy positions.

Note. Subordinate company fire support officers develop quick-fire plans to support their respective commander's mission. In quick-fire planning the fire support officer assigns targets (and possibly a schedule of fires) to the most appropriate fire support means available to support the mission. In this type of fire support planning, the available time usually does not permit evaluation of targets on the target list and consolidation with targets from related fire support agencies. (See ATP 3-09.42 for additional information on quick-fire planning.)

4-130. The platoon leader and subordinate leaders plan for sustainment to ensure the platoon's freedom of action and prolong endurance through unit trains (for example, company trains and battalion field trains). (See chapter 8 for additional information on unit sustainment operations during the attack.)

4-131. Protection increases a platoon's survivability and their ability to maintain combat power. Planning for protection involves identifying enemy capabilities (for example, fires, aviation, and CBRN) and the necessary actions to counter or mitigate those threats. Emphasis on protection increases during preparation and continues throughout execution. Protection is a continuing activity; it integrates all protection capabilities to protect the force before, during, and after the attack. (See paragraphs 4-40 to 4-47 on the challenges to protect the force during the offense.)

PREPARE

4-132. During preparation for the attack, the platoon leader and subordinate leaders must continue to analyze the most current information from organic and higher echelon information collection efforts. When possible, even in the most fluid of situations, attacks are best organized and coordinated in a currently established TAA, or one that is hastily established due to a change of mission.

4-133. Unless already in a TAA or patrol base, the attacking platoon moves into one to prepare for future operations. The unit moves with as much secrecy as possible, normally at night and along routes preventing or degrading the enemy's capabilities to visually observe or otherwise detect the movement. It avoids congesting its TAA or patrol base and occupies it for the minimal possible time. While in the TAA or patrol base, the platoon is responsible for its own protection activities, such as local security and electromagnetic emissions control and masking. (See chapter 7 for additional information on establishing a patrol base.)

4-134. The platoon leader and subordinate leaders continue their TLP and priorities of work to the extent the situation and mission allow before moving towards their attack position(s) and/or LD and if established their LC. These preparations include but are not necessarily limited to:

- Protecting the force.
- Conducting task organization.
- Performing reconnaissance.
- Refining the plan.

- Briefing the troops.
- Conducting rehearsals, to include test firing (test firing dependent on the situation, though not conducted past the LD) of weapons.
- Moving logistics and medical support forward.
- Promoting adequate rest for both leaders and Soldiers.
- Positioning the force for subsequent action.
- PCCs and PCIs.

4-135. Preparation for an attack against a moving enemy force is limited because the opportunity to attack the enemy at the appropriate time and place depends on the enemy's movement. This forces the platoon to focus the preparation on executing fires and movement and maneuver actions regarding the enemy's location, or possible locations, and disposition when engaged. The platoon leader prioritizes each location to ensure the platoon prepares for the most likely engagements first. The platoon leader ensures all subordinate leaders understand their role at each location and the decision point for execution at each location. Leaders rehearse actions, as possible, in each location against various enemy conditions to promote flexibility and initiative consistent with the commander's intent. Battle drills, SOPs, and rehearsals against likely enemy actions are essential for success in preparation for an attack.

4-136. As part of TLP, a leader's reconnaissance (see paragraph 7-15) of the actual terrain should be conducted with all key leaders when this will not compromise the attack or result in excessive risk to the leaders on the reconnaissance. Modern information systems can enable leaders to conduct a virtual reconnaissance when a physical reconnaissance is not practical. If a limited-visibility attack is planned, they should reconnoiter the terrain at night.

EXECUTE

4-137. During execution, the platoon leader maneuvers the platoon to close with and destroy or defeat enemy forces, seize, and secure terrain, or both. The following paragraphs discuss executing a company attack (specific to platoon scheme of maneuver) using the four-step tactical framework of the offense (find, fix, finish, and follow through) discussed earlier in this chapter.

Note. Regarding the enemy's plan, a defending enemy generally establishes a security zone around or forward of main enemy forces to make early contact with friendly attacking forces. This helps the enemy determine the attacking force's capabilities, intent, chosen COA, and delays their approach. The enemy leader uses the security zone to strip away friendly reconnaissance forces and hide the enemy's main body dispositions, capabilities, and intent. The goal is to compel the friendly attacking force to conduct movement to contact against the enemy defending force while knowing the exact location of attacking friendly forces.

Find the Enemy

4-138. Gaining and maintaining contact with the enemy (when the enemy is determined to hide, deny, or break contact) is vital to the success of an attack. As platoons moves from the LD and cross the PLD towards the objective and as current intelligence and relevant combat information on the enemy is updated, the enemy situation generally becomes clearer. Platoon leaders and subordinate leaders, through actions on contact (see paragraph 2-48), rapidly develop the situation in accordance with the company commander's plan and intent for the attack.

Note. The fictional scenario, used as discussion vehicles in figure 4-8 on page 4-49, figure 4-9 on page 4-52, and figure 4-10 on page 4-54, illustrate different ways an Infantry rifle platoon can accomplish its mission within the conduct of a higher echelon's attack. For discussion purposes, the fictional scenario will primarily focus on the actions of Company C, the battalion's main effort. These three examples illustrate potential challenges confronting the platoon leader in accomplishing an attack (specifically platoon breach and platoon assault) but are not intended to be prescriptive of how the Infantry rifle platoon performs any particular operation. This same scenario drives the attack techniques used in ATP 3-21.10 and ATP 3-21.20. ATP 3-21.8 focuses on the techniques used to perform missions, functions, or tasks in support of the Infantry rifle company within an Infantry battalion's attack.

Fix the Enemy

4-139. Prior to the attack, battalion support by fire positions (SBF-3 and SBF-4, reference figure 4-8 on page 4-49) once established fix enemy actions prior to (initially), on, and surrounding the company objective. Additional operations to fix and/or *disrupt*—a tactical mission task in which a unit upsets an enemy's formation or tempo and causes the enemy force to attack prematurely or in a piecemeal fashion (FM 3-90)—the enemy includes preparatory fires and deception operations to deceive the enemy. By fixing and/or disrupting the enemy, the objective of the attacker's main effort is isolated to prevent the enemy from maneuvering to reinforce the enemy targeted for destruction. Massing forces in one place by using economy of force measures in other areas allows the attacker to mass the effects of overwhelming combat power against a portion of the enemy. The following example, see figure 4-8 on page 4-49, used for discussion purposes, represents one way to fix enemy actions prior to the assault. In the scenario, as assault forces close on the objective, Company C commander employs attached Assault Platoon 3 (with attached company mortar section) in such a way that the commander is best able to maximize isolation of the two points of breach (see chapter 6 section I) locations forward of Objective Fox North.

Note. On order Company D (-), with Assault Platoon 1 and Assault Platoon 2, battalion mortar Section A, and the battalion tactical command post, conducts approach march vicinity Phase Line Jim, moves on Axis Orange (not illustrated) into battalion Support by Fire Position 3 and Support by Fire Position 4 (illustrated in figure 4-8, figure 4-9 on page 4-52, and figure 4-10 on page 4-54). (See ATP 3-21.10 and ATP 3-21.20 for additional information on the scenario at the company and battalion echelons, respectively.)

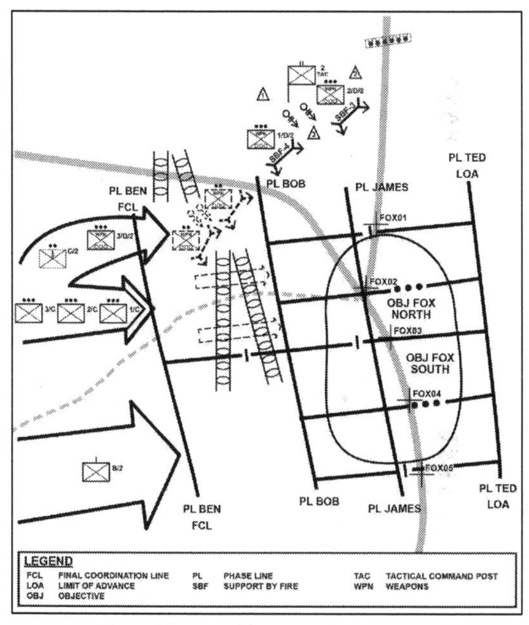

Figure 4-8. Attack (support by fire), example

Finish the Enemy

4-140. During the assault, the attacking force maneuvers to gain positional advantage to seize, retain, and exploit the initiative while avoiding the enemy's defensive strength. The attacker employs tactics defeating the enemy by attacking through a point of relative weakness, such as a flank or the rear. The key for success is to strike hard and fast, overwhelm a portion of the enemy force, and quickly transition to the next objective or phase, thus maintaining the momentum of attack without reducing the pressure.

Movement From the Line of Departure to the Probable Line of Deployment

4-141. The platoon conducts tactical movement (approach march) once it crosses the LD. It moves aggressively and as quickly as the terrain and enemy situation allow. It moves forward using the appropriate movement formation and movement technique assisted by the fires (when required) of supporting units. Fire and movement are coordinated closely. Suppressive fires facilitate friendly movement, and friendly movement facilitates more fires. Whenever possible, the attacking platoon uses avenues of approach avoiding strong enemy defensive positions, takes advantage of all available cover and concealment, placing the unit on the flanks or rear of the defending enemy. Where cover and concealment are not available, the unit uses obscurants to conceal its movement.

Actions at the Probable Line of Deployment, Assault Position, or Final Coordination Line

4-142. The attacking platoon maintains the pace of its advance as it approaches its PLD. The attacking platoon splits into one or more assault and support forces once it reaches the PLD if not previously completed. Assaulting forces usually plan to pass through the assault position and deploy to their final assault formation at final coordination line (see paragraph 4-114). All forces supporting the assault force should be set in their support and attack by fire positions before the assault force begins its assault. The platoon leader synchronizes the occupation of support by fire positions with the maneuver of the supported attacking element to limit the vulnerability of forces occupying these positions. Leaders use unit tactical SOPs, battle drills, prearranged signals, engagement area, and TRP to control direct and indirect fires from these supporting positions and normally employs restrictive fire lines (RFLs) between converging forces.

Conduct a Breach

4-143. To conduct a breach successfully, the platoon applies the breaching fundamentals of suppress, obscure, secure, reduce, and assault (see chapter 6 section I for information on the fundamentals). The support force sets the conditions; the breach force reduces, clears, and marks the required number of lanes through the enemy's tactical obstacles to support the maneuver of the assault force. The leader must clearly identify the conditions allowing the breach force to proceed to avoid confusion. For example, one condition is the use of a minimum safe line when a friendly element, such as a breach force, is moving toward an area of indirect fires. As the breach force approaches the minimum safe line, observers call for fires shift (or cease), allowing the breach force to move safely in the danger area (see paragraph A-29). From the PLD, the assault force maneuvers against or around the enemy to take advantage of support force's efforts to suppress the targeted enemy positions.

Note. For discussion purposes, the platoon breach and assault will primarily focus on the actions of Company C, the battalion's main effort.

4-144. The following example, see figure 4-9, used for discussion purposes, represents one way to conduct a platoon breach prior to the company's assault on Objective Fox North. In the scenario, as breach forces close on the points of breach, Company C commander establishes local support by fire positions (with attached Assault Platoon 3) to support First Platoon's breach. Assault Platoon 3 (with Company C mortar section to engage targets on the far side of the breach) moves by section on the company's left flank. As Company C's First Platoon advances, Assault Platoon 3, on order or as First Platoon starts to receive effective fires in route to the obstacle begins to suppress the enemy overwatching the obstacle to *isolate*—a tactical mission task in which a unit seals off an enemy, physically and psychologically, from sources of support and denies it freedom of movement (FM 3-90)—the points of breach, west of Phase Line Bob. Once the points of breach locations are isolated by Assault Platoon 3, First Platoon moves forward with attached engineers to conduct two explosive breaches of the wire obstacle. The effect of the enemy wire obstacle, west of Objective Fox North, is to block any eastward movement along the trail network. First and Second Squads of First Platoon establish local support by fire positions to support engineer breaches. Once the obstacle breach is open, attached engineers create and mark lanes, and guide Third Squad, Fire Team A (under control of the Third Squad leader, with one attached machine gun) and Third Squad, Fire Team B (under the control of the First Platoon, Weapons Squad Leader, with one attached machine gun) through the obstacle to establish two footholds on the far side of the two breaches. Once footholds are established, fire teams employ two local support by fire positions (SFB-A and SBF-B, reference figure 4-9 on page 4-52) to support Second and Third Platoons' assault through the breach and movement to the platoon's PLD for the assault of Objective Fox North. (See chapter 6 section I for additional information on actions on the objective involving support, breach, and assault elements in the attack.)

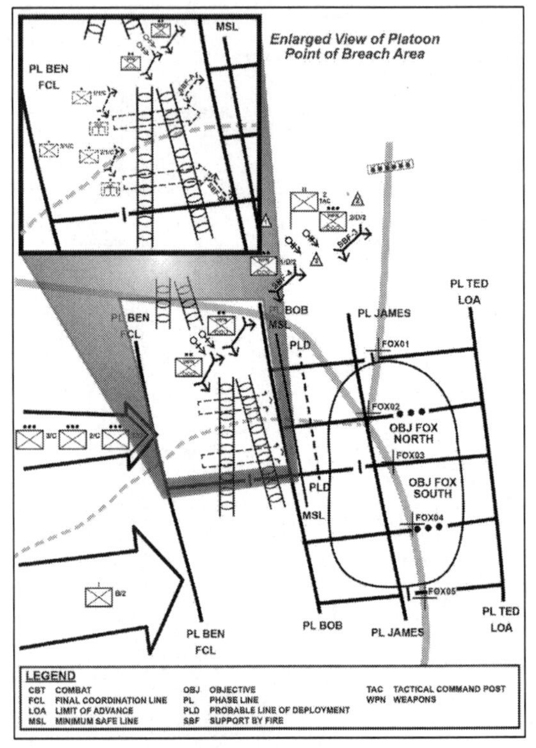

Figure 4-9. Attack (platoon breach), example

4-145. In the scenario, once conditions are set for the assault, Second and Third Platoons maneuver through the lanes established by First Platoon. (See figure 4-10 on page 4-54.) During the establishment of Support by Fire Positions A and B. Company C commander coordinates with Company D for Support by Fire Position 4 to the north to cease-fire. Support by Fire Position 3 continues to support Second Platoon and Third Platoon (assault forces) maneuver to seize Objective Fox North, assault forces clear to LOA, Phase Line Ted. Simultaneously, Company B (guiding off Company C's rate of advance to LOA, Phase Line Ted) assaults with two platoons abreast (not illustrated), seizes Objective Fox South, assault forces clear to LOA, Phase Line Ted.

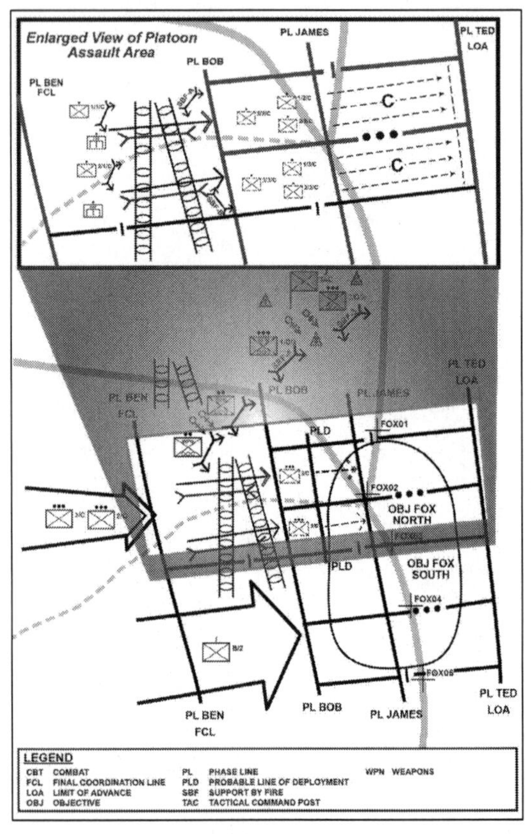

Figure 4-10. Attack (platoon assault), example

Actions on the Objective

4-146. The effects of overwhelming and simultaneous application of fire, movement, and shock action characterize the final assault. This violent assault destroys or defeats and drives the enemy from the objective area. Small units conduct the final assault while operating under the control of the appropriate echelon command post.

4-147. Assaulting forces usually plan to pass through the assault position and deploy to their final assault formation at the final coordination line. The *final coordination line* is a phase line close to the enemy position used to coordinate the lifting or shifting of supporting fires with the final deployment of maneuver elements (ADP 3-90). Ideally when established and utilized, the platoon's assault element occupies the assault position without the enemy detecting the platoon's elements. Preparations in the assault position may include preparing Bangalore, other breaching equipment, or demolitions; fixing bayonets; lifting or shifting direct fires; or preparing generated obscuration (for example, smoke pots and hand grenades).

4-148. If the platoon is detected as it nears its assault position, indirect fire suppression is required on the objective and the support element increases its volume of fire. If the platoon needs to make last-minute preparations, then it occupies the assault position. If the platoon does not need to stop, it passes through the assault position, treating it as a PLD and assaults the objective. Sometimes, a platoon must halt to complete preparation and to ensure synchronization of friendly forces. Once the assault element moves forward of the assault position, the assault continues. If the assault element stops or turns back, the element could sustain excessive casualties.

4-149. Infantry leaders employ all direct and indirect fire support means to destroy and suppress the enemy and sustain the momentum of attack. By carefully synchronizing the effects of indirect-fire systems and available close air support (CAS), leaders improve the likelihood of success. Fires are planned in series or groups to support maneuver against enemy forces on or near the geographical objective. As leaders shift artillery fires and obscurants from the objective to other targets, the assault element moves rapidly across the objective. The support element must not allow its suppressive fires to lapse. These fires isolate the objective and prevent the enemy from reinforcing or counterattacking. They also destroy escaping enemy forces and systems.

Follow Through

4-150. Once an enemy is finished, actions by the platoon are not complete. Small remaining elements of the enemy may require the platoon to destroy them in detail. If all enemy forces are not *neutralized*—a tactical mission task in which a unit renders the enemy incapable of interfering with an operation (FM 3-90)—the platoon maintains constant pressure to keep them off balance while capitalizing on successful tactical actions.

4-151. Consolidate is the process of organizing and strengthening a newly captured position so it can be defended. Normally, the attacking unit tries to exploit its success regardless the type of assault. In some situations, however, the unit may have to consolidate its gains. Consolidate may vary from a rapid repositioning of forces and security elements on the objective, to a reorganization of the attacking force, to the organization and detailed improvement of positions for defensive missions.

4-152. Consolidate consists of actions taken to secure the objective and defend against an enemy counterattack. Platoon leaders use TLP to plan and prepare for this phase of operation. They ensure the unit is ready to conduct the following actions (see paragraph 4-116 for a further discussion) that usually are part of consolidate:

- Establish security beyond the objective by securing areas that may be the source of enemy direct fires or enemy artillery observation.
- Prepare for and assist the passage of follow-on forces (if required).
- Continue to improve security by conducting other necessary defensive actions. These defensive actions include engagement area development, direct fire planning, and battle position preparation.
- Adjust FPFs and register targets along likely mounted and dismounted avenues of approach.
- Protect the obstacle reduction effort.
- Secure enemy detainees.
- Prepare for enemy counterattack.

4-153. Reorganization usually is conducted concurrently with consolidate. It consists of actions taken to prepare units for follow-on operations. As with consolidate, unit leaders plan and prepare for reorganization as they conduct TLP. (See paragraph 4-118 for a further discussion.) Platoon leaders ensure the following actions are conducted:

- Provide essential medical treatment and evacuate casualties, as necessary.
- Treat and evacuate wounded detainees and process the remainder of detainees.
- Cross-level personnel and adjust task organization as required to support the next phase or mission.
- Conducts resupply operations, including rearming and refueling.
- Redistribute ammunition.
- Conduct required maintenance.
- Continue improvement of defensive positions, as necessary.

4-154. For all attacks, Infantry units should and must plan to exploit success. However, at the conclusion of an engagement, the platoon leader may be forced to defend. For short defenses, units make use of existing terrain to enhance their survivability. If a longer defense is envisioned, engineer assets immediately should refocus their efforts on providing survivability support (fighting positions and similar activities). Engineer assets should do this even as they sustain mobility and integrate countermobility into the planned defensive mission. The Infantry leader considers the higher commander's concept of operations, friendly capabilities, and enemy situation when making the decision to defend or continue the offense.

ASSESS

4-155. Assessment refers to the continuous monitoring and evaluation of the current situation, particularly the enemy, and progress of an operation. Platoon leaders constantly assess the actual progress of the operation against the plan and look for variance in the form of emerging threats or opportunities. Assessment precedes and guides every operations process activity and concludes each operation or phase of an operation. It involves a comparison of forecasted outcomes to actual events. Assessment entails three tasks:

- Continuously assessing the enemy's reactions and vulnerabilities.
- Continuously monitoring the situation and progress of operation towards the commander's desired end state.
- Evaluating the operation against measures of effectiveness and measures of performance.

4-156. Upon receiving the mission, platoon leaders perform an initial assessment of the situation and the mission variables of METT-TC (I), focusing on the unit's role in the larger operation, and allocating time for planning and preparing. The two most important products from this initial assessment should be at least a partial restated mission, and a timeline. Platoon leaders issue their initial warning order (WARNORD) on this first assessment and time allocation. Assessment continues during planning, updated as needed throughout preparation and execution based on information gained through the platoon leader's continuous assessment.

4-157. Assessment of risk assists the platoon leader in making informed decisions on changing task organization, shifting priorities of effort and support, and future operations. Effectiveness entails making accurate assessments and good decisions about how to fight the enemy. Vital supporting concepts are TLP, actions on contact, and risk management (see chapter 2). Leaders use the assessment process to generate combat power.

4-158. Specific to execution, assessment involves a deliberate comparison of forecasted outcomes to actual events, using indicators to judge progress toward attaining desired end state conditions. Assessments help platoon leaders to adjust plans based on changes in the situation, when the operation is complete, and when to transition into the next cycle of the operations process. Platoon leaders constantly assess the actual progress of the operation against the plan through the following (although not all-inclusive) questions:

- Is there new or changing intelligence from high headquarters?
- Are there new priority intelligence requirements (PIRs) and/or information requirements (and were current PIRs answered correctly)?
- Is the platoon synched with the plan, for example:
  - Is the movement faster or slower than planned?
  - Are the obstacle types and locations templated correctly?
  - Is planned ammunition consumption correct?
  - Is combat power still adequate to accomplish the mission?

- Is the threat assessment made during planning and preparation still correct, for example:
  - Is the enemy stronger or weaker than templated?
  - Are preparatory fires adequate?
  - Are enemy locations templated correctly?
  - Are there still local enemy remnants or unengaged counterattacking forces?
- Is there an ability, in terms of opportunity, to seize or regain the initiative, for example:
  - Is there terrain to gain a better position of advantage over the enemy?
  - Is there an opportunity to catch the enemy by surprise or prior to the employment of enemy forces?
  - Are there unplanned but now-identified enemy reinforcements or capabilities on the way to, or on the objective that need to be engaged?

VARIATIONS OF THE ATTACK

4-159. Variations of the attack are ambush, counterattack, raid, and spoiling attack. (See ATP 3-21.10 for additional information.) The commander's intent and analysis of the mission variables of METT-TC (I) determines which variation of the attack to employ. Each variation of attack, except for a raid, can be conducted as either a hasty or a deliberate operation. As variations of an attack, they share many of the planning, preparation, and execution considerations of attack.

AMBUSH

4-160. An ambush is an attack by fire or other destructive means from concealed positions on a moving or temporarily halted enemy. An ambush stops, denies, or destroys enemy forces by maximizing the element of surprise. Ambushes can employ direct fire systems as well as other destructive means, such as command-detonated mines, indirect fires, and supporting nonlethal effects. They may include an assault to close with and destroy enemy forces. In an ambush, ground objectives do not have to be seized and held.

4-161. The three common forms of a small-unit ambush are point, area, and antiarmor ambushes. In a point ambush, a unit deploys to attack a single kill zone. In an area ambush, a unit deploys into two or more related point ambushes. Units smaller than a platoon normally do not conduct an area ambush. In an antiarmor ambush, a unit focuses on moving or temporarily halted enemy armored vehicles. (See chapter 7 for additional information.)

COUNTERATTACK

4-162. A *counterattack* is a variation of attack by a defending force against an attacking enemy force (FM 3-90). (For example, within a company's area defense, a platoon or squad may conduct a counterattack from its subsequent battle position to retake its prior primary battle position, see paragraphs 5-172 to 5-173.) The leader directs a counterattack to defeat or destroy enemy forces, exploit an enemy weakness such as an exposed flank, or to regain control of terrain and facilities after an enemy success. Units executing a counterattack can use all or part of the defending force, to include their reserve when given as a priority planning mission, against an enemy attacking force.

4-163. A unit conducts a counterattack to seize the initiative from the enemy through offensive action. A counterattacking force may maneuver to isolate and destroy a designated enemy force, or it can be an attack by fire into an engagement area to defeat or destroy an enemy force, restore the original position, or block an enemy penetration.

4-164. To be decisive, the counterattack occurs when the enemy is overextended, dispersed, and disorganized during the attack. All counterattacks should be rehearsed in the same conditions they will be conducted. During planning and preparation, leaders consider planning the commitment criteria and triggers for conducting a counterattack. In execution, leaders must be alert to those criteria and triggers emerging during an operation. If they do emerge, the platoon leader immediately launches the planned counterattack. Once committed, that force becomes the main effort.

Raid

4-165. A raid is a limited-objective, deliberate operation entailing swift penetration of hostile terrain. A raid is not intended to hold territory; and it requires detailed intelligence, preparation, and planning. (See chapter 7 for additional information.) The Infantry platoon and squad conducts raids as part of a larger force to accomplish several missions, including the following:

- Capture prisoners, installations, or enemy materiel.
- Capture or destroy specific enemy command and control locations.
- Destroy enemy materiel or installations.
- Obtain information concerning enemy locations, dispositions, strength, intentions, or methods of operation.
- Confuse the enemy or disrupt the enemy's plans.
- Liberate friendly personnel.

Spoiling Attack

4-166. A *spoiling attack* is a variation of an attack employed against an enemy preparing for an attack (FM 3-90). The spoiling attack usually employs armored, attack helicopter, or fire support elements to attack enemy assembly positions in front of a friendly commander's main line of resistance or battle positions.

4-167. The objective of a spoiling attack is to disrupt the enemy's offensive capabilities and timelines while destroying targeted enemy personnel and equipment, not to seize terrain and other physical objectives. Two conditions must be met to conduct a survivable spoiling attack:

- The spoiling attack's objective must be obtainable before the enemy is able to respond to the attack in a synchronized and coordinated manner.
- The force conducting the spoiling attack must be prevented from becoming over extended, fixed, or having insufficient mobility to break contact and /or withdraw.

4-168. Infantry forces conduct a spoiling attack whenever possible during friendly defensive missions to strike an enemy force while it is in a TAA or attack position(s) preparing for its own offensive mission or is stopped temporarily. For example, a platoon may conduct an infiltration behind enemy lines to impair (platoon ambush or raid) an enemy's ability to organize for its operation.

OPERATIONS DURING LIMITED VISIBILITY

4-169. Effective use of advanced optical sights and equipment during limited visibility attacks enhances the ability of squads and platoons to achieve surprise, hit targets, and cause panic in a lesser-equipped enemy. Advanced optics and equipment allow the Infantry Soldier to see farther and with greater clarity.

USE OF ILLUMINATION, NIGHT VISION, AND INFRARED LASERS

4-170. Night vision devices provide good visibility in all but pitch-black conditions but do somewhat limit the Soldier's field of view. Leaders must understand that night vision devices emit light at the eyecups, and many have an infrared illuminator which can emit light that is visible to a peer threat. Night vision devices are also visible to other night vision devices. Infantry platoons have night vision equipment mounted on the helmet of each Soldier and weapon-mounted and handheld devices to identify and designate targets.

4-171. Infantry leaders and Soldiers have an increased ability to designate and control fires during limited visibility. There are three types of advanced optics and equipment for use in fire control:

- Target designators. Leaders can designate targets with greater precision using infrared laser pointers to designate targets and sectors of fire and to concentrate fire. The leader lazes a target for Soldiers to place their fires. The Soldiers then use their weapon's aiming lights to engage the target.
- Aiming lights. Soldiers with aiming lights have greater accuracy of fires during limited visibility. Each Soldier in the Infantry platoon is equipped with an aiming light for the individual weapon. Aiming lights work with the individual Soldier's helmet-mounted night vision goggles. It puts an infrared light on the target at the point of aim.
- Target illuminators. Leaders can designate larger targets using target illuminators. Target illuminators are essentially infrared light sources that

light the target, making it easier to acquire effectively. Leaders and Soldiers use the infrared devices to identify enemy or friendly personnel and then engage targets using their aiming lights.

4-172. Illuminating rounds fired to burn on the ground can mark objectives. This helps the platoon orient on the objective but may adversely affect night vision devices.

4-173. Leaders plan but do not always use illumination during limited visibility attacks. Battalion commanders normally control conventional illumination but may authorize the company commander to do so. If the commander decides to use conventional illumination, the commander should not call for it until the assault is initiated or the attack is detected. It should be placed on several locations over a wide area to confuse the enemy as to the exact place of the attack. It should be placed beyond the objective to help assaulting Soldiers see and fire at withdrawing or counterattacking enemy Soldiers. Infrared illumination is a good capability to light the objective without lighting it for enemy forces without night vision devices. This advantage is degraded when used against a peer threat with the same night vision capabilities.

4-174. The platoon leader, squad leaders, and vehicle commanders must know unit tactical SOP and develop sound COAs to synchronize the employment of infrared illumination devices, target designators, and aiming lights during their assault on the objective. These include using luminous tape or chemical lights to mark personnel and using weapons control restrictions.

4-175. The platoon leader may use the following techniques to increase control during the assault:

- Use no flares, grenades, or obscuration on the objective.
- Use mortar or artillery rounds to orient attacking units.
- Use a base squad or fire team to pace and guide others.
- Reduce intervals between Soldiers and squads.

4-176. Like a daylight attack, indirect and direct fires are planned for a limited visibility attack but are not executed unless the platoon is detected or is ready to assault. Some weapons may fire before the attack and maintain a pattern to deceive the enemy or to help cover noise made by the platoon's movement. This is not done if it will disclose the attack.

4-177. Obscuration further reduces the enemy's visibility, particularly if the enemy has night vision devices. The FO fires obscuration rounds close to or on enemy positions, so it does not restrict friendly movement or hinder the reduction of obstacles. Employing obscuration on the objective during the assault may make it hard for assaulting Soldiers to find enemy fighting positions. If enough thermal sights are available, obscuration on the objective may provide a decisive advantage for a well-trained platoon.

Note. If the enemy is equipped with night vision devices, leaders must evaluate the risk of using each technique and ensure the mission is not compromised by the enemy's ability to detect infrared light sources.

DELIBERATE BATTLEFIELD OBSCURATION

4-178. Battlefield obscuration missions can occur during both the offense and the defense and can be highly effective. Firing obscuration on enemy positions can degrade the vision of gunners and known or suspected observation posts, preventing them from seeing or tracking targets and, thereby, reducing their effectiveness. When employed for its incendiary effect, white phosphorous (WP) can cause confusion and disorientation by degrading the enemy's command and control capabilities; while friendly units retain the ability to engage the enemy using thermal sights and DA Form 5517 (*Standard Range Card*). Enemy vehicles become silhouetted as they emerge from the obscuration. If obscuration employment is planned and executed correctly, this occurs as the enemy reaches the trigger line. (See ATP 3-11.50 for additional information.)

Note. Phosphorous obscuration is used in instantaneous burst munitions (for example, artillery shells and mortar cartridges), with the showers of burning phosphorous particles being highly incendiary. The projectile has an incendiary-producing effect and is ballistically similar to the high explosive projectile. Normally, shell WP is employed for its incendiary effect. The projectile can also be used for screening, spotting, and signaling purposes.

Projected Obscuration

4-179. Projected obscuration is an obscurant produced by artillery or mortar munitions, naval gunfire, helicopter-delivered rockets or, potentially, weapon grenade launchers. The advantage of using projected obscurant munitions is that they can place obscurants directly on the target without becoming decisively engaged. The disadvantage of projected obscuration is that most projected obscurant devices and munitions are lethal; they cannot be used on or near friendly forces. Ideal military applications for projected obscuration systems are to produce protection effects by obscuring enemy forces at distant locations and to mark distant targets for destruction by lethal fire or protection obscuration to initiate protection screens forward of an attacking force.

4-180. Obscuration employed to screen friendly forces and their interests is delivered in areas between friendly and enemy forces or in a friendly AO to degrade enemy ground or aerial detection, observation, and engagement capabilities. Units use screening obscuration to attack enemy target acquisition and guidance systems by placing obscuration effects between the friendly unit and enemy sensors. There are three visibility categories for visual screening obscuration that the supported unit commander uses to establish the visibility requirement for an obscuration mission. The visibility categories include obscuration blanket, haze, and curtain:

- Obscuration blanket. An obscuration blanket is a dense horizontal concentration of smoke covering an area of ground with visibility inside the concentration less than 50 meters. The blanket is established over and around friendly areas to protect them from air visual observation and visual precision bombing attacks. An obscuration blanket may hamper the operations of friendly forces by restricting movement and activity within the screen. It provides maximum concealment.

- Obscuration haze. An obscuration haze is obscuration placed over friendly areas to restrict adversary observation with fire, but not dense enough to disrupt friendly operations within the screen.
- Obscuration curtain. An obscuration curtain is a vertical development of smoke that reduces the enemy's ability to clearly identify what is occurring on the other side of the cloud. Visual recognition depends on the curtain width and obscurant density. It is placed between friendly and enemy positions to prevent or degrade enemy ground observation of friendly positions. Since the obscuration curtain is not placed directly on friendly troops, it will not hamper friendly operations.

Planning and Preparation Considerations

4-181. Obscuration missions are important functions for grenade launchers and mortars. Obscuration missions must be planned well and prepared for in advance so that the grenade launchers or mortar carriers are loaded with enough obscuration rounds.

4-182. Atmospheric stability, wind velocity, and wind direction are the most important factors when planning target effects for obscuration and WP mortar rounds. The effects of atmospheric stability can determine whether projected obscuration is effective at all or, if effective, how much ammunition is needed. The considerations are—

- During unstable conditions, grenade and mortar obscuration rounds are almost ineffective—the obscuration does not spread but often climbs straight up and quickly dissipates.
- Under stable conditions, both grenade and mortar obscuration rounds are effective.
- The higher the humidity, the better the screening effects of grenade and mortar obscuration rounds.

4-183. The terrain in the target area affects obscuration rounds. If the terrain in the target area is swampy, rain-soaked, or snow-covered, then burning obscuration rounds may not be effective. These rounds produce obscuration by ejecting felt wedges soaked in phosphorous. These wedges then burn on the ground, producing a dense, long-lasting cloud. If the wedges fall into mud, water, or snow, they can extinguish. Shallow water can reduce the obscuration produced by these rounds by as much as 50 percent.

4-184. The vehicle mounted grenade launchers (from the assault platoon of the weapons company) can provide a screening, incendiary, marking, and casualty-producing effect. It produces a localized, instantaneous obscuration cloud by scattering burning WP particles. The 120-millimeter (mm) heavy mortar and 81-mm medium battalion mortar WP and red phosphorous rounds produce a long lasting and wide area obscuration screen and can be used for incendiary effects, such as marking, obscuring, screening, and casualty producing. The 60-mm lightweight company mortar WP round can be used as a screening, signaling, and incendiary agent. All mortar obscuration rounds can be used as an aid in target location and navigation.

Employment Considerations

4-185. When used correctly obscurants can significantly reduce the enemy's effectiveness both in daytime and at night. Use obscurants to reduce the ability of the enemy to deliver effective fires, to hamper hostile operations, and to deny the enemy information on friendly positions and maneuvers. Obscurant reduces the effectiveness of laser beams and inhibits electro optical systems including some night vision devices. Additional usage and purpose include:

- Obscuration to:
  - Defeat flash to bang ranging and restrict the enemy's counterfire program.
  - Obscure adversary observation posts and reduce their ability to provide accurate target location for enemy fire support assets.
  - Obscure enemy direct fire weapons and lasers.
  - Instill apprehension and increase enemy patrolling.
  - Slow adversary vehicles to blackout speeds.
  - Increase control problems by preventing effective visual signals and increasing radio traffic.
  - Defeat night observation devices and reduce the capability of most infrared devices.
- Obscuration curtain for:
  - Deceptive screens. Smoke draws fire. Deceptive screens cause the enemy to disperse their fires and expend ammunition.
  - Flank screens. Smoke may be used to screen exposed flanks.
  - Areas forward of the objective. Smoke helps the maneuver units consolidate on the objective unhindered by enemy ground observers.
  - Gap crossing operations. Screening the primary crossing site denies the enemy information. Deceptive screens deceive the enemy as to the exact location of the main crossing.
  - Obstacle breaching. The enemy is denied the ability to observe breaching unit activities.

Chapter 5
Defense

The Infantry rifle platoon conducts defensive operations to defeat enemy attacks, gain time, control key terrain, protect critical infrastructure, secure the population, and economize forces. Most importantly, the platoon sets conditions to transition to the offense. Defensive operations alone are not decisive unless combined with offensive operations to surprise the enemy, attack enemy weaknesses, and pursue or exploit enemy vulnerabilities. Even within the conduct of the Infantry rifle company defense, the Infantry rifle platoon exploits opportunities to conduct offensive actions within its area of operations (AO) to deprive the enemy of the initiative and create the conditions to assume the offense. Other reasons for conducting defensive operations include, retain decisive terrain, or deny a vital area to the enemy, attrition or fix the enemy as a prelude to the offense, counter surprise action by the enemy, or to increase the enemy's vulnerability by forcing the enemy commander to concentrate subordinate forces. This chapter covers the conduct of the defense, common defensive planning considerations, types of defensive operations, engagement area development, and fighting position construction.

SECTION I – CONDUCT OF THE DEFENSE

5-1. The Infantry platoon and squad use the defense to *occupy*—a tactical mission task in which a unit moves into an area to control it without enemy opposition (FM 3-90)—and prepare positions and mass the effects of fires on likely enemy avenues of approach or mobility corridors. *Avenue of approach* is a path used by an attacking force leading to its objective or to key terrain. Avenues of approach exist in all domains (ADP 3-90). Mobility corridors are areas that are relatively free of obstacles where a force will be canalized due to terrain restrictions allowing military forces to capitalize on the principles of mass and speed. While the offense is the most decisive type of combat operation, the defense is the stronger type. The inherent strength of the defense is the defender's ability to occupy positions before an attack and use the available time to improve those defenses.

CHARACTERISTICS OF THE DEFENSE

5-2. Successful defenses share the following characteristics: disruption, flexibility, maneuver, mass and concentration, operations in-depth, preparation, and security. Defenses are aggressive. Defenders use all available means to disrupt attacking enemy forces and isolate them from mutual support to defeat them in detail. Defenders seek to increase their freedom of maneuver while denying it to attackers. Defenders use every opportunity to transition to the offense, even if only temporarily. As attackers' losses increase, they falter, and the initiative shifts to the defenders. Defenders strike swiftly when the attackers reach their decisive point(s). Surprise and speed enable counterattacking forces to seize the initiative and overwhelm the attackers' concentration, flexibility, maneuver, and operations in-depth. (See FM 3-96 for additional information.)

DISRUPTION

5-3. Defenders disrupt attackers' tempo and synchronization with actions designed to prevent them from massing combat power. Disruptive actions attempt to unhinge the enemy's preparations and, ultimately, enemy's attack. Methods include defeating or misdirecting enemy reconnaissance forces, breaking up their formations, isolating their units, and attacking (for example, conduct aggressive combat patrols [see chapter 7]) or disrupting their systems.

FLEXIBILITY

5-4. The defense requires flexible plans. Planning focuses on preparation in-depth, use of reserves, and ability to shift the main effort. Defenders add flexibility by designating supplementary positions and designing counterattack plans. Planned flexibility only works when leaders recognize—and act on—the emerging conditions that necessitate that planned flexibility. Defenders can have a flexible plan but if they do not act because they are wedded to the base plan, they still lose the opportunity to regain the offensive element of conventional warfare.

MANEUVER

5-5. Maneuver allows the defender to take full advantage of AO and to mass and concentrate when desirable. Maneuver, through movement in combination with fire, allows the defender to achieve a position of advantage over the enemy to accomplish the mission. It also encompasses defensive actions such as security force operations in the security area and security patrol activities in the main battle area and rear area.

MASS AND CONCENTRATION

5-6. Defenders seek to mass the effects of overwhelming combat power where they choose and shift it to support the main effort. To obtain an advantage at decisive points, defenders economize and accept risk in some areas; retain and, when necessary, reconstitute a reserve; and maneuver to gain local superiority at the point of decision.

Defenders accept risk in some areas to mass effects elsewhere. Obstacles, security forces, and fires can assist in reducing risk.

Operation In-Depth

5-7. *Operations in depth*—the simultaneous application of combat power throughout an area of operations (ADP 3-90)—improves the chances for success while minimizing friendly casualties. Quick, violent, and simultaneous action throughout the depth of the defender's AO can hurt, confuse, and even paralyze an enemy force just as it is most exposed and vulnerable. Such actions weaken the enemy's will and do not allow all early enemy successes to build the confidence of the enemy's Soldiers and leaders. In-depth planning prevents the enemy from gaining momentum in the attack. Synchronization of main and supporting efforts facilitate mission success.

Preparation

5-8. The defense has inherent strengths. The defender arrives in the AO before the attacker and uses the available time to prepare. These preparations multiply the defense's effectiveness. Preparations end only when the defenders retrograde or begin to fight. Until then, preparations are continuous. Preparations in-depth continues, even as the close fight begins.

Security

5-9. Security helps deceive the enemy as to friendly locations, strengths, and weaknesses. It also inhibits or defeat enemy reconnaissance. Security measures (for example, security patrols and counterreconnaissance [see paragraphs 7-183 to 7-184]) provide early warning and disrupt enemy attacks early and continuously.

Note. *Counterreconnaissance* is a tactical mission task that encompasses all measures taken by a unit to counter enemy reconnaissance and surveillance efforts (FM 3-90). Counterreconnaissance is not a distinct mission, but a component of all security operations and local security measures. It prevents hostile observation of a force or area. It involves both active and passive elements and includes combat action to destroy or repel enemy reconnaissance units and surveillance (both air and ground) assets.

DEFENSIVE OPERATIONS

5-10. A *defensive operation* is an operation to defeat an enemy attack, gain time, economize forces, and develop conditions favorable for offensive or stability operations (ADP 3-0). The three types of defensive operations are area defense (see section III), mobile defense (see note on page 5-59), and retrograde (see section IV). Each contains elements of the others and usually contains both static and dynamic aspects.

5-11. These three operations have significantly different concepts and pose significantly different problems. Each defensive operation is dealt with differently when planning and executing the defense. As with offensive operations, defensive operations can result in non-physical effects, such as those generated in the *information environment*—the aggregate of social, cultural, linguistic, psychological, technical, and physical factors that affect how humans and automated systems derive meaning from, act upon, and are impacted by information, including the individuals, organizations, and systems that collect, process, disseminate, or use information (JP 3-04).

5-12. For example, the use of deception in support of operations security can be highly effective at gaining time and tactical deception can support the economization of forces. Infantry forces use the three types of defensive operations to deny enemy forces advantages, each has its own focus—

- Area defense focuses on terrain.
- Mobile defense focuses on the movement of enemy forces.
- Retrograde focuses on the movement of friendly forces.

5-13. Infantry rifle platoons serve as the primary maneuver element or terrain-controlling units for the Infantry rifle company. They can defend within an AO or battle position or serve as a security force or a reserve as part of the Infantry rifle company's or Infantry battalion's coordinated defense. (See ATP 3-21.10 and ATP 3-21.20 for additional information.)

5-14. As part of a defense, the platoon can (although not all-inclusive) defend, delay, withdraw, counterattack, block, fix, turn, or perform security operations. The Infantry platoon usually defends, as part of the Infantry rifle company's defense in the main battle area. It conducts the defense to achieve one or more of the following:

- Gain time.
- Retain essential terrain.
- Support other operations.
- Preoccupy the enemy in one area while friendly forces attack in another.
- Wear down enemy forces at a rapid rate while reinforcing friendly operations.

COMMON DEFENSIVE CONTROL MEASURES

5-15. The defending leader controls defensive operations by using control measures to provide the flexibility needed to respond to changes in the situation and allow the defender to concentrate combat power at the decisive point(s). Defensive control measures within the organization of forces of an AO includes designating the security area, the main battle area, within the close area, with its associated forward edge of the battle area, and rear area with its echeloned support areas. The defender can use battle positions and additional direct fire control and fire support coordination measures in addition to those control measures to synchronize the employment of combat power. The defender designates a *disengagement line(s)*—a phase line located on identifiable terrain that, when crossed by the enemy, signals to defending elements that it is time to displace to their next position (ADP 3-90)—to trigger the displacement of subordinate forces. (See FM 3-96 for additional information.)

Security Area

5-16. A *security area* is that area occupied by a unit's security elements and includes the areas of influence of those security elements (ADP 3-90). It may be located as necessary to the front, flanks, or rear of a protected unit, facility, or location. Forces in a security area furnish information on an enemy force; delay, deceive, and disrupt that enemy force; and conduct counterreconnaissance.

Main Battle Area

5-17. The *main battle area* is the area where the commander intends to deploy the bulk of their unit to defeat an attacking enemy (FM 3-90). The defender's major advantage is the ability to select the ground on which the battle takes place. The defender positions subordinate forces in mutually supporting positions in-depth to absorb enemy penetrations or *canalize*—a tactical mission task in which a unit restricts enemy movement to a narrow zone (FM 3-90)—them into prepared engagement area(s), defeating the enemy's attack by concentrating the effects of overwhelming combat power. The natural defensive strength of positions determines the distribution of forces in relation to both frontage and depth. In addition, defending units typically employ field fortifications and obstacles to improve the terrain's natural defensive strength.

Battle Handover Line

5-18. The *battle handover line* is a designated phase line where responsibility transitions from the stationary force to the moving force and vice versa (ADP 3-90). The common higher commander of the two forces establishes the battle handover line (BHL) after consulting with both forces. The stationary defender determines the exact location of the line. The BHL is forward of the forward edge of the battle area in the defense or the forward line of own troops (FLOT) in the offense and utilized during a passage of lines (forward or rearward).

Battle Positions

5-19. A *battle position* is a defensive location oriented on a likely enemy avenue of approach (ADP 3-90). Units as large as battalion task forces and as small as squads or sections use battle positions. They may occupy the topographical crest of a hill, a forward slope, a reverse slope, or a combination of all areas. The platoon leader selects positions based on terrain, enemy capabilities, and friendly capabilities. The leader can assign all or some subordinates battle positions within the AO.

5-20. There are five types of battle positions: primary, alternate, supplementary, subsequent, and strong point. When assigning battle positions within a company defense, the commander always designates the primary battle positions. Platoon leaders will generally define exactly where primary battle positions are location and designate alternate, supplementary, and subsequent positions as time and other resources permit, and if the terrain or situation requires them. (See figure 5-1 on page 5-6.) Before assigning a strong point, the commander ensures that the strong point force has sufficient time and resources to construct the position.

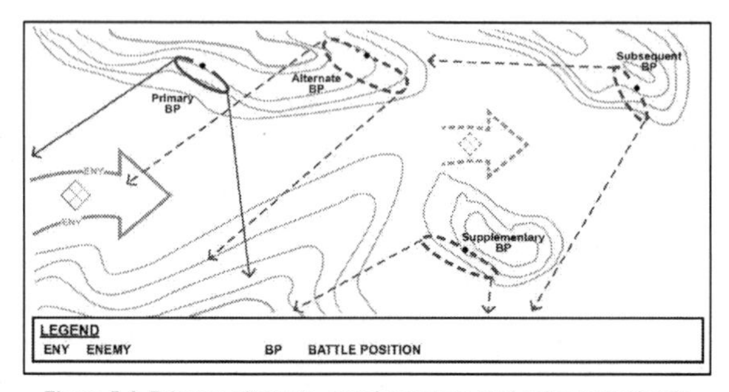

Figure 5-1. Primary, alternate, supplementary, and subsequent battle positions

5-21. A *primary position* is the position that covers the enemy's most likely avenue of approach into the assigned area (FM 3-90). It is the best position from which to accomplish the defensive mission, such as the overwatch of an engagement area to prevent enemy penetration.

5-22. An *alternate position* is a defensive position that the commander assigns to a unit or weapon system for occupation when the primary position becomes untenable or unsuitable for carrying out the assigned task (FM 3-90). When established the alternate position covers the same area as the primary position. The correct positioning of an alternate position improves survivability by providing good cover and concealment for movement to the alternate position and back to the primary position. The position is located slightly to the front, flank, or rear of the primary battle position. The position may be positioned forward of the primary battle position during limited visibility operations. An alternate position is employed to supplement or support positions with weapons of limited range.

> ***Note.*** Supplementary and subsequent positions are a way to achieve both flexibility and depth within the defense; however, they use must be planned for with displacement criteria and then rehearsed. All leaders should know when and why squads and/or the platoon will move to one of those positions.

5-23. A *supplementary position* is a defensive position located within a unit's assigned area that provides the best sectors of fire and defensive terrain along an avenue of approach that is not the primary avenue where the enemy is expected to attack (FM 3-90). It is assigned when more than one avenue of approach runs into a unit's AO. For example, an avenue of approach into a platoon's AO from one of its flanks could require the company to direct its platoons to establish supplementary positions to allow the platoons to engage enemy forces traveling along an avenue. The platoon leader assigns supplementary positions when the platoon must cover more than one avenue of approach.

5-24. A *subsequent position* is a position that a unit expects to move to during the course of battle (FM 3-90). Subsequent positions can have primary, alternate, and supplementary positions associated with them.

5-25. A *strong point* is a heavily fortified battle position tied to a natural or reinforcing obstacle to create an anchor for the defense or to deny the enemy decisive or key terrain (ADP 3-90). A strong point implies retention of terrain to control key terrain and blocking, fixing, or canalizing enemy forces. Defending units require permission from the higher headquarters to withdraw from a strong point. All combat, maneuver enhancement, and sustainment assets within the strong point require fortified positions. In addition, extensive protective and tactical obstacles are required to provide an all-around defense. Primary positions cover the enemy's most likely avenue of approach into the area. (See figure 5-2.)

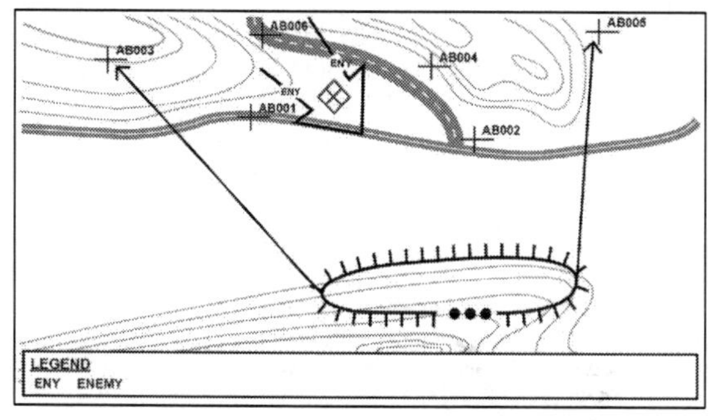

Figure 5-2. Platoon strong point battle position

Note. A minimally effective strong point typically requires a 1-day effort from an engineer unit the same size as the unit defending the strong point. (See ATP 3-21.10 for additional information.)

FORWARD EDGE OF THE BATTLE AREA

5-26. The *forward edge of the battle area* is the foremost limits of a series of areas in which ground combat units are deployed to coordinate fire support, the positioning of forces, or the maneuver of units, excluding areas in which covering or screening forces are operating (JP 3-09.3). When the defender defends forward in an area defense the unit employs the majority of its combat forces near the forward edge of the battle area.

FORWARD LINE OF OWN TROOPS

5-27. The *forward line of own troops* is a line which indicates the most forward positions of friendly forces in any kind of military operation at a specific time (FM 3-90). The FLOT normally identifies the forward location of covering or screening forces. In the defense, it may be beyond, at, or short of the forward edge of the battle area. It does not apply to small, long-range reconnaissance assets and similar stay-behind forces. Friendly forces forward of the FLOT may have a restrictive fire support coordination measure, such as a restrictive fire area, placed around them to prevent fratricide incidents.

TACTICAL FRAMEWORK OF THE DEFENSE

5-28. Usually as part of a larger force, the Infantry rifle platoon conducts the defense performing several integrated and overlapping activities. As in the offense, defensive operations are typically executed in a four-step tactical framework. The framework is for discussion purposes only and is not the only way of executing defensive operations. Normally the first two steps are supporting efforts, while the finish step is the main effort. Follow through is usually a sequel or branch to the plan based upon the situation. These steps may not occur sequentially; they may occur simultaneously. These steps are—

- Find the enemy. Intel drives fires and maneuver.
- Fix the enemy. Prevent repositioning or reinforcement making them easier to destroy.
- Finish the enemy. Mass available combat power to accomplish the mission.
- Follow through. Defeat in detail, consolidate, reorganize, and transition.

SECTION II – COMMON DEFENSIVE PLANNING CONSIDERATIONS

5-29. Planning a defensive operation is a complex effort requiring detailed planning and extensive coordination. In the defense, synchronizing the effects of the Infantry rifle platoon's capabilities enables the platoon leader to apply overwhelming combat power against selected advancing enemy forces to disrupt the enemy's plan. As an operation evolves, the platoon leader knows a shift to main and supporting efforts is a probability to press the fight and keep the enemy off balance. Warfighting functions provide the Infantry leader a means and structure for planning, preparing, executing, and assessing the defense. The following paragraphs discuss the synchronization and coordination of activities within each warfighting function critical to the success of the Infantry rifle platoon and squad. Additionally, this section addresses environmental planning considerations in the defense. (See FM 3-96 for additional information.)

COMMAND AND CONTROL

5-30. The first step is the expression of the platoon leader's vision of anticipated enemy actions integrated with the Infantry rifle company's intelligence preparation of the operational environment (known as IPOE). The Infantry battalion and company IPOE should not differ significantly, giving the Infantry platoon and squad a clear understanding of how the Infantry battalion and company commanders envision the enemy fight and plan for the operation. The company commander refines the IPOE to focus on the details of the operation in the company AO. The platoon leader refines the IPOE to focus on the details of the mission in the platoon AO. The Infantry battalion commander usually defines where and how the Infantry battalion will defeat or destroy the enemy. The Infantry rifle company commander and platoon leader then describe a vision of how their units will execute their portion of the battalion fight.

5-31. The platoon's leaders will locate where they can best provide command and control to subordinate units and Soldiers. Understanding the enemy's likely scheme of maneuver allows the platoon leader to select those decisive point(s) where best to position leaders and key weapons in order to mass their effects on enemy forces. When a platoon command post is established within the defense, it is located where the platoon leader determines it is best able to provide communications with higher, lower, adjacent, and supporting units; to assist the platoon leader in planning, coordinating, and issuing platoon orders, and to support continuous operations by the platoon. In static positions (for example, a battle position[s]), a stationary command post location may be designated by the platoon leader where field expedient antennas and/or land wire are employed to allow communications to be established with company, and subordinate static units, observation posts, and/or patrols within the platoon's AO. Accurate primary, alternate, contingency, and emergency (known as PACE) communications plan (see paragraph 5-104) is crucial to a leader's situational awareness. The PACE plan designates the order in which a leader will move through available communications systems until contact can be established with the desired element.

MOVEMENT AND MANEUVER

5-32. Movement and maneuver considerations involve the movement and employment of direct fire weapons on the battlefield. In the defense, weapons positioning is critical to the Infantry platoon's and squads' success. Weapons positioning enables the platoon to mass fires at critical points on the battlefield and shift fires as necessary. The platoon leader exploits the strengths of the weapon systems while minimizing the platoon's exposure to enemy observation and fires.

DEPTH AND DISPERSION

5-33. Dispersing positions laterally and in-depth helps to protect the force from enemy observation and fires. The positions are established in-depth, allowing sufficient maneuver space within each position to establish in-depth placement of weapon systems and Infantry elements. Engagement areas are established to provide for the massing of fires at critical points on the battlefield. Sectors of fire are established to distribute and shift fires throughout the extent of the engagement area. Once the direct fire plan is determined, fighting positions are constructed in a manner to support the fire support and obstacle plan.

Flank Positions

5-34. Flank positions enable a defending force to fire on an attacking force moving parallel to the defender's forces. A flank position provides the defender with a larger and more vulnerable target while leaving the attacker unsure of the defense's location. Major considerations for employment of a flank position are the defender's ability to secure the flank and the ability to achieve surprise by remaining undetected.

Disengagement Criteria

5-35. *Disengagement criteria* are the protocols that specify those circumstances where a friendly force must break contact with the direct fire and observed indirect fire to avoid becoming decisively engaged or to preserve friendly combat power (FM 3-90). It dictates to subordinate elements the circumstances, in which they will displace to alternate, supplementary, or subsequent positions. The criteria are tied to an enemy action, such as an enemy unit advancing past a certain phase line. They also are linked to the friendly situation. For example, the criteria might depend on whether artillery or an overwatch element can engage the enemy. Disengagement criteria are developed for each specific situation.

Displacement Criteria

5-36. Displacement criteria define triggers for planned withdrawals, passage of lines, or reconnaissance handovers (battle handover for security operations) between units. As with engagement, disengagement, and bypass criteria, the conditions and parameters set in displacement criteria integrate the leader's intent with tactical feasibility. Conditions and parameters are event-driven, time-driven, or enemy-driven. Displacement criteria conditions and parameters are key rehearsal events due to the criticality of identifying the triggers to anticipate during the mission. An example of event-driven conditions and parameters are associated priority intelligence requirements (PIRs) being met, enemy contact not expected in the area, and observed named area of interest (NAI) or avenue of approach denied to the enemy. Time-driven conditions and parameters ensure the time triggers are met (for example, latest time information is of value). An observation post compromised by threat or local civilian contact is a threat-driven condition. Failure to dictate conditions and parameters of displacement, nested within the higher scheme of maneuver, results in mission failure.

Disengagement and Displacement Planning

5-37. Disengagement and displacement allow the platoon to retain its flexibility and tactical agility in the defense. For example, retrogrades (see section IV) are conducted to improve a tactical situation or preventing a worse situation from developing. Platoons usually conduct retrogrades as part of a larger force but may conduct independent retrogrades (withdrawal) as required during defensive operations. The ultimate goals of disengagement and displacement are to enable the platoon to avoid being fixed or decisively engaged by the enemy. The overarching factor in a displacement is to maintain a mobility advantage over the enemy. The platoon leader must consider several important factors in displacement planning. These factors include, among others:

- The enemy situation, for example, an enemy attack with one company-size enemy unit might prevent the platoon from disengaging.
- Disengagement criteria (see paragraph 5-35).
- Availability of direct fire suppression that can support disengagement by suppressing or disrupting the enemy.
- Availability of cover and concealment, indirect fires, and obscurants to assist disengagement.
- Obstacle integration, including situational obstacles.
- Positioning of forces on terrain that provides an advantage to the disengaging elements such as linear obstacles.
- Identification of displacement routes and times when disengagement or displacement will take place. Routes and times are rehearsed.
- The size of the friendly force that must be available to engage the enemy in support of the displacing unit.

5-38. While disengagement and displacement are valuable tactical tools, they can be extremely difficult to execute in the face of a rapidly moving enemy force. The platoon leader must establish clear disengagement criteria, and it must be rehearsed. In execution, leaders carefully evaluate the situation and are alert to those criteria. The platoon leader makes a displacement decision early enough to ensure that it is feasible and does not result in unacceptable personnel or equipment losses.

5-39. For example, in a delay from alternate positions, First Platoon in an area defense occupies delaying positions in-depth. As First Squad engages the enemy vicinity Phase Line Dog, Second Squad occupies its next position in-depth adjacent to Third Squad and prepares to assume responsibility for the operation. When the enemy reaches Phase Line Cat, First Squad disengages and passes around (preferred method) or through Second Squad. First Squad then moves to its next position and prepares to reengage the enemy while Second Squad and Third Squad take up the fight.

Direct Fire Suppression

5-40. The attacking enemy force must not be allowed to bring direct and indirect fires to bear on a disengaging friendly force. Direct fires from the base-of-fire element, employed to suppress or disrupt the enemy, are the most effective way to facilitate disengagement. The platoon may receive base of direct fire support from another element in the company, but in most cases, the platoon establishes its own base-of-fire element. Having an internal base of fire requires the platoon leader to sequence the displacement of forces.

Cover and Concealment

5-41. The platoon and subordinate squads use covered and concealed routes when moving to alternate, supplementary, or subsequent positions. Regardless of the degree of protection the route itself affords, the platoon and squads should rehearse the movement both day and night prior to contact. Rehearsals increase the speed at which they can conduct the move and provide an added measure of security. The platoon leader allocates available time to rehearse movement in limited visibility and degraded conditions.

Indirect Fires and Obscurants

5-42. Artillery or mortar fires assist the platoon during disengagement. Suppressive fires slow the enemy and cause the enemy to seek cover. Smoke obscures the enemy's vision, slows the progress, or screens the defender's movement out of the battle position or along the displacement route.

Obstacle Integration

5-43. Obstacle efforts create opportunities to engage enemy forces with direct and indirect fires. Leaders should plan to mass fires on the places that obstacles will force enemy forces to go, not on the obstacles themselves. Ideally, obstacles should be positioned far enough away from the defender that enemy elements could be engaged on the far side of the obstacle while keeping the defender out of range of the enemy's massed direct fires. One or two well-constructed obstacles are more valuable than more obstacles constructed to marginal standards.

Countermobility

5-44. *Countermobility* is a set of combined arms activities that use or enhance the effects of natural and man-made obstacles to prevent the enemy freedom of movement and maneuver (ATP 3-90.8). To succeed in the defense, defenders reinforce the terrain to prevent the enemy from gaining a position of advantage. Reinforcing the terrain focuses on existing and reinforcing obstacles. These categories of obstacles provide a way for viewing the obstacle classification described in paragraphs 5-45 to 5-64 from the countermobility perspective. (See ATP 3-21.10 for additional information on countermobility operations in the defense.)

Existing Obstacles

5-45. Existing obstacles are inherent aspects of the terrain that impede movement and maneuver. Existing obstacles may be natural (rivers, mountains, wooded areas) or man-made (enemy explosive and nonexplosive obstacles and structures, including bridges, canals, railroads, and their associated embankments). Although not specifically designed or intended as an obstacle, structures may pose as an obstacle based on existing characteristics or altered characteristics that result from combat operations or a catastrophic event. Structures, such as bridges and overpasses, present an inherent impediment to the enemy's mobility based on weight and clearance restrictions. Existing obstacles are shown on the combined obstacle overlay developed as part of the IPOE. As described in ATP 3-34.80, geospatial engineering is critical in accurately predicting the effects that existing obstacles will have on enemy and friendly movement and maneuver.

Reinforcing Obstacles

5-46. Reinforcing obstacles are those man-made obstacles that strengthen existing terrain to achieve a desired effect. For U.S. forces, reinforcing obstacles on land consist of land mines, networked munitions, and demolition and constructed obstacles. The basic employment principles for reinforcing obstacles are—

- Support the maneuver commander's plan.
- Integrate with observation and fires.
- Integrate with other obstacles.
- Employ in-depth.
- Employ for surprise.

5-47. Land mines. A land mine is a munition on or near the ground or other surface area that is designed to explode by the presence, proximity, or contact of a person or vehicle. Land mines can be employed in quantities within a specific area to form a minefield, or they can be used individually to reinforce nonexplosive obstacles. Land mines can be further qualified as antivehicle or antipersonnel. They can be air-, artillery-, or ground-delivered. Land mines fall into two general categories:

- Persistent. A persistent mine is a land mine, other than nuclear or chemical, that is not designed to self-destruct. It is designed to be emplaced by hand or mechanical means and can be buried or surface emplaced.
- Nonpersistent. A nonpersistent mine remains active for a predetermined period of time until self-destruction, self-neutralization, or self-deactivation renders the mine inactive. This means that it is capable of self-destructing or self-deactivating.

Note. As of 31 January 2020, U.S. forces are not authorized to employ persistent or nondetectable land mines.

5-48. Networked munitions. *Networked munitions* are remotely controlled, interconnected, weapon systems designed to provide rapidly emplaced ground-based countermobility and protection capability through scalable application of lethal and nonlethal means (JP 3-15).

5-49. Demolition obstacles. Demolition obstacles are created using explosives. Examples include bridge or other structure demolition (rubble) and road craters. (See ATP 3-90.8 appendix B for additional information on demolition obstacles.)

5-50. Constructed obstacles. Constructed obstacles are created without the direct use of explosives. Examples include wire obstacles, antivehicle ditches, or similar construction that involves the use of heavy equipment.

5-51. Reinforcing obstacles are categorized as tactical and protective. They are employed as part of the movement and maneuver and protection warfighting functions. (See ATP 3-90.8 for additional information.) In the defense, leaders use reinforcing obstacles to:

- Slow the enemy's advance to give the platoon or squad more time to mass fires on the enemy.
- Protect defending units.
- Canalize the enemy into places where they can easily be engaged.
- Separate the enemy's armor from its infantry.
- Strengthen areas that are lightly defended.

Tactical Obstacles

5-52. The primary purposes of tactical obstacles are to shape enemy maneuver and to maximize the effects of fires. Tactical obstacles directly attack the ability of a force to move, mass, and reinforce; therefore, they affect the tempo of operations. Leaders integrate obstacles into the scheme of movement and maneuver/scheme of maneuver to enhance the effects of fires. Preexisting obstacles that a unit reinforces and integrates with observation and fires may become tactical obstacles. The types of tactical obstacles are clearly distinguished by the differences in execution criteria. The three types are—directed obstacles, situational obstacles, and reserved obstacles.

5-53. A *directed obstacle* is an obstacle directed by a higher commander as a specified task to a subordinate unit (ATP 3-90.8). The higher commander directs and resources these obstacles as specified tasks to a subordinate unit. Unit leaders plan, prepare, and execute directed obstacles during the preparation of the AO. Most tactical obstacles are directed obstacles.

5-54. A *situational obstacle* is an obstacle that a unit plans and possibly prepares prior to starting an operation but does not execute unless specific criteria are met (ATP 3-90.8). They are preplanned obstacles as part of a target area of interest and provide the commander flexibility for emplacing tactical obstacles based on battlefield development. Situational obstacles can incorporate any type of obstacle that can be employed within the time required. Situational obstacles are used to get inside the decision cycle of the enemy or react promptly to unexpected situations. Situational obstacles help units retain flexibility to quickly adapt in changing environments. Situational obstacles support the maneuver commander's synchronization matrix, decision support template, or branches and sequels. The following should be considered in addition to the obstacle employment principles:

- Identifying the need for a situational obstacle.
- Planning for the appropriate resources.
- Planning for the obstacle.
- Identifying obstacle execution triggers.
- Withholding the execution of the obstacle until needed.

Note. The U.S. ban on persistent land mines has resulted in a greater reliance on scatterable mines to fulfill tactical obstacle requirements. The limited duration of scatterable mines (after their activation) makes them the most suitable for use as situational obstacles.

5-55. A *reserved obstacle* are obstacles of any type, for which the commander restricts execution authority (ATP 3-90.8). The commander specifies the unit that is responsible for preparing, guarding, and executing reserved-obstacle emplacement. Units execute them only on command of the designating commander or based on specific criteria identified by the commander. The purpose of a reserved obstacle is to retain control over the mobility along an avenue of approach. Commanders use reserved obstacles when failure to maintain control over the mobility along an avenue of approach will have disastrous effects on the current battle or future operations.

Note. The three definitions above can apply to any type of obstacle. For example, the Modular Pack Mine System can be employed in support of disengagement, to either block a key displacement route once the displacing unit has passed through it or close a lane through a tactical obstacle. The location of obstacles in support of disengagement depends on mission variables of METT-TC (I) mission, enemy, terrain and weather, troops and support available, time available, civil considerations, and informational considerations.

Protective Obstacles

5-56. Protective obstacles have two roles: defense or security. Infantry platoons plan and construct their own protective obstacles. For best effect, protective obstacles are tied into existing or tactical reinforcing obstacles. The platoon can use mines and wire, or it might receive additional materiel from the company, for example, Classes IV or V. In planning protective obstacles:

- The platoon leader evaluates the potential threat to the platoon's position. Then, employs the best system for that threat.
- Protective obstacles usually are located beyond hand grenade distance (40 to 100 meters) from the Soldier's fighting position and may extend out 300 to 500 meters to tie into tactical obstacles and existing restricted terrain. (As with tactical obstacles, the platoon leader should plan protective obstacles in depth and try to maximize the range of the platoon's weapons.)
- The platoon leader considers preparation time, the burden on the logistical system, the Soldiers' loads, and the risk of loss of surprise.

5-57. The three types of wire obstacles (see figure 5-3 on page 5-16) are protective, tactical, and supplementary:

- Protective wire can be a complex obstacle providing all-around protection of a platoon perimeter. It also might be a simple wire obstacle on the likely dismounted avenue of approach into a squad ambush position. Command detonated M18 claymores (see TC 3-22.23) can be integrated into the protective wire or used separately.

- Tactical wire is positioned to increase the effectiveness of the platoon's fires. Usually, it is positioned along the friendly side of the medium machine gun final protective line (FPL). Tactical minefields also may be integrated into these wire obstacles or used separately.
- Supplementary wire obstacles can break up the line of tactical wire. This helps prevent the enemy from locating friendly weapons (particularly the medium machine guns) by following the tactical wire.

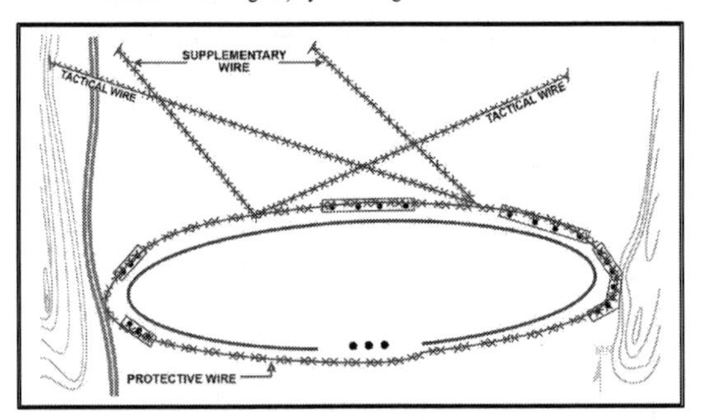

Figure 5-3. Wire obstacles

> ***Note.*** See ATP 3-90.8 for information on nuisance and phony obstacles. Nuisance obstacles impose caution on opposing forces. They disrupt, delay, and sometimes weaken or destroy follow-on echelons. Phony obstacles deceive the attacking force concerning the exact location of real obstacles. They cause the attacker to question its decision to breach and may cause the attacker to expend reduction assets wastefully.

5-58. Certain actions must occur when a protective obstacle (same actions apply for a tactical obstacle) is turned over or transferred to ensure that obstacle effectiveness or integration is not degraded. After an obstacle is completed, the emplacing unit (for example, the platoon) conducts obstacle transfer with the owning unit (for example, the company). The details for obstacle transfer (including who, what, when, and where) are established between the emplacing and owning units during obstacle siting. The owning unit may subsequently transfer ownership of an obstacle to another unit as part of a relief or unit rotation. (A withdrawal may involve obstacle removal.) Obstacle transfer ensures that the commander who is gaining ownership of the obstacle is familiar with the obstacle characteristics and features and that the responsibilities for maintaining obstacle integration are understood. In addition to the information presented in obstacle records, demolition target folders, and transfer reports (if applicable), leaders who conduct obstacle transfer should consider providing the following information to the gaining unit leader:

- Features or characteristics of the obstacle, such as the—
  - Obstacle data, including obstacle number, grid location, and emplacement date and time.
  - Dimensions and composition.
  - Anchor points.
  - Lane-marking data.
  - Intended effect.
  - Strengths and weaknesses.
- Updates on friendly and enemy activities near the obstacle.
- Areas for enemy observation and likely points of breach.
- Fire control measures.
- Detailed instructions, along with demonstrations as necessary, for—
  - Closing lanes and gaps.
  - Repairing and maintaining obstacles.
  - Removing, clearing, dismantling, disabling, or disarming obstacles.
  - Reusing or properly disposing of recovered obstacle material.

Note. Tracking obstacle removal and obstacles remaining on the battlefield is just as critical as tracking obstacle emplacement. (See ATP 3-90.8 for additional information.)

Obstacle Intent

5-59. Obstacle intent includes the target and desired effect (clear task and purpose) and the relative location of the obstacle group. The purpose influences many aspects of the operation, from selecting and designing obstacle sites to conducting the defense. Normally, the company commander designates the purpose of an obstacle group. When employing obstacles, the platoon leader considers the following principles:

- Support the tactical plan. Ensures obstacles supplement combat power to decrease the mobility of the enemy and provide security for the platoon. While considering enemy avenues of approach, the platoon leader also considers the platoon's own movement requirements, such as routes for resupply, withdrawal, counterattacks, patrols, and observation posts.
- Tie in. Ensures reinforcing obstacles are tied in with existing or natural obstacles. Ensures fire support plan supports obstacle plan, based on estimated enemy's rate of movement and formation, amount of time it will take the enemy to breach or bypass specific obstacles, and how long it will take to process and fire each target. This allows for the creation of effective

triggers that quickly synchronize fires and achieve the desired effects against enemy formations.

- Covered by observation and fire. Ensures that all obstacles are covered by observation and fire. This reduces the enemy's ability to remove or breach the obstacles out of contact.
- Constructed in-depth. Emplaces obstacles so that each new obstacle encountered by the enemy attrits the enemy force or equipment and causes a desired and controlled reaction. Proper use of obstacles in-depth wears the enemy down and significantly increases the overall effect.
- Employed for surprise. Avoids an obvious pattern of obstacles that would divulge locations of units and weapons. The platoon leader must avoid readily discernable, repetitive obstacle patterns.

5-60. The company commander assigns obstacle groups and tells the platoon leader and supporting engineers the intent of the obstacles against the enemy, and then resources the groups accordingly. Obstacle intent includes the elements of target, effect, and relative location:

- Target identifies the targeted enemy force size, type, echelon, avenues of approach, or any combination of these.
- Effect describes how to affect the enemy's maneuver with obstacles and fires. Tactical obstacles block, turn, fix, or disrupt. Obstacle effect integrates the obstacles with direct and indirect fires.
- Relative location is the placement of the obstacle effect to occur against the targeted enemy force. Initiates an obstacle integration process after identifying where on the terrain the obstacle will most decisively affect the enemy.

5-61. For example, the battalion commander might say, "deny the enemy access to the battalion's flank by turning the northern, enemy mechanized Infantry company into Engagement Area Crow to allow A Company (specifically First Platoon and Second Platoon) to mass their fires to destroy the enemy." Scatterable mine systems and submunitions are the main means of constructing tactical obstacles. These systems, with their self- and command-destruct capabilities, are flexible, and they aid in rapid transitions between defensive and offensive operations. They do this better than other constructed obstacles. The force constructs conventional minefields and obstacles only for a deliberate, long-term defense. In those cases, the company and platoons usually are augmented with assets from the brigade engineer battalion of the Infantry brigade combat team (IBCT) and/or a divisional engineer battalion. Table 5-1 shows the symbols for each obstacle effect, and it describes the purpose and characteristics of each.

Table 5-1. Tactical obstacle effects

APPLICATION	DESCRIPTION	PURPOSE	FIRES AND OBSTACLES MUST:	OBSTACLE CHARACTERISTICS
DISRUPT	• The arrows indicate the direction of enemy advance. • The length of the arrows indicate where the enemy is slowed or allowed to bypass.	• Breakup enemy formations. • Interrupt the enemy's timetable and C2. • Cause premature commitment of breach assets. • Cause the enemy to piecemeal his attack.	• Cause the enemy to deploy early. • Slow part of the enemy formation while allowing part to advance unimpeded.	• Do not require extensive resources. • Difficult to detect at long range.
TURN	• The heel of the arrow is the anchor point. • The direction of the arrow indicates the desired direction of the turn.	• Force the enemy to move in the direction desired by the the friendly commander.	• Prevent the enemy from bypassing or breaching the obstacle belt. • Maintain pressure on the enemy force throughout the turn. • Mass direct and indirect forces at the anchor point of the turn.	• Tie into impassable terrain at the anchor point. • Consist of obstacles in depth. • Provide a subtle orientation relative to the enemy's approach.
FIX	• The arrow indicates the direction of enemy advance. • The irregular part of the arrow indicates where enemy advance is slowed by obstacles.	• Slow an attacker within an area so the attacker can be destroyed. • Generate the time necessary for the friendly force to disengage.	• Cause the enemy to deploy into attack formation before encountering the obstacles. • Allow the enemy to advance slowly in an EA or AO. • Make the enemy fight in multiple directions once the enemy is in the EA or AO.	• Arrayed in depth. • Span the entire width of the avenue of approach. • Must not make the terrain appear impenetrable.
BLOCK	• The vertical line indicates the limit of enemy advance and where the obstacle ties into severely restricted terrain. • The horizontal line shows the depth of the obstacle effort.	• Stop an attacker along a specific avenue of approach. • Prevent an attacker from passing through an AO or EA. • Stop the enemy from using an avenue of approach and force the enemy to use another avenue of approach.	• Prevent the enemy from bypassing or penetrating through the belt. • Stop the enemy's advance. • Destroy all enemy breach efforts.	• Must tie into impassable terrain. • Consistent of complex obstacles. • Defeat the enemy's mounted and dismounted breaching effort.
Direction of enemy attack ⟶				

LEGEND
AO AREA OF OPERATION C2 COMMAND & CONTROL EA ENGAGEMENT AREA

Obstacle Lanes

5-62. Leaders identify mobility requirements to determine the need for lanes or gaps to be left in obstacles. Mission variables of METT-TC (I) are considered to determine if these areas remain open, closed on order, or closed with the defender able to open the lane. The on-off-on feature of networked munitions makes them well suited for lanes. Units use lanes or gaps to allow patrols (combat and reconnaissance), *quick reaction forces*—commander designated forces to respond to threat attacks or emergencies (FM 3-90), quick response forces, or tactical combat forces, and counterattack forces to enter and leave through the defense.

> ***Note.*** A quick response force is a dedicated force on a base with adequate tactical mobility and fire support designated to defeat Level I and Level II threats and shape Level III threats until they can be defeated by a tactical combat force or other available response forces (see FM 3-90).

5-63. Lanes and gaps are weak points in protective obstacles, so units consider allocating increased direct and indirect fires to cover them. The lanes and gaps must be disseminated to all levels and overwatched as a protective measure to prevent fratricide. Obstacles that are used for rapid lane closure are often demolition obstacles or mines, for example, units may use a cratering charge to close a lane.

5-64. When the platoon is responsible for actions related to lanes through obstacles. Its duties can include marking lanes in an obstacle, reporting locations of the start and ends of each lane, operating contact points, providing guides for elements passing through the obstacle, and planning and rehearsing lane closure. (See ATP 3-90.8 for additional information.)

MOBILITY

5-65. Mobility operations in the defense ensure the ability to reposition forces, delay, and counterattack. Initially during defensive preparations, mobility operations focus on the ability to resupply, reposition, and conduct rearward and forward passage of forces, materiel, and equipment. Once major defensive preparations are complete, the focus normally shifts to supporting the local reserve and counterattack(s), and the higher headquarters counterattack or reserve. Priorities set by the company may specify routes for improvement in support of such missions. Normally, most engineer assets go to survivability and countermobility. At a set time or trigger, engineers disengage from obstacle and survivability position construction and start preparing for focused mobility missions. The defender analyzes the scheme of maneuver, obstacle plan, fire support plan, and terrain to determine mobility requirements. (See ATP 3-21.10 for additional information on mobility operations in the defense.) Critical considerations may include:

- Lanes and gaps in the obstacle plan.
- Lane closure plan and subunit responsibility.
- Route reconnaissance, improvement, and maintenance.

SECTOR

5-66. A *sector* is an operational area assigned to a unit in the defense that has rear and lateral boundaries and interlocking fires (FM 3-0). The non-bounded side is open towards the enemy. A higher echelon headquarters uses fire support coordination and maneuver control measures such as battle positions and trigger lines to synchronize subordinate units. Higher headquarters are responsible for synchronizing operations forward of the main battle and security areas or coordinated fire line. Units use sectors to synchronize and coordinate engagement areas and allow for mutually supporting fields of fire, which do not require coordination between adjacent units. Units treat everything behind the FLOT as an AO. Sectors can be further subdivided as needed. (See FM 3-90 for additional information on sectors.)

RESERVE ROLE

5-67. If the platoon, or squad are designated in a reserve role, positioning the reserve in a location where it can react to several contingency plans is vital to success. The designating commander considers terrain, trafficability of roads, potential engagement area(s), and probable points of enemy penetrations and commitment time when positioning the reserve. Additionally, reserve forces must understand that they may be a primary target for enemy destruction through fires or CBRN attacks and must be prepared to react to this targeting. The reserve force must understand, plan for, and rehearse multiple assigned planning priorities, and must time, and provide those times to the designating commander, for all anticipated movements.

5-68. The battalion commander can have a single reserve, or if the terrain dictates, the subordinate commanders can designate their own reserve (generally a squad or platoon). Commanders must ensure reserves are positioned in a covered and concealed position and that they cannot be trapped in these reserve positions by the enemy's scatterable mines. Information concerning reserves may be considered essential element of friendly information and protected from enemy reconnaissance. The commander or platoon leader (when the platoon designates a reserve) might choose to position the reserve forward initially to deceive the enemy, or to move the reserve occasionally to prevent it from being targeted by enemy indirect fires.

INTELLIGENCE

5-69. The Infantry platoon leader never has all the information needed about the enemy. Therefore, the platoon leader obtains or develops the best possible IPOE products, conducts continuous reconnaissance, and integrates new and updated intelligence products and assessments and combat information throughout the operation. The leader requests information through the company from the battalion staff to answer platoon information requirements. (See ATP 3-21.10 for additional information.)

5-70. As with all tactical planning, IPOE is a critical part of defensive planning. It helps the platoon leader define where to concentrate combat power, where to accept risk, and where to plan potential decisive points in the operation. To aid in the development of a flexible defensive plan, the IPOE must present all feasible enemy courses of action (COAs). The essential areas of focus are—

- Analyze terrain and weather.
- Determine enemy force size and likely COAs with associated decision points.
- Determine enemy vulnerabilities and high-value targets (generally developed during step 4 of the battalion's IPOE, see chapter 2).
- Impact of civilian population on the defense.

5-71. The platoon leader, in coordination with the commander, base determinations of how and where to defeat the enemy on potential future enemy locations, the terrain, and forces available. The company may define a defeat mechanism (destroy, dislocate, disintegrate, and isolate, see ATP 3-21.10) including the use of single or multiple counterattacks to achieve success. The platoon leader analyzes the platoon's role in the company fight and determines how to achieve success.

FIRES

5-72. The Infantry platoon leader plans and executes fires in a manner, which achieves the intended task and purpose of each target. The platoon's scheme of fires serves to:

- Slow and disrupt enemy movement.
- Prevent the enemy from executing breaching operations.
- Destroy or delay enemy forces at obstacles using massed or precision fires.
- Suppress enemy support by fire elements.
- Defeat attacks along avenues of approach with final protective fires (FPFs).
- Disrupt the enemy to enable friendly forces to displace or counterattack.
- Obscure enemy observation or screen friendly movement during disengagement and counterattacks.
- Provide obscurants to separate enemy echelons or to silhouette enemy formations to facilitate direct fire engagement.
- Provide illumination, as necessary.
- Execute suppression of enemy air defense missions to support aviation operations.

5-73. In developing the fire plan, the platoon leader evaluates the indirect fire systems available to provide support. Considerations when developing the plan include tactical capabilities, weapons ranges, and available munitions. These factors help the platoon leader and forward observer (FO) determine the best method for achieving the task and purpose for each target in the fire plan. The Infantry rifle company fire support Soldiers contribute significantly to the platoon fight. Positioning is critical. The platoon leader, in coordination with the company fire support officer, selects positions providing the platoon FO with unobstructed observation of the AO, ensuring survivability.

SUSTAINMENT

5-74. In addition to the sustainment functions (see chapter 8) required for all missions, the platoon leader's planning process includes pre-positioning of ammunition caches, identifying the positioning of company trains, and Classes IV and V supply points, and explosive mine supply points.

5-75. The platoon leader's mission analysis may reveal the platoon's ammunition requirements during an upcoming mission exceed its basic load. This requires the platoon to coordinate with the company to pre-position ammunition caches. The platoon usually positions ammunition caches at alternate or subsequent positions. The platoon also may dig in these caches and guard them to prevent their capture or destruction.

5-76. The Infantry company trains usually operate 500 to 1,000 meters or one terrain feature to the rear of the company in covered and concealed positions close enough to the platoons to provide responsive support (support distance) but out of enemy direct fire. The company trains conduct evacuation (of those wounded in action, weapons, and equipment) and resupply as required. The company first sergeant or executive officer positions the trains and supervises sustainment operations with the platoon. It is the

Infantry company commander's, along with the first sergeant's, responsibility to ensure all subordinate units know the locations of battalion combat and field trains as well as the company casualty collection point (CCP), battalion aid station, and casualty evacuation (CASEVAC) and medical evacuation (MEDEVAC) procedures. The platoon leader's analysis determines the measures for every mission. The platoon sergeant responses to whatever system of supply management the company establishes.

PROTECTION

5-77. Because the platoon defends to conserve combat power for use elsewhere or later, the platoon leader must *secure*—a tactical mission task in which a unit prevents the enemy from damaging or destroying a force, facility, or geographical location (FM 3-90)—the force. The leader enables security, by means of providing information about the activities and resources of the enemy, through the employment of reconnaissance and surveillance within its assigned AO. The leader may employ counterreconnaissance (security patrols) and establish observation posts to counter enemy reconnaissance and surveillance efforts. The platoon leader integrates platoon reconnaissance and surveillance efforts, and security patrols with those of the company information collection effort.

ACTIVE AND PASSIVE AIR DEFENSES

5-78. In the face of an enemy air threat, the platoon usually has only active and passive (with its organic weapons) air defenses. Air defense units are usually not assigned below brigade combat team level. However, air defense assets may be located near the platoon and may provide area coverage. If air defense elements are assigned, the higher commander with the advisement of the air defense leader determines likely enemy air avenues of approach, and plans positions accordingly. The higher commander establishes priorities for protection that may include the company command post, mortar positions, and logistics assets. A stinger section may support maneuver elements from positions to the rear of defensive battle positions or along offensive attack routes as operations transition. Stinger teams on the early warning net can warn supported units of an air attack. (See ATP 3-01.8 for additional information.)

5-79. The platoon leader and subordinate leaders ensure all active and passive air defense measures are well planned and implemented, to include scanning for any hostile small-unmanned aircraft systems. (See Battle Drill-React to Aircraft While Dismounted, appendix E.) Passive measures include use of concealed routes and tactical assembly areas (TAAs), movement on secure routes, movement at night, increased intervals between elements of the columns, dispersion between Soldiers, using terrain to mask movement, and slowing or stopping movement to eliminate dust signature. Air guard duties assigned to specific Soldiers during dismounted (or mounted) movements and marches give each a specific search area. For movements and marches, seeing the enemy first gives the unit time to react. Leaders understand that scanning for long periods decreases the Soldier's ability to identify enemy aircraft. During extended or long movements and marches, Soldiers are assigned air guard duties in shifts. Units take passive measures unless or until threat aircraft are within range of assigned or attached weapons, at which time when threatened they take active measures. Active measures primarily involve engaging threat aircraft with a high volume of fires with the appropriate lead applied. Active measures include use of organic and attached weapons according to the OPORD and unit SOP. (See paragraphs 4-41 to 4-44 for additional information on active and passive air defense.)

Note. Counterrocket and artillery batteries, and countermortar fire elements may be located in or near an assigned AO to support a mission. Battery sensors detect incoming rockets, artillery, and mortar shells and may be used to detect Groups 1 and 2 unmanned aircraft systems. The battery's fire control system predicts the flight path of incoming rockets and shells, prioritizes targets, and activates in the supported AO warning system according to established ROE. Exposed elements within the AO then can take cover and provide cueing data that allows the battery's weapon system to defeat the target before the target can impact the area. Commanders clearly define command and support relationships between counterrocket, artillery, and mortar elements and supported units during planning. (See ATP 3-01.60 for additional information on counterrocket, artillery, and mortar fires and ATP 3-01.81 for additional information on counter-unmanned aircraft system measures.)

Chemical, Biological, Radiological, and Nuclear Passive Defense

5-80. Operationally, chemical, biological, radiological, and nuclear (CBRN) passive defense enables the unit's ability to continue military operations in a CBRN environment while minimizing the vulnerability of the force to the degrading effects of CBRN threats and hazards. Tactical-level doctrine is organized around the key activities of CBRN protection and mitigation. (See ATP 3-11.32 and ATP 3-21.10 appendix H for additional information on the CBRN environment and passive defense activities.)

Chemical, Biological, Radiological, and Nuclear Protection

5-81. CBRN protection measures are taken to keep CBRN threats and hazards from having an adverse effect on Soldiers, equipment, and facilities. Tasks that enable CBRN protection include the following:

- Employ individual protective equipment and other CBRN protective equipment.
- Establish CBRN alarm conditions.
- Exercise personal hygiene and force health protection programs.
- Utilize shielding or protective cover.

5-82. CBRN protection is an integral part of all operations. Other forms of protection involve sealing or hardening positions, protecting Soldiers, assuming appropriate mission-oriented protective posture (MOPP) levels (see table 5-2), reacting appropriately to the CBRN attack, maintaining dispersion, overhead cover (known as

OHC), employment of alarms upwind, and having CBRN-trained personnel for alarms, detection, and monitoring capabilities. Individual protective items include the protective mask, joint service lightweight integrated suit technology (also known as JSLIST), overboots, and gloves. The higher-level commander above the IBCT establishes the minimum level of protection. Subordinate units may increase this level as necessary but may not decrease it. Regardless of the directed MOPP level, in the event of an actual or suspected chemical attack, Soldiers will immediately don their protective mask and then the remainder of their protective equipment to come to MOPP 4.

5-83. The JSLIST provides protection for 45 days with up to 6 launderings or up to 120 days after being removed from packaging. The JSLIST still provides protection for up to 24 hours after it is contaminated.

Table 5-2. Mission-oriented protective posture levels

Level/ Equipment	*MOPP Ready*	*MOPP0*	*MOPP1*	*MOPP2*	*MOPP3*	*MOPP4*	*Mask Only*
Mask	Carried	Carried	Carried	Carried	Worn	Worn	Worn ***
JSLIST	Ready*	Available**	Worn	Worn	Worn	Worn	
Overboots	Ready*	Available**	Available**	Worn	Worn	Worn	
Gloves	Ready*	Available**	Available**	Available**	Available **	Worn	
Helmet Cover	Ready*	Available**	Available**	Worn	Worn	Worn	

Notes. *Items available to Soldier within 2 hours with replacement available within 6 hours.
**Items must be positioned within arms-reach of the Soldier.
***Never "mask only" if nerve or blister agents are used in area of operation.

Legend: JSLIST–joint service lightweight integrated suit technology; MOPP–mission-oriented protective posture

5-84. Passive measures can be used to monitor for the presence of CBRN hazards. Depending on the threat and probability of use, periodic or continuous techniques are used. An area array of CBRN detectors and/or monitors can be positioned within a given area for detection and early warning of a CBRN incident.

Note. CBRN environment is an operational environment that includes CBRN threats and hazards and their potential effects. CBRN environment conditions can be the result of deliberate enemy action or the result of an industrial accident. CBRN threats include the intentional employment of, or intent to employ CBRN weapons. A CBRN hazard is a substance that could create adverse effects due to an accidental or deliberate release and dissemination. CBRN hazards may result from weapons of mass destruction employment. (See ATP 3-21.10 appendix H or FM 3-11 for additional information on CBRN hazards.)

Chemical Hazards

5-85. A chemical hazard is any chemical manufactured, used, transported, or stored that can cause death or other harm through toxic properties of those materials, including chemical agents and chemical weapons prohibited under the chemical weapons convention as well as toxic industrial chemicals. This includes—

- Chemical weapons or toxic chemicals specifically designed as a weapon.
- Chemical agents or chemical substance that are intended for use in military operations to kill, seriously injure, or incapacitate, mainly through physiological effects.
- Toxic industrial chemicals are chemicals developed or manufactured for use in industrial operations or research.

5-86. A chemical weapon is a munition or device, specifically designed to cause death or other harm through the toxic properties of specified chemicals. Chemicals are released as a result of the employment of such munition or device; any equipment specifically designed for use directly in connection with the employment of munitions or devices.

5-87. A chemical agent is a chemical substance that is intended for use in military operations to kill, seriously injure, or incapacitate mainly through its physiological effects. Chemical hazards cause casualties, degrade performance, slow maneuver, restrict terrain, and disrupt operations (see table 5-3). They can cover large areas and may be delivered as liquid, vapor, or aerosol. They can be delivered by various means including artillery, mortars, rockets, missiles, aircraft spray, bombs, land mines, and covert means. (See Battle Drill-React to a Chemical Agent Attack, appendix E.)

Table 5-3. Characteristics of chemical agents

Agent	*Nerve*	*Blister*	*Blood*	*Choking*
Protection	Mask and IPE	Mask and IPE	Mask	Mask
Detection	JCAD, M256A2, CAM, and M8 and M9 paper	JCAD, M256A2, CAM, and M8 and M9 paper	JCAD, M256A2	Odor (freshly mowed hay)
Symptoms	Difficult breathing, drooling, nausea, vomiting, convulsions, and blurred vision	Burning eyes, stinging skin, irritated nose	Convulsions and coma	Coughing, nausea, choking, headache, and tight chest
Effects	Incapacitates	Blisters skin, damages respiratory tract	Incapacitates	Floods and damages lungs
First Aid	ATNAA and CANA DECON	Treat for 2nd and 3rd degree burns	None	Keep warm and avoid movement
Decontamination	RSDL and flush eyes with water	RSDL and flush eyes with water	RSDL	RSDL

Legend: ATNAA–antidote treatment nerve agent auto-injector; CAM–chemical agent monitor; CANA–convulsive antidote nerve agent; DECON–decontamination; IPE–individual protection equipment; JCAD–joint chemical agent detector; RSDL–reactive skin decontamination lotion

Unmasking Procedures

5-88. During an actual or suspected chemical attack, Soldiers will go to MOPP 4. Once masked, Soldiers will not unmask until there is no remaining chemical threat, and only under the approval of their commander. Soldiers will use the unmasking procedures found in paragraphs 5-89 through 5-91 to determine if it is safe to unmask.

Note. Persistent chemicals present a long-duration threat while non-persistent chemicals disperse within hours, depending on the weather.

Unmasking with M256/M256A1/A2 Kit

5-89. If an M256/M256A1/A2 detector kit is available, use it before conducing unmasking procedures. The kit does not detect all agents; therefore, proper unmasking procedures, which take approximately 15 minutes, must still be used. If all tests with the kit (including a check for liquid contamination using M8 detector paper) have been performed and the results are negative, use the following procedures:

- The senior person should select one or two Soldiers to start the unmasking procedures. If possible, they move to a shady place; bright, direct sunlight can cause pupils in the eyes to constrict, giving a false symptom.
- The selected Soldiers unmask for 5 minutes, then clear and reseal their masks.
- Observe the Soldiers for 10 minutes. If no symptoms appear, leaders will report the results of their unmasking procedures and request permission from higher headquarters to signal ALL CLEAR.
- Watch all Soldiers for possible delayed symptoms. Always have first-aid treatment immediately available in case it is needed.

Note. Time to complete two simultaneous M256/M256A1/A2 detector kits, including using M8 detector paper for liquid, is approximately 20 minutes. Unmasking procedures with the M256/M256A1/A2 detector kit will take approximately 35 minutes to complete.

Unmasking without M256/M256A1/A2 Kit

5-90. If an M256/M256A1/A2 kit is not available, the unmasking procedures take approximately 35 minutes. When a reasonable amount of time has passed after the attack, find a shady area. Use M8 paper to check the area for possible liquid contamination. Conduct unmasking using these procedures:

- The senior person selects one or two Soldiers. They take a deep breath and break their mask seals, keeping their eyes wide open.
- After 15 seconds, the Soldiers clear and reseal their masks. Observe them for 10 minutes.
- If no symptoms appear, the same Soldiers break their seals, take two or three breaths, and clear and reseal their masks. Observe them for 10 minutes.

- If no symptoms appear, the same Soldiers unmask for 5 minutes, and then re-mask.
- If no symptoms appear in 10 minutes, leaders will report the results of their unmasking procedures and request permission from higher headquarters to signal ALL CLEAR. Continue to observe all Soldiers in case delayed symptoms develop.

All-Clear Signal

5-91. Leaders initiate the request for ALL CLEAR after testing for contamination proves negative. Once approved, units pass the all-clear signal by word of mouth through their chain of command. The commander designates the specific all-clear signal and includes it in the unit SOP or the OPORD. If required, standard sound signals may be used, such as a continuous, sustained blast on a siren, vehicle horn, or similar device. When ALL CLEAR is announced on the radio, the receiving unit must authenticate the transmission before complying.

Decontamination

5-92. During continuous operations in areas of contamination, decontamination is essential in preventing casualties and severe combat degradation. The platoon gains maximum benefit from the available time and decontamination resources by observing these considerations:

- The platoon should execute decontamination as soon as possible and as far forward as possible.
- Decontamination should be conducted only to the extent necessary to ensure the platoon's safety and operational readiness.
- Decontamination priorities with regard to unit safety and mission accomplishment should be strictly observed.

5-93. These principles are consistent with doctrine that places the burden of decontamination at battalion or company level. For this reason, the platoon must develop a thorough SOP covering decontamination methods and priorities, using all available assets to the maximum extent possible. Paragraphs 5-94 to 5-101 address immediate and operational levels (see table 5-4) of decontamination. (See ATP 3-11.32 for additional information on decontamination procedures and higher level [thorough/clearance] decontamination.)

Table 5-4. Immediate and operational levels of decontamination tasks

Levels	Purpose	Tasks	Best Start Time	Performed By
Immediate	- Saves lives - Stops agent from penetrating - Limits agent spread	Skin decontamination	Before 1 minute	Individual
		Personal wipe down	Within 15 minutes	Individual or buddy
		Operator wipe down	Within 15 minutes	Individual or crew
		Spot decontamination	Within 15 minutes	Individual or crew
Operational	- Continues operations in a contaminated environment - Limits agent spread	MOPP gear exchange CCS and/or CCA	Within 6 hours	Contaminated unit
		Vehicle wash down	Within 6 hours (CARC)	Battalion or decontamination unit
			Or within 1 hours (non-CARC)	
Legend: CARC–chemical agent resistance coating; CCA–contamination control area; CCS–contamination control station; MOPP–mission-oriented protective posture				

Immediate Decontamination

5-94. Immediate decontamination minimizes casualties and limits the spread or transfer of contamination. This action is carried out by the contaminated individual and the purpose is to save lives and reduce penetration of agent into surfaces. This may include decontamination of Soldiers, clothing, and equipment (see paragraphs 5-95 to 5-98). Immediate decontamination limits casualties and permits the use of individual equipment and key systems.

5-95. Skin decontamination is a basic survival skill and should be performed within 1 minute of being contaminated. Decontaminate the eyes by flushing with water as soon as possible following contamination. Soldiers use their own reactive skin decontamination kit to decontaminate their skin and conduct personal wipe down (see paragraph 5-88). The skin decontamination kit should be stored in the individual's mask carrier or if issued, in the individual equipment carrier bag.

5-96. Personal wipe down decontamination is performed within 15 minutes of contamination to remove contamination from individual equipment with a M295 decontamination kit. Use a chemical detector or detector paper to locate some chemical agents. Use a radiac meter to locate radiological contamination; and then brush, wipe, or shake it off.

5-97. Teams and crews use joint chemical agent detectors to monitor potential chemical contamination and then use the M100 Sorbent Decontamination System for immediate decontamination of chemically contaminated surfaces, for example, weapon systems and communications equipment. In the event of chemical or biological contamination of optics, crews conduct tactical decontamination of vision blocks and optics including machine gun and close combat missile system (known as CCMS) optics, using the M334 decontamination kit, individual equipment which augments the M295.

5-98. Operator wipe down decontamination is done within 15 minutes of contamination of surfaces that operators need to touch or contact to operate the equipment. Radiological contamination in the form of dust particles may be wiped, scraped, or brushed off. During operator wipe down, the unit or activity should use individual equipment decontamination kits.

Operational Decontamination

5-99. Operational decontamination allows a force to continue fighting and sustain momentum after being contaminated and eliminating or reducing the duration that MOPP equipment should be use.

5-100. Operational decontamination is carried out by the contaminated unit under the control of the Infantry battalion (with assistance from a higher headquarters decontamination team with an M26 Joint Service Transportable Decontamination System or support CBRN unit decontamination organization). Operational decontamination is restricted to the specific parts of contaminated, operationally-essential equipment and material to minimize contact hazards and to sustain operations.

5-101. MOPP gear exchange should be performed as mission allows, but within 24 hours of being contaminated. A MOPP gear exchange allows a unit to remove the gross contamination from Soldiers and equipment and a return to an increased operational readiness in the pursuit of mission accomplishment.

SURVIVABILITY

5-102. *Survivability* is a quality or capability of military forces which permits them to avoid or withstand hostile actions or environmental conditions while retaining the ability to fulfill their primary mission (ATP 3-37.34), which can be enhanced through various means and methods. One way the platoon can enhance survivability when existing terrain features offer insufficient *cover*, protection from the effects of fires (FM 3-96), and *concealment*, protection from observation or surveillance (FM 3-96), is to alter the physical environment to improve cover and concealment. Similarly, natural, or artificial materials may be used as camouflage to confuse, mislead, or evade the enemy. Together, these are called *survivability operations*—those protection activities that alter the physical environment by providing or improving cover, camouflage, and concealment (ATP 3-37.34). Although survivability encompasses capabilities of military forces both while on the move and when stationary, survivability operations focus more on stationary capabilities—constructing fighting and protective positions and hardening facilities.

5-103. Survivability construction includes fighting positions, protective positions, and hardening. These are prepared to protect Soldiers, weapon systems, and when attached and required, vehicles. Positions can be constructed and reinforced with OHC to increase the survivability of dismounts and crew-served weapons against shrapnel from airbursts. In addition, the Infantry platoon and squad may use digging assets for ammunition caches at alternate, supplementary, or subsequent positions. All leaders must understand survivability plans and priorities. Typically, at platoon level the engineer platoon leader creates a leader's card, which enables the platoon leader to track the survivability effort. Typically, the platoon sergeant is designated to enforce the plan and priorities and ensures the completion status is reported to the company and disseminated within the platoon, including attachments. (See section VI for a detailed discussed on fighting position construction.)

ELECTROMAGNETIC PROTECTION

5-104. One of the most recognized electromagnetic protection (EP) techniques at the small-unit tactical level is the establishment of a PACE plan. PACE plans (see paragraphs 2-34 to 2-39) should use a mix of systems that operate across a wide range of the electromagnetic spectrum to present multiple dilemmas to enemy electromagnetic warfare (EW) operators. The greatest vulnerability in a PACE plan is the failure to understand and rehearse the plan at the lowest possible level. Triggers must be established to move between PACE communications systems. This must be rehearsed continuously at all echelons during troop leading procedures (TLP). (See paragraph 2-42 for additional information on EP.)

SECTION III – AREA DEFENSE

5-105. An *area defense* is a type of defensive operation that concentrates on denying enemy forces access to designated terrain for a specific time rather than destroying the enemy outright (ADP 3-90). The focus is on retaining terrain where the bulk of the defending force positions (main battle area) itself in mutually supporting prepared positions. Defenders maintain their positions and control the terrain between these positions and focus their fires into engagement area (see paragraph 5-204), which reserve units can supplement. The leader uses the reserve force to reinforce fires, add depth, block penetrations, restore positions, counterattack to destroy enemy forces, or seize the initiative.

ORGANIZATION OF FORCES

5-106. The Infantry rifle platoon leader organizes an area defense around the static framework of the defensive positions seeking to destroy enemy forces by interlocking fire or local counterattacks. The platoon leader (in coordination with the company commander) has the option of selecting from the two methods of positioning units, defending in-depth or defending forward (see paragraphs 5-114 to 5-118). The platoon's ability to defend in depth depends on the threat, task organization of the company, and nature of the terrain. When the leader defends forward within an AO, the force is organized so that most of the available combat power is committed early in the defensive effort. In an area defense, the leader organizes the defending force to accomplish information collection (especially reconnaissance and surveillance), security, main battle area, reserve, and sustainment missions.

5-107. The platoon normally conducts defensive operations within the Infantry rifle company as part of an Infantry battalion's area defense to disrupt or destroy an enemy force, control terrain, or protect a friendly force. The orientation of the platoon defense is determined from the mission analysis. The platoon's purpose in the mission statement clearly focuses the platoon on the enemy, the terrain, or a friendly force. The platoon may take part in many differing roles as part of a larger force: as part of the security force (in the security area, see paragraph 5-16) and main body (in the main battle area, see paragraph 5-17), conducting information collection (reconnaissance and surveillance, and security operations), occupying a battle position(s) in the main battle area, or as a reserve or part of a reserve. The platoon defense normally focuses on reconnaissance, surveillance, security, defensive operations, and actions by the reserve.

5-108. The platoon uses numerous security techniques in concert with deliberate well-planned defensive positions, incorporating such things as ambushes, reconnaissance, and security patrolling (movement to contact). The platoon leader analyzes how the terrain effects where the enemy can go and where to emplace forces from the platoon (within the security area and main battle area) that deny the enemy the ability to compromise the company and battalion defenses.

5-109. Platoon missions within the security area consist of security, reconnaissance, surveillance, and counterreconnaissance missions. Depending on the specific missions assigned, the platoon may simply observe and report, engage with indirect fires, and/or engage with direct fire (the least preferred) weapons. Security operations are crucial throughout the security area initially, to support the preparation of the defense; early in the fight, to disrupt the enemy attack and/or to identify the main effort; and in the main battle area, to support the higher commander's decision-making process. The platoon leader considers activities that must be conducted in the security area that inhibit the enemy compromising the defense. Combat patrols, ambushes, security patrols, and raids should be considered in concert with the preparation of defensive positions. The combat power applied to these combat patrols are affected by METT-TC (I) and the higher commander's intent.

5-110. Platoon missions conducted in the main battle area are oriented on enemy destruction, terrain retention, and force protection. Normally, the decisive fight occurs in the main battle area; therefore, the main effort is located there. Units tasked with security missions, supporting efforts, or reserve missions must support the main effort in the commander's concept for conducting the defense.

Note. The reserve is used to seize and maintain the initiative. Although the reserve does not have an assigned mission that directly supports the main effort, the reserve is positioned to be employed at the decisive time and place to ensure the success of the defense.

5-111. The platoon uses numerous local security techniques in concert with deliberate well planned defensive positions, incorporating such things as ambushes, reconnaissance, and security patrolling (movement to contact). The platoon leader analyzes how the terrain effects where the enemy can go and where to emplace forces from the platoon (within the security area and main battle area) that *deny*—a task to hinder or prevent the enemy from using terrain, space, personnel, supplies, or facilities (ATP 3-21.20)—the enemy the ability to compromise the company and battalion defenses.

5-112. Sustainment missions in an area defense requires a balance between establishing forward supply stocks of ammunition, barrier material, and other supplies in sufficient amounts, and having the ability to move the supplies in conjunction with enemy advances. Maintenance and medical support, with their associated repair parts and medical supplies, must also be balanced between being employed forward and having the ability to move. Echeloned company trains (echeloned trains include battalion combat and field trains) provide the company commander with the flexibility to execute platoon sustainment missions in support of the area defense. Company trains (see chapter 8) are located between 500 and 1,000 meters away from forward platoon combat missions. By placing at least one terrain feature between it and the enemy, the company trains will be out of the enemy's direct fire weapons. In some circumstances, it may be necessary to emplace cache's forward of the company trains, and to the rear of the platoon battle positions or company command post in a defilade position, with cover when possible, for easier access throughout the duration of the defense.

5-113. A platoon may perform all these operations at the same time in its own AO; or it may be tasked to do one or more of them for a larger unit. (See ATP 3-21.10 and ATP 3-21.20 for additional information.)

POSITION SELECTION

5-114. The two methods for the positioning of units within an area defense are defense in-depth and forward defense. The Infantry platoon is expected to be able to do both. While the Infantry company commander usually selects the type of area defense to use, the higher commander often defines the general defensive scheme for the Infantry rifle company and platoon. The specific mission may impose constraints such as time, security, and retention of certain areas that are significant factors in determining how the rifle company and platoon will defend. Regardless of the method of positioning, platoon will typically defend from battle positions.

DEFENSE IN-DEPTH

5-115. Defense in-depth reduces the risk of the attacking enemy quickly penetrating the defense. The enemy is unable to exploit a penetration because of additional defensive positions employed in-depth. (See figure 5-4 on page 5-34.) The in-depth defense provides more space and time to defeat the enemy attack. Platoons can achieve depth in positions through planning supplementary or subsequent battle positions (see paragraphs 5-22 and 5-23, respectively). When platoons are tasked to defend, the expectation is that they plan for depth by planning for multiple battle positions in depth within their assigned AO.

5-116. The Infantry platoon uses a defense in-depth when—

- The mission allows the platoon to fight throughout the depth of the AO.
- The terrain does not favor a defense well forward, and better defensible terrain is available deeper in the AO.
- Sufficient depth is available in the AO.
- Cover and concealment forward in the AO are limited.
- Weapons of mass destruction may be used.

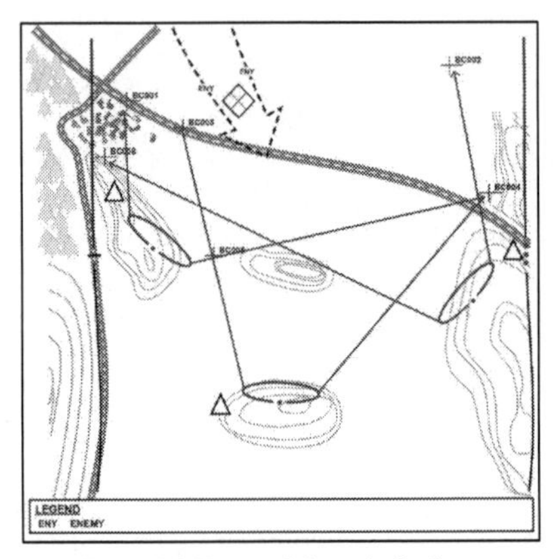

Figure 5-4. Platoon defense in-depth

Forward Defense

5-117. The intent of a forward defense is to prevent enemy penetration of the defense. (See figure 5-5.) Due to lack of depth, a forward defense is least preferred. The platoon deploys most of its combat power into forward defensive positions near the forward edge of the battle area. While the company defense may lack depth, the platoon and squads must build depth into the defense at their levels. Achieving depth usually comes with a cost in terms of reduced frontage but is essential to an effective defense. The leader fights to retain the forward position and may conduct counterattacks against enemy penetrations or destroy enemy forces in a forward engagement area. Often, counterattacks are planned forward of the forward edge of the battle area to defeat the enemy.

5-118. The platoon uses a forward defense when—

- Terrain forward in the AO favors the defense.
- Strong existing natural or man-made obstacles, such as river or a rail line, are located forward in AO.
- The assigned AO lacks depth due to location of the area or facility to be protected.
- Cover and concealment in rear portions of the AO are limited.
- Directed by higher headquarters to retain or initially control forward terrain.

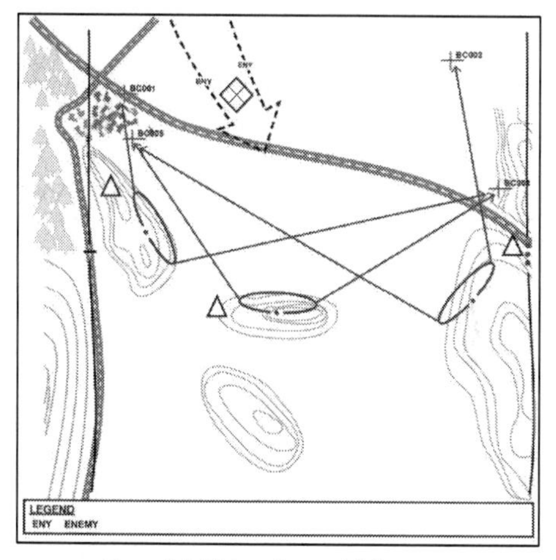

Figure 5-5. Platoon forward defense

VARIATIONS OF THE AREA DEFENSE

5-119. The Infantry rifle platoon usually defends using one of three variations of the area defense: defense of a linear obstacle, perimeter defense, or reverse slope. The platoon also can defend using a combination of these variations. (See ATP 3-21.10 for additional information.)

DEFENSE OF A LINEAR OBSTACLE

5-120. A platoon may conduct a defense of a linear obstacle along or behind a linear obstacle within an area defense. Linear obstacles such as mountain ranges or river lines generally favor a forward defense. It is extremely difficult to deploy in strength along the entire length of a linear obstacle. The defending leader must conduct economy of force measures in some areas.

5-121. Within an area defense, the leader's use of a defense in-depth accepts the possibility the enemy may force a crossing at a given point. The depth of the defense should prevent the enemy from rapidly exploiting its success. It also diffuses the enemy's combat power by forcing the enemy to contain bypassed friendly defensive positions in addition to continuing to attack positions in greater depth.

5-122. This variation of an area defense may be used when defensible terrain is available in the forward portion of the platoon's AO, or to take advantage of a major linear natural obstacle. This technique allows interlocking and overlapping observation and fields of fire across the platoon's front. (See figure 5-6.) The bulk of the platoon's combat power is well forward. Sufficient resources must be available to provide adequate combat power to detect and stop an attack. The platoon relies on fighting from well-prepared mutually supporting positions. It uses a high volume of direct and indirect fires to stop the attacks. The main concern when fighting this variation of an area defense is the lack of flexibility and the difficulty of both seizing the initiative, seeking out enemy weaknesses, and the difficulty of developing depth within the defense. Obstacles, indirect fires, and contingency plans are vital to this maneuver. The platoon depends upon surprise, well-prepared positions, and deadly accurate fires to defeat the enemy. The platoon will normally not have the resources to establish a separate reserve, the company's reserve in this case is generally small, and perhaps a squad from one of the three platoons.

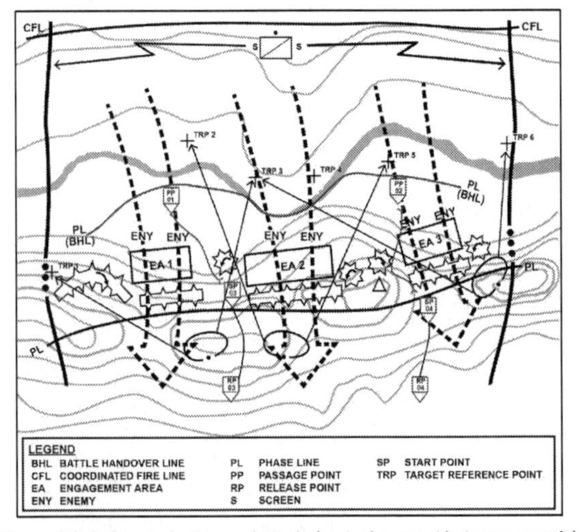

Figure 5-6. Defense of a linear obstacle (mutual support between squads)

5-123. Minefields and other obstacles are positioned and covered by fire to slow the enemy and inflict casualties. Engaging the enemy at long-range by supporting fires disrupts the momentum of the enemy's attack. Use fires from mortars, machine guns, and small arms as the enemy comes into range. If the defense is penetrated, block the advance with the reserve and shift fire from the forward squads onto the enemy flanks. Then, counterattack with the reserve or a less committed squad with intense fires. The purpose is to destroy isolated or weakened enemy forces and regain key terrain.

5-124. The counterreconnaissance effort is critical when fighting to deny the enemy the locations of the platoon's forward positions. If the enemy locates the forward positions, the enemy will concentrate combat power where desired while fixing the rest of the platoon to prevent their maneuver to disrupt the enemy attack. This effort might be enhanced by initially occupying and fighting from alternate positions (see paragraph 5-22) forward of the primary positions. This tactic enhances the security mission and deceives the enemy reconnaissance that may get through the security force.

PERIMETER DEFENSE

5-125. A perimeter defense is a defense oriented in all directions. (See figure 5-7 on page 5-38.) The platoon uses it for self-security, and to protect other units located within the perimeter. The platoon can employ a perimeter defense in urban or woodland terrain. The platoon might be called upon to execute the perimeter defense under a variety of conditions including:

- When it must conserve or build combat power to execute offensive operations or patrolling missions.
- When it must hold critical terrain in an AO.
- When its defense is not tied in with adjacent units.
- When it has been bypassed and isolated by the enemy and must defend in place.
- When it is used in a reserve position, in a TAA or patrol base, on a follow-on decentralized platoon operation during resupply or when the platoon is otherwise isolated.
- When it is directed to concentrate fires into two or more adjacent avenues of approach.

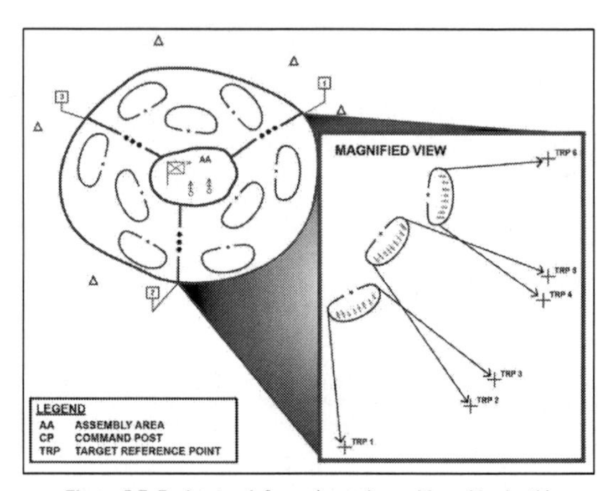

Figure 5-7. Perimeter defense (squads positioned in-depth)

> *Note*. The actions addressed in paragraph 5-126 will generally apply to the defense of a linear obstacle and reverse-slope defense.

5-126. Preparing a perimeter defense is like preparing any other position defense, but the platoon must disperse in a circular configuration for all-round security. The actual shape depends on the terrain. The following actions constitute setting up a perimeter defense:

- The platoon leader assigns squads to cover the most likely approach and prepares alternate and supplementary positions within the perimeter.
- Javelins (or M3 Multi-role, Antiarmor, Antipersonnel Weapon System [known as MAAWS]) cover likely armor approaches.
- The platoon may use hide positions and move forward to fire as the enemy appears. The platoon leader assigns several firing positions. If there are few positions for them, they are assigned a primary position and are dug in.
- Snipers (when attached) or squad designated marksman should cover likely or suspected enemy positions or observation posts.
- Snipers (when attached) and squad designated marksmen also should be used to observe or overwatch areas where civilians congregate.

- Keep attached mortars near the center of the perimeter so their minimum range does not restrict their ability to fire in any direction.
- Mortars should dig in and have covered ammunition storage bunkers.
- If possible, hold one or more rifle teams in reserve.
- The platoon leader assigns a primary position to the rear of the platoon, covering the most dangerous avenues of approach, and may assign the rifle squad supplementary positions since the platoon is prepared to fight in all directions.
- Prepare obstacles in-depth around the perimeter.
- Plan direct and indirect fire as for any type of defense.
- Plan and use direct and indirect fire support from outside the perimeter when available.
- Counter enemy probing attacks by area fire weapons (artillery, mortars, claymores, and grenade launchers) to avoid revealing the locations of fighting positions (rules of engagement [ROE]-dependent).
- If the enemy penetrates the perimeter, the reserve destroys, and then blocks the penetration:
  - It also covers friendly Soldiers during movement to alternate, supplementary, or subsequent positions.
  - Even though the platoon's counterattack ability is limited, it must strive to restore its perimeter.
- Sustainment elements may support from within the perimeter or from another position.
- Supply and evacuation might be by air. Consider the availability of landing zones and drop zones (protected from enemy observation and fire) when selecting and preparing the position.
- Position machine gun teams with principal direction of fires (known as PDFs) or FPLs. (Build the defense by first positioning key weapons, and then squads/teams.)
- Position riflemen where they can protect machine guns.
- Identify dead space and position grenadiers.
- Position riflemen with depth, overlap/interlock fires, and stand off to grenade range.
- Ensure gaps/seams are closed between elements through adjacent unit coordination (left-to-right, front-to-rear).

5-127. The Y-shaped perimeter defense is a variation of the perimeter defense that uses the terrain effectively. This defense is used when the terrain, cover and concealment, or fields of fire do not support the physical positioning of the squads in a circular manner. The Y-shaped perimeter defense is so named because the squad's battle positions are positioned on three different axes radiating from one central point. (See figure 5-8 on page 5-40.)

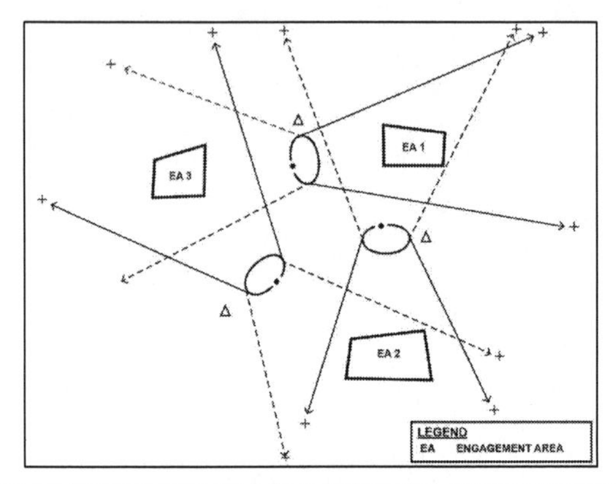

Figure 5-8. Y-shaped perimeter defense (primary and alternate engagements)

5-128. The Y-shaped variation is still a perimeter defense because it is effective against an attack from any direction. The Y-shaped defense provides all-round perimeter fires without having to position Soldiers on the perimeter. It is likely to be most effective in mountainous terrain, but it also may be used in a dense jungle environment due to limited fields of fire. All the fundamentals of a perimeter defense previously discussed apply, with the following adjustments and special considerations:

- Although each squad battle position has a primary orientation for its fires, each squad must be prepared to reorient to mass fires into the engagement areas to its rear.
- When there is no identified most likely enemy approach, or in limited visibility, each squad may have half its Soldiers oriented into the engagement areas to the front and half into the engagement areas to the rear. Ideally, supplementary individual fighting positions are prepared, allowing Soldiers to reposition when required to mass fires into one engagement area.
- When a most likely enemy avenue of approach is identified, the platoon leader may adjust the normal platoon orientations to concentrate fires (see figure 5-9) for the following reasons:
  - This entails accepting risk in another area of the perimeter.
  - The platoon security plan should compensate for this with additional observation posts, patrols, or other measures.
- The positioning of the platoon command post, reserve, or any sustainment assets is much more difficult due to a lack of depth within the perimeter.

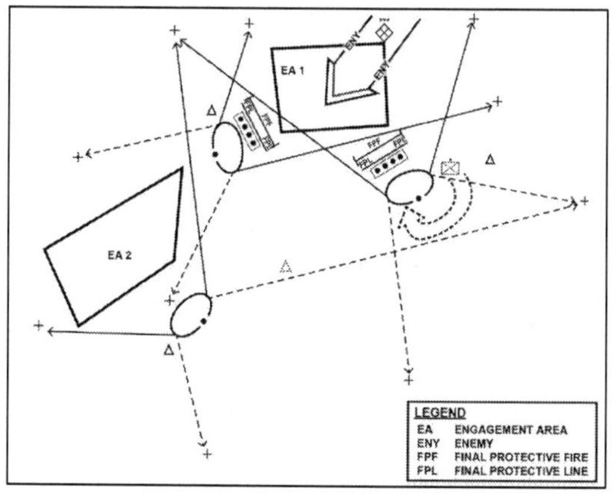

Figure 5-9. Modified Y-shaped perimeter defense

5-129. The most difficult aspect of the Y-shape perimeter defense is the fire control measures required. To fight this defense without casualties from fratricide, the leaders must ensure the limits of fire for each weapon do not allow fires into the adjacent squad's positions. In a mountainous environment, firing downward into the engagement area may make this simpler. Some measures to consider include:

- Position medium machine guns near the apex of the "Y" to allow an FPL that covers the platoon front while firing away from the adjacent units.
- Cover the areas of the engagement areas closest to the apex with claymores, non-persistent mines, or obstacles to reduce the need for direct fires in these areas.
- Identify those positions at most risk to fratricide and prepare the fighting position to protect the Soldier from fires in this direction.
- The loss of one squad position may threaten the loss of the entire platoon. To prevent this, plan and rehearse immediate counterattacks with a reserve or the least committed squad.
- Consider allowing the enemy to penetrate well into the engagement areas and destroy the enemy as in an ambush.
- Be aware that if a Y-shape defense is established on the prominent terrain feature and the enemy has the ability to mass fires, the enemy may fix the platoon or squad with direct fires and destroy it with massed indirect fires.

REVERSE-SLOPE DEFENSE

5-130. An alternative to defending on the forward slope of a hill or a ridge is to defend on a reverse slope. (See figure 5-10.) In such a defense, the platoon is deployed on terrain that is masked from enemy direct fire and ground observation by the crest of a hill. Although some units and weapons might be positioned on the forward slope, the crest, or the counterslope (a forward slope of a hill to the rear of a reverse slope), most forces are on the reverse slope. The key to this defense is control the crest by direct fire.

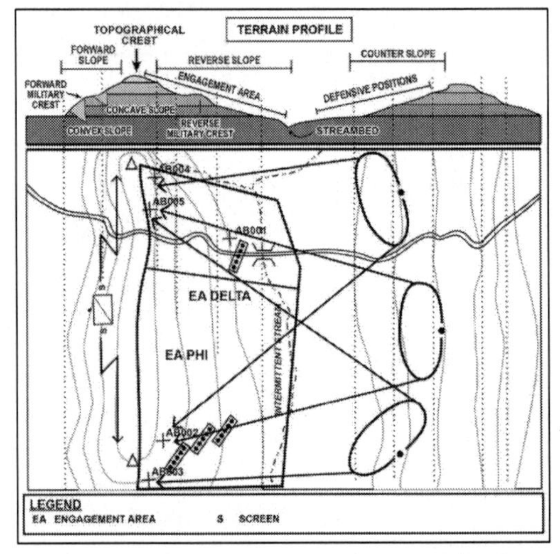

Figure 5-10. Reverse-slope defense (slope terminology)

5-131. Another reverse-slope variation available when line of sight restrictions exist to a unit's direct front is to organize a system of reverse slope defenses firing to the oblique defilade, each covering the other. (See figure 5-11.) In this example, platoon main battle area positions were unable to engage targets directly to their front but could cover each other using oblique defilade. Line of sight restrictions can be obstacles, terrain, and vegetation driven. This system of reverse slope defenses protects defenders from enemy frontal and flanking fires and from fires coming from above the main defensive area.

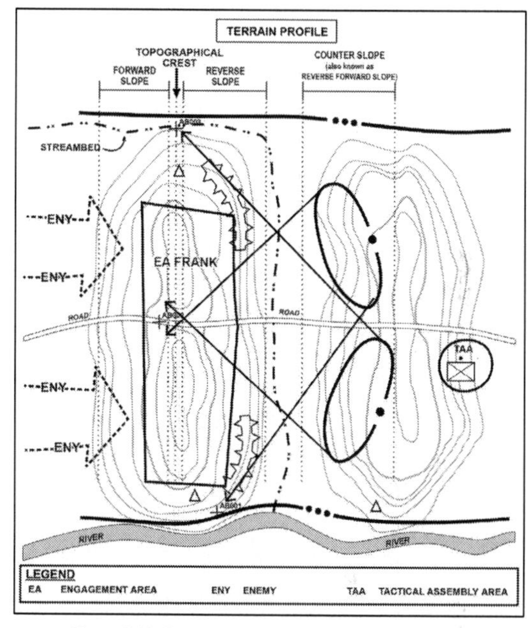

Figure 5-11. Reverse-slope defense (oblique defilade)

5-132. To succeed, the platoon leader prevents the attacker from conducting a detailed observation of the reverse-slope defense through the deployment of aggressive patrols (combat and reconnaissance [see chapter 7]) and security operations (see chapter 6 section VII) forward of the topographical crest. The platoon establishes observation posts and combat outposts on, or forward of the topographical crest. This allows long-range observation over the entire front, and indirect fire coverage of forward obstacles. Observation posts may vary in size and should include FOs. During limited visibility, the numbers may increase to improve security. Additional planning fundamentals to a defense on a reverse slope include:

- Positioning forward squads so they block enemy approaches and exploit existing obstacles. Plans should—

  - Permit surprise fire on the crest and on the approaches around the crest.
  - Have rear and OHC to protect friendly Soldiers from fratricide while in forward fighting positions.
- Positioning observation posts on the crest or the forward slope of the defended hill. Plans should—
  - Increase observation posts and patrols to prevent infiltration at night.
  - Consider attaching medium machine guns to observation posts.
- Positioning the squad in-depth or reserve where it can provide the most flexibility, support the forward squads by fire, protect the flanks and the rear of the platoon, and counterattack, if necessary. It might be positioned on the counterslope to the rear of the forward squad if that position allows it to fire and hit the enemy when the enemy reaches the crest of the defended hill.
- Positioning the platoon command post to the rear where it will not interfere with the reserve or supporting units. Plans should consider that—
  - The platoon leader may have an observation post on the forward slope or crest and another on the reverse slope or counterslope.
  - The observation post is used on the forward slope or crest before the battle starts when the platoon leader is determining the enemy's intentions.
  - During the fight, the platoon leader moves the observation post on the reverse slope or counterslope.
- Planning indirect fire well forward of, on, and to the flanks of the forward slope, crest, reverse slope, and counterslope.
- Planning FPFs on the crest of the hill to control the crest and stop assaults.
- Reinforcing existing obstacles.
- Planning protective obstacles on the reverse slope—just down from the crest where it can be covered by fire—can slow the enemy's advance and hold the enemy under friendly fire.
- Planning for counterattacks and plans to drive the enemy off the crest by fire, if possible.
- Planning to drive the enemy off by fire and movement.

5-133. The platoon leader may position all or part of the defending platoon to employ a reverse-slope defense. This technique allows subordinate units to concentrate direct fires into a relatively small area while being protected from the enemy's direct observation and supporting fires. The defender can destroy the enemy that is isolated forward through surprise and concentrated fires. Maintaining control of the forward slope is essential for success. Control of the forward slope is accomplished by using dominating terrain behind the defender or by using stay behind forces, such as an observation post, which can observe and call for fire on the attacker. The platoon leader may choose to employ a reverse slope defense when—

- Enemy fire makes the forward slope untenable.
- Lack of cover and concealment on the forward slope makes it untenable.
- The forward slope has been lost or not yet been gained.

- The forward slope is exposed to enemy direct fire weapons fired from beyond the effective range of the defender's weapons. Moving to the reverse slope removes the attacker's standoff advantage.
- The terrain on the reverse slope provides better fields of fire than the forward slope.
- Surprising and deceiving the enemy as to the true location of the platoon's defensive positions is essential.
- Enemy weapon systems have overmatch in range and lethality.

5-134. When executing a reverse slope defense, it is important for the platoon leader to consider emplacing overwatching elements forward of the topographic crest and on the flanks to protect the main defensive positions of the defending platoon. This is especially desirable when overwatching elements can observe and place fires on the crest and forward slope. Additional areas of special emphasis include:

- A direct and indirect fire support plan to prevent the enemy's occupation and using the crest of the hill.
- The use of observation posts or reconnaissance elements on the forward slope to provide observation across the entire front and security to the main battle positions.
- A counterattack plan specifying measures necessary to clear the crest or regain it from the enemy.
- Fire support to destroy, disrupt, and attrit enemy forces on the forward slope.

5-135. The forward edge of positions should be at the maximum range allowable by the available fields of fire allowing the defender time to place well-aimed fire on the enemy before the enemy reaches friendly positions. The platoon establishes observation posts on or forward of the topographical crest. This allows long range observation over the entire front and indirect fire coverage of forward obstacles. Observation posts usually are provided by the unit owning the terrain being observed and may vary in size from a few Soldiers to a reinforced squad. They should include FOs. At night, their number should be increased to improve security. Additional considerations the platoon leader may apply when defending on a reverse slope include:

- Observation of the enemy is more difficult.
- Soldiers in this position see forward no farther than the crest. This makes it hard to determine exactly where the enemy is as they advance, especially when visibility is poor.
- Observation posts must be placed forward of the topographic crest for early warning and long-range observation.
- Withdraw from the position might be more difficult.
- Fields of fire are usually short.
- Obstacles on the forward slope can be covered only with indirect fire or by units on the flanks of the platoon unless some weapon systems are placed forward initially.
- If the enemy gains the crest, they can assault downhill. This may give the enemy a psychological advantage.

- If observation posts are insufficient or improperly placed, the defenders might have to fight an enemy who suddenly appears in strength at close range.
- It is exceedingly difficult to break contact meaning that a reverse slope engagement is decisive resulting in one or both forces being severely attritted.
- The attacking force retains the initiative as long as it stays behind the crest of the front slope.

AREA DEFENSE SEQUENCE OF EVENTS, EXAMPLE

5-136. The Infantry rifle platoon may assume a defensive mission following an attack of its own or in anticipation of an enemy attack. Most defenses will follow a sequence of events similar to the example sequence addressed in paragraphs 5-137 to 5-173, as part of a larger defense. This sequence of events is used for discussion purposes and is not the only way to sequence an area defense. Some events may be performed simultaneously or in an order different from shown in the example sequence. For example, events (such as reconnaissance and surveillance, security, and occupation) may recur several times during the conduct of defense. With any defense, the platoon leader and platoon subordinate leaders understand events will vary depending on the mission variables of METT-TC (I) and events will overlap.

LEADER'S RECONNAISSANCE

5-137. Before occupying any position, to include those in the forward security area, leaders at all echelons conduct reconnaissance (building on the company commander's reconnaissance plan) of their assigned position(s). The reconnaissance effort is as detailed as possible regarding the mission variables of METT-TC (I). Reconnaissance can consist of a simple map reconnaissance, or a more detailed leader's reconnaissance and initial layout of the new position. When feasible, the platoon leader and subordinate leaders conduct a leader's reconnaissance of the complete AO to develop plans based on their view of the actual terrain. When available and dependent upon the situation, aviation assets may be used to conduct the leader's reconnaissance.

5-138. An effective leader's reconnaissance allows the platoon to rapidly occupy defensive positions and immediately start priorities of work. Participants in the platoon's reconnaissance include (the company commander when possible) the platoon leader, subordinate squad leaders, and security personnel. The goals are, but not limited to, identification of enemy avenues of approach, engagement areas, sectors of fire, the tentative obstacle plan, indirect fire plan, observation post locations, command post locations, mortar firing point, and logistics element location. The platoon leader develops a plan for the leader's reconnaissance that includes the following:

- Provisions for security.
- Areas to reconnoiter.
- Priorities and time allocated for the reconnaissance.
- Considerations for fire support, communications, and CASEVAC and MEDEVAC plans.
- Contingency plan if the reconnaissance is compromised.

ESTABLISH SECURITY

5-139. When the Infantry rifle platoon is part of a larger unit's area defense, the higher-level commander (for example, the company or battalion) establishes a forward security area before the unit moves to defend. Even with this forward security area established, subordinate platoons of the company must still provide for their own local security, especially over large geographical areas or in complex terrain and noncontiguous assigned area.

Security Operations Within the Security Area

5-140. In order to prevent the enemy from observing and interrupting defensive preparations and identifying unit positions, the higher-level commander (normally the battalion commander) establishes the security area well forward of the planned main battle area for the battalion, but within indirect fire and communications range. When the battalion commander is unable to push the security area forward to achieve this objective, the battalion may have to hold its positions initially, as it transitions and then withdraws units to the defensive main battle area, establishing a forward security force in the process. The rifle platoon may be called upon to act as anyone of these forces in the conduct of the defense.

5-141. In a contiguous assigned area, the battalion commander normally organizes and defines the security area forward of the forward edge of the battle area and assigns company assigned area to prevent gaps in the battalion area defense. In a noncontiguous assigned area, the battalion commander normally organizes and defines the security area forward of the main battle area, or along likely avenues of approach. In a noncontiguous assigned area, individual companies will have more responsibility for independent security area actions. The platoon leader ensures the independent security area actions at all levels within the platoon align with the security area plan for the company.

Security Operations Within the Main Battle Area

5-142. As the security forces forward of the battalion's main battle area are arrayed, the company commander and platoon leaders plan security operations within the main battle area, to prevent enemy reconnaissance, reduction of obstacles, targeting of friendly positions, and other disruptive actions. Subordinate platoons and squads secure obstacles, battle positions, command posts, and sustainment sites (company trains) throughout the AO.

Active and Passive Measures to Avoid Detection

5-143. Security in the defense includes all active and passive measures taken to avoid detection by the enemy, deceive the enemy, and deny enemy reconnaissance elements accurate information on friendly positions. The two primary tools available to the platoon leader are observation posts and patrols. In planning for security in the defense, the platoon leader considers the military aspects of terrain observation and fields of fire, avenues of approach, key terrain, obstacles and cover, and concealment. The leader uses the map to identify terrain that will protect the platoon from enemy observation and fires, while providing observation and fires into the engagement area. The leader uses intelligence and combat information updates to increase situational understanding, reducing the possibility of the enemy striking at a time or in a place for which the platoon is unprepared.

5-144. An observation post provides the primary security in the defense. Observation posts provide early warning of impending enemy contact by reporting direction, distance, and size. It detects the enemy early and sends accurate reports to the platoon. The platoon leader establishes observation posts along the most likely enemy avenues of approach into the position or into the AO. Leaders ensure that observation posts (normally dismounted) have communication with the platoon.

5-145. Early detection reduces the risk of the enemy overrunning the observation post. Observation post may be equipped with a Javelin command launch unit (known as CLU); class 1 unmanned aircraft system; seismic, acoustic, or frequency detecting sensors to increase its ability to detect the enemy. They may receive infrared trip flares, infrared parachute flares, infrared M320 rounds, and even infrared mortar round support to illuminate the enemy. The platoon leader weighs the advantages and disadvantages of using infrared illumination when the enemy is known to have night vision devices that detect infrared light. Although infrared and thermal equipment within the platoon enables the platoon to see the observation post at a greater distance, the platoon observation post should not be positioned beyond the supporting range of the platoon's small-arms weapons.

5-146. To reduce the risk of fratricide further, observation posts use Global Positioning Systems (GPSs) to navigate to the exit and entry point in the platoon's position. The platoon leader submits an observation post location to the company commander to ensure a no fire area is established around each observation post position.

5-147. Platoons actively patrol in the defense. Patrols enhance the platoon's ability to fill gaps in security between observation posts. The platoon leader forwards the tentative patrol route to the commander to ensure it does not conflict with other elements within the company. The commander forwards the entire company's patrol routes to the battalion. This allows the operations and intelligence staff officers to ensure all routes are coordinated for fratricide prevention and no gaps are present. The patrol leader may use a GPS to enhance basic land navigational skills as the leader tracks the patrol's location on a map.

5-148. During the defense, the platoon leader will either set a security level or comply with the security posture established by the company commander, this increases or decreases depending on enemy situation. For example, 25 percent security, 50 percent security, or up to 100 percent security at stand-to (30 minutes before and after begin morning nautical twilight or end evening nautical twilight) or when the situation dictates. Local security patrols continually inspect the integrity of emplaced obstacles and clear dead space and identify signs of enemy reconnaissance activities within the platoon's AO. Unit alert plans ensure all positions are check periodically, observation posts are relieved periodically, and at least one leader (more depending on the situation) is always alert.

Occupation and Priorities of Work

5-149. When the battalion or other higher echelon establishes a security area, a subordinate company or platoon within the battalion may be identified as a security force to deploy forward as remaining forces occupy and prepare positions in the main battle area. Security of the main battle area is critical during occupation to ensure units within the main battle area avoid detection and maintain combat power for the actual engagements with the defense.

Plan of Occupation

5-150. The plan of occupation for the platoon must be thoroughly understood to maximize the time available for occupation and preparation of the defense. When occupying/establishing the security area, the battalion commander may lead with the scout platoon to conduct reconnaissance and establish observation post along the forward edge of the security area. Company commanders must ensure that subordinate forces understand security force locations forward of the main defenses.

5-151. Within the security area, platoons within the company may be called to assist or augment forward security forces from the scout platoon or conduct security force operations in support of the battalion. During occupation of the security area, security forces establish observation posts forward, and from these observation posts, the security forces use long-range fires to hinder the enemy's preparations, to reduce the force of the enemy's initial blows, and to start the process of wresting the initiative from the enemy. Throughout the security area, security forces position and reposition (though most movement should be limited to avoid detection) to—

- Prevent enemy observation of defensive preparations.
- Defeat infiltrating enemy reconnaissance forces.
- Prevent the enemy from delivering direct fires or observed indirect fires into the battalion area defense.
- Provide early warning of the enemy's approach.

5-152. Within the main battle area, and depending on the situation, commanders at all levels may send a subordinate force to initially secure positions prior to the main body's arrival. The mission of this force is to continue to conduct reconnaissance and surveillance of key terrain and obstacles, guide and provide local security as the defenses main body occupies the defense and initiates priorities of work. As all elements within the main battle area establish local security, priorities of work continue and includes but is not limited to refining the plan, positioning of forces, preparing positions, constructing obstacles, planning and synchronizing fires, positioning logistics, and conducting inspections and rehearsals. To aid in operations security and to reduce vulnerability, the platoon leader analyzes the benefits of dispersion against the requirements and resources for the security area. Usually, the greater the dispersion between platoons (or companies within the battalion defense) the larger the security area.

Preparation and Conduct of a Defense

5-153. Priority of work is a set method of controlling the preparation and conduct of a defense. Tactical SOPs should describe priority of work including individual duties. The platoon leader changes priorities based on the situation. All leaders in the platoon should have a specific priority of work for their duty position. Although listed in sequence, several tasks are performed at the same time. An example leader and Soldier priority of work sequence is as follows:

- Post local security.
- Set and then maintain a designated level of security.
- Establish the platoon's reconnaissance and surveillance.
- Position Javelins (or M3 MAAWS) and machine guns (designate FPLs and FPFs); assign sectors of fire.
- Position subordinate units and other assets.
- Prepare hasty fighting positions.
- Establish platoon command post.
- Clear fields of fire and prepare standard range cards and sector sketches.
- Adjust indirect fire FPFs. The firing unit fire direction center should provide a safety box clearing of all friendly units before firing adjusting rounds.
- Emplace early warning devices.
- Modify hasty fighting positions to deliberate fighting positions as the enemy situation allows.
- Install wire communications, if applicable.
- Emplace obstacles and mines.
- Mark (or improve marking for) target reference points (TRPs) and direct fire-control measures.
- Improve primary fighting positions such as OHC.
- Prepare alternate and supplementary positions.
- Conduct maintenance (specifically, weapons and communications equipment).
- Establish sleep and rest plan.
- Reconnoiter movements and routes (for example, trafficability and timing).
- Rehearse engagements and disengagements or displacements.
- Adjust positions and control measures as required.
- Stockpile ammunition, food, and water.
- Dig trenches between positions.
- Continue to improve positions (for example, replaced camouflage concealment if it changes color or dries out).

Note. Leaders carefully manage security levels to achieve priorities of work.

Duties and Areas of Responsibilities During Defensive Preparation

5-154. Many platoon leader duties and responsibilities during the defense can be delegated to subordinates, but the platoon leader ensures they are done. This includes:

- Ensuring local security and assigning observation post responsibility.
- Conducting a leader's reconnaissance with the platoon sergeant and selected leaders.
- Confirming or denying significant deductions or assumptions from the mission analysis.
- Understanding/confirming the initial terrain analysis (for example, avenues of approach, potential gaps and seams, and dead space) made prior to arrival.
- Confirming/revising the direct fire plan, to include engagement area, sectors of fire, position key weapons, and fire control measures based on the reality of the terrain.
- Designating primary, alternate, supplementary, and subsequent positions supporting the direct fire plan, for platoons, sections, and supporting elements.
- Requiring squads to conduct coordination.
- Integrating indirect fire plan and obstacles to support the direct fire plan.
- Designating the platoon command post location (for example, collocating with a squad) and positioning key weapons.
- Checking the platoon command post and briefing the platoon sergeant on the situation and logistics requirements.
- Upon receipt of the squads' sector sketches, making two copies of the platoon defensive sector sketch and fire plan, retaining one copy, and forwarding the other copy to the company. (See figure B-7 on page B-13 for an example of a platoon defense sector sketch.)
- Confirming the direct fire plan and squad positions before digging starts.
- Coordinating with the left and right units.
- Checking with the company commander for all changes or updates in the orders.
- Finishing the security, deception, counterattack, and obstacle plans.
- Walking the platoon positions after they are dug.
- Confirming clear fields of fire and complete coverage of the platoon's entire AO by all key weapons.
- Looking at the defensive plan from an enemy point of view, conceptually and physically.
- Checking dissemination of information, interlocking fires, and dead space.
- Ensuring immediate correction of deficiencies.
- Ensuring rehearsals are conducted and obstacle locations reported.
- Ensuring routine report status of and completion of defensive preparations to the company.

5-155. Platoon sergeant's duties and responsibilities include:

- Establishing the platoon command post and ensures wire communications link the platoon, squads, and attached elements, if applicable.
- Establishing CCPs, platoon logistics release points (known as LRPs), and detainee collection points, and locating company-level points.
- Briefing squad leaders on the platoon command post location, logistics plan, and routes between positions.
- Assisting the platoon leader with the sector of fire and sector sketch.
- Requesting and allocating pioneer tools, barrier materiel, rations, water, and ammunition.
- Walking the positions with the platoon leader. Supervising emplacement of squads, key weapons, check standard range cards, and sector sketches.
- Establishing routine security or alert plans, radio watch, rest plans, and briefing the platoon leader.
- Supervising continuously and assisting the platoon leader with other duties as assigned.
- Selecting a straddle trench location and ensuring it is properly marked (see ATP 4-25.12).

5-156. The squad leader—

- Emplaces local security.
- Confirms positioning and assigned sectors of fire within the squad.
- Confirms positioning and assigned sectors of fire for the CCMS and medium machine gun teams.
- Positions and assigns sectors of fire for automatic rifleman, grenadiers, and riflemen.
- Establishes command post and wire communications.
- Confirms designated FPL and FPFs (for example, changes to locations, directions, or systems).
- Clears fields of fire and prepares standard range cards.
- Prepares squad standard range card and sector sketch.
- Digs fighting positions.
- Establishes communication and coordination within the platoon and adjacent units.
- Coordinates with adjacent units. Reviews sector of fire and sector sketch.
- Emplaces AT and claymores, then wire and other obstacles.
- Marks or improves marking for TRPs and other fire control measures.
- Improves primary fighting positions and adds OHC (stage 2, see paragraph 5-237).
- Prepares supplementary and alternate positions (same procedure as the primary position).
- Establishes sleep and rest plans.

- Distributes and stockpiles ammunition, food, and water.
- Digs trenches to connect positions.
- Continues to improve positions, construct revetments, replace camouflage, and add to OHC.

5-157. The FO—

- Assists the platoon leader in planning the indirect fires to support defensive missions.
- Advises the platoon leader on the status of all firing units, and on the use of obscurants or illumination.
- Coordinates with the Infantry company fire support officer, firing units, and squad leaders to ensure the fire plan is synchronized and fully understood.
- Ensures the indirect fire plan is rehearsed and understood by all.
- Ensures all FPFs are adjusted as soon as possible.
- Develops an observation plan.
- Coordinates and rehearses all repositioning of observers within the platoon AO to ensure they can observe targets or areas of responsibility.
- Develops triggers.
- Reports information collection activities.
- Ensures redundancy in communications.

Adjacent Unit Coordination

5-158. The goal of adjacent unit coordination is to ensure unity of effort in accomplishment of the platoon mission. Items adjacent units coordinate include—

- Unit positions, including locations of vital leaders' call signs and frequencies.
- Locations of observation posts and patrols.
- Overlapping fires (to ensure direct fire responsibility is clearly defined).
- Coverage of exploitable gaps or seams between units.
- TRPs (to ensure identification/coordinated method of target handoff).
- Coverage of terrain features (roads, trails, and dead space) that lay along boundaries.
- Alternate, supplementary, and subsequent battle positions.
- Indirect fire information.
- Obstacles (location and type).
- Air defense considerations, if applicable.
- Routes to be used during occupation and repositioning.
- Sustainment considerations.

Coordination

5-159. In the defense, coordination ensures that units provide mutual support and interlocking fires. In most circumstances, the platoon leader conducts face-to-face coordination (the preferred method due to reduced signal emissions) to facilitate understanding and to resolve issues effectively. However, when time is extremely limited, digital coordination may be the only means of sending and receiving this information. The platoon leader should send and receive the following information using the radio or command and control system before conducting face-to-face coordination:

- Location of key leaders.
- Location of fighting positions.
- Location of observations posts and withdrawal routes.
- Location and types of obstacles.
- Location of *contact points*—in land warfare, points on the terrain, easily identifiable, where two or more units are required to make contact (JP 3-50). Point control measures—designated military symbols are generally numbered from left to right.
- Location of *coordination points*—are points that indicate a specific location for the coordination of tactical actions between adjacent units (FM 3-90).
- Location, activities, and passage plan for scouts and other units forward of the platoon's position.
- Platoon's digital sector sketch.
- Location of all Soldiers and units operating in and around the platoon's AO.

5-160. Current techniques for coordination hold true for units that are digitally equipped. If a digitized and a nondigitized unit are conducting adjacent unit coordination, face-to-face is the preferred method. The leader of the digitized unit has the option to enter pertinent information about the nondigitized unit into command and control systems for later reference. The digitally equipped platoon leader should show the adjacent unit leader the digital sector sketch. If face-to-face coordination is not possible, leaders share pertinent information by radio.

Note. The fictional scenario, used as discussion vehicles in figures 5-12 and 5-13 on pages 5-55 and 5-57, illustrate different ways an Infantry rifle platoon can accomplish its mission within the conduct of a higher echelon's area defense. These two examples focus on potential challenges confronting the platoon leader in establishing a battle position within its company's area defense but are not intended to be prescriptive of how the Infantry rifle platoon performs any particular operation. This same scenario drives the area defense techniques used in ATP 3-21.10 and ATP 3-21.20. ATP 3-21.8 focuses on the techniques used to perform missions, functions, or tasks in support of the Infantry rifle Company B along the enemy avenue of approach in COA 1 in both figures.

Security Area Engagement

5-161. When the platoon is part of the higher echelon's security area defense, the platoon leader integrates engagements within the higher echelon's defense. Within the security area defense, planned fires usually include FOs and fire support teams (FISTs) executing indirect fire targets on a primary enemy avenue of approach. Security forces within the defense can come in the form of the scout platoon, sniper squad, subordinate units within the Infantry rifle company, or a combination of all. (See figure 5-12.) This can be in support of the higher headquarters' scheme of fires using IBCT or higher echelon artillery, or in support of the battalion, and company scheme with the use of organic mortars and allocated artillery fires.

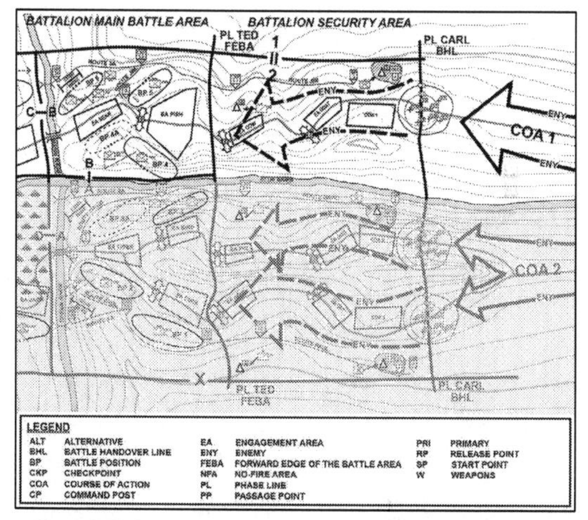

Figure 5-12. Area defense (security area engagements), example

5-162. The scheme of fires within the security area combined with the use of tactical obstacles serve to disrupt the enemy and canalize the enemy in the engagement areas, and to force the enemy to commit enemy engineer assets prior to the main battle area engagement. Tactical obstacles (for example, situational obstacles) and fire support tasks are planned and triggered relative to specific enemy COAs. They are essential, allowing for more effective engagements within the security area. Forward security forces employed forward may cover these situational obstacles with direct fires or indirect fires prior to their withdrawal to positions within the main battle area.

5-163. A platoon may support a higher echelon or its own scheme of maneuver by fighting a delay through the depth of the security area and into the main battle area. The purpose may be to take advantage of restrictive avenues of approach, to set the conditions for a counterattack, or to avoid a decisive engagement until favorable conditions are set. As security forces complete the rearward passage of lines, main battle area forces assume control of the battle at the BHL. Battle handover from forward security forces to forward main battle area forces requires firm, clear arrangements—

- For assuming command of the action.
- For coordinating direct and indirect fires.
- For the security force's rearward passage of lines.
- For closing lanes in obstacles.
- For movement of the security force with minimal interference to main battle area actions.

5-164. As security area engagements transition into the main battle area, security area forces withdraw to battle positions within the main battle area and counterattack or reserve positions. Security area forces may remain in place to continue to attrit the enemy as it continues on its avenue of approach or may move to a flank or to the rear of the main battle area to provide security. (See ATP 3-21.10 for additional information.)

MAIN BATTLE AREA ENGAGEMENT

5-165. The platoon leader seeks to defeat, disrupt, or neutralize the enemy's attack forward of or within the main battle area. The platoon leader integrates direct and indirect fires with the obstacle plan, local counterattacks, and, when established, reserve forces to destroy the enemy in designated engagement areas or to force the enemy transition to a retrograde or hasty defense. The leader uses planned engagement areas and battle positions to attack the enemy throughout the depth of the defense. Direct fire control measures allow the platoon to focus, shift, or distribute their fires as necessary. However, fire support may be limited to critical points and times. Control measures allow the leader to rapidly concentrate the use of combat power at the decisive point(s), provide flexibility to respond to changes, and allocate responsibility of terrain and obstacles to synchronize the employment of combat power.

5-166. As attacking forces reach the forward edge of the battle area, the enemy will try to find weak points in the defense and attempt to force a passage, possibly by a series of probing attacks. Forward elements engage the enemy's lead forces as the enemy attack develops along identified enemy avenues of approach. (See figure 5-13.) The platoon arrays forces and establishes engagement areas using obstacles and fires to canalize enemy forces. When shaping efforts allow for the canalization of the enemy, the enemy advance slows, and the increased density of forces present good targets for defensive fires within engagement areas. The maximum effects of these simultaneous and sequential fires, brought to bear at this stage of the battle, enable the destruction of the attacking enemy force. Additional considerations and key leader duties may include:

- Focusing and shifting fires between direct fire control measures depending on enemy movement and controlling rates of fire.
- Decisions on displacement to alternate, subsequent, and supplementary positions.
- Reporting enemy situation and alerts to emerging threats, opportunities, or changing nature of the situation.

- Reaction to enemy fires (for example, artillery and/or aviation) and CBRN attacks.
- Reports to higher, monitoring stockage levels, and cross leveling or resupply.
- CASEVAC and MEDEVAC procedures.
- Criteria to commitment the reserve.

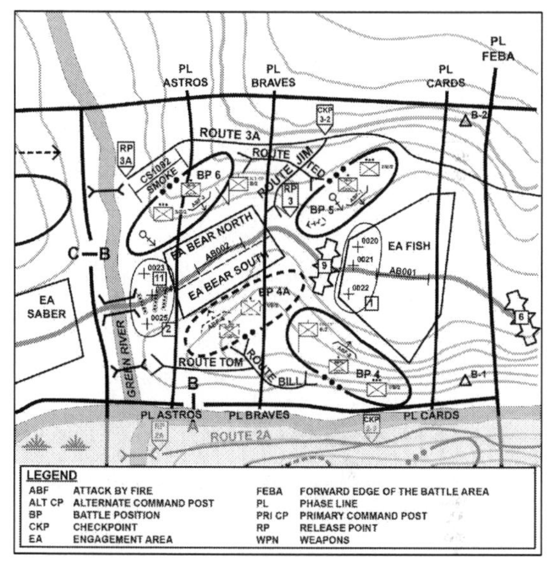

Figure 5-13. Main battle area (platoon engagements), example

FOLLOW THROUGH

5-167. During the planning for the defensive operation, the platoon leader must discern from the company OPORD what the potential follow-on missions are and begin to plan how to achieve them. During this planning, the leader determines the possible timeline and location for defeat in detail, consolidate, reorganize, and transition which best facilitates future operations and provides adequate protection.

5-168. The platoon leader and platoon subordinate leaders plan and prepare for consolidate during TLP. The following actions are usually a part of consolidate:

- Eliminate remaining enemy resistance in battle positions, noncontiguous assigned areas, and the security area.
- Reestablish, reoccupy fighting positions and adjust unit boundaries.
- Reestablish local security including OPs and patrols to deny enemy direct or indirect fire on defensive positions.
- Continue to improve security by conducting other necessary defensive actions. These defensive actions include engagement area development, direct fire planning, and battle position preparation.
- Adjust FPFs and register targets along likely mounted and dismounted avenues of approach.
- Close gaps or lanes in obstacles; repair or replace damaged or breached obstacles and replace expended mines.
- Secure enemy detainees.
- Prepare for enemy counterattack.

5-169. Reorganization usually is conducted concurrently with consolidate. It consists of actions taken to prepare the unit for follow-on tasks. As with consolidate, the platoon leader and platoon subordinate leaders plan and prepare for reorganization during TLP. During reorganization, these small-unit leader ensures the following actions are taken:

- Provide essential medical treatment and evacuate casualties, as necessary.
- Treat and evacuate wounded detainees and process the remainder of detainees.
- Reestablish key leaders using the designated succession of command.
- Reestablish manning of key weapons and equipment: CCMS, machine guns, grenade launchers, radio telephone operators.
- Reestablish communications within the organization to higher and to adjacent units.
- Cross-level personnel and adjust forces as required to support operations.
- Conducts resupply operations.
- Redistribute ammunition.
- Conduct required maintenance.
- Continue improving defensive positions, as necessary.

5-170. At the conclusion of an engagement, the platoon may continue the defense, or if ordered, transition to the offense. The platoon leader considers the higher commander's concept of operations, friendly capabilities, and enemy situation when making this decision. All missions should include plans for exploiting success or assuming a defensive posture.

5-171. A defending unit may transition from defensive operations to the retrograde as a part of continuing operations. A retrograde (see paragraph 5-174) usually involves a combination of a delay, withdrawal, and retirement that may occur simultaneously or sequentially. As in other missions, the leader's concept of operations and intent drive planning for the retrograde. Each form of retrograde has its unique planning considerations, but considerations common to all retrogrades are risk, the need for synchronization and security.

5-172. A company commander may order a defending platoon to conduct a hasty operation (for example, counterattack) or participate in a movement to contact. As part of a reserve force, the platoon and squad may execute a counterattack to destroy exposed enemy elements and free decisively engaged friendly elements. A base-of-fire element suppresses or fixes the enemy force while the counterattack element moves on a concealed route to firing positions from which it can engage the enemy in the flank and rear. The counterattack force must maneuver rapidly to its firing position, often fighting through enemy flank security elements, to complete the counterattack before the enemy can bring follow-on forces forward to influence the fight.

5-173. Execution of the counterattack is an attack by part or all of a defending force against an enemy attacking force, for such specific purposes as regaining ground lost or cutting off or destroying enemy advance units, and with the general objective of denying to the enemy the attainment of the enemy's purpose in attacking. Planning and preparation considerations for counterattack vary depending on the purpose and location of the operation. For example, the counterattack may be conducted forward of friendly positions, requiring a force not in contact or the reserve force to move around friendly elements and through their protective and tactical obstacles. In other situations, the platoon leader may use a counterattack to block, fix, or contain a penetration within its AO. In any case, the counterattacking force conducts the counterattack as an enemy-oriented task.

Note. The *mobile defense* is a type of defensive operation that concentrates on the destruction or defeat of the enemy through a decisive attack by a striking force (ADP 3-90). Mobile defenses focus on defeating or destroying the enemy by allowing enemy forces to advance to a point where they are exposed to a decisive counterattack by the *striking force*—a dedicated counterattack force in a mobile defense constituted with the bulk of available combat power (ADP 3-90). The *fixing force*—a force designated to supplement the striking force by preventing the enemy from moving from a specific area for a specific time (ADP 3-90)—holds attacking enemy in position, to help channel attacking enemy forces into ambush areas, to retain areas from which to launch the striking force.

Mobile defenses require an AO of considerable depth. Commanders shape their battlefields causing enemy forces to overextend their lines of communications, expose their flanks, and dissipate their combat power. Commanders move friendly forces around and behind enemy forces to cut off and destroy them. Divisions or larger formations usually conduct a mobile defense because of their ability to fight multiple engagements throughout the width, depth, and height of their AO, while simultaneously resourcing the striking, fixing, and reserve forces. Typically, the striking force in a mobile defense consists of one-half to two-thirds of the defender's combat power. The striking force decisively engages the enemy as it becomes exposed in attempts to overcome the fixing force.

The Infantry rifle platoon's missions as part of a larger force in a mobile defense are similar to the missions in area defense. The platoon is generally part of the fixing force. As part of the fixing force, the platoon defends within its assigned AO, although the AO might be larger than usual. Weapons system from within the Infantry weapons company of the Infantry battalion may be attached to the platoon to enable better coverage of the larger AO and to engage enemy armor heavy forces from fixed positions within the area defense. (See ATP 3-21.10 for additional information.)

SECTION IV – RETROGRADE

5-174. A *retrograde* is a type of defensive operation that involves organized movement away from the enemy (ADP 3-90). The enemy may force a retrograde or the commander may execute it voluntarily. In either case, the higher commander of the force executing the operation must approve retrograding. The three variations of the retrograde are delay, withdrawal, and retirement.

5-175. Retrogrades are conducted to improve a tactical situation or preventing a worse situation from developing. Platoons usually conduct retrogrades as part of a larger force but may conduct independent retrogrades (withdrawal) as required. *Retrograde movement* is any movement to the rear or away from the enemy (FM 3-90). Retrograde operations can accomplish the following:

- Resist, exhaust, and defeat enemy forces.
- Draw the enemy into an unfavorable situation.
- Avoid contact in undesirable conditions.
- Gain time.
- Disengage a force from battle for use elsewhere for other missions.
- Reposition forces, shorten lines of communication, or conform to movements of other friendly units.
- Secure favorable terrain.

DELAY

5-176. A *delay* is when a force under pressure trades space for time by slowing down the enemy's momentum and inflicting maximum damage on enemy forces without becoming decisively engaged (ADP 3-90). Delays allow units to trade space for time, avoiding decisive engagement and safeguard its forces. Ability of a force to trade space for time requires depth within the AO assigned to the delaying force. The amount of depth required depends on several factors, including the—

- Amount of time to be gained.
- Relative combat power of friendly and enemy forces.
- Relative mobility of forces.
- Nature of terrain.
- Ability to shape the AO with obstacles and fires.
- Degree of acceptable risk.

5-177. Delays succeed by forcing the enemy to concentrate forces to fight through a series of defensive positions. Delays must offer a continued threat of serious opposition, forcing the enemy to repeatedly deploy and maneuver. Delaying forces displace to subsequent positions before the enemy can concentrate sufficient resources to engage decisively and defeat delaying forces in current positions. The length of time a force can remain in position without facing danger of becoming decisively engaged is primarily a function of relative combat power and METT-TC (I). (See ATP 3-21.10 for additional information.) Delays gain time to—

- Allow friendly forces to establish a defense.
- Cover withdrawing forces.
- Protect friendly force's flanks.
- Allow friendly forces to counterattack.

PARAMETERS OF THE DELAY

5-178. Parameters of the delay are specified in the order for a delay mission. First, leaders direct one of two alternatives: delay within the AO or delay forward of a specified line or terrain feature for a specified time. A *delay line* is a phase line over which an enemy is not allowed to cross before a specific date and time or enemy condition (FM 3-90). The delay line is depicted as part of the graphic control measure. The second parameter in the order must specify acceptable risk. Acceptable risk ranges from accepting decisive engagement to hold terrain for a given time maintaining integrity of the delaying force. The order must specify whether the delaying force may use the entire AO or must delay from specific battle positions. A delay using the entire AO is preferable, but a delay from specific positions may be required to coordinate two or more units.

ALTERNATE OR SUCCESSIVE POSITIONS

5-179. Leaders normally assign subordinate unit contiguous operations that are deeper than they are wide. Leaders use obstacles, fires, and movement throughout the depth of assigned AO. If the leader plans the delay to last only a short time or the AO depth is limited, delaying units may be forced to fight from a single set of positions. If the leader expects the delay to last for longer periods, or sufficient depth is available, delaying units may delay from either alternate or successive positions.

5-180. In both techniques, delaying forces normally reconnoiter subsequent positions before occupying them if possible, and post guides on one or two subsequent positions. Additionally, in executing both techniques, it is critical the delaying force maintains contact with the enemy between delay positions. Advantages and disadvantages of the two techniques are summarized in table 5-5.

Table 5-5. Advantages and disadvantages of delay techniques

Method of Delay	*Use When*	*Advantages*	*Disadvantages*
Delay from Subsequent Positions	• Area of operations is wide. • Forces available do not allow themselves to be split.	• Masses fires of all available combat elements.	• Limited depth to the delay positions. • Less available time to prepare each position. • Less flexibility.
Delay from Alternate Positions	• Area of operations is narrow. • Forces are adequate to be split between different positions.	• Allows positioning in-depth. • Provides best security on most dangerous avenue of approach. • Allows more time for equipment and Soldier maintenance. • Increases flexibility.	• Requires more forces. • Requires continuous maneuver coordination. • Requires passage of lines. • Engages only part of the force at one time. • Risk losing contact with enemy between delay positions.

5-181. The alternate position technique normally is preferred when adequate forces are available, and AO have sufficient depth. Delays from alternate positions, two or more units in a single AO occupy delaying positions in-depth. (See figure 5-14.) As the first unit engages the enemy, the second occupies the next position in-depth and prepares to assume responsibility for the operation. The first force disengages and passes around or through the second force. It then moves to the next position and prepares to re-engage the enemy while the second force takes up the fight.

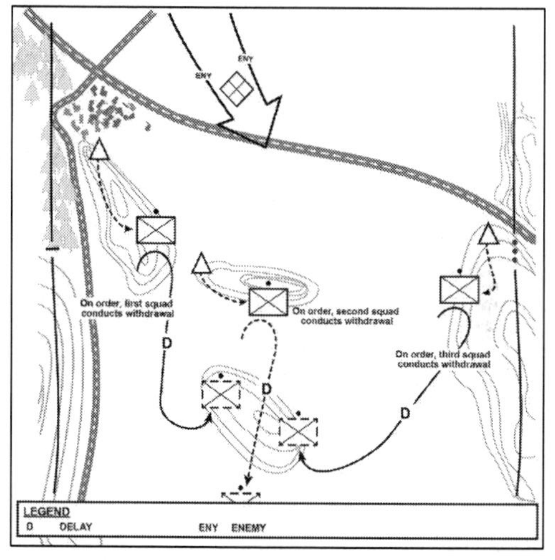

Figure 5-14. Delay from alternate positions

5-182. Delays from subsequent positions are used when assigned AO are so wide available forces cannot occupy more than a single tier of positions. (See figure 5-15 on page 5-64.) Delays from subsequent positions must ensure all delaying units are committed to each of the series of battle positions or across the AO on the same phase line. Most of the delaying force is located well forward. Mission dictates the delay from one battle position or phase line to the next. Delaying unit movement is staggered so not all forces are moving at the same time.

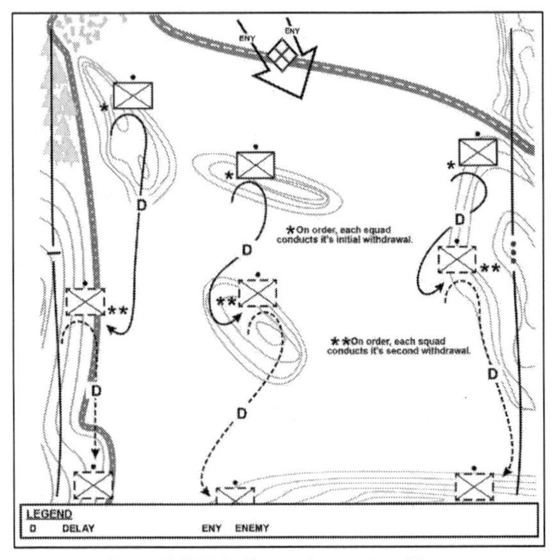

Figure 5-15. Delay from subsequent positions

WITHDRAWAL

5-183. *Withdraw* is to disengage from an enemy force and move in a direction away from the enemy (ADP 3-90). Withdrawal is a planned retrograde operation, which a force in contact disengages from an enemy force and moves away from the enemy. Although the leader avoids withdrawing from action under enemy pressure, it is not always possible. Withdrawal is used to preserve the force or release it for a new mission.

5-184. Withdrawals are inherently dangerous. They involve moving units to the rear and away from what is usually a stronger enemy force. The heavier the previous fighting and closer the contact with the enemy, the more difficult the withdrawal. Units usually confine rearward movement to times and conditions when the advancing enemy force cannot observe the activity or easily detect the operation. Operations security is critical, particularly during the initial stages of a withdrawal when most of the functional and sustainment forces displace. (See ATP 3-21.10 for additional information.)

PLANNING A WITHDRAWAL

5-185. The leader plans and coordinates a withdrawal in the same manner as a delay. METT-TC (I) applies differently because of differences between a delay and withdrawal. A withdrawal always begins under the threat of enemy interference. Because the force is most vulnerable when the enemy attacks, the leader plans for a withdrawal under pressure. The leader then develops contingencies for a withdrawal without pressure. In both cases, the leader's main considerations are to—

- Plan a deliberate break from the enemy.
- Displace the main body rapidly, free of enemy interference.
- Safeguard withdrawal routes.
- Retain sufficient maneuver, functional/multifunctional support and sustainment capabilities throughout the operation supporting forces in contact with the enemy.

ASSISTED OR UNASSISTED

5-186. Withdrawals may be assisted or unassisted. They may or may not take place under enemy pressure. These two factors combined produce four variations. Figure 5-16 depicts the mission graphic for a withdrawal and withdrawal under enemy pressure. The withdrawal plan considers which variation the force currently faces.

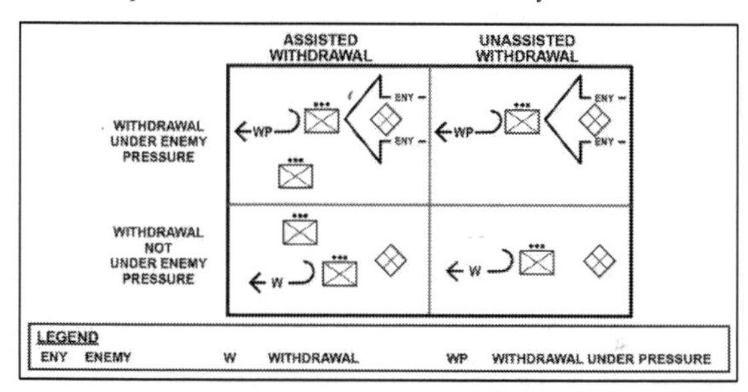

Figure 5-16. Types of withdrawals

5-187. Leaders prefer to conduct a withdrawal while not under enemy pressure and without assistance. Actions by the enemy, as well as additional coordination needed because of the presence of an assisting unit, complicate the operation.

5-188. During an assisted withdrawal, the assisting force occupies positions to the rear of the withdrawing unit and prepares to accept control of the situation. Both forces closely coordinate the withdrawal. A withdrawing force can receive assistance from another force in the form of—

- Additional security for the area through which the withdrawing force will pass.
- Information concerning withdrawal routes (reconnaissance and trafficability).
- Forces to secure choke points or key terrain along withdrawal routes.
- Elements to assist in movement control, such as traffic control post.
- Required maneuver, direct fire support and sustainment, which can involve conducting a counterattack to assist the withdrawing unit in disengaging from the enemy.

5-189. During an unassisted withdrawal, the withdrawing unit establishes routes and develops plans for the withdrawal. It establishes the security force as a rear guard while the main body withdraws. Sustainment and protection forces usually withdraw first, followed by combat forces. As the unit withdraws, the detachment left in contact (DLIC) disengages from the enemy and follows the main body to its final destination. A *detachment left in contact* is an element left in contact as part of the previously designated security force while the main body conducts its withdrawal (FM 3-90). It simulates—as nearly as possible—the continued presence of the main body until it is too late for an enemy force to react by conducting activities such as electronic transmissions or attacks. The DLIC requires specific instructions about what to do when the enemy force attacks and when and under what circumstances to delay or withdraw. If the DLIC must disengage from the enemy force, it uses the same techniques as in the delay (see paragraph 5-176).

5-190. A *stay-behind operation* is an operation in which a unit remains in position to conduct a specified mission while the remainder of the force withdraws or retires from an area (FM 3-90). While both a DLIC and a stay-behind force delay and disrupt the enemy, the biggest difference is the level of stealth. A stay-behind force delays the enemy through small scale engagements such as raids and ambushes. A stay-behind force only masses its combat power when combat power ratios and the tactical situation are in its favor. A DLIC engages the enemy and tries to represent itself as a larger force, so its presence must be known as it fights in a very conventional manner. Both forces seek to link-up with friendly forces when they accomplish their commander's intent. (See ATP 3-21.10 for additional information.)

Withdrawal Not Under Pressure

5-191. In a withdrawal not under pressure, platoons may serve as or as part of the DLIC. A DLIC is used to deceive the enemy into thinking that the entire force is still in position. As the DLIC, the platoon—

- Repositions squads and weapons to cover the company's withdrawal.

- Repositions a squad in each of the other platoon positions to cover the most dangerous avenue of approach into the position.
- Continues the normal operating patterns of the company and simulates company radio traffic.
- Covers the company withdrawal with planned direct and indirect fires if the company is attacked during withdrawal.
- Withdraws by echelon (squads or teams) once the company is at its next position.

5-192. In a withdrawal not under pressure, the platoon leader may designate one squad to execute the DLIC mission for the platoon or constitute the DLIC using elements from the remaining rifle squads with the platoon sergeant as the DLIC leader. Figure 5-17 shows an example of an unassisted withdrawal, not under pressure.

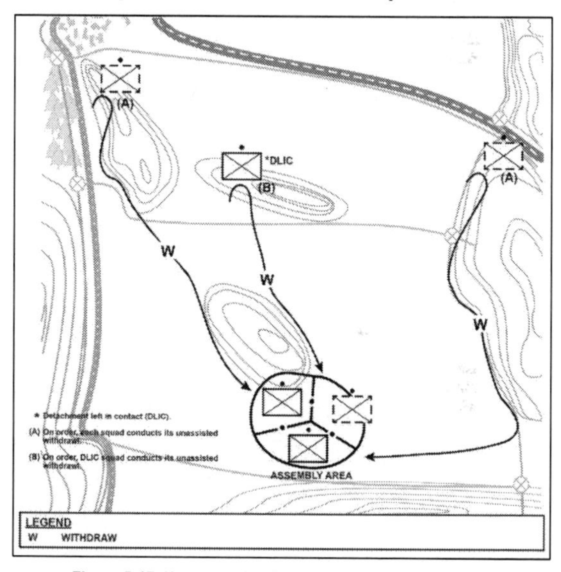

Figure 5-17. Unassisted withdrawal (not under pressure)

Withdrawal Under Pressure

5-193. In a withdrawal under enemy pressure, all units withdraw simultaneously when available routes allow, using delaying tactics to fight their way to the rear. When simultaneous withdrawal of all forces is not practical, (for example, when the platoon cannot prepare and position the security force, it conducts a fighting withdrawal) the platoon leader decides the order of withdrawal. Soldiers and squads not in contact are withdrawn first to provide suppressive fire and to allow Soldiers and squads (or teams) in contact to withdraw. Several factors influence this decision:

- Subsequent missions.
- Availability of transportation assets and routes.
- Disposition of friendly and enemy forces.
- Level and nature of enemy pressure.
- Degree of urgency associated with the withdrawal.

Methods of Disengagement

5-194. Based on the OPORD, the unit may be required to retain defensive positions and to remain and fight for a certain amount of time, or it may be required to *disengage*—a tactical mission task in which a unit breaks contact with an enemy to conduct another mission or to avoid becoming decisively engaged (FM 3-90)—and displace to subsequent positions. A platoon, as part of a company, may disengage to defend from another battle position, prepare for a counterattack, delay, withdraw, or prepare for another mission.

Fire and Movement

5-195. Fire and movement to the rear is the basic method for disengaging. All available fires are used to slow the enemy and allow platoons to move away. The company commander may move subordinate platoons and mass fires to stop or slow the enemy advance before beginning the movement away from the enemy.

5-196. Using bounding overwatch, a base of fire is formed to cover platoons or squads moving away from the enemy. One platoon or squad acts as the base of fire, delaying the enemy with fire or retaining terrain blocking the advance, while other platoons or squads disengage (see figure 5-18).

5-197. Moving platoons or squads get to their next position and provide a base of fire to cover the rearward movement of forward platoons and squads. Fire and movement is repeated until contact with the enemy is broken, the platoons pass through a different base-of-fire force, or the platoons are in position to resume their defense.

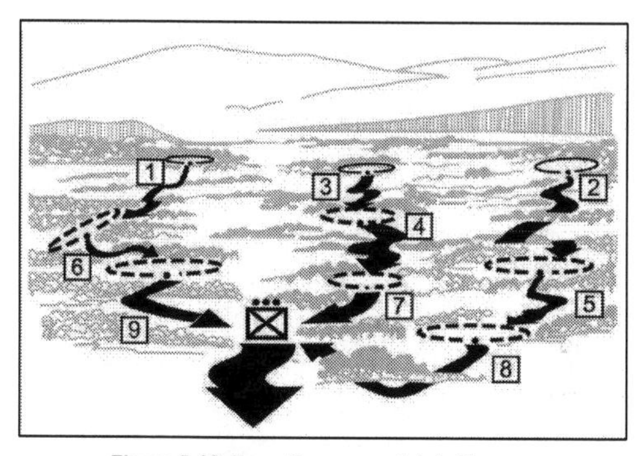

Figure 5-18. Bounding overwatch to the rear

Plan for Disengagement

5-198. Tactics used by the platoon to disengage from the enemy differ according to the company commander's plan for disengagement, how the platoon is deployed, and other factors. The following actions apply in all cases:

- Maximum use is made of the terrain to cover rearward movement. Squads back out of position and move, attempting to keep a terrain feature between them and the enemy.
- Rapid movement and effective base of fire enhance mobility and are key to a successful disengagement.

5-199. Plans for disengagement may be part of any defensive plan. When squads are separated, there are three ways they can disengage: by squads and fire teams; by thinning the lines when they must cover their own movement; or simultaneously when they are covered by another force.

Squads and Fire Teams

5-200. When the rifle platoon must cover their own movement, (for example) two squads may stay in position as a base of fire. The third squad and weapons squad then move to the rear (crew-served weapons move based on the platoon leader's assessment of when they could best move). The squads left in position must fire into the entire element's sector to cover the movement of the other squad(s). Sectors of fire are adjusted for better coverage of the element's sector. The moving squad may displace by squad or as fire teams depending on the enemy's effective fires or because there are two squads covering their movement. The squads left in position sequentially disengage. Movement to the rear by alternating squads continues until contact is broken.

Thinning the Lines

5-201. When disengaging by thinning the lines, selected Soldiers from each fire team (usually one Soldier from each fighting position) disengage and move to the rear (see figure 5-19). The Soldiers still in position become the base of fire to cover the movement.

Figure 5-19. Disengagement by thinning the lines

Simultaneous Disengagement

5-202. Squads disengage simultaneously when they are covered by another force. Simultaneous disengagement is favored when rapid movement is critical, when the disengaging element is adequately covered by overwatching fires, when the enemy has not closed on the rifle squad or cannot fire effectively at it, and when there are obstacles to delay the enemy. Simultaneous disengagement is used when rifle squads are able to move before the enemy can close on their position.

RETIREMENT

5-203. A *retirement* is when a force out of contact moves away from the enemy (ADP 3-90). Retirement is a form of retrograde, which a force not in contact with the enemy moves away from the enemy. A retiring unit organizes for combat but does not anticipate interference by enemy ground forces. Typically, another unit's security force covers the tactical movement of one formation as the unit conducts a retirement. However, enemy forces may attempt to interdict the retiring force through unconventional forces, air strikes, air assault, or long-range fires. The leader anticipates enemy actions and organizes the unit to fight in self-defense. The leader usually conducts retirement to reposition forces for future operations or to accommodate the current concept of operations. Units may utilize tactical road marches (see chapter 6) where security and speed are the most important considerations. (See ATP 3-21.10 for additional information.)

SECTION V – ENGAGEMENT AREA

5-204. The *engagement area* is an area where the commander masses effects to contain and destroy an enemy force (FM 3-90). *Contain* is a tactical mission task in which a unit stops, holds, or surrounds an enemy force (FM 3-90). *Destroy* is a tactical mission task that physically renders an enemy force combat-ineffective until reconstituted (FM 3-90). The platoon leader uses natural and man-made obstacles to canalize the attacking force into an engagement area where they can be destroyed by direct and indirect fires. The key to success is the platoon leader's ability to integrate the obstacles, fires and direct fires plans within the engagement area to achieve the Infantry platoon's and squads' tactical purposes. (See ATP 3-21.10 and ATP 3-21.20 for additional information.)

KEY ENGAGEMENT CONSIDERATIONS

5-205. *Engagement criteria* are protocols that specify those circumstances for initiating engagement with an enemy force (FM 3-90). Engagement criteria may be restrictive or permissive in nature. For example, the platoon leader may instruct a subordinate squad leader not to engage an approaching enemy unit until the enemy commits to an avenue of approach. The platoon leader establishes engagement criteria in the direct fire plan in conjunction with engagement priorities and other direct fire control measures to mass fires and control fire distribution.

5-206. *Engagement priority* identifies the order in which the unit engages enemy systems or functions (FM 3-90). The platoon leader assigns engagement priorities (see paragraph A-61) based on the type or level of threat at different ranges to match organic weapon system capabilities against enemy vulnerabilities. Engagement priorities are situationally dependent and used to distribute fires rapidly and effectively. Subordinate elements can have different engagement priorities but will normally engage the most dangerous targets first, followed by targets in-depth or specialized systems, such as engineer vehicles.

5-207. A *target reference point* is a predetermined point of reference, normally a permanent structure or terrain feature that can be used when describing a target location. (JP 3-09.3). The platoon may designate TRPs (see paragraph A-41) to define unit or individual sectors of fire and observation, usually within the engagement area. TRPs, along with trigger lines, designate the center of an area where the platoon leader plans to distribute or converge the fires of all weapons rapidly. TRPs allow leaders to distribute, shift, or focus fires throughout the width and depth of an engagement area: for instance, a squad sector distributes fires between TRP 1 and TRP 2. A machine gun PDF may be established using a TRP to focus their fires. A squad may shift fires by shifting from TRP 1 in their primary sector to TRP 3 in their secondary sector. Trigger lines, when added, serve as a method of control. Once designated, TRPs may also constitute indirect fire targets.

5-208. A *trigger line* is a phase line located on identifiable terrain used to initiate and mass fires into an engagement area at a predetermined range (FM 3-90). The platoon leader can designate one trigger line (see paragraph A-62) for all weapon systems or separate trigger lines for each weapon or type of weapon system. The leader specifies the engagement criteria for a specific situation. The criteria may be either time- or event-driven, such as a certain number or certain types of vehicles to cross the trigger line before initiating engagement. The leader can use a time-based fires delivery methodology or a geography-based fires delivery.

ENGAGEMENT AREA DEVELOPMENT

5-209. Engagement area development is a complex function demanding parallel planning and preparation if the Infantry platoon and squad are to accomplish the tasks for which they are responsible. Despite this complexity, engagement area development resembles a drill, and the platoon leader and subordinate leaders use an orderly, standard set of procedures. The steps of engagement area development are not a rigid sequential process. Some steps may occur simultaneously to ensure the synergy of combined arms (asterisks [*] denote steps that occur simultaneously). The platoon leader integrates these steps within the TLP and the IPOE (see chapter 2 for additional information). Beginning with evaluation of mission variables of METT-TC (I), the development process—

- Identifies likely enemy avenues of approach.
- Determines most likely enemy COA.
- Determines where to kill the enemy.
- Positions subordinate forces and weapon systems.*
- Plans and integrates obstacles.*
- Plans and integrates fires.*
- Rehearses the execution of operations in the engagement area.

IDENTIFY LIKELY ENEMY AVENUES OF APPROACH

5-210. This step begins during the IPOE (see paragraph 2-4) and creating the platoon graphic terrain analysis overlay (GTAO), see paragraph 2-67, that is similar to the modified combined obstacle overlay (MCOO) prepared by the battalion intelligence staff officer (S-2) (see paragraph 2-6). Identifying avenues of approach (see paragraph 5-1) is important because all COAs that involve maneuver depend on available avenues of approach. During defensive operations, it is important to identify avenues of approach that support threat offensive capabilities when identifying the most likely enemy COA. See paragraphs 2-63 to 2-66 for information on the classification of severely restricted terrain, restricted terrain, and unrestricted terrain, reference the MCOO. Results of IPOE are validated during leader's reconnaissance and preparation of the defense.

Note. The primary analytic tools used to aid in determining the effects specific to terrain, are the MCOO and the terrain effects matrix. The company commander expands upon the products received from the battalion to further identify the enemy avenues of approach.

5-211. Avenues of approach consist of a series of mobility corridors (see paragraph 5-1) through which a maneuvering force must pass to reach its objective. (See figure 5-20 on page 5-74.) Avenues of approach must provide ease of movement and enough width for dispersion of a force large enough to affect the outcome of the operation significantly. Avenues of approach are developed by identifying, categorizing, and grouping mobility corridors and evaluating avenues of approach. Additional procedures and considerations when identifying the enemy's likely avenues of approach include:

- Conducting initial reconnaissance. If possible, do this from the enemy's perspective along each avenue of approach into the AO or engagement area.
- Identifying *key terrain*—is an identifiable characteristic whose seizure or retention affords a marked advantage to either combatant (ADP 3-90) and *decisive terrain*—key terrain whose seizure and retention is mandatory for successful mission accomplishment (ADP 3-90). This includes locations affording positions of advantage over the enemy, as well as natural obstacles and choke points restricting forward movement.
- Determining which avenues provide cover and concealment for the enemy while allowing the enemy to maintain its tempo.
- Determining what terrain, the enemy is likely to use to support each avenue.
- Evaluating lateral routes adjoining each avenue of approach.

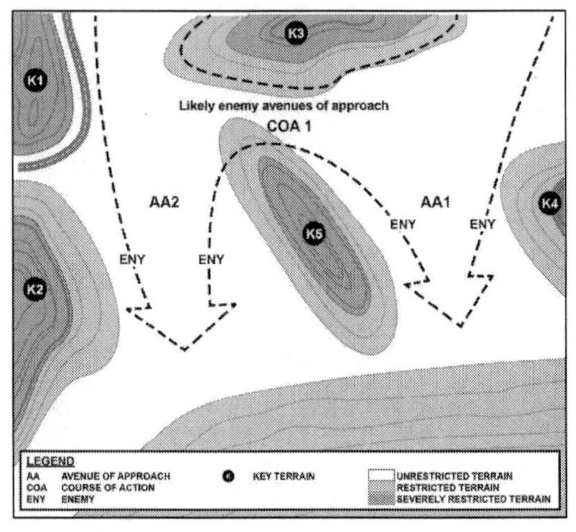

Figure 5-20. Likely enemy avenues of approach

IDENTIFY MOST LIKELY ENEMY COURSE OF ACTION

5-212. This step is completed after step 4 (determine threat COAs) of the IPOE process (see paragraph 2-4) identifies and describes threat COAs that can influence the platoon's AO and area of interest. Outputs from the S-2 cell during step 4 includes situation templates (known as SITEMPs) and COA statements for each COA. Procedures and considerations in determining the enemy's most likely COA (enemy scheme of maneuver) include:

- Determining how the enemy will structure the attack.
- Determining how the enemy will use reconnaissance and surveillance assets. Will the enemy attempt to infiltrate friendly positions?
- Determining where and when the enemy will change formations and establish support by fire positions.
- Determining where, when, and how the enemy will conduct the assault or breaching operations.
- Determining where and when the enemy will commit follow-on forces.
- Determining the enemy's expected rates of movement.

- Assessing the effects of enemy combat multipliers and anticipated locations/areas of employment.
- Determining what reactions, the enemy is likely to have in response to projected friendly actions.

DETERMINE WHERE TO KILL THE ENEMY

5-213. Generally, leaders at the platoon level decide where to kill the enemy within an engagement area assigned by a higher headquarters; however, leaders at the platoon level must assess the best place to kill the enemy within the assigned AO by understanding the terrain effects and the enemy they will face.

5-214. An analysis of the terrain, enemy scheme of maneuver and avenues of approach enables the platoon leader to understand how the enemy will enter, and flow through the platoon's AO. Within the engagement area (either assigned or identified) the platoon leader determines through this analysis where exactly to mass fires. For example, is it where the enemy stops to cross a stream; is it where the terrain forces the enemy to cross in front of the platoon laterally giving it a flank shot; is it where the platoon can fire at a point where it can achieve enfilade fires; or is a chokepoint where the platoon can drop a wire obstacle to create a place to mass fires on a stationary enemy? Only after this analysis can the platoon leader accurately identify the most advantageous places to concentrate fires within the engagement area and then identify TRPs to focus, distribute, and shift fires within those places.

5-215. The following steps apply in identifying and marking where the enemy engagement (see figure 5-21 on page 5-76) is to occur:

- Identify TRPs matching the enemy's scheme of maneuver allowing the Infantry platoon and squad to identify where it will engage enemy forces through the depth of the AO.
- Identify and record the exact location of each TRP.
- In marking TRPs, use thermal sights to ensure visibility at the appropriate range under varying conditions, including daylight and limited visibility (darkness, smoke, dust, or other obscurants).
- Determine how many weapon systems will focus fires on each TRP to achieve the desired end state.
- Determine which element will mass fires on each TRP.
- Establish engagement areas around TRPs.
- Develop the direct fire planning measures necessary to focus fires at each TRP.

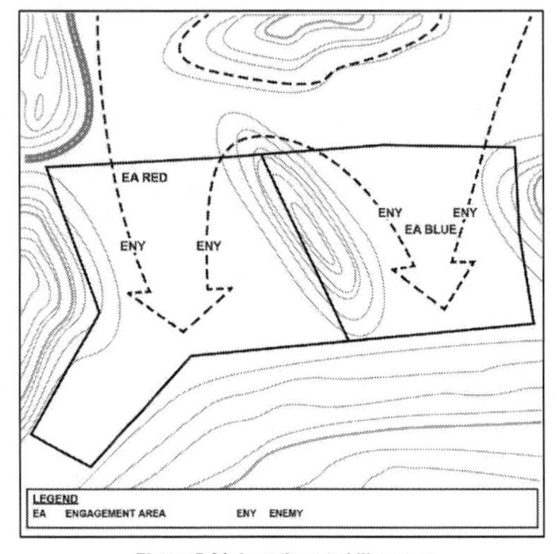

Figure 5-21. Locations to kill enemy

POSITION SUBORDINATE FORCES AND WEAPON SYSTEMS

5-216. The platoon leader does not normally position subordinate forces and weapon systems before completing steps 1, 2, and 3 of engagement area development. With an understanding of how the enemy will move through the engagement area, the optimal places to kill enemy forces within the engagement area, and tentative TRPs designated, the platoon leader plans where to emplace machine guns, antitank (AT) missiles, and squad battle positions to most effectively cover those places with direct fires. The platoon leader plans the use of primary and secondary sectors of fire, primary and alternate, subsequent and supplementary battle positions, and direct fire control measures to enable the platoon to focus, shift, and distribute direct fires throughout the EA. Additionally, emplacement of forces and weapon systems needs to support the desired obstacle effect to maximize the effectiveness of the engagement area. The following steps apply in selecting and improving battle positions and emplacing the Infantry platoon and squads, antiarmor weapon systems, crew-served weapon systems, vehicles (when attached), and individual fighting positions (see figure 5-22 on page 5-78):

- Position key weapons systems first; ensure they can observe likely target locations.
- Select tentative platoon/squad battle positions capable of massing fires in likely target locations; ensure they can observe tentative TRPs from the prone position.

Note. When possible, select battle positions while moving in the engagement area. Using the enemy's perspective enables the Infantry leader to assess survivability of positions.

- Conduct a leader's reconnaissance of tentative battle positions.
- Move along the avenue of approach into the engagement area (when possible) to confirm selected positions are tactically advantageous.
- Confirm and mark the selected battle positions.
- Ensure battle positions do not conflict with those of adjacent units and are tied in with adjacent positions.
- Select primary, alternate, and supplementary fighting positions to achieve the desired effect for each TRP.
- Ensure leaders position weapon systems so each TRP is covered by the required number of weapon systems and squads.
- Stake weapons system positions and vehicles (when attached) according to unit SOPs so engineers can dig in the positions.
- Confirm all weapons system positions.
- When established, mark and rehearse route(s) and trigger(s) for the defending force to move from its hide position(s) (concealed from enemy observation) to its OP(s) and/or battle position(s).

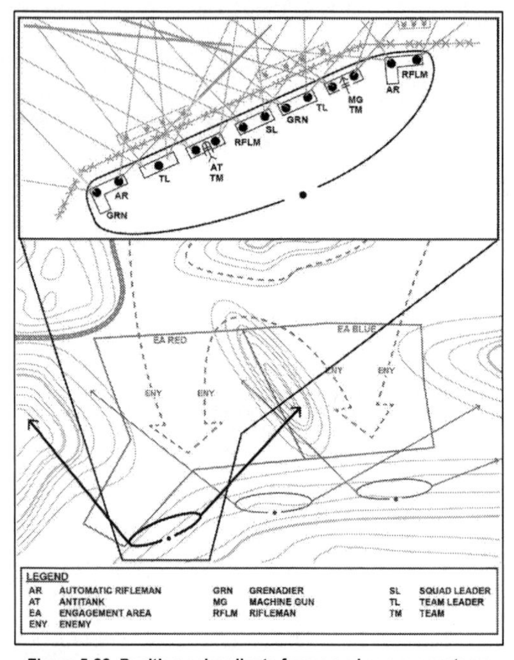

Figure 5-22. Position subordinate forces and weapon systems

PLAN AND INTEGRATE OBSTACLES

5-217. The platoon leader, as directed by the company commander, with the assistance of the engineer leader (when attached) determine an obstacle's intent relative to its location, and effect to block, disrupt, fix, or turn attacking enemy forces into planned engagement areas. Countermobility planning includes the positioning of protective obstacles to prevent the enemy from closing with defensive battle positions, and integrating mines, wire, indirect fires, and direct fires in conjunction with placement.

The following apply in planning and integrating obstacles (see figure 5-23 on page 5-80) during defensive missions:

- Platoon leader must understand what obstacle effort the platoon has been tasked to emplace by higher headquarters, whether an obstacle group or a specific directed obstacle. This will take priority.
- Obstacle effort is time-intensive; platoons must immediately start emplacing obstacles. This may mean emplacing lower priority obstacles first while higher priority obstacles are still being planned.
- Obstacles shape the engagement area by achieving specific effects: denying enemy access to terrain (block), forcing the enemy to change direction (turn), significantly delaying their movement (fix), or limiting their ability to move freely or maintain their movement formations (disrupt).
  - *Block* is an obstacle effect that integrates fire planning and obstacle effort to stop an attacker along a specific avenue of approach or prevent the attacking force from passing through an engagement area (FM 3-90).
  - *Turn* is an obstacle effect that integrates fire planning and obstacle effort to divert an enemy formation from one avenue of approach to an adjacent avenue of approach or into an engagement area (FM 3-90).
  - *Fix* is an obstacle effect that focuses fire planning and obstacle effort to slow an attacker's movement within a specified area, normally an engagement area (FM 3-90).
  - *Disrupt* is an obstacle effect that focuses fire planning and obstacle effort to cause the enemy to break up its formation and tempo, interrupt its timetable, commit breaching assets prematurely, and attack in a piecemeal effort (FM 3-90).
- Protective obstacles are positioned close to friendly unit positions to disrupt an enemy assault.
- Tactical wire obstacles support the defensive scheme, for instance by running along a machine gun FPL.
- Determine the obstacle or obstacle groups' intent with the engineer platoon leader confirming the target, relative location, and effect. Ensure intent supports the commander's scheme of maneuver.
- In conjunction with the engineer platoon leader, identify, site, and mark the obstacles within the obstacle group, tie reinforcing obstacles into natural obstacles.
- Ensure coverage of all obstacles with observation and direct and/or indirect fires. Ensure squads can observe and fire on obstacles but also on the terrain the enemy will use when making contact with those obstacles.
- Assign responsibility for guides and lane closure as required.
- According to METT-TC (I), assist engineer platoons in emplacing obstacles, securing Class IV/V point, and securing obstacle work sites.
- Coordinate engineer disengagement criteria, actions on contact, and security requirements with the engineer platoon leader at the obstacle work site.
- Assess what effect the obstacle will have on the enemy.

- Understand resources required to construct each obstacle effect based off the enemy they are facing.
- Understand how the terrain supports the obstacle effect, examples include block, disrupt, fix, and turn.

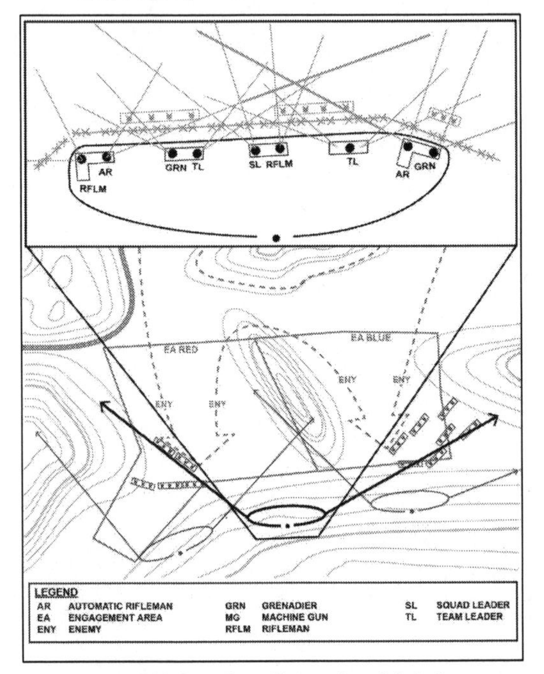

Figure 5-23. Plans for and integration of obstacles

PLAN AND INTEGRATE FIRES

5-218. The platoon leader, as directed by the company commander and with the assistance of the fire support officer, determines the purpose of fires and where that purpose is best achieved. This plan provides the most effective fires (specifically direct and indirect fires) resources and mitigate the risk of fratricide as the attacking enemy nears the designated engagement areas while supporting air assets conduct army aviation and close air support (CAS) attacks. The following apply in planning and integrating fires (see figure 5-24 on page 5-82):

- Determine the purpose of fires.
- Determine where purpose will best be achieved.
- Establish the observation plan that includes—
  - Redundancy for each target.
  - Observers who will include the FIST, as well as members of maneuver elements with direct fire support execution responsibilities.
- Establish triggers based on enemy movement rates.
- Obtain accurate target locations using organic target location devices or survey/navigational equipment.
- Refine target locations to ensure coverage of obstacles.
- Plan final protection fire.
- Request critical friendly zone for protection of maneuver elements and no-fire areas for protection of observation posts and forward positions.
- When able, leaders walk the terrain with their FOs to ensure the observers can fire targets at pre-planned triggers. This also ensures the observers can provide bottom-up refinement through the fires planning process.

Note. Leaders at the platoon level provide bottom-up refinement on the effectiveness of the higher fire support plan. Leaders at the platoon level have a better appreciation of the terrain within the engagement area because of their proximity to the engagement area.

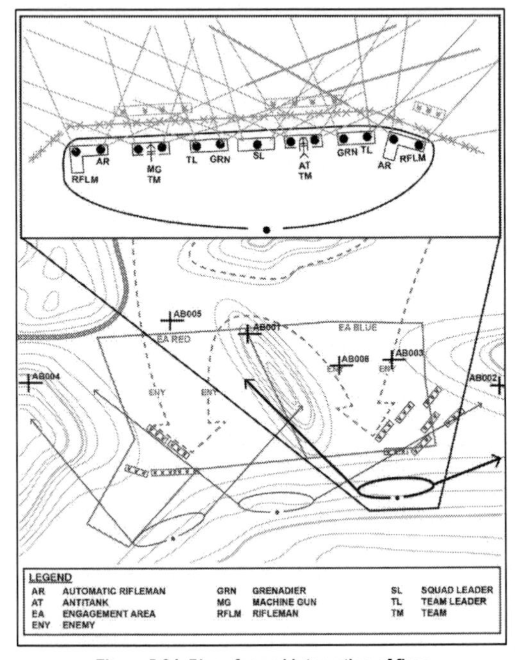

Figure 5-24. Plans for and integration of fires

REHEARSALS

5-219. The purpose of rehearsals is to ensure every leader and Soldier understands the plan and elements are prepared to cover their assigned areas with direct and indirect fires. The rehearsal should cover—

- Rearward passage of security forces (as required).
- Closure of lanes (as required).
- Movement from the hide position (when used) to the battle position.

- Use of fire commands, triggers, and maximum engagement lines to initiate direct and indirect fires.
- Shifting of fires to refocus and redistribute fire effects.
- Disengagement criteria.
- Identification of displacement routes and times.
- Preparation and transmission of critical reports using radio and digital systems, as applicable.
- Assessment of the effects of enemy weapon systems.
- Displacement to alternate, supplementary, or subsequent battle positions.
- Cross leveling or resupply of Class V.
- CASEVAC and MEDEVAC procedures.
- Evacuation of casualties.
- Friendly passage of lines or reducing own obstacles to counterattack or pursue enemy.

5-220. The platoon leader should coordinate rehearsals with higher headquarters to ensure there are no conflicts with other units. Coordination leads to efficient use of planning and preparation time for all units involved with the operation. It eliminates dangers of misidentifying friendly forces in the rehearsal area, which could result in fratricide.

SECTION VI – FIGHTING POSITION CONSTRUCTION

5-221. Whether the defending unit is in a defensive perimeter or on a deliberate ambush line, the unit seeks cover from fire and concealment from observation. From the time the unit prepares and occupies a fighting position, the unit continues to improve its positions. How far the unit gets depends on how much time it has, regardless of whether it is a hasty position or a well-prepared one with OHC.

CHARACTERISTICS OF A FIGHTING POSITION

5-222. Defending units occupying immediately establish hasty fighting positions and designate sectors of fire. Depending on time available, units continue to improve their positions through depth and adding frontal, flank, rear, and OHC.

LEADER RESPONSIBILITIES

5-223. Leaders ensure assigned fighting positions enable Soldiers to engage the enemy within the Soldiers' assigned sector of fire to the maximum effective range of the Soldiers' weapon (terrain dependent) with grazing fire and minimal dead space. Leader responsibilities include:

- Protect Soldiers.
- Plan and select fighting position sites.
- Provide materials.
- Supervise construction.

- Inspect periodically.
- Request technical advice from engineers as required.
- Improve and maintain unit survivability continuously.
- Determine if need to build up or build down OHC.

SECTORS OF FIRE AND WEAPONS EMPLACEMENT

5-224. Though fighting positions provide maximum protection for Soldiers within the fighting position, the primary consideration is always given to sectors of fire and effective weapons employment. Weapon systems are sited where natural or existing positions are available, or where terrain will provide the most protection while maintaining the ability to engage the enemy. To maximize the capabilities of a weapon system, always consider the best use of available terrain and how to modify it to provide the best sectors of fire.

COVER

5-225. The inclusion of natural cover afforded by the terrain enhances protection. Cover protects against small arms fire and indirect fire fragments. The different types of cover, frontal, flank, rear, and OHC, are used to make fighting positions. Positions are connected by tunnels and trenches to move between positions for engagements or resupply, while remaining protected. Natural covers such as rocks, trees, logs, and rubble are best, because it is hard for the enemy to detect. When natural cover is unavailable, use dirt removed from the hole to construct the fighting position. Improve the effectiveness of dirt as a cover by putting it in sandbags. Fill sandbags only three quarters full to alleviate gaps in the cover.

CONCEALMENT

5-226. Concealment hides the position from enemy detection. When a position can be detected, it can be hit by enemy fire. Natural, undisturbed concealment is better than man-made concealment. When digging a position, Soldiers try not to disturb the natural concealment around the position. Unused dirt (not used for front, flank, or rear cover which must be concealed) from the hole is placed behind the position and camouflage. Natural, undisturbed concealment material that does not have to be replaced (rocks, logs, live bushes, and grass) is best, but avoid using so much camouflage that the position looks different from its surroundings. Man-made concealment must blend with its surroundings so that is cannot be detected and must be replaced if it changes color or dries out. Protection through concealment requires the position to be undetectable from a minimum distance of at least 35 meters.

PRIORITY OF WORK

5-227. Leaders establish priorities of work that may include but is not limited to posting local security, preparing appropriate fighting positions for all weapon systems, assigning sectors of fire, emplacing antiarmor assets such as Javelins (or M3 MAAWS) and protective obstacles (mines/wire), and rehearsals. Building fighting positions is woven into the defender's priority of work (see paragraphs 5-149 to 5-160) and is directly limited by levels of security established (see paragraphs 5-139 to 5-148), and that the platoon leader will develop a plan for any internal or external resources (for example, pioneer tools and engineering assets).

Note. Firing from fighting positions or enclosures is prohibited for the M3 MAAWS. (See TC 3-22.84 for additional information on ammunition firing restrictions.)

HASTY AND DELIBERATE FIGHTING POSITIONS

5-228. Fighting positions protect Soldiers by providing cover from direct and indirect fires and concealment through positioning and proper camouflage. Because the battlefield conditions confronting Soldiers are never standard, no single standard fighting position design fits all tactical situations. The two types of fighting position are hasty and deliberate. The one constructed depends on time and equipment available, and the required level of protection. Fighting positions are designed and constructed to protect the individual Soldier and the Soldier's weapons system.

HASTY FIGHTING POSITION

5-229. Hasty fighting positions, used when there is little time for preparation, should be behind whatever cover is available. However, the term hasty does not mean that there is no digging. If a natural hole or ditch is available, use it. This position should give frontal cover from enemy direct fire but allow firing to the front and the oblique. When there is little or no natural cover, hasty positions provide as much protection as possible.

5-230. A shell crater, which is 2 to 3 feet (0.61 to 1 meter) wide, offers immediate cover (except for overhead) and concealment. Digging a steep face on the side toward the enemy creates a hasty fighting position. A small crater position in a suitable location can later develop into a deliberate position.

5-231. A skirmisher's trench is a shallow position that provides a hasty prone fighting position. When immediate need for shelter from enemy fire, and there are no defilade firing positions available, lie prone or on the side, scrape the soil with an entrenching tool, and pile the soil in a low parapet (wall) between the body and the enemy. To form a shallow, body-length pit quickly, use this technique in all but the hardest ground. Orient the trench so it is oblique to enemy fire. This keeps the silhouette low and offers some protection from small-caliber fire.

5-232. The prone position is a further refinement of the skirmisher's trench. It serves as a good firing position and provides better protection against the direct fire weapons than the crater position or the skirmisher's trench. The position should be about 18 inches deep and use the dirt from the hole to build cover around the edge of the position (see figure 5-25 on page 5-86).

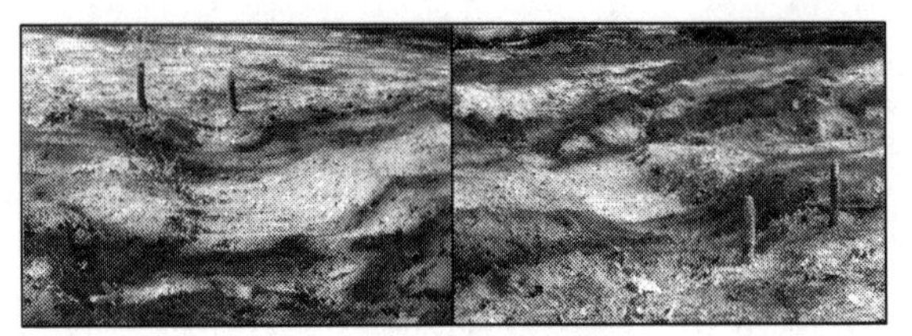

Figure 5-25. Prone position

Deliberate Fighting Positions

5-233. Deliberate fighting positions are modified hasty positions prepared during periods of relaxed enemy pressure. Positions must provide Soldiers protection from direct and indirect fires. Soldier construct parapets retaining walls 39 inches (minimum) thick (length of M4 rifle [stock extended] or length of M7 rifle [stock extended/without suppressor] plus 6 inches) and 10 to 12 inches (length of a bayonet) to the front, flank, and rear. Whether building up, or building down, OHC should be a minimum of 18 inches thick (length of an extended entrenching tool) to protect against overhead burst. Leaders identify requirements for additional OHC based on threat capabilities.

> ***Note.*** The following fighting position dimensions use the M4 rifle (or M7 rifle when fielded [stock extended/without suppressor]) with an approximant length of 33 inches/stock extended.

Two-Soldier Fighting Position with Built-Up Overhead Cover

5-234. Soldiers prepare a two-Soldier position in four stages. Leaders inspect the position at each stage before the Soldier moves to the next stage.

Stage 1 – Establish Sectors and Decide Whether to Build Overhead Cover Up or Down (H+0 to H+.5 Hours)

5-235. Leaders establish sectors of fire (see figure 5-26) and decide whether to build OHC up or down. Leaders consider the mission variables of METT-TC (I) to decide on the most appropriate fighting position to construct. For example, due to more open terrain the leader may decide to use built-down OHC.

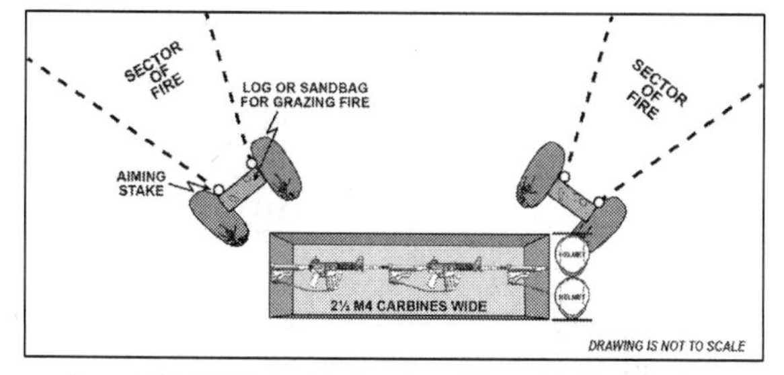

Figure 5-26. Establishment of sectors of fire and building method

5-236. Leader's checklist—

- Site position:
  - Check fields of fire from prone.
  - Assign sectors of fire (primary and secondary).
  - Emplace aiming and limiting stakes.
  - Emplace grazing fire logs or sandbags.
  - Decide whether to build OHC up or down, based on potential enemy observation of position.
- Prepare:
  - Scoop out elbow holes.
  - Trace position outline.
  - Clear primary and secondary fields of fire.
- Inspect:
  - Site location tactically sound.
  - Low profile maintained.
  - OHC material requirements identified according to paragraphs 5-260 to 5-266.

Stage 2 – Place Supports for Overhead Cover Stringers and Construct Parapet Retaining Walls (H+.5 to H+1.5 Hours)

5-237. During stage 2, Soldiers place supports for OHC stringers and construct parapet retaining walls. Stringers are long horizontal timbers, beams, poles, or pickets to support the roof (see table 5-6 on page 5-94). The parapet retaining wall supports the stringers and protects the front and rear edges of the roof. For an example, see figure 5-27 on page 5-88.

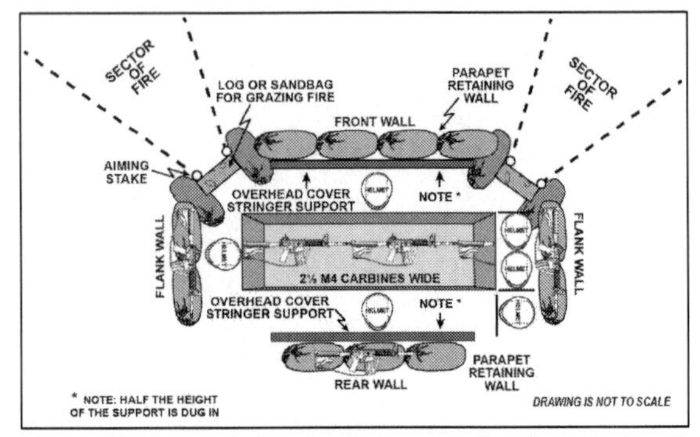

Figure 5-27. Placement of overhead cover supports and construction of retaining walls

5-238. Leader's checklist—

- Prepare:
  - OHC supports to front and rear of position:
    1. Ensure at least 1 foot (about 1 helmet length) or 1/4 cut, whichever is greater, distance from the edge of the hole to the beginning of the supports needed for the OHC.
    2. If logs or cut timber are used, secure in place with strong stakes about 2 to 3 inches in diameter and 18 inches long (short U-shaped picket will work).
    3. Dig in stringer support (front and rear) about 1/2 the height.
  - Front retaining wall–at least 10 inches high (2 filled sandbags in-depth) and approximately 2 and 1/3 M4s/M7s long.
  - Rear retaining walls–at least 10 inches high and 1 and 1/3 M4s/M7s long.
  - Flank retaining walls:
    1. At least 10 inches high and 1 and 1/3 M4s/M7s long.
    2. Start digging hole; use soil to fill sandbags for walls.
- Inspect preparations:
  - Setback for OHC supports–minimum of 1 foot or 1/4 depth of cut.
  - Front retaining wall–at least 10 inches high (2 filled sandbags in-depth) and approximately 2 and 1/3 M4s/M7s long.
  - Rear retaining walls–at least 10 inches high and 1 andl/3 M4s/M7s long.
  - Flank retaining walls–at least 10 inches high and 1 and 1/3 M4s/M7s long.

Stage 3 – Dig Position, Build Parapets, and Place Stringers for Overhead Cover (H+1.5 to H+6 Hours)

5-239. During stage 3, Soldiers dig position, build parapets, and place stringers. For an example, see figure 5-28.

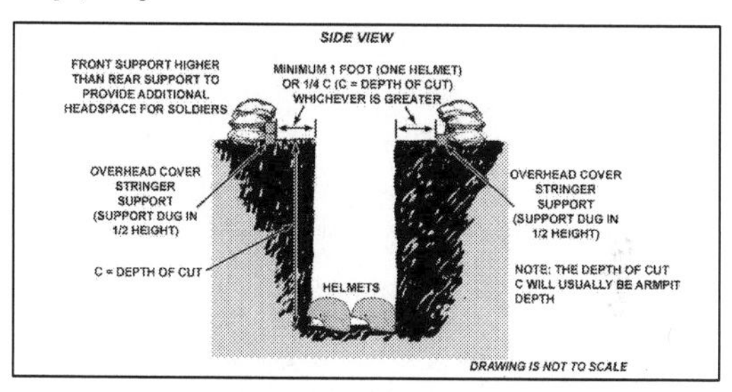

Figure 5-28. Dig position and build supports

5-240. Leader's checklist—

- Prepare:
  - Dig position:
  1. Maximum depth is armpit deep (if soil condition permits).
  2. Use spoil from hole to fill parapets in order of front, flanks, and rear.
  3. Dig two grenade sumps in the floor (one on each end). If the enemy throws a grenade into the hole, kick or throw it into one of the sumps. The sump will absorb most of the blast. The rest of the blast will be directed straight up and out of the hole. Dig the grenade sumps as wide as the entrenching tool blade; at least as deep as an entrenching tool and as long as the position floor is wide (see figure 5-29 on page 5-90).
  4. Dig a storage compartment in the bottom of the back wall; the size of the compartment depends on the amount of equipment and ammunition to be stored (see figure 5-30 on page 5-90).

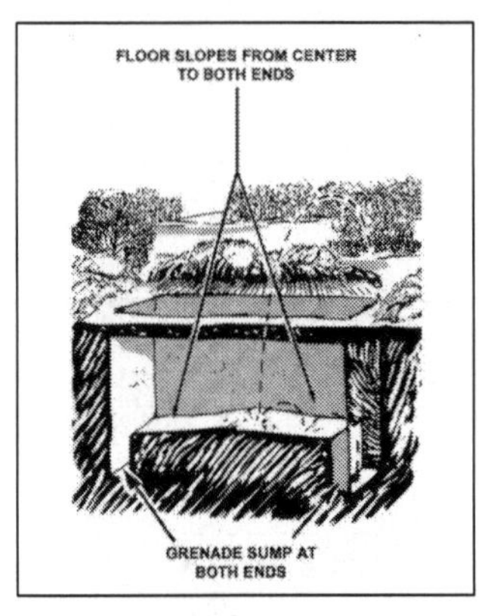

Figure 5-29. Grenade sumps

Figure 5-30. Storage compartment

5. Dig walls vertically.
6. If site soil properties cause unstable soil conditions, construct revetments, and consider sloping walls.
7. Sloped walls: First dig vertical hole and then slope walls at a 1:4 ratio (move 1 foot horizontally for each 4 feet vertically).

- Install revetments to prevent wall collapse and/or cave-in:

1. Required in unstable soil conditions.
2. Use plywood or sheeting material and pickets to revet walls.
3. Tie back pickets and/or posts.

- Place OHC stringers (see figure 5-31, reference paragraph 5-261):

1. Stringer: 2 inches x 4 inches, 4 inches x 4 inches, or pickets (U-facing down).
2. Standard OHC stringer length 8 feet (allows for length if sloping of walls occurs).
3. L equals stringer length and H equals stringer spacing.
4. Second layer of sandbags in front and rear retaining walls removed to place stringers. Replace these sandbags on top of stringers once stringers are properly positioned.

- Inspect:
  - Stringers firmly rest on structural support.
  - Stringer spacing based on values found on paragraph 5-261.
  - Lateral bracing placed between stringers at OHC supports.
  - Revetments built in unstable soil to prevent wall cave-in; walls sloped if needed.

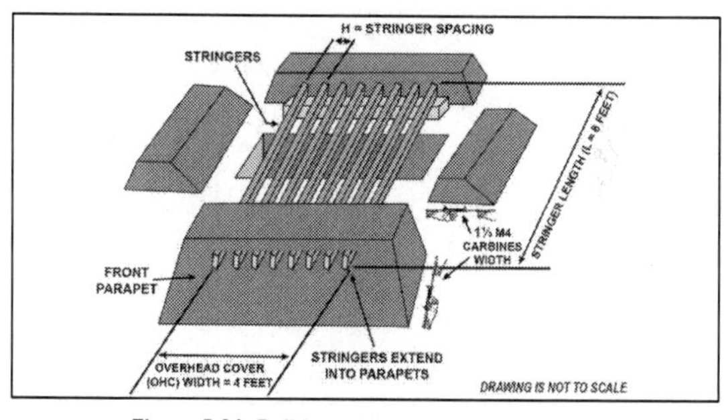

Figure 5-31. Build parapets and place stringers

Stage 4 – Install Overhead Cover and Camouflage (H+6 to H+11 Hours)

5-241. During stage 4, Soldiers install OHC and camouflage (they continue to re-camouflage throughout occupation as required) their position. See figure 5-32.

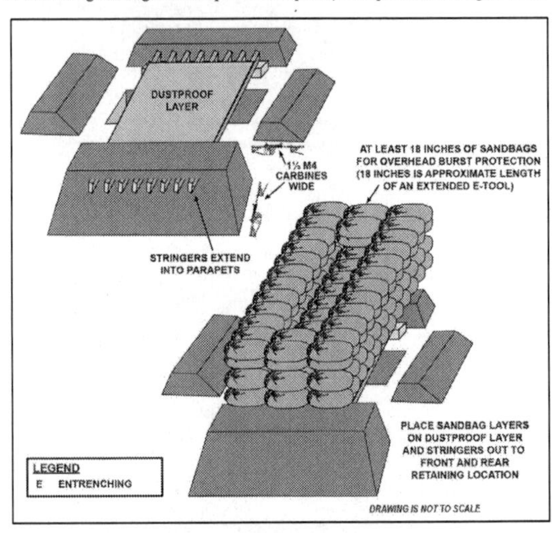

Figure 5-32. Install built-up overheard cover

5-242. Leader's checklist—

- Prepare:
  - Install OHC:

1. Use plywood, sheeting mat, or fighting position cover for dustproof layer (could be boxes, plastic panel, interlocked U-shaped pickets. Standard dustproof layer is 4 feet by 4 feet sheets of 3/4-inch plywood centered over dug position.
2. Nail plywood dustproof layer to stringers.
3. Use minimum of 18 inches of sand-filled sandbags for overhead burst protection (4 layers). As a minimum, these sandbags must cover an area that extends to the sandbags used for the front and rear retaining walls.
4. Use plastic or a poncho for waterproofing layer.

5. Fill center cavity with soil dug from hole and surrounding soil (see figure 5-33).
   - Camouflage position—use surrounding topsoil and camouflage screen systems.
- Inspect:
  - Dustproof layer-plywood or panels.
  - Sandbags filled 75 percent capacity.
  - Burst layer of filled sandbags at least 18 inches deep.
  - Waterproof layer in place.
  - Camouflage in place.
  - Position undetectable at 35 meters.
  - Soil used to form parapets, used to fill cavity, or spread to blend with surrounding ground.

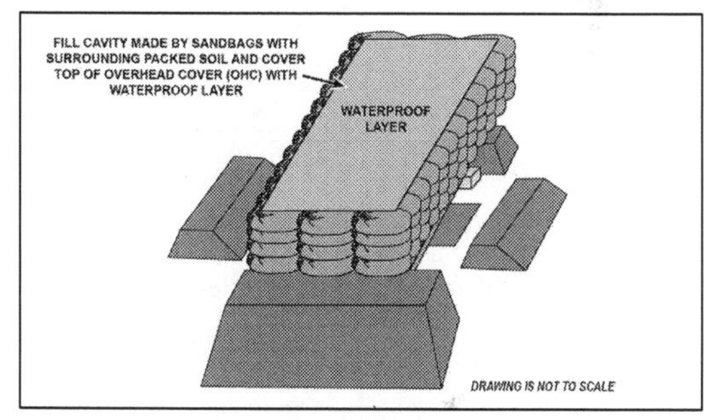

Figure 5-33. Two-Soldier fighting position with built-up overhead cover

5-243. Table 5-6 on page 5-94 lists a summary of considerations for the construction of a deliberate fighting position.

Table 5-6. Considerations for construction

DO	DO NOT
Construct to standard.	Fail to supervise.
Ensure adequate material is available.	Use sand for structural support.
Dig down as much as possible.	Use sandbags for structural support of a position.
Maintain, repair, and improve positions continuously.	Overfill sandbags.
Inspect and test position safety daily, after heavy rain, and after receiving direct and indirect fires.	Put Soldiers in marginally safe positions.
Revet walls in unstable and sandy soil.	Take short cuts.
Revet walls if staying in positions for extended time.	Build above ground unless necessary.
Interlock sandbags for double wall construction and corners.	Forget lateral bracing on stringers.
Check stabilization of wall bases.	Forget to camouflage.
Fill sandbags about 75 percent.	Drive vehicles within 6 feet.
Use common sense.	
Use soil to fill sandbags, fill overhead cover cavity, or spread to blend with surroundings.	

Two-Soldier Fighting Position with Built-Down Overhead Cover

5-244. A two-Soldier fighting position with built-down OHC lowers the profile of the fighting position (see figure 5-34), which aids in avoiding detection. Leader's checklist remains the same as stated previously for a built-up OHC fighting position. However, there are three major differences and/or concerns as stated in paragraphs 5-245 through 5-247.

5-245. Maximum height. The maximum height above ground for a built-down OHC should not exceed 12 inches. Parapets may be used up to a maximum height of 12 inches. Leaders must ensure that Soldiers taper OHC portions and parapets above ground surface to conform to the natural lay of the ground.

5-246. Adequate fighting space. The position is a minimum of 3 and 1/2 M4s/M7s in length. This provides Soldiers adequate fighting space between the end walls of the fighting position and the built-down OHC. This requires an additional 2.5 hours to dig.

5-247. Firing platform construction. When firing, Soldiers must construct a firing platform in the natural terrain upon which to rest their elbows. They position the firing platform to allow the use of the natural ground surface as a grazing fire platform.

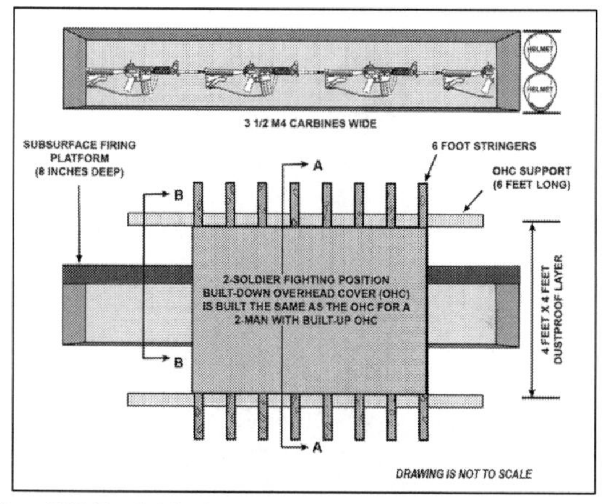

Figure 5-34. Two-Soldier fighting position with built-down overhead cover (overhead view)

5-248. Figure 5-35 on page 5-96 view A depicts a fighting position in which no parapets are used, and the OHC is flush with the existing ground surface. View B depicts a Soldier utilizing a firing platform.

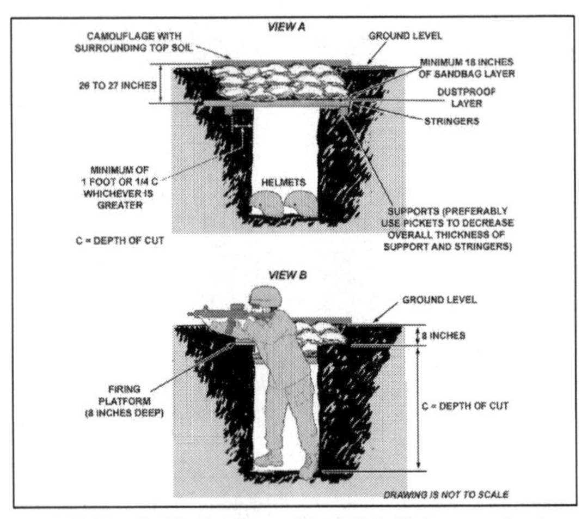

Figure 5-35. Two-Soldier fighting position with built-down overhead cover (side view)

One-Soldier Fighting Position

5-249. When a one-Soldier fighting position is required, (for example, for an ammunition bearer in a machine gun team), except for its size, a one-Soldier position is built the same way as a two-Soldier fighting position. The hole of a one-Soldier position is only large enough for the Soldier and personal equipment. The one-Soldier fighting position does not have the security of a two-Soldier position; therefore, it must allow a Soldier to shoot to the front or oblique from behind frontal cover.

Machine Gun Fighting Position

5-250. Machine gun fighting positions are usually positioned on a flank, so the gun's designated FPL lies obliquely to the position, firing across the platoon's front (see figure 5-36). When terrain does not support an FPL, the gun is laid on a PDF. A PDF or FPL is always the primary sector of fire and is fired from the tripod. To establish the orientation of the machine gun fighting position, the crew first positions the tripod on the designated PDF or FPL and then identifies the secondary sector of fire. The fighting position is then built around these two points.

5-251. Two Soldiers (gunner and assistant gunner) are required to work the weapon system. Therefore, the hole is shaped so both the gunner and assistant gunner can get to the gun and fire it from either side of the frontal protection. The gun's height is reduced by digging the tripod platform down as much as possible. However, the platform is dug to keep the gun traversable across the entire sector of fire. The tripod is used on the side with the primary sector of fire, and the bipod legs are used on the side with the secondary sector. When changing from primary to secondary sectors, the machine gun is moved but the tripod stays in place. When a machine gun has only one sector of fire, the team digs only half of the position.

5-252. With a three-Soldier crew for a machine gun, the (ammunition bearer) digs a one-Soldier fighting position (see paragraph 5-249) to the flank. From this position, the Soldier can see and shoot to the front and oblique. Usually, the one-Soldier position is on the same side of the machine gun as its FPL or PDF. From that position, the Soldier can observe and fire into the machine gun's secondary sector and, at the same time, see the gunner and assistant gunner. The ammunition bearer's position is connected to the machine gun position by a crawl trench so that the ammunition bearer can bring ammunition to the gun or replace the gunner or the assistant gunner.

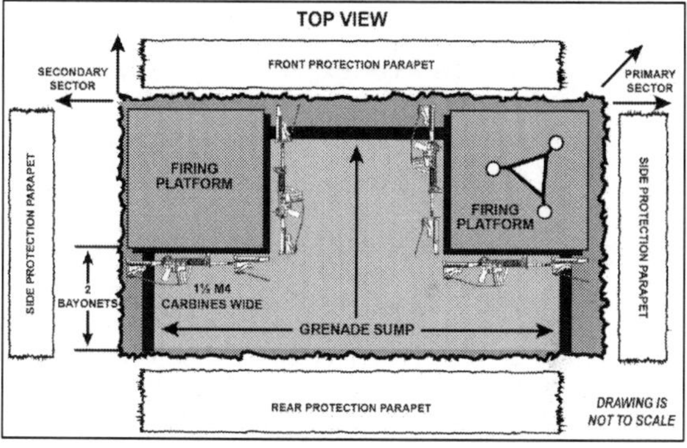

Figure 5-36. Machine gun fighting position with firing platforms

Stage 1 – Establish Sectors (Primary and Secondary) and Outline Position (H+0 to H+.5 Hours)

5-253. Leader's checklist—

- Site position:
  - Check fields of fire from prone.
  - Assign sectors of fire (primary and secondary) and FPL or PDF.
  - Emplace aiming stakes.
  - Emplace grazing fire logs or sandbags.
  - Decide whether to build OHC up or down based on potential enemy observation of position.
- Prepare:
  - Mark position of the tripod legs where the gun can be laid on the FPL or PDF.
  - Trace position outline to include location of two distinct firing platforms.
  - Clear primary and secondary fields of fire.
- Inspect:
  - Site location tactically sound.
  - Low profile maintained.
  - OHC material requirements identified according to paragraphs 5-261 to 5-265.

Stage 2 – Dig Firing Platforms and/or Place Supports for Overhead Cover Stringer and/or Construct Parapet Retaining Walls (H + .5 to H + 1.0 Hours)

5-254. Leader's checklist—

- Prepare:
  - OHC supports to front and rear of position-OHC centered in position and supports placed according to steps established for stage 2 of the two-Soldier fighting position.
  - Rear retaining walls–Construction steps same as those established for stage 2 of the two-Soldier fighting position.
  - Start digging hole:
    1. First dig firing platforms 6 to 8 inches deep and then position machine gun to cover primary sector of fire.
    2. Use soil to fill sandbags for walls.
- Inspect:
  - Setback for OHC supports–minimum of 1 foot or 1/4 depth of cut.
  - Dig firing platforms to a depth of 6 to 8 inches.
  - Before moving to stage 3, position the machine gun to cover sector of fire.
  - Ensure machine gun can traverse sector freely.

Stage 3 – Dig Position, Build Parapets, and Place Stringers for Overhead Cover (H+1.0 to H+7 Hours)

5-255. Leader's checklist—

- Prepare:
  - Dig position:
    1. Complete digging the position to a maximum depth if armpit deep around the firing platforms.
    2. Use soil from hole to fill parapets in order of front, flanks, and rear.
    3. Dig grenade sumps and slope floor toward them.
  - Install revetments to prevent wall collapse and/or cave-in if needed–follow same steps established for two-Soldier fighting position.
  - Place OHC stringers (see figure 5-31 on page 5-91, reference paragraph 5-240):
    1. Follow same steps established for two-Soldier fighting position.
    2. Minimum stringer length is 8 feet.
- Inspect:
  - Depth of cut is at least armpit deep (if soil conditions permit).
  - If unstable soil conditions exist, then revet walls.
  - Stringers firmly rest on structural support.
  - Lateral bracing of stringers placed between stringers at OHC supports.

Stage 4 – Install Overhead Cover and Camouflage (H+7 to H+12 Hours)

5-256. Leader's checklist—

- Prepare:
  - Install OHC—for a machine gun position that is built the same as OHC for a two-Soldier fighting position.
  - Camouflage position:
    1. Use surrounding topsoil and camouflage screen systems.
    2. Ensure no enemy observation within 35 meters of position.
    3. Use soil from hole to fill sandbags, OHC cavity, or spread to blend in with the surrounding ground.
- Inspect:
  - Dustproof layer—plywood or panels.
  - Sandbags filled 75 percent capacity.
  - Burst layer of filled sandbags at least 18-inches deep.
  - Waterproof layer in place.
  - Camouflage in place.
  - Position undetectable at 35 meters.
  - Soil used to form parapets, used to fill cavity, or spread to blend with surrounding ground.

Close-Combat Missile Fighting Position

5-257. The CCMS available to the Infantry rifle platoon is the Javelin Missile System (medium). The organization of Javelin CCMSs within the Infantry rifle platoon includes two Javelins teams within each Infantry weapons squad of the platoon. The AT4 CCMS system is available within each Infantry rifle squad. Paragraphs 5-258 and 5-259 discuss close combat missile fighting positions for the AT4 and Javelin.

AT4 Position

5-258. The AT4 is fired from any of the previously described fighting positions. However, all Soldiers must be alert to the hazards posed by backblast, particularly in terms of positioning fighting positions in depth, when firing from a two-Soldier position, or the risk of backblast deflection from walls, parapets, large trees, or other objects to the rear of the weapon. The front edge of a fighting position is a good elbow rest to help the Soldier steady the weapon and gain accuracy. Stability is better if the body is leaning against the position's front or sidewall.

Standard Javelin Fighting Position with Overhead Cover

5-259. The standard Javelin fighting position has cover to protect Soldiers from direct and indirect fires (see figure 5-37). The position is prepared the same as the two-Soldier fighting position with two additional steps. First, the back wall of the position is extended and sloped rearward, which serves as storage area. Secondly, the front and side parapets are extended twice the length as the dimensions of the two-Soldier fighting position with the Javelin's primary and secondary seated firing platforms added to both sides.

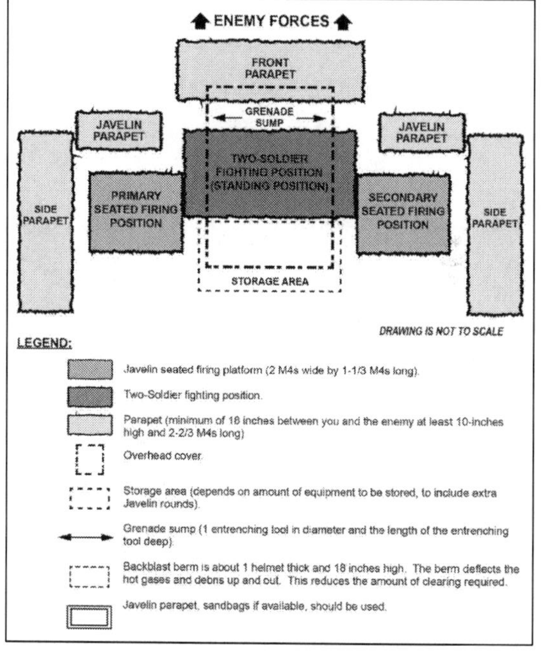

Figure 5-37. Standard Javelin fighting position

Note. When a Javelin is fired, the muzzle end extends 6 inches beyond the front of the position, and the rear launcher extends out over the rear of the position. As the missile leaves the launcher, stabilizing fins unfold. Soldiers must keep the weapon at least 6 inches above the ground when firing to leave room for the fins. OHC that would allow firing from beneath it is usually built if the backblast area is clear.

CLASS IV MATERIAL REQUIREMENTS FOR STANDARD OVERHEAD COVER

5-260. The Infantry rifle platoons works within the plan and assigned priorities of work to maximize the timely completion of all fighting positions. As the platoon begins preparations, the battalion establishes combat and field trains and begins to work the defensive logistical concerns that will assist the establishment of the defense, and the coordination with company trains to meet the platoon's Class IV requirements (see paragraphs 5-261 to 5-266).

Note. Typically, each Soldier carries five to seven sandbags which is the starting set for building up the position.

5-261. Lateral supports and stringers:

- Cut timber:
  - 4 inches by 4 inches by 6 feet; Used for supports (need total of 3 each [2 inches front and 1 in rear]).
  - 4 inches by 4 inches by 8 feet; Used for stringers (need total of 9).
- Stringers spaced 6 inches on center.
- Single U-shaped picket beams:
  - Use pickets or cut timber for supports (6-foot length).
  - If used for supports, need 6 each 6-foot pickets for supports (3 front and 3 rear).
  - Need 11 each 8-foot pickets for stringers spaced 5 inches on center; open side facing down.

5-262. Dustproof layer:

- One (1) sheet of 3/4- or 1-inch plywood.
- (4 feet by 4 feet) for two-Soldier position and two sheets (4 feet by 4 feet) for machine gun position.
- See summary of materials (reference paragraph 5-266) for other materials to be used.

5-263. Sandbags:

- Front retaining wall, 12 each.
- Flank and rear retaining walls, 6 each times 3 equals 18 each.
- OHC: 76 each sandbag completely covers first layer while next 3 layers form perimeter only. These 3 layers form a cavity to be filled with surrounding soil to meet minimum 18-inch cover. Reference figure 5-32 on page 5-92 to visualize this process. Plan on one 75 percent full sandbag to be 10 inches by 15 inches by 5 inches.
- Position needs (limiting stakes and grazing fire, for example) 10 each.
- Total sandbags equal 116 each.

5-264. Waterproofing. Use poncho or plastic sheeting.

5-265. Revetments. 4 feet by 6 feet sheets of plywood and 3 each 6-foot U-shaped pickets for each front and rear wall.

5-266. In a combat situation, the Soldier may need to improvise construction of a survivability position by using materials not normally associated with the construction. Examples of field-expedient construction materials are listed in table 5-7.

Table 5-7. Field-expedient material

	Wall Revetment		*Wall Construction (Building Up)*
	Sheet metal.		55-gallon drums filled with sand.
	Corrugated sheet.		Shipping boxes/packing material.
	Plastic sheeting.		Expended artillery shells filled with sand.
	Plywood.		Prefabricated concrete panels.
	Air mat panels.		Prefabricated concrete traffic barriers.
	Air force air load pallets.		Sand grid material.
	Overhead Cover Stringers		*Stand-alone Positions*
	Single pickets.		Prefabricated concrete catch basins.
	Double pickets.		Military vans.
	Railroad rails.		Conex or shipping containers.
	"I" beams.		Large diameter pipe/culvert.
	2" (inch)-diameter pipe and larger.		Steel water tanks.
	Timbers 2 inches by 4 inches, 4 inches by 4 inches, and larger.		Vehicle hulks.
	Reinforced concrete beams.		Prefabricated concrete catch basins.
	55-gallon drums cut longitudinally in half.		
	Culverts cut in half.		*Aiming Stakes*
	Precast concrete panels 6 to 8 inches thick.		2-foot pickets.
	Airfield panels.		Wooden tent poles.
	Single pickets.		
			Limiting Stakes
			2-foot pickets.
			Wooden tent poles.
			Filled sandbags.

Chapter 6

Tactical Enabling Operations and Activities

As in all operations, Infantry platoons and squads conduct a variety of tactical *enabling operations*—an operation that sets the friendly conditions required for mission accomplishment (FM 3-90)—and activities in support of offensive and defensive operations. This chapter addresses breaching, relief in place, passage of lines, linkup tactical deception, security, security operations, and troop movement. (See chapter 7 section IV for additional information on reconnaissance operations.)

SECTION I – BREACHING

6-1. A *breach* is a tactical mission task in which a unit breaks through or establishes a passage through an enemy obstacle (FM 3-90). As a tactical mission task, a breach is an action by a friendly force conducted to allow maneuver despite the presence of obstacles. During maneuver, the commander attempts to bypass and avoid obstacles and enemy defensive positions to the maximum extent possible to maintain tempo and momentum. Breaching enemy defenses and obstacle systems is normally the last choice to prevent loss of personnel, time, and/or equipment. A breach is a synchronized operation under the control of the maneuver commander in contact with the obstacle or enemy defense. The breach begins when friendly forces detect an obstacle and begin to apply the breaching fundamentals and ends when battle handover has occurred between follow-on forces and a unit conducting the breach. See appendix E, Battle Drill–Breach a Mined Wire Obstacle for information on the tactical mission task-breach.

Note. A *tactical mission task* is the specific activity a unit performs while executing a tactical operation or form of maneuver (FM 3-90). A breach as defined as a tactical mission task is an action by a friendly force.

6-2. A *breach* is a synchronized combined arms activity under the control of the maneuver commander conducted to allow maneuver through an obstacle (ATP 3-90.4). The breaching tenets (intelligence, breaching fundamentals, breaching organization, mass, synchronization) apply when conducting a breach, whether it is a hasty or deliberate operation. Breaching activities include the *reduction*—the creation of lanes through a minefield or obstacle to enable passage of the attacking ground force (JP 3-15)—of minefields, other explosive hazards, and other obstacles. Breaching at battalion and above (and to a lesser degree at company level) require significant combat engineering support to accomplish. (See ATP 3-21.10 and ATP 3-90.4 for additional information.)

BREACHING FUNDAMENTALS

6-3. Suppress, obscure, secure, reduce, and assault are the breaching fundamentals applied to ensure success when breaching against a defending enemy. These obstacle reduction fundamentals will always apply, but they may vary based on the mission variables of METT-TC (I) mission, enemy, terrain and weather, troops and support available, time available, civil considerations, and informational considerations.

SUPPRESS

6-4. Suppression is a tactical task used to employ direct or indirect fires on enemy personnel, weapons, or equipment to prevent or degrade enemy fires and observation of friendly forces. During breaching operations, the initial purpose of suppression is to enable the approach of the breaching force to the obstacle. Then it is to protect, for example, the platoon and/or squad reducing and maneuvering through an obstacle. Suppression is a mission-critical task performed during breaching operations. Achieving suppression is usually the required condition that triggers subsequent breaching actions. Fire control measures ensure all fires are synchronized with other actions supporting the platoon's breach of the obstacle. Although suppressing the enemy overwatching the obstacle is the mission of the support force, the breach force should always plan to provide additional (local) suppression against an enemy the supporting force cannot suppress. Effective suppression serves to isolate enemy forces in the *breach area*—a defined area where a breach occurs (ATP 3-90.4). The breach area includes the point of breach, reduction area, far side objective, and point of penetration. (See figure 6-1.)

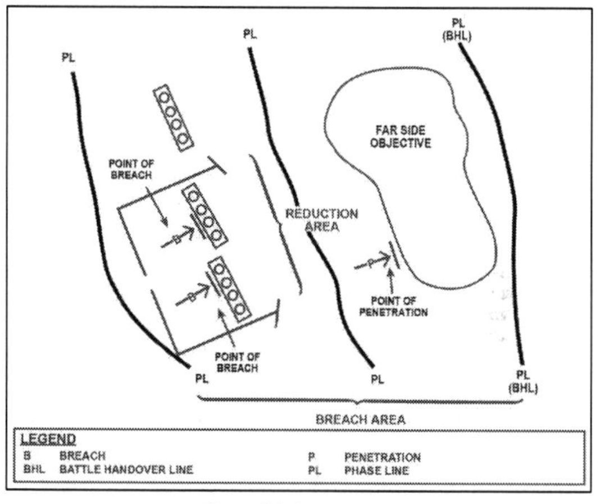

Figure 6-1. Breach area

OBSCURE

6-5. Obscuration must be employed to protect forces conducting obstacle reduction and passage of assault forces. Obscuration hampers enemy observation and target acquisition by concealing friendly activities and movement. Obscuration smoke deployed on or near the enemy's position minimizes its vision. Screening obscurants employed between the *reduction area*—a number of adjacent points of breach that are under the control of the breaching commander (ATP 3-90.4)—and the enemy conceals movement and reduction activities. It also degrades enemy ground and aerial observations. Obscuration must be planned carefully to provide maximum degradation of enemy observation and fires, but it must not degrade friendly fires and control significantly.

SECURE

6-6. Friendly forces secure reduction areas to prevent the enemy from interfering with obstacle reduction and the passage of the assault force through lanes created during the reduction. Security must be effective against outposts and fighting positions near the obstacle and against overwatching units as necessary. The far side of the obstacle must be secured by fires or be occupied before attempting efforts to reduce the obstacle. The attacking unit's higher headquarters is responsible for isolating the breach area by fixing adjacent units, attacking enemy reserves in-depth, and providing counterfire support.

6-7. Identifying the extent of the enemy's defenses is critical before selecting the appropriate technique to secure the *point of breach*—the location at an obstacle where the creation of a lane is being attempted (ATP 3-90.4). If the enemy controls the point of breach and cannot be suppressed adequately, the force must secure the point of breach before it can reduce the obstacle.

6-8. The breach force must be resourced with enough maneuver assets to provide local security against the enemy forces the supporting force cannot engage sufficiently. Elements within the breach force securing the reduction area also may be used to suppress the enemy once reduction is complete. The breach force also may need to assault to the far side of the breach and provide local security so the assault element can seize its initial objective.

Reduce

6-9. Breaching forces reduce an obstacle to create and mark lanes through, over, or around that obstacle to allow the attacking force to accomplish its mission. The number and width of lanes created varies with the enemy situation, the assault force's size, composition, and scheme of maneuver. The lanes must allow the assault force to pass through the obstacle rapidly. The breach force will reduce, proof (if required), mark, and report lane locations and lane-marking method (if changed from the predetermined method) to higher command headquarters. Follow-on units will reduce or clear the obstacle when required. Reduction cannot be accomplished until suppression and obscuration are in place, the obstacle has been identified, and the point of breach is secure.

Assault

6-10. The assault force's primary mission is to seize terrain on the far side of the obstacle to prevent the enemy from placing or observing direct and indirect fires on the reduction area. If planned, the battle handover with follow-on forces occurs.

BREACHING ORGANIZATION

6-11. The platoon leader, when tasked, organizes the platoon to accomplish breaching fundamentals quickly and effectively. This generally requires the leader to organize as either one or part of the support, breach, or assault force with the necessary assets to accomplish its role.

SUPPORT FORCE

6-12. The support force's primary responsibility is to eliminate the enemy's ability to place direct or indirect fire on friendly forces in the reduction area. The support force must—

- Isolate the reduction area with fires and establish a support by fire position to destroy, fix, or suppress the enemy. Depending on METT-TC (I), this may be the weapons squad or the entire platoon.
- Mass-control direct and indirect fires to suppress the enemy and to neutralize weapons able to bring fires on the breach force.
- Control obscuring smoke to prevent enemy-observed direct and indirect fires.

BREACH FORCE

6-13. The breach force assists in the passage of the assault force by detecting, creating, proofing (if necessary), marking and reporting lanes. The breach force is a combined-arms force. It may include engineers, reduction assets, and enough maneuver forces to provide additional suppression and local security. The entire Infantry platoon or squad may be part of the breach force. The breach force may apply portions of the following breaching fundamentals as it reduces an obstacle.

Suppress

6-14. The breach force must be allocated enough maneuver forces to provide additional suppression against various threats, including:

- Enemy direct-fire systems that cannot be observed and suppressed by the support force due to the terrain or the masking of the support force's fires by the breach force as it moves forward to reduce the obstacle.
- Counterattacking and/or repositioning forces that cannot be engaged by the support force.

Obscure

6-15. When required, the breach force employs smoke pots, vehicle mounted smoke, handheld smoke, or indirect fire obscurants. These type obscurants provide for self-defense and cover lanes as the assault force passes through them. As the breach force starts their breach, they generally assume control of obscuration fires from the support force as they are in a better condition to assess its effects.

Secure

6-16. The breach force secures itself from threat forces providing close in protection of the obstacle. The breach force also secures the lanes through the tactical obstacles once they are created to allow safe passage of the assault force.

Reduce

6-17. The breach force performs its primary mission by reducing the obstacle. To support the development of a plan to reduce the obstacle, the composition of the obstacle system must be an information requirement. If the obstacles are formidable, Infantry platoons and squads will be augmented with engineers to conduct reduction. Without engineers and special equipment such as Bangalore torpedoes and line charges, minefields must be probed.

ASSAULT FORCE

6-18. The breach force assaults through the point of breach to the far side of an obstacle and seizes the foothold. The assault force's primary mission is to destroy the enemy and seize terrain on the far side of the obstacle to prevent the enemy from placing direct fires on the created lanes. The assault force may be tasked to assist the support force with suppression while the breach force reduces the obstacle.

6-19. The assault force must be sufficient in size to seize the *point of penetration*—the location, identified on the ground, where the commanders concentrate their efforts to seize a foothold on the far side objective (ATP 3-90.4). Combat power is allocated to the assault force to achieve a minimum 3:1 ratio on the point of penetration. The breach and assault assets may maneuver as a single force when conducting breaching operations as an independent company team conducting an attack.

6-20. If the obstacle is defended by a small enemy force, assault and breach forces' missions may be combined. This simplifies command and control and provides more immediate combat power for security and suppression.

6-21. Fire control measures are essential because support and breach forces may be firing on the enemy when the assault force is committed. Suppression of overwatching enemy positions must continue, and other enemy forces must remain fixed by fires until the enemy has been destroyed. The assault force must assume control for direct fires on the assault objective as support and breach force fires are ceased or shifted. Table 6-1 illustrates the relationship between the breaching organization and breaching fundamentals.

Table 6-1. Relationship between breaching organization and breaching fundamentals

Breaching Organization	*Breaching Fundamentals*	*Responsibilities*
Support force	Suppress. Obscure.	Suppress enemy direct fire systems covering the reduction area. Initiates obscuring smoke. Prevent enemy forces from repositioning or counterattacking to place direct fires on the breach force.
Breach force	Suppress (provides additional suppression). Obscure (provides additional obscuration in the reduction area). Secure (provides local security). Reduce.	Refines/controls obscuration. Create, proof, and mark the necessary lanes in an obstacle. Secure the near side and far side of an obstacle. Defeat forces placing immediate direct fires on the reduction area. Report the lane status/location.
Assault force	Assault. Suppress (if necessary).	Destroy the enemy on the far side of an obstacle if the enemy is capable of placing direct fires on the reduction area. Assist the support force with suppression if the enemy is not suppressed. Be prepared to breach follow-on and/or protective obstacles after passing through the reduction area.

DETAILED REVERSE PLANNING

6-22. The platoon leader along with the platoon sergeant and squad leaders develop the breaching plan using the reverse planning sequence in figure 6-2 on page 6-8, when planning for a protective obstacle breach. The platoon leader can plan to breach wire, minefields, trenches, and craters. The following considerations must be made:

- Reverse planning begins with actions on the objective.
- Actions on the objective begin from the designated point of penetration.
- The planned point of penetration determines the point of breach.
- The size and composition of support, breach, and assault forces are based on the results of reverse planning.
  - Actions on the objective drive the size and composition of the assault force.
  - The ability of the enemy to interfere with the reduction of the obstacle determines the size and composition of the security element in the breach force.
  - The ability of the enemy to mass fires on the point of breach determines the amount of suppression and size and composition of the support force.

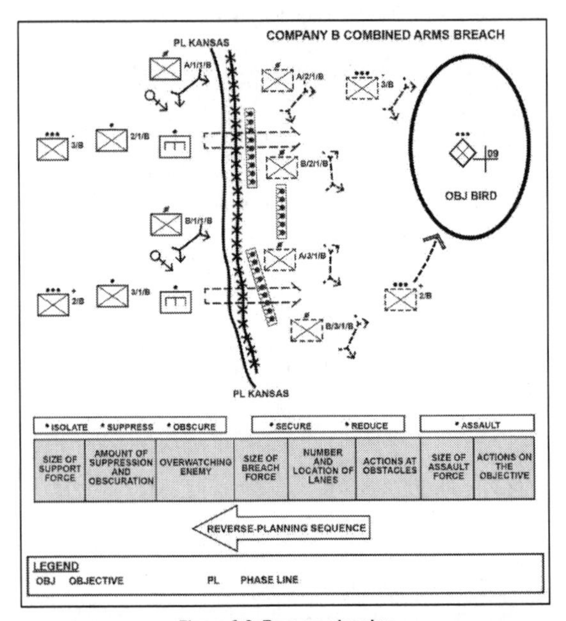

Figure 6-2. Reverse planning

SECTION II – RELIEF IN PLACE

6-23. A *relief in place* is an operation in which, by direction of higher authority, all or part of a unit is replaced in an area by the incoming unit and the responsibilities of the replaced elements for the mission and the assigned zone of operations (Army uses the term, area of operations) are transferred to the incoming unit (JP 3-07.3). The incoming unit continues the operations as ordered. Normally, the Infantry rifle platoon conducts a relief in place as part of a larger operation, primarily to maintain the combat effectiveness of committed forces. (See ATP 3-21.10 for additional information.)

VARIATIONS OF RELIEF IN PLACE

6-24. The commander directing the relief, in coordination with subordinate elements during planning, determine the most appropriate variation of a relief in place. The three variations of a relief are sequential, simultaneous, or staggered.

SEQUENTIAL

6-25. A *sequential relief in place* occurs when each element within the relieved unit is relieved in succession, from right to left or left to right, depending on how it is deployed (ADP 3-90). This technique is the most deliberate and time-consuming; however, it minimizes confusion and maintains the best command and control, and readiness posture. A sequential relief involves sequentially relieving maneuver platoons of the company one at a time. Separate routes to the rear of relieved platoons' locations are planned for each platoon (or squad) and placed on the operations overlay. Routes are labeled sequentially and correspond to the order in which the platoon executes them during the relief. When the lead platoon reaches its release point, squads are guided into the positions they will occupy. Crews exchange standard range card and fire support information. Once the relief occurs, relieved units move to the rear to occupy their next location. When the lead platoon is in position, the next platoon moves along its designated route(s) to relieve its counterpart. This process repeats until each platoon has been relieved.

SIMULTANEOUS

6-26. A *simultaneous relief in place* occurs when all elements are relieved at the same time (ADP 3-90). Simultaneous relief takes the least time to execute but is more difficult to control and more easily detected by the enemy. A simultaneous relief involves relieving maneuver platoons of the company at the same time. Separate routes to the rear of the relieved platoons' locations are planned and labeled for each platoon and placed on the operations overlay. When relieving platoons reach their release points, squads are guided to the positions they are occupying. Crews exchange standard range card and fire support information, and the relieved unit then moves to the rear to its next location.

STAGGERED

6-27. A *staggered relief in place* occurs when a commander relieves each element in a sequence determined by the tactical situation, not its geographical orientation (ADP 3-90). As with a sequential relief, staggered reliefs can occur over a significant amount of time. Separate routes to the rear of the relieved platoons' locations are planned and labeled for each platoon and placed on the operations overlay. When relieving platoons reach their release points, squads are guided to the positions they are occupying. Information exchanges and transfers of supplies are the same as in the other two techniques. Once the relief occurs, relieved units move to the rear to occupy their next location. When the first platoon to move is in position, the next platoon is identified to move along its designated route(s) to relieve its counterpart: thereby repeating the relief process. This process repeats until each platoon has been relieved.

PLANNING AND PREPARATION

6-28. Once ordered to conduct a relief in place, the platoon leader of the relieving unit contacts the leader of the unit to be relieved. The collocation of unit command posts also helps achieve the level of coordination required. If the relieved unit's forward elements can defend the area of operations (AO), the relieving unit executes the relief in place from the rear to the front. This facilitates movement and terrain management.

6-29. When planning for a relief in place, the platoon leader takes the following actions:

- Issues a warning order (WARNORD) immediately.
- Accompanies platoon key leaders in the advance party to conduct detailed reconnaissance and coordination.
- As the relieving unit, adopts the outgoing unit's normal pattern of activity as much as possible.
- As the relieving unit, determines when the platoon will assume responsibility for outgoing unit's position.
- As the relieving unit, collocates with the relieved unit's headquarters.
- Maximizes operations security to prevent the enemy from detecting the relief operation, to include minimizing signatures on electromagnetic spectrum.

Note. When possible, conduct the relief at night or under other limited visibility conditions.

- As the unit being relieved, plans for transfer of excess ammunition, wire, and other materiel of tactical value to the incoming unit.
- Controls movement by reconnoitering, designating, and marking routes, and providing guides.

6-30. The incoming and outgoing unit leaders meet to exchange tactical information, conduct a joint reconnaissance of the area, and complete other required coordination. The two leaders carefully address passage of command and jointly develop contingency actions to deal with enemy contact during the relief. This process usually includes coordination of the following information:

- Location of individual fighting positions (to include hide, alternate, and supplementary positions) and vehicles (if attached). Leaders should verify fighting positions both by conventional map and using command and control systems that are available.
- The enemy situation.
- The outgoing unit's tactical plan, including graphics, platoon and company direct and indirect fire plans, sector sketches, and key weapons' standard range cards.

- Direct and indirect fire support coordination, including indirect fire plans and time of relief for supporting artillery and mortar units.
- Types of weapon systems being replaced.
- Time, sequence, and relief technique.
- Location and disposition of obstacles, and time when the leaders will transfer responsibility.
- Supplies and equipment to be transferred.
- Movement control, route priority, and placement of guides.
- Command and signal information.
- Visibility considerations.

Note. Units conduct relief on the radio nets (though limited) of the outgoing unit to facilitate control during the relief.

EXECUTION

6-31. When executing a relief in place, the linkup (see section IV) method chosen prior to the intermediate tactical assembly area (TAA) (when used) usually involves conducting linkup at a predetermined contact point(s). During movement to the contact point(s), the platoon leader monitors the progress and execution of the linkup to ensure that established positive control measures are followed or adjusted as required. Contact points must be readily recognizable and posted on overlays. When possible, the moving force should halt short of the contact point and send a smaller force (patrol or team) forward to pinpoint the contact point. After the patrol makes contact with the force in position at the contact point, the patrol may leave a portion of patrol at the contact point then move back with the remaining members of the patrol to guide the subordinate unit back to conduct linkup. Following linkup, the relieving force is guided to a TAA, or depending on the situation, guided directly into relief position(s). (See figure 6-3 on page 6-12.)

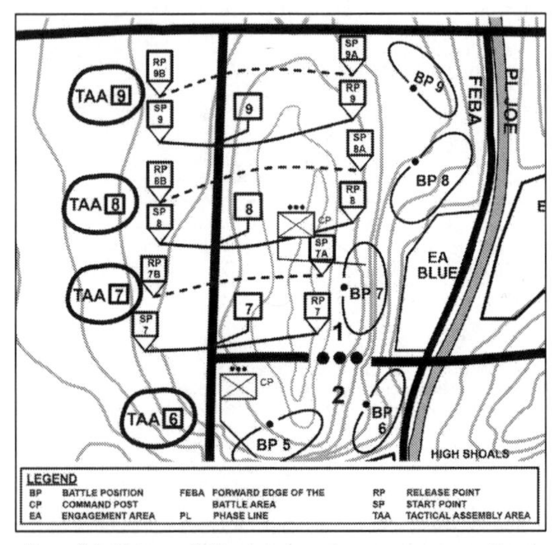

Figure 6-3. Platoon relief in place (squads occupy same positions)

6-32. During the relief, the outgoing leader retains responsibility of the AO and mission. The outgoing leader exercises operational control over those subordinate elements of the incoming unit that have completed their portion of the relief. Responsibility passes to the incoming leader when all elements of the outgoing unit are relieved and adequate communications are established.

SECTION III – PASSAGE OF LINES

6-33. A *passage of lines* is an operation in which a force moves forward or rearward through another force's combat positions with the intention of moving into or out of contact with the enemy (JP 3-18). A unit conducts a passage of lines to continue an attack, conduct a counterattack or retrograde; pass-through security or main battle forces; and anytime one unit cannot bypass another unit's position. A passage of lines may involve engagement with the enemy shortly before or after the completion of the operation, and usually involves a battle handover. (See ATP 3-21.10 for additional information.)

6-34. Passage of lines occur under two basic conditions. A *forward passage of lines* occurs when a unit passes through another unit's positions while moving toward the enemy (ADP 3-90). A *rearward passage of lines* occurs when a unit passes through another unit's positions while moving away from the enemy (ADP 3-90).

PLANNING AND PREPARATION

6-35. A passage of lines is a complex operation requiring close supervision and detailed planning, coordination, and synchronization between the unit conducting the passage and the unit being passed. The headquarters ordering the passage of lines is responsible for planning and coordination; however, specific coordination tasks are normally delegated to subordinate units. Terrain management is critical to successful completion of a passage of lines. At least two units are occupying and concentrated on the same terrain. The commander and subordinate leaders of both units at all levels must understand their respective plans and be flexible in their execution. Terrain is controlled through the sharing of a common operational picture and overlays that contain—

- Assigned/assembly area(s) (AA).
- Attack positions(s).
- Battle handover line (BHL).
- Primary and alternate routes.
- Checkpoint data.
- Friendly and enemy unit locations and status.
- Passage points and lanes.
- Fire support coordination measures.
- Friendly and enemy obstacle types and locations.
- Sustainment locations and descriptions (for example, ambulance exchange points and casualty collection points [CCPs]).
- Contact points.

6-36. A passage of lines may require either the reduction of some obstacles or the opening and closing of lanes through friendly obstacles. The passing unit should task the attached engineer officer if available to coordinate with the stationary unit engineer or stationary commander. At a minimum, this coordination must address the following:

- Location and status of friendly and enemy tactical obstacles.
- Routes and locations of lanes and bypasses through friendly and enemy obstacles.
- Transfer of obstacle and passage lane responsibilities.

6-37. As with any activity involving transferred combat responsibility from one unit to another, the complex nature of a passage of lines involves risk. The passage of lines is either hasty or deliberate. In a hasty passage of lines, leaders use verbal orders. In a deliberate passage of lines, both the stationary and moving force have time to—

- Review fire support observation plan, target execution, communication linkages, and mutual support.
- Confirm fire support coordination measures.

- Review routes and positioning.
- Locations and descriptions of obstacles, lanes, bypasses, and markings.
- Locations of any stockpiles, especially engineer stockpiles.
- Responsibility for closing passage lanes after the passage of lines is complete.
- Air defense weapons locations, early warning communications, air threat, and weapons control status (WCS).
- Passage point recognition procedures.
- Route management, contact points, checkpoints, and use of guides.
- Locations for and movement of sustainment units.
- Locations of aid stations, ambulance exchange points, and casualty evacuation (CASEVAC) procedures.

6-38. Units plan their passages of lines to maintain enemy contact and provide constant fires on enemy forces. They reduce risk and ensure synchronization through detailed planning and centralized execution. The need for positive control increases during the passage because of the intermingling of passing and stationary forces. The passage requires close coordination, clearly understood control measures, and liaison among all headquarters and echelons involved in the passage. Clear identification of the moment or event that causes one force to assume responsibility for the assigned area from another is vital to successfully performing this task. The passage of lines requires effective communication between passing and stationary forces. Units build redundancy of communication signals and means into their passage plans. They also designate contact points to ensure effective communication between these forces at the lowest possible tactical level for example, near and far recognition signals.

6-39. During rehearsals for a passage of lines, passing forces rehearse when and where to move, as well as how to execute required coordination. Passing forces' rehearsals address necessary branches and sequels in case enemy forces attack them during the execution of their passages to prevent degrading friendly maneuver.

FORWARD PASSAGE OF LINES

6-40. In a forward passage, the passing unit first moves to a TAA or an attack position behind the stationary unit. Designated liaison personnel move forward to linkup with guides and confirm coordination information with the stationary unit. Guides then lead the passing elements through the *passage lane*—a lane through an obstacle that provides safe passage for a passing force (FM 3-90).

6-41. The platoon conducts a forward passage by employing tactical movement. It moves quickly, using appropriate dispersal and formations whenever possible, and keeping radio traffic to a minimum. The passing unit holds its fire until it passes the BHL or the designated fire control measure unless the leader has coordinated fire control with the stationary unit. The stationary unit will hold its fire once the passing unit crosses the BHL or designated fire control measure unless the leader has coordinated fire control with the passing unit. Once clear of passage lane restrictions, the platoon consolidates at a rally point or continues conducting tactical movement according to its orders. (See figure 6-4.)

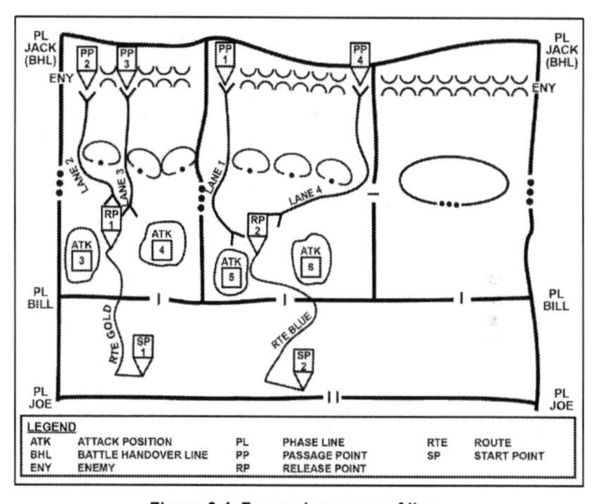

Figure 6-4. Forward passage of lines

REARWARD PASSAGE OF LINES

6-42. Because of the increased chance of fratricide during a rearward passage, coordination of recognition signals and direct fire restrictions are critical. Rehearsals and training can help reduce fratricide. The passing unit contacts the stationary unit while it is still beyond direct fire range and conducts coordination as discussed previously. Near recognition signals and location of the BHL are emphasized. Both passing and stationary units can employ additional fire control measures, such as a restrictive fire line (RFL), to minimize the risk of fratricide. (See figure 6-5 on page 6-16.)

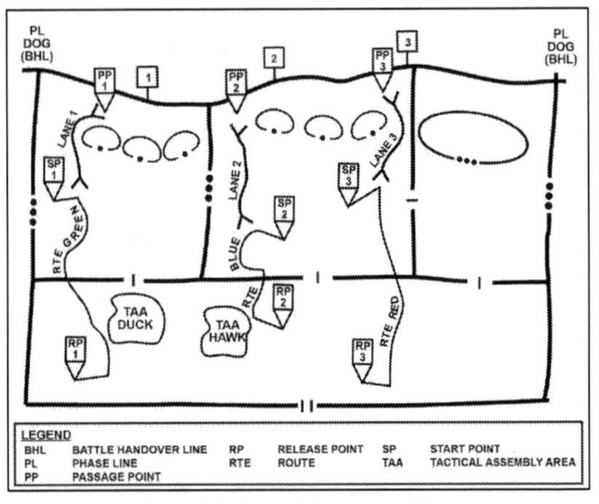

Figure 6-5. Rearward passage of lines

6-43. Following coordination, the passing unit continues tactical movement toward the passage lane. The passing unit is responsible for its security until it passes the BHL. If the stationary unit provides guides, the passing unit can conduct a short halt to linkup at a contact point and coordinate with them. The passing unit moves quickly through the passage lane to a designated location behind the stationary unit. See table 6-2 for stationary unit and passing unit responsibilities.

Table 6-2. Stationary and passing unit responsibilities

Stationary Unit	*Passing Unit*
Clears lanes or reduces obstacles along routes.	May assist with reducing obstacles.
Provides obstacle and friendly units' locations.	Provides order of movement and scheme of maneuver.
Clears and maintains routes up to the battle handover line (BHL).	May assist with maintaining routes.
Provides traffic control for use of routes and lanes.	Augments the traffic control capability of the stationary unit as required.
Provides security for passage up to the BHL.	Maintains protection measures.
Identifies locations for the passing unit to use as a tactical assembly area (TAA) and attack positions.	Reconnoiters from its current location to its designated TAA and attack positions.
Provides the passing unit previously coordinated or emergency logistics assistance within its capability.	Assumes full responsibility for its own sustainment support forward of the BHL.
Controls all fires in support of the passage.	Positions artillery to support the passage.

SECTION IV – LINKUP

6-44. A *linkup* is a type of enabling operation that involves the meeting of friendly ground forces, which occurs in a variety of circumstances (FM 3-90). It happens when an advancing force reaches an objective area previously seized by an airborne or air assault; when an encircled element breaks out to rejoin friendly forces or a force comes to the relief of an encircled force, and when converging maneuver forces meet. Both forces may be moving toward each other, or one may be stationary. (See GTA 07-04-008 [*Infantry Reference Card for Linkup Operations*] for additional information.)

PLANNING AND PREPARATION

6-45. Whenever possible, joining forces exchange as much information as possible before starting an operation. The headquarters ordering the linkup establishes—

- A common operational picture.
- Command relationship and responsibilities of each force before, during, and after linkup.
- Coordination of direct and indirect fire support before, during, and after linkup, including control measures.
- Linkup method.
- Recognition signals and communication procedures to use, including pyrotechnics, armbands, vehicle markings, gun-tube orientation, panels, colored obscurant (smoke), lights, and challenge and passwords.
- Operations to conduct following linkup.

6-46. The higher headquarters who orders the linkup, in coordination with the units linking up, establishes control measures for the linkup—

- Assigns each unit an AO defined by left and right boundaries and an RFL also acts as a limit of advance (LOA).
- Establishes a no fires area around one or both units and establishes a coordinated fire line beyond the area where the unit's linkup.
- Establish no fire areas to prevent delivery of fire support across RFLs or boundaries where they can impact friendly forces.

6-47. The coordinated fire line allows available fires to attack enemy targets quickly as they approach the area where the linkup is to occur. The linkup forces use the linkup points established by the higher headquarters to make physical contact with each other. The higher headquarters designates alternate linkup points since enemy action may interfere with the primary linkup points. Control measures are adjusted during the operation to provide for freedom of action as well as positive control.

EXECUTION

6-48. There are two linkup methods. The preferred method is when the moving force has an assigned LOA near the other force and conducts the linkup at predetermined contact points. Units then coordinate additional operations. The other method is used during highly fluid mobile operations when the enemy force escapes from a potential *encirclement*—where one force loses its freedom of maneuver because an opposing force is able to isolate it by controlling all ground lines of communications and reinforcement (FM 3-90), or when one of the linkup forces is at risk and requires immediate reinforcement. In this method, the moving force continues to move and conduct long-range recognition via radio or other measures, stopping only when it makes physical contact with the other force.

6-49. The platoon and squad can conduct linkup activities independently or as part of a larger force. Within a larger unit, the platoon may lead the linkup force. The linkup consists of three phases. The following actions are critical to the execution of a linkup.

Note. See ATP 3-21.10 and ATP 3-21.20 for information regarding situations unique to the conduct of defensive operations that include denial operations, defending encircled, breakout from encirclement, and stay-behind operations.

PHASE 1 – FAR RECOGNITION SIGNAL

6-50. During this phase, the forces conducting a linkup establish both radio and when possible digital communications before reaching direct fire range. The lead element of each linkup force should monitor the radio frequency of the other friendly force.

Phase 2 – Coordination

6-51. Before initiating movement to the linkup point, the forces must coordinate necessary tactical information including the following:

- The known enemy situation.
- Command and control systems.
- Number of friendly personnel; type and number of friendly vehicles.
- Disposition of stationary forces (if either unit is stationary).
- Routes to the linkup point and rally point, if any.
- Direct and indirect fire control measures.
- Near recognition signals.
- Communications information.
- Sustainment responsibilities and procedures.
- Finalized location of the linkup point and rally points, if any.
- Special coordination, such as those covering maneuver instructions or requests for medical support.

Phase 3 – Movement to the Linkup Point and Linkup

6-52. All units or elements involved in the linkup enforce strict fire control measures to help prevent fratricide. Moving or converging forces must easily recognize linkup points and RFL. Linkup elements take the following actions:

- Conduct far recognition using radios or command and control systems.
- Conduct short-range (near) recognition using the designated signal.
- Complete movement to the linkup point.
- Establish local security at the linkup point.
- Conduct additional coordination and linkup activities, as necessary.

Note. The following scenario (see figure 6-6 on page 6-20) is used for discussion purposes and is not the only way of conducting a linkup operation.

The mission determines that Company A is the stationary unit and First Platoon, Company B is the moving unit.

1. Company A, the first unit to arrive, establishes linkup, Rally Point 1 based on the mission variables of METT-TC(I) mission, enemy, terrain and weather, troops and support available, time available, civil considerations and informational considerations. Company A contacts higher echelon immediately; then contacts other linkup units.
2. Company A sends a contact team to Linkup Point 1.
3. Company A's contact team establishes the linkup point by:
    (a) Clearing the area.
    (b) Pinpointing the linkup point.
    (c) Marking the linkup point by using the agreed-upon recognition signal.
    (d) Moving to a covered and concealed position; observing the linkup site.
    (e) Informing higher echelon.
4. First Platoon, Company B repeats Steps 1 and 2 (Rally Point 2). When First Platoon's contact team arrives vicinity Linkup Point 1 and identifies the recognition signal, it initiates the far recognition signal. Company A's contact team returns the far recognition signal. First Platoon's contact team moves to the Company A's contact team position where it exchanges near-recognition signals and conducts coordination.
5. First Platoon's contact team guides Company A's contact team to the First Platoon's unit linkup, Rally Point 2.
6. The Company A contact team leads First Platoon to Company A's unit linkup, Rally Point 1. The units are linked up.

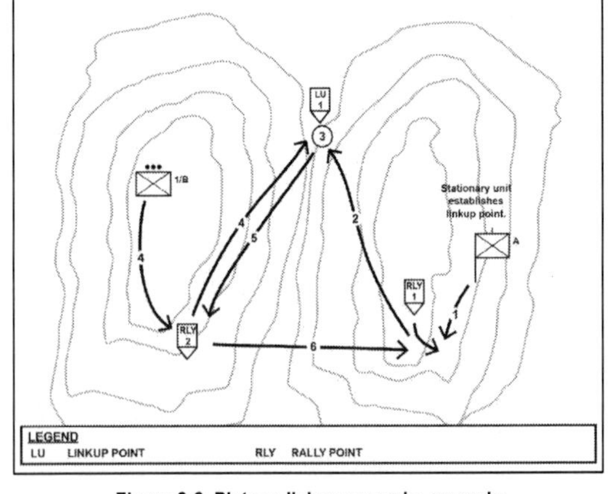

Figure 6-6. Platoon linkup scenario, example

SECTION V – TACTICAL DECEPTION

6-53. Surprise, a principle of war, is a combat multiplier that amplifies the effects of the other principles of war (see FM 3-0). Its effective use allows friendly units to strike at a time and place or in a manner that the enemy is unprepared for, which induces shock and causes hesitation. Every echelon works to achieve surprise in an operation and only by multiple echelons working together is achieved by surprise. The easiest way to achieve surprise is to use deception. This section provides a brief overview of the considerations unique to tactical deception operations. (See ATP 3-21.10 and FM 3-90 for additional information.)

6-54. *Tactical deception* is a friendly activity that causes enemy commanders to take action or cause inaction detrimental to their objectives (FM 3-90). Tactical deception operations are designed to support the commander's plan by enhancing the probability of mission success. It allows units to create windows of opportunity by doing the unexpected and causing the target to react to the unit's operation. Vital to the unit's ability to execute tactical deception is through maintaining operational security and ensuring the realism. The purpose of tactical deception is to—

- Gain the initiative.
- Reduce the overall operational risk.
- Preserve combat power.

6-55. Leaders conduct tactical deception at every echelon with either the means they have or with assistance from their higher echelon. *Deception means*—methods, resources, and techniques that can be used to convey information to the deception target (JP 3-13.4)—are the tools that friendly commanders use to accomplish tactical deception. Leaders normally employ deception plans to supplement several means that aim to mislead multiple types of enemy sensors. This increases credibility and the likelihood of deceiving the deception target. Tactical deception means provide the signatures, associations, and profiles of friendly alleged activities to the enemy. Typically, the means are found in either the coordinating instructions of an order or through commander's guidance. There are two tactical deception means categories are physical and technical:

6-56. Physical means are resources, methods, and techniques used to convey information normally derived from direct observation or active sensors by the deception target. (Most physical means also have technical signatures visible to sensors that collect scientifically or electronically.) Examples might include—

- Movement of the platoon.
- Decoy equipment and devices.
- Security measures.
- Tactical actions.
- Reconnaissance, surveillance, and security activities.

6-57. Technical means are resources, methods, and techniques used to convey or deny selected information or signatures. Technical means may be applied with corresponding physical means or alone to replicate something physical that is absent from visual contact. Examples of technical means might include—

- The establishment of communications networks and interactive transmissions that replicate a specific unit type, size, or activity.
- Emission or suppression of chemical or biological odors associated with a specific capability or activity.
- Organic capabilities that disrupt an enemy sensor or affect data transmission.

6-58. Variations of tactical deception are missions that a higher headquarters can assign to a subordinate unit. A unit may be told to conduct one of the two variations of tactical deception: feints and demonstrations. The selected variation and its use depend on the unit's understanding of the current situation as well as the commander's desired end state. All other deception guidance should be issued in the coordinating instructions or done to accomplish the commander's intent.

6-59. Units should consider cost of tactical deception in terms of resource expenditure. For tactical deception to appear real, units must dedicate adequate resources. The cost depends on the variation of tactical deception (demonstration or feint) and its objective. The units should also measure costs in risk and flexibility. For example, it may be very risky for the success of a unit's main effort to rely solely on the success of a planned demonstration. Should the demonstration not produce the expected enemy reaction, it could cause the main effort to fail. Flexibility is built into the plan by using branches, sequels, or executable deceptions.

6-60. Units have two options to disseminate the tactical deception plan to their subordinates. It is imperative that the unit maintains operational security for the success of the operation. The first option is for the unit's leadership to know the true task and purpose of the organization, subordinates may receive a different task and purpose in order to maintain operational security. The second option is for the unit to provide complete briefings to their subordinates. In both options, caution must be exercised to ensure that deception details are not mentioned in their operation orders (OPORDs). Operations security must be recognized as a vital element of the variation of the tactical deception.

6-61. A *demonstration* is a variation of tactical deception used as a show of force in an area where a unit does not seek a decision and attempts to mislead an adversary (FM 3-90). The commander uses demonstrations and feints in conjunction with other military deception activities. The commander generally attempts to deceive the enemy and induce the enemy commander to move reserves and shift fire support assets to locations where they cannot immediately affect the friendly main effort or take other actions not conducive to the enemy's best interests during the defense. The commander must synchronize the conduct of these variations of an attack with higher and lower echelon plans and operations to prevent inadvertently placing another unit at risk. Both variations are always shaping efforts, but a feint will require more combat power and usually requires ground combat units for execution.

6-62. A *feint* is a variation of tactical deception that makes contact solely to deceive the adversary as to the location, time of attack or both (FM 3-90). The principal difference between a feint and a demonstration is that in a feint the commander assigns the force an objective limited in size, scope, or some other measure. The force conducting the feint makes direct fire contact with the enemy but avoids decisive engagement. The planning, preparing, and executing considerations for demonstrations and feints are the same as for the other attack. The commander assigns the operation to a subordinate unit and approves plans to assess the effects generated by the feint, to support the operation.

SECTION VI – SECURITY

6-63. *Security* is measures taken by a military unit, activity, or installation to protect itself against all acts designed to, or which may, impair its effectiveness (JP 3-10). Maintaining security is a constant theme of tactical movement (see section VIII). Security can prevent enemy surprise. Security requires everyone to concentrate on the enemy. This means leaders and Soldiers must be proficient in the basics of tactical movement (see chapter 3). Failure to attain proficiency diverts attention away from the enemy, thereby directly reducing the platoon's ability to fight.

PLANNING AND PREPARATION

6-64. During planning and preparation for movement, leaders analyze the enemy situation, determine known and likely enemy positions, and develop possible enemy courses of action (COAs) and contact. After first considering the enemy and terrain, leaders determine what security measures to emplace during tactical movement.

ENEMY

6-65. Leaders must decide whether they are going to move aggressively to make contact, or stealthily to avoid contact. Either way, leaders must anticipate enemy contact throughout. If possible, leaders should avoid routes with obvious danger areas such as built-up areas, roads, trails, and known enemy positions. If these places cannot be avoided, risk management should be conducted to develop ways to reduce danger to the platoon or squad. If stealth is desired, the route should avoid contact with local inhabitants, built-up areas, and natural lines of drift.

6-66. Movement techniques help the leader manage the amount of security the platoon has during movement. Traveling is the least secure and used when contact is unlikely. Traveling overwatch is used when contact is likely but not imminent. Bounding overwatch is used when contact is imminent. The leader establishes the probable line of deployment (PLD) to indicate where the transition from traveling overwatch to bounding overwatch should occur. When in contact with the enemy, the platoon transitions from movement to maneuver (fire and movement) while the leader conducts actions on contact. (See figure 6-7 on page 6-24.)

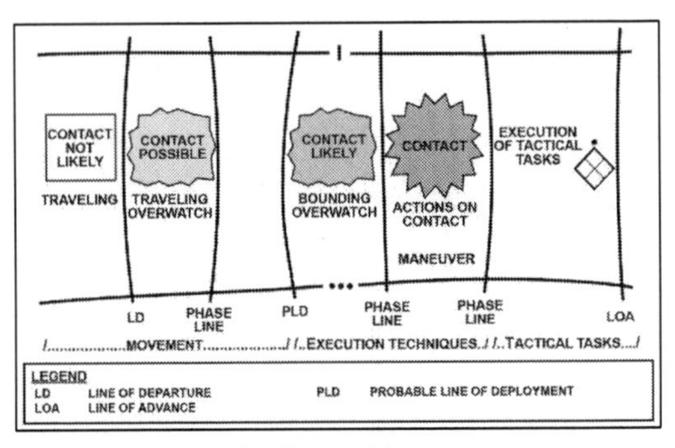

Figure 6-7. Movement to maneuver

TERRAIN

6-67. Platoon and squads enhance their own security during movement using covered and concealed terrain; the use of the appropriate movement formation and movement technique; the actions taken to secure danger areas during crossing; the enforcement of noise, light, and emissions control (for example, thermal and electronic) discipline; and use of proper individual camouflage techniques. When planning and preparing for movement, leaders must consider how terrain affects security while simultaneously considering METT-TC (I). Some missions may require the platoon or individual squad to move on other than covered and concealed routes. While leaders may not be able to prevent the unit's detection, they can ensure it moves on the battlefield in a time and place for which the enemy is unprepared. Particularly when moving in the open, leaders must avoid predictability and continue to use terrain to their advantage.

EXECUTION

6-68. During execution, leaders enforce camouflage discipline (Soldiers and their equipment). Leaders ensure the camouflage used by their Soldiers is appropriate to the terrain and season. Platoon standard operating procedures (SOPs) specify elements of camouflage, noise and light discipline and emissions control; security halts; and actions at security halts.

CAMOUFLAGE, NOISE, AND LIGHT DISCIPLINE AND EMISSIONS CONTROL

6-69. The platoon is visible to enemy forces and target acquisition capabilities on every spectrum, including visible light, sound, and across the electromagnetic spectrum. The platoon must take steps to eliminate or minimize all these signatures to prevent targeting. Natural or artificial materials may be used as camouflage to confuse, mislead, or evade the enemy. Leaders ensure all elements abide by strict execution of noise (this includes carried equipment) and light discipline and emissions control (for example, thermal and electronic). Infrared lights can provide additional illumination when necessary; however, peer threat forces can easily detect it. Radios and command and control systems all emit detectable signatures and must be used carefully; for instance, minimizing transmissions, reduced power settings, masking signals with terrain, and so forth, can help reduce the likelihood of detection. Cell phones must not be used under any circumstances.

DISMOUNTED SECURITY HALTS

6-70. Units conducting tactical movement frequently make temporary halts. These halts range from brief to extended periods. For short halts, platoons use a cigar-shaped perimeter intended to protect the force while maintaining the ability to continue movement. When the platoon leader decides not to resume tactical movement immediately, the platoon leader transitions the platoon to a perimeter defense. The perimeter defense is used for longer halts or during lulls in combat.

Note. During short halts, Soldiers use a brief relief bag or cat-hole latrine. The cat-hole latrine is dug about 1 foot deep and wide, and it is completely covered and packed down after use. In temporary AAs (1 to 3 days), the straddle trench latrine is used unless permanent facilities are provided. (See ATP 4-25.12 for additional information on basic field sanitation.)

Cigar-Shaped Perimeter

6-71. When the unit halts, if terrain permits, Soldiers should move off the route and face out to cover the same sectors of fire they were assigned while moving, allowing passage through the center of the formation. This results in a cigar-shaped perimeter. Actions by subordinate leaders and their Soldiers occur without an order from the leader. Soldiers initially take a knee; however, if the platoon does not begin movement quickly, they automatically move to a covered position and take up a prone position facing out. If the halt lasts more than a few minutes, leaders reposition their Soldiers as necessary to take advantage of the best cover, concealment, and fields of fire.

Perimeter Defense

6-72. When operating independently, the platoon uses a perimeter defense during extended halts, resupply, issuance of platoon orders, or lulls in combat. Normally the unit first occupies a short halt formation. Then after conducting a leader's reconnaissance of the position and establishing security, the unit moves into the perimeter defense.

Mounted Security Halt

6-73. When vehicles are attached, the platoon (squad when moving independently) employs the coil, herringbone, and triangle "Y" formations to maintain 360-degree security when stationary. The coil provides all-round security and observation when the platoon is stationary. The platoon also uses the coil for tactical refueling, resupply, and issuing orders. When in a coil, leaders post security. (See figure 6-8.)

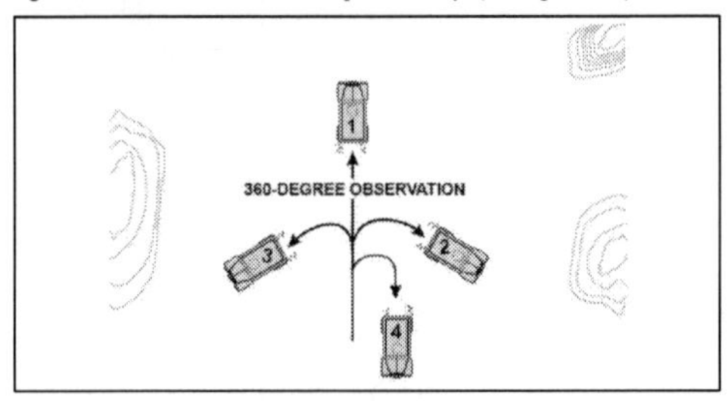

Figure 6-8. Coil formation

6-74. The patrol leader uses the herringbone and triangle during temporary halts or when getting off a road to allow another unit to pass. It lets the patrol move to covered and concealed positions off a road or from an open area and establishes all-round security without issued detailed instructions. The truck commander repositions their vehicles as necessary to take advantage of the best cover, concealment, and fields of fire. Fire team members dismount and establish security. (See figures 6-9 and 6-10.)

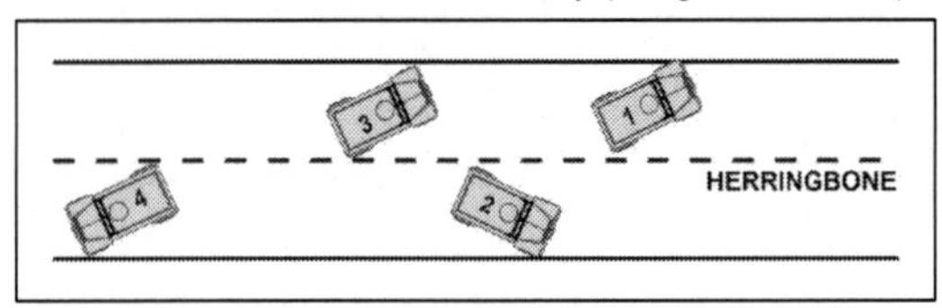

Figure 6-9. Herringbone formation

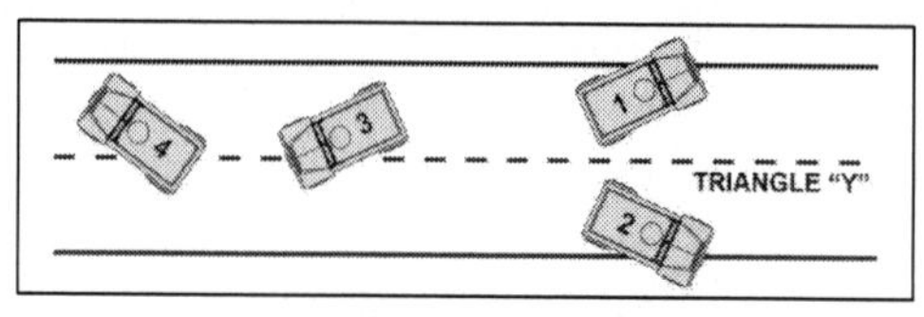

Figure 6-10. Triangle Y formation

ACTIONS AT HALTS

6-75. Table 6-3 lists the standard actions taken at halts by the Soldier, squad leader, and platoon leader.

Table 6-3. Actions at halts

*Soldier Actions**	*Squad Leader (or Section Leader) Actions*	*Platoon Leader Actions*
• Moves to as much of a covered and concealed position as available. • Visually inspects and physically clears immediate surroundings (a roughly 5- to 25-meter radius around the position). • Establishes a sector of fire for assigned weapon (using 12 o'clock as the direction the Soldier is facing out, the Soldier's sector of fire ranges from 10 o'clock to 2 o'clock). • Determines observation and field of fire. Identifies dead space in field of fire. • Identifies obstacles and determines enemy avenues of approach (both mounted and dismounted). • Identifies the dominant ground in immediate surroundings. • Coordinates the actions with the Soldiers on left and right. (*These actions occur without leader prompting.)	• Adjusts perimeter: ▪ If operating independently, the squad leader establishes 360-degree, three-dimensional security. ▪ Attempts to find terrain that anchors position. ▪ If operating as part of a platoon, the squad leader arrays teams to best fit into the platoon leader's defensive scheme, based on the platoon leader's guidance. • Visually inspects and physically clears (if required) the squad's immediate surroundings (about 35 meters, the distance within hand grenade range). • Ensures squad's individual sectors of fire overlap with each other, creating a seamless perimeter with no gaps of fire coverage. • Identifies dead space and adjusts grenadiers accordingly. • Identifies obstacles and the likely enemy avenue of approach (mounted and dismounted). • Identifies the dominant ground in area of operations. • Coordinates responsibilities and sectors with the units on left and right.	• Adjusts perimeter: ▪ If operating independently, establishes 360-degree, three-dimensional security. ▪ If operating as part of another organization, arrays squads to best fit into the controlling commander's defensive scheme. ▪ Supervises the emplacement of the weapons squad's weapon systems. • Dispatches an element (usually a fire team) to visually inspect and physically clear the platoon's immediate surroundings (an area out to small arms range, roughly 100 to 300 meters depending on terrain). • Ensures the squads' sectors of fire overlap with each other, creating a seamless perimeter with no gaps of fire coverage. • Identifies dead space not covered and requests indirect fire support to overwatch dead space in the area of operations. • Identifies obstacles and the likely enemy avenue of approach (mounted and dismounted). • Identifies the dominant ground in area of operations. • Coordinates with the units on the left and right.

SECTION VII – SECURITY OPERATIONS

6-76. *Security operations* are those operations performed by commanders to provide early and accurate warning of enemy operations, to provide the forces being protected with time and maneuver space within which to react to the enemy, and to develop the situation to allow commanders to effectively use their protected forces (ADP 3-90). They prevent surprise, reduce uncertainty, and provide early warning of enemy activities. This section addresses the general security considerations, fundamentals of security, and types of security operations.

Note. The main difference between the performance of security operations and reconnaissance operations is that security operations orient on the friendly force, area, or facility, while reconnaissance operations (see chapter 7) orient on the enemy and terrain.

GENERAL CONSIDERATIONS

6-77. Security operations are not to be confused with the more general term of security (see section VI). Security includes a wide range of activities such as preventing unauthorized access into secure areas or establishing perimeter security around a base camp. Security is inherent in all operations and is always the first priority of work. Security is the responsibility of every Soldier and unit whereas security operations are distinct missions. Security operations are supporting efforts. As a supporting effort, economy of force is often a condition associated with the performance of security operations and essential for preserving friendly force combat power. The following activities (local security, observation posts, and combat outposts) are general security considerations that are applicable to all security operations and units.

INFANTRY SMALL UNIT SECURITY OPERATIONS

6-78. Security operations, assigned to Infantry small units (also called security forces), protect their higher organization from observation, indirect fires, harassment, and surprise. At the same time, they provide information about the size, composition, location, and movement of enemy forces including information about the terrain and populations within their higher organization's assigned area. Effective security operations can also draw enemy forces into exposed positions, trade space for time, allow the higher organization to concentrate forces elsewhere, deceive the enemy, attrite enemy forces, and hold, deny, or control key terrain. Security forces must be prepared to destroy enemy reconnaissance efforts and fight for information to seize, retain, or exploit the initiative.

LOCAL SECURITY

6-79. *Local security* is the low-level security activities conducted near a unit to prevent surprise by the enemy (ADP 3-90). All units are responsible for their own local security.

Local security is not an operation of its own. It includes any local measure taken by units that protect against enemy actions. It involves avoiding enemy detection or deceiving the enemy about friendly positions and intention. Local security provides immediate protection to friendly forces and is typically performed by a unit for self-protection, but it may also be provided by another unit when the security requirements are greater than that unit's security capabilities. Local security may include countermobility (see paragraph 5-44) and survivability activities (see paragraph 5-102) as well as the use of active and passive measures (see paragraph 5-143) to provide local security.

OBSERVATION POSTS

6-80. An *observation post* is a position from which observations are made or fires are directed and adjusted (FM 3-90). All observation posts should possess appropriate communications. Commanders task subordinate units to perform reconnaissance and combat patrols to cover gaps between observation posts. Units place restrictive fire support coordination measures (such as no-fire area or restrictive fire area) around observation post locations to prevent fratricide. Dismounted Infantry observation posts provide maximum stealth but lack the speed of displacement, optics, and weapons of mounted ones. It takes a minimum of two Soldiers to establish an observation post, and then they can only operate effectively for no more than 12 hours. Observation posts manned for more than 12 hours require a minimum of an Infantry squad to ensure continuous operation. Combat and reconnaissance patrols (see chapter 7) cover dead space and the area between observation posts. In addition, under limited-visibility conditions, units can establish observation posts as listening posts to take advantage of the increased auditory acuteness that occurs when Soldier vision is degraded.

COMBAT OUTPOSTS

6-81. A *combat outpost* is a reinforced observation post capable of conducting limited combat operations (FM 3-90). Using combat outposts is a technique for employing security forces in restrictive terrain that precludes mounted security forces from covering the area. Leaders use combat outposts when enemy forces infiltrating into and through the security area could overrun smaller observation posts. Leaders use a combat outpost to extend the depth of the security area, to keep friendly forward observation posts in place until they can observe the enemy force's main body, or to secure friendly forward observation posts that enemy forces might encircle.

6-82. While the mission variables of METT-TC (I) determine the size, location, and number of combat outposts a unit establishes, a reinforced platoon typically occupies each combat outpost. A combat outpost requires enough resources to accomplish its missions, but it should not seriously deplete the strength of the main body (see chapter 5, main battle area). Combat outposts are usually located far enough forward of the protected force to preclude enemy ground reconnaissance elements from observing the actions of the protected force.

6-83. Units organize the combat outpost to provide an all-around defense (see chapter 5, perimeter defense) to withstand a superior enemy force. When the enemy force has significant armored capability, commanders may give combat outposts more antitank

(AT) weapons (see appendix D). Forces operating combat outposts conduct aggressive patrolling (combat and reconnaissance), engage and destroy enemy reconnaissance elements, and engage the enemy main body before their extraction. Leaders plan to extract friendly forces from their outposts before enemy forces overrun them.

FUNDAMENTALS OF SECURITY OPERATIONS

6-84. Five fundamentals of security operations establish the framework for security operations. These fundamentals, listed in paragraphs 6-85 to 6-89, provide a set of principles that remind leaders of the inherent characteristics required to execute security operations. These fundamentals include provide early and accurate warning, provide reaction time and maneuver space, orient on the force, area, or facility to be secured, perform continuous reconnaissance, and maintain enemy contact. (See FM 3-96 for additional information.)

PROVIDE EARLY AND ACCURATE WARNING

6-85. The platoon (when task in a security force role) detects, observes, and reports threat forces that can influence the company's or higher headquarters main body. Early detection and warning through rapid reporting enables the protected commander to make timely and well-informed decisions to apply forces relative to the threat. As a minimum, security forces should operate far enough from the protected force to prevent enemy ground forces from observing or engaging the protected force with direct fires.

PROVIDE REACTION TIME AND MANEUVER SPACE

6-86. The platoon (security force) provides the protected force with enough reaction time and maneuver space to respond effectively to likely enemy actions by operating at a distance from the protected force and by offering resistance (within its capabilities and mission constraints) to enemy forces. Providing the platoon with an area of operations that has sufficient depth to operate enhances its ability to provide reaction time and maneuver space to the protected force.

ORIENT ON THE FORCE, AREA, OR FACILITY TO BE SECURED

6-87. While reconnaissance forces orient on the enemy, security forces orient on the protected force by understanding their scheme of maneuver and follow-on mission. The platoon (security force) focuses all its actions on protecting and providing early warning operating between the protected force and known or suspected enemy.

PERFORM CONTINUOUS RECONNAISSANCE

6-88. Reconnaissance fundamentals are implicit in all security operations. The platoon (security force) continuously seeks the enemy and reconnoiters key terrain. The platoon uses continuous reconnaissance to find the enemy (gain and maintain enemy contact), develop the situation, report rapidly and accurately, retain freedom of maneuver to provide early and accurate warning, and provide reaction time and maneuver space to the protected force. The platoon conducts zone, area, or route reconnaissance (see chapter 7 section IV for additional information) along with using observation posts and combat outpost (see paragraphs 6-80 and 6-81, respectively) to detect enemy movement or enemy preparations for action and to learn as much as possible about the terrain with the ultimate goal to determine the enemy's COA and to assist the protected force in countering it.

MAINTAIN ENEMY CONTACT

6-89. Once the platoon (security force) makes enemy contact, it does not break contact unless the main force commander specifically directs it. However, the individual security asset that first makes contact does not have to maintain that contact if the entire security force maintains contact with the enemy. The security force commander (or platoon leader) ensures that subordinate security assets hand off contact with the enemy from one security asset to another in this case. The platoon must continuously collect information on the enemy's activities to assist the protected main body in determining potential and actual enemy COAs and to prevent the enemy from surprising the protected force. Depth in space and time enables the platoon to maintain continuous visual contact, to use direct and indirect fires, and to maneuver freely.

TYPES OF SECURITY OPERATIONS

6-90. Security operations provide the protected force with varying levels of protection and are dependent upon the size of the unit conducting the security operation. All security operations provide protection and early warning to the protected force. The four types of security operations are screen, guard, cover, and area security.

Note. Security operations conducted in the security area by one force or a subordinate element of a force that provides security for the larger force are screen, guard, and cover. The screen, guard, and cover security operations, respectively, contain increasing levels of combat power and provide the main body with increasing levels of security. The more combat power in the security force means less combat power for the main body. Normally, the commander designates a security area in which security forces provide the protected force with reaction time and maneuver space to preserve freedom of action. (See chapter 5 for additional information on security area and main battle area operations within the defense.)

SCREEN

6-91. *Screen* is a type of security operation that primarily provides early warning to the protected force (ADP 3-90). Screens provide less protection than guards or covers. Screen missions are defensive in nature and accomplished by establishing a series of observation posts and patrols to ensure observation of the assigned sector. The screen force gains and maintains enemy contact consistent with the fundamentals and destroys or repels enemy reconnaissance units by conducting counterreconnaissance (see paragraph 5-9). A unit performing a screen observes, identifies, and reports enemy actions. Generally, a screening force engages and destroys enemy reconnaissance elements within its capabilities—augmented by indirect fires, Army aviation, and/or close air support (CAS)—but otherwise fights only in self-defense. If a significant enemy force is expected or a significant amount of time and space is needed to provide the required degree of protection, the commander assigns and resources a guard mission instead of a screen. A screen is appropriate to cover gaps between forces, exposed flanks, or the rear of stationary and moving forces. (See FM 3-90 for additional information.)

GUARD

6-92. *Guard* is a type of security operation conducted to protect the main body by fighting to gain time while preventing enemy ground observation of and direct fire against the main body (ADP 3-90). A guard differs from a screen in that a guard force contains sufficient combat power to defeat, cause the withdrawal of, or fix the lead elements of an enemy ground force before it can engage the main body with direct fire. A guard force routinely engages enemy forces with direct and indirect fires. A screening force, however, primarily uses indirect fires or CAS to destroy enemy reconnaissance elements and slow the movement of other enemy forces. A guard force uses all means at its disposal to prevent enemy forces from penetrating to a position to observe and engage the main body. It operates within the range of the main body's fire support weapons, deploying over a narrower front than a comparable-sized screening force to permit concentrating combat power. The three types of guard operations are advance, flank, and rear guard. Commanders can assign a guard mission to protect either a stationary or a moving force. The guard force commander normally conducts the guard mission as an area defense, a delay, a zone reconnaissance, or a movement to contact mission in the security area to provide reaction time and maneuver space to the main body. (See FM 3-90 for additional information.)

COVER

6-93. *Cover* is a type of security operation done independent of the main body to protect them by fighting to gain time while preventing enemy ground observation of and direct fire against the main body (ADP 3-90). The biggest difference between a guard force and a covering force is that a covering force is able to operate independently of the main body, while a guard force relies on indirect support from the main body. A division covering force is normally a reinforced BCT. It performs reconnaissance or other security missions. If the division assigned area is narrow enough, an adequately reinforced combined arms battalion, reconnaissance squadron, or Stryker battalion may perform a cover mission. At both corps and division echelons, the amount of reinforcement provided to the covering force determines the distance and time it can operate away from the main body. These reinforcements typically revert to their parent organizations once the covering mission is complete. BCTs and battalions typically organize a guard force instead of a covering force because their resources are limited. (See FM 3-90 for additional information.)

Area Security

6-94. *Area security* is a type of security operation conducted to protect friendly forces, lines of communications, installation routes and actions within a specific area (FM 3-90). Area security operations occur during all types of operations. They allow commanders to provide protection to critical assets without a significant diversion of combat power. Protected forces range from echelon headquarters through artillery and echelon reserves to the sustaining base. Protected installations can be part of the sustaining base, or they can constitute part of the area's critical infrastructure. Operations in noncontiguous assigned areas require units to emphasize area security.

6-95. During the offense, various military organizations may be involved in conducting area security operations in an economy-of-force role to protect lines of communications, convoys, or critical fixed sites and radars. Route security operations are defensive in nature and are terrain oriented. A route security force may prevent an enemy force from impeding, harassing, or destroying lines of communications. Establishing a movement corridor for traffic along a route or portions of a route is an example of route security operations. (See ATP 3-21.10 for additional information.)

SECTION VIII – TROOP MOVEMENT

6-96. *Troop movement* is the movement of Soldiers and units from one place to another by any available means (FM 3-90). This is inherent in all military operations. Successful movement places Soldiers and equipment at their destination at the proper time, ready for combat. Infantry small units perform troop movements using different methods such as dismounted and mounted movement. The method employed depends on the situation, the size and composition of the moving unit, the distance the unit must cover, the urgency of execution, and the condition of the troops. This section addresses the types of troop movement and the methods of troop movement.

TYPES OF TROOP MOVEMENT

6-97. The ability of small Infantry leaders to posture their units in the right starting location to conduct operations depends on their ability to move their forces. The essence of battlefield agility is the capability to conduct rapid and orderly troop movement to concentrate combat power at decisive points and times. Units can expect the enemy to attempt to deny freedom of movement throughout their assigned area. The two types of troop movement are nontactical and tactical movement.

Nontactical Movement

6-98. *Nontactical movement* is a movement in which troops and vehicles are arranged to expedite their movement and conserve time and energy when no enemy ground interference is anticipated (FM 3-90). Units only conduct nontactical movements in secure areas. Examples of nontactical movements include rail and highway movement in the continental United States. Once units deploy into a theater of operation, they do not normally conduct nontactical movements.

Tactical Movement

6-99. A *tactical movement* is a movement in which troops and vehicles are arranged to protect combat forces during movement when a threat of enemy interference is possible (FM 3-90). Units maintain security against enemy attacks from both the air and ground and prepare to take immediate action against enemy ambushes, although they do not expect contact with significant enemy ground forces. During movement, the moving force employs security measures, even when contact with enemy ground forces is not expected. During a tactical movement, units are always prepared to take immediate action. There are three methods of a tactical movement that units of all types can conduct: approach march, forced march, and tactical road march.

Approach March

6-100. An *approach march* is the advance of a combat unit when direct contact with the enemy is intended (FM 3-90). Units employ an approach march when they know the approximate location of enemy forces. The approach march terminates in a march objective—such as an attack position, TAA, assault position—or it can be used to transition to an attack, for example, when the unit is no longer conducting a movement to contact. During approach marches, units use movement formations and movement techniques (see chapter 3) to balance security and speed throughout the operation. (See ATP 3-21.10 for additional information concerning approach marches.)

Forced March

6-101. A *forced march* is a march longer or faster than usual or in adverse conditions (FM 3-90). Forced marches require speed, exertion, and an increase in the number of hours marched each day beyond normal standards. Soldiers cannot sustain forced marches for more than a short period. In a forced march, the platoon may not halt as often or for as long as recommended for maintenance, rest, and feeding, or when vehicles are attached, fuel. Leaders must understand that immediately following a long and fast march, Soldiers experience a temporary deterioration in their physical condition. The combat effectiveness and cohesion of the unit also temporarily decreases, and the plan must accommodate stragglers. (See ATP 3-21.18 for additional information concerning forced marches.)

Tactical Road March

6-102. A *tactical road march* is a rapid movement used to relocate units within an assigned area to prepare for combat operations (FM 3-90). The primary consideration of the tactical road march is rapid movement. Typically, Infantry small units executes a tactical road march using the dismounted method of troop movement (see paragraph 6-104). Based on the mission variables of METT-TC (I) a unit can execute a mounted method of troop movement (see paragraph 6-105). In both methods, the moving force employs security measures, even when contact with enemy ground forces is not expected. During a tactical road march, the leader is always prepared to take immediate action if the enemy attacks. (See ATP 3-21.10 for additional information concerning tactical road marches.)

Note. The organization for a tactical road march and forced march is the march column. A *march column* is all march serials using the same route for a single movement under control of a single commander (FM 3-90). The subordinate element for a march column is a march serial. A *march serial* is a subdivision of a march column organized under one commander (FM 3-90). An example is a battalion serial formed from a brigade-sized march column. The subordinate element for the march serial is a march unit. A *march unit* is a subdivision of a march serial (FM 3-90). It moves and halts under the control of a single commander who uses voice and visual signals. An example of a march unit is a company from a battalion-sized march serial. Units organize a march column into four elements: reconnaissance, quartering party, main body, and trail party. Organized march columns employ three ground movement techniques: open column, close column, and infiltration march. All techniques use scheduled halts to control and sustain the road march. (See ATP 3-21.20 and ATP 3-21.10 for additional information.)

METHODS OF TROOP MOVEMENT

6-103. There are five methods of troop movements that units can execute. Generally, units can execute these methods in combination. For example, a unit can initially execute a mounted movement and then transition to a dismounted movement. Additionally, each method can be applied to nontactical or tactical movement. These methods are—dismounted movement, mounted movement, air movement, rail movement (see ATP 4-14 for additional information concerning rail movement), and water movement (see TC 3-21.76 for additional information concerning waterborne operations).

DISMOUNTED MOVEMENT

6-104. A *dismounted movement* is a movement of troops and equipment mainly by foot, with limited support by vehicles (FM 3-90). Dismounted movement is also called foot marches. Foot marches are characterized by combat readiness (because all Soldiers can immediately respond to enemy attack without the need to dismount), ease of control, adaptability to terrain, slow rate of movement, and increased Soldier fatigue. Foot marches do not depend on the existence of roads. (See ATP 3-21.18 for additional information concerning dismounted movement.)

MOUNTED MOVEMENT

6-105. A *mounted movement* is the movement of troops and equipment by combat and tactical vehicles (FM 3-90). The speed of the march and the increased supplies that can accompany the unit characterize this movement method. Infantry maneuver units cannot move themselves with organic truck assets and require assistance from transportation elements to conduct mounted movement. (See ATP 3-21.10 for additional information concerning mounted movement.)

Air Movement

6-106. *Air movement* is air transport of units, personnel, supplies, and equipment including airdrops and air landings (JP 3-36). Planning for air movements is like other missions. In addition to the normal planning process, however, air movement planning must cover specific requirements for air infiltration and exfiltration. They include—

- Coordinate with the supporting aviation units.
- Plan and rehearse with the supporting aviation unit before the mission, if possible. If armed escort accompanies the operation, the platoon leader and company commander, as well as the assault or general support aviation unit, should ensure aircrews are included in the planning and rehearsals.
- Gather as much information as possible, such as the enemy situation, in preparation of the mission.
- Plan and coordinate (commonly accomplished at battalion and above echelons) joint suppression of enemy air defense.
- Plan and coordinate (commonly accomplished at battalion and above echelons) different ingress and egress routes, covering the following:
  - Planned insertion and extraction points.
  - Emergency extraction rally points.
  - Lost communications extraction points.

6-107. Planned extraction points and emergency extraction rally points require communications to verify the preplanned pickup time or coordinate an emergency pickup time window. Planning must include details for extraction when communications between higher headquarters and unit are lost. The lost communications extraction point involves infiltration teams moving to the emergency extraction point after two consecutive missed communications windows and waiting up to 24 hours for pickup. (See FM 3-99 for additional information concerning air movement.)

Water Movement

6-108. Use of inland and coastal waterways may add flexibility, surprise, and speed to tactical movements. Use of these waterways also increases the load-carrying capability of dismounted Infantry rifle units. See TC 3-21.76 chapter 12 for additional information on tactical movement by watercraft (specifically, the combat rubber raiding craft).

Note. During tactical movement by watercraft and crossing water obstacles, leaders identify weak or non-swimmers and pair them with good swimmers in their squads.

6-109. When platoons or squads must move by, into, though, or out of rivers, lakes, streams, or other bodies of water, they treat the water obstacle as a danger area. While on the water, the platoon is exposed and vulnerable. (See TC 3-21.76 for information on waterborne operations.) Common techniques during the employment of waterborne operations include:

- To offset the detection challenges, the platoon—
  - Moves during limited visibility.
  - Disperses.
  - Moves near the shore to reduce the chances of detection.
- When moving in more than one boat, the platoon—
  - Maintains tactical integrity and self-sufficiency.
  - Cross-loads essential Soldiers and equipment.
  - Ensures the radio is with the leader.
- If boats are not available, several other techniques can be used such as—
  - Swimming.
  - Poncho rafts.
  - Air mattresses.
  - Waterproof bags.
  - A 7/16-inch rope used as a semisubmersible, one-rope bridge, or safety line.
  - Water wings (made from a set of trousers).

This page intentionally left blank.

Chapter 7
Patrols and Patrolling

A *patrol* is a detachment sent out by a larger unit to conduct a specific mission that operates semi-independently and returns to the main body upon completion of mission. Patrolling fulfills the Infantry's primary function of finding the enemy to engage them or report their disposition, location, and actions. Patrols act as ground sensors or early warnings for larger units and the planned action determines the type of patrol. This chapter provides an overview on patrolling by the Infantry rifle platoon and squad and discusses in detail combat and reconnaissance patrols.

SECTION I – OVERVIEW

7-1. When a patrol is made up of a single unit, such as a rifle squad sent out on a reconnaissance patrol, the squad leader is responsible and designated as the patrol leader. If a patrol is made up of mixed elements from several units, then the senior officer or noncommissioned officer is designated as the patrol leader. This temporary title defines the role and responsibilities during the mission. The patrol leader may designate an assistant, normally the next senior Soldier in the patrol, and subordinate element leaders the patrol leader requires.

Note. In this chapter, the patrol leader is the person in charge of the patrol. In a platoon-size element, that person would most likely be the platoon leader. The assistant patrol leader (known as APL) is the second person in charge of the patrol. In a platoon-size element, that person most likely is the platoon sergeant.

COMMON CONSIDERATIONS

7-2. The leader of any patrol, regardless of the type or the tactical task assigned, has an inherent responsibility to plan and prepare for possible enemy contact while patrolling. Patrols are always assigned a tactical mission. After the mission, the patrol returns to the main body, the patrol leader reports to the commander and describes the patrol's actions, observations, and condition.

Purpose of Patrolling

7-3. The planned action determines if the patrol is combat, or reconnaissance focused. Regardless of the type of patrol, the unit needs a clear task and purpose. There are several specific purposes which can be accomplished by patrolling—

- Gathering information on the enemy, on the terrain, or on the populace.
- Regaining contact with the enemy or with adjacent friendly forces.
- Engaging the enemy in combat to destroy them or inflict losses.
- Providing unit security.
- Protecting essential infrastructure or bases.
- Deterring and disrupting insurgent or criminal activity.

Organization of Patrols

7-4. A patrol is organized to perform a specific mission. It must be prepared to secure itself, navigate accurately, identify and cross danger areas, and reconnoiter the patrol objective. If it is a combat patrol, it must be prepared to support by fire, breach obstacles, and assault the objective. If it is a reconnaissance patrol, it must reconnoiter and conduct surveillance of an objective area while securing itself. Additionally, a patrol must be able to conduct detailed searches as well as deal with casualties and detainees.

7-5. The patrol leader identifies those tasks that must be or will likely be conducted during the patrol and decides which elements will perform which tasks. Where possible, the leader should maintain squad and fire team integrity. Squads and fire teams may perform more than one task during the time a patrol is away from the main body, or it may be responsible for only one task. The leader must plan carefully to identify and assign all required tasks in the most efficient way.

Note. A patrol can consist of a unit as small as a fire team but are usually squad and platoon sized. For larger combat tasks such as for a raid, the patrol is sometimes a company (see ATP 3-21.10).

7-6. The headquarters element of a patrol (for example, platoon size) normally consists of the patrol leader (platoon leader) and a radiotelephone operator (known as RTO). The platoon sergeant may be designated as the APL. Additional teams and individuals common to all patrols generally include:

- Forward observer (FO), combat medics, and combat lifesavers (CLSs).
- Aid and litter teams responsible for locating, treating, and evacuating casualties.
- Detainee teams responsible for processing detainees, according to the five Ss (search, silence, segregate, speed, and safeguard) and leader's guidance. These teams also may be responsible for accounting for and controlling recovered personnel.
- Recorder designated to record all information collected during the patrol.

- Compass and pace Soldiers to navigate using terrain association, dead reckoning, or a combination of both (see chapter 3 section I).

RALLY POINTS

7-7. A rally point is a place designated by the patrol leader where the patrol (or individual Soldiers) move(s) to reassemble and reorganize if it becomes dispersed. Forces conducting a patrol, or an infiltration or exfiltration commonly use this control measure. Despite these different types of rally points (see paragraph 7-10), the occupation and actions occurring are generally the same.

Selection of Rally Points

7-8. The patrol leader physically reconnoiters routes to select rally points whenever possible and determines what actions will occur there. The leader selects tentative points if there is only time for a map reconnaissance. Routes are confirmed by the leader through actual inspection as the patrol moves through them.

7-9. When occupying a rally point, leaders use a perimeter defense to ensure all-around security. Rally points used to reassemble the unit after an event are likely to be chaotic scenes and will require immediate actions by whatever Soldiers happen to arrive. These actions and other considerations are listed in table 7-1.

Table 7-1. Actions and considerations at a rally point

Rally Points	*Soldier Actions at Rally Points*	*Other Considerations*
Select a rally point that— • Is easily recognized. • Is large enough for the unit to assemble. • Is defensible for a short time. • Is away from normal movement routes and natural lines of drift. • Designate a rally point by one of the following three ways: - Physically occupy it for a short period. - Use hand and arm signals (either pass by at a distance or walk through). - Radio communication.	• Establish security. • Reestablish the chain of command. • Account for personnel and equipment status. • Determine how long to wait until continuing the unit's mission or linkup at a follow-on rally point. • Complete last instructions.	• Travel time and distance. • Maneuver room needed. • Adjacent unit coordination requirements. • Line of sight and range requirements for communication equipment. • Trafficability and load bearing capacity of the soil (especially when mounted). • Ability to prevent surprise from the enemy. • Energy expenditure of Soldiers and their condition at the end of the movement.

Types of Rally Points

7-10. The most common types of rally points used during movement and crossing danger areas are initial, en route, reentry, and nearside and far side. An objective rally point (known as ORP) is most used prior to actions on or near the objective area during combat and reconnaissance patrols. Soldiers must know which rally point to move to at each phase of the patrol mission. They should know what actions are required there and how long they are to wait at each rally point before moving to another. The following is a description of each rally point:

- Initial rally point. A place inside of friendly lines where a unit may assemble and reorganize if it makes enemy contact during the departure of friendly lines or before reaching the first en route rally point. It is normally selected by the commander of the friendly unit.
- En route rally point. A point where the leader designates en route rally points based on the terrain, vegetation, and visibility.
- Reentry rally point. A point located out of sight, sound, and small-arms weapons range of the friendly unit through which the patrol will return. This also means the reentry rally point should be outside the final protective fires (FPFs) of the friendly unit.
- Nearside and far side rally points. These rally points are on the near and far side of danger areas. If the patrol makes contact while crossing the danger area and control is lost, Soldiers on either side move to the rally point nearest them. They establish security, reestablish the chain of command, determine their personnel and equipment status, continue the patrol mission, and linkup at the ORP.
- ORP. A point out of sight, sound, and small-arms range of the objective area (reconnaissance or combat). It normally is in the direction the patrol plans to move after completing its actions on the objective. The ORP is tentative until the objective is pinpointed. Actions at or from the ORP include—
  - Issuing a final fragmentary order (FRAGORD).
  - Disseminating information from reconnaissance if contact was not made.
  - Making final preparations before continuing operations.
  - Accounting for Soldiers and equipment after actions at the objective are complete.
  - Reestablishing the chain of command after actions at the objective is complete.

Note. Isolated Soldiers still able to function on their own will execute their isolated Soldier guidance designated in the patrol plan. (See ATP 3-21.10 for additional information.)

Occupation of an Objective Rally Point

7-11. The occupation of an ORP is somewhat different (from the other rally points listed in paragraph 7-10) in that it is established in much the same way as a tactical assembly area (TAA) (see paragraph 4-61), and a patrol base (see paragraph 7-17). As this rally point is typically near the infiltrating (see paragraphs 4-16 to 4-20) unit's objective, each element within the patrol is assigned a sector within the perimeter of the ORP and is responsible for its security. Prior to occupation the patrol leader conducts a leader's reconnaissance of the ORP. Once the leader's reconnaissance is completed and as the patrol occupies the ORP, the patrol leader prepares for a second leader's reconnaissance (see paragraph 7-15) of the patrol's objective. As before the patrol leader assembles the appropriate leaders and personnel for this leader's reconnaissance. Once all elements have closed within the ORP, the patrol begins the leader's reconnaissance of the patrol's objective. The patrol leader can emplace appropriate surveillance, and both the left, right, and rear security elements while on the leader's reconnaissance. The ORP is tentative until the objective is pinpointed. The patrol occupies an ORP (see figure 7-1 for example sequence on page 7-6) in the following general sequence:

- The patrol halts beyond sight and sound of the ORP (200 to 400 meters in good visibility, 100 to 200 meters in limited visibility).
- The patrol establishes a security halt (see paragraph 6-70) according to the unit SOP.
- After issuing a five-point contingency plan (see paragraph 7-15) to the APL, the patrol leader moves forward from the security halt with a reconnaissance element to conduct a leader's reconnaissance of the ORP.
- For a squad-sized patrol, the patrol leader moves forward with a compass Soldier and one member of each fire team to confirm the location of the ORP.
  - After physically clearing the ORP location, the patrol leader leaves two Soldiers at the 6 o'clock position facing in opposite directions.
  - The patrol leader issues a contingency plan and returns with the compass Soldier to guide the patrol forward.
  - The patrol leader guides the patrol forward into the ORP, with one team occupying from 3 o'clock through 12 o'clock to 9 o'clock, and the other occupying from 9 o'clock through 6 o'clock to 3 o'clock.
- For a platoon-sized patrol, the patrol leader, RTO, weapons squad leader, two assistant gunners, a team leader, a squad automatic weapon gunner, and riflemen conduct the leader's reconnaissance to confirm the ORP.
  - After physically clearing the ORP location, the patrol leader leaves selected Soldiers in the ORP at the 10, 2, and 6 o'clock positions. (For example, the weapons squad leader establishes machine gun positions marked by the assistant gunners at the 10 and 2 o'clock positions and positions the squad automatic weapon gunner at the 6 o'clock position.)
  - The patrol leader leaves a leader (generally the weapons squad leader or team leader) at the ORP, issues a contingency plan, then returns with selected Soldier(s) to guide the patrol forward.
  - The patrol leader guides the patrol forward into the ORP: with the first squad in the order of movement (base squad), occupying from 10 to 2 o'clock; trail squads occupy from 2 to 6 o'clock and 6 to 10 o'clock, respectively; and patrol headquarters element occupying the center of the triangle (or circle perimeter [in whichever fits the terrain]).

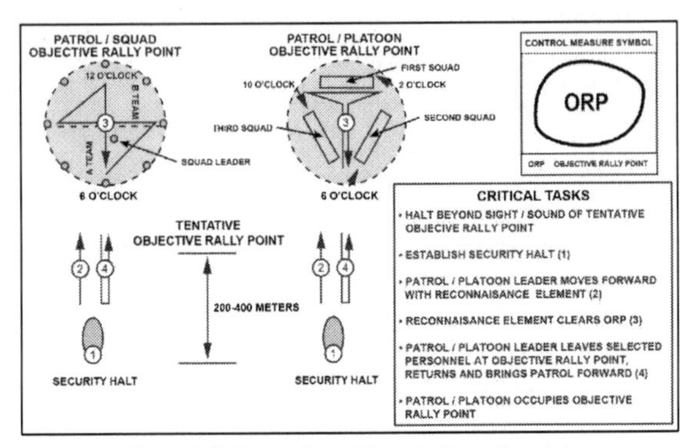

Figure 7-1. Occupation of the objective rally point

MOUNTED PATROLS

7-12. An analysis of METT-TC (I) mission, enemy, terrain and weather, troops and support available, time available, civil considerations, and informational considerations variables determines whether the patrol is mounted or dismounted. Some factors to consider when determining which mode to use include—

- Mission, especially where distance and speed are factors.
- Onboard visibility, navigation, and communication.
- Firepower and protection.
- Stealth and surprise.
- Terrain.

Additional Considerations

7-13. The planning and coordination required for dismounted patrols apply to vehicle-mounted patrols. The patrol leader additionally considers the following:

- Organize and orient vehicle gunners and vehicle commanders to maintain all-round security and, for urban areas, high-low security. Carefully consider leader locations in each vehicle and within the convoy.
- Rehearse mounted battle drills, reaction to contact, roll over drills, and mounting and dismounting in contact. If water hazards are present, include evacuation of the vehicle under these conditions. Include drivers in all rehearsals.

- Plan alternate routes to avoid civilian traffic and roadblocks.
- Remember that four is generally the minimum number of vehicles to conduct an operation. If one vehicle is disabled or destroyed, it can be recovered while the others provide security. Unit tactical standard operating procedures (SOPs) determine the number of vehicles required.
- Plan for actions required if a vehicle breaks down and must be repaired or recovered. Review self-recovery procedures. Plan actions in case a vehicle gets stuck and cannot be recovered. Plan actions for catch-ups and breaks in contact.
- Establish primary, alternate, contingency, and emergency (known as PACE) communications plan (see paragraphs 2-34 to 2-39).
- Secure external gear to prevent theft.
- Plan for heavy civilian vehicle and pedestrian traffic.
- Conduct a map reconnaissance and identify likely chokepoints, ambush sites (intersections), and overpasses.
- Plan primary and alternate routes to avoid potential hazards.
- Drive offensively, unpredictably, but within rules of engagement (ROE) restrictions.
- Avoid stopping; it can create a potential kill zone.
- Learn the characteristics of the vehicle, to include how high a vehicle can clear curbs and other obstacles, its turning radius, its high-speed maneuverability, and its estimated width (especially with slat armor).

Patrols with Mounted and Dismounted Phases

7-14. The mounted patrol normally moves to a dismount point (often the designated ORP) and conducts the same actions on the objective as a dismounted patrol. If possible, the vehicles establish a support by fire position to cover the objective, establish blocking positions, provide security, or otherwise support the actions of the dismounted element. The dismounted element conducts its required part of the mission and returns to the vehicles, remounts, and continues mission. Types of combat patrols that are especially suited for mounted movement include antiarmor ambushes and reconnaissance and security patrols covering long distances.

LEADER'S RECONNAISSANCE

7-15. The patrol leader reconnoiters the objective just before an attack or prior to sending elements forward to locations where they will support by fire. The leader confirms the condition of the objective, gives each subordinate leader a clear picture of the terrain where they will move, and identifies parts of the objective they must seize or suppress. The leader's reconnaissance patrol can consist of the patrol leader or representative, the leaders of major subordinate elements, and (sometimes) security personnel and unit guides. The leader can use the memory aid (see figure 7-2 on page 7-8) to help in remembering the five-point contingency plan (going, others, time, what, actions [known as GOTWA]) which is used when anybody separates from the main body.

G:	Going – where is the leader going?
O:	Others – are others going with the leader and who?
T:	Time (duration) – how long will the element be gone?
W:	What procedures do we take if the leader fails to return?
A:	Actions – what actions does the departing element and main body plan to execute on enemy contact?

Figure 7-2. Five-point contingency plan

7-16. A patrol leader should conduct a leader's reconnaissance when time or the situation allows. The plan includes a leader's reconnaissance of the objective once subordinate elements establish the ORP. During a reconnaissance, the leader pinpoints the objective, selects positions for the elements, squads, and teams and adjusts the plan based upon observation of the objective. Each type of patrol requires different tasks during the leader's reconnaissance, and the leader takes different elements depending upon the patrol's mission. The leader ensures the objective remains under continuous surveillance once deciding to return to the ORP. The leader designates a rally point and plans for adequate time to return to the ORP, complete the plan, disseminate information, issue orders and instructions, and allow the subordinate units to make additional preparations. (See table 7-2.)

Table 7-2. Leader's reconnaissance (tasks and control measures)

Leader Tasks to Be Performed	*Control Measures Identified*
Pinpoint Objective	Trigger Points
Determine Enemy Situation	Left and Right Limits
Identify Kill Zone/Objective	Limit of Advance
Identify Assault Position(s)	
Identify Support by Fire Position(s)	
Identify/Establish Surveillance Position	
Identify/Emplace Security Positions	
Identify Friendly Positions and Routes	
Confirm Plan	

Note. A leader's reconnaissance may alert the enemy a patrol is in the area by evidence of movement or noise before the patrol begins its mission. The leader must weigh the risk with that of only conducting a map reconnaissance.

PATROL BASE ACTIVITIES

7-17. A patrol base is a security perimeter, which is set up when a squad or platoon is conducting patrol halts for an extended period. A patrol base should not be occupied for more than a 24-hour period (except in emergency). A patrol never uses the same patrol base twice. The following activities at a minimum should be taken into consideration:

- Use.
- Site selection.
- Planning consideration.
- Security measures.
- Occupation.
- Priorities of work.

USE

7-18. Patrol bases typically are used to—

- Avoid detection by eliminating movement.
- Hide a unit during a long-detailed reconnaissance.
- Perform maintenance on weapons, equipment, eat, and rest.
- Plan and issue orders.
- Reorganize after infiltrating an enemy area.
- Establish a base from which to execute several consecutive or concurrent operations.

SITE SELECTION

7-19. The leader selects the tentative site from a map or by aerial reconnaissance. The site's suitability must be confirmed and secured before the unit moves into it. Plans to establish a patrol base must include selecting an alternate patrol base site. The alternate site is used if the first site is unsuitable or if the patrol must unexpectedly evacuate the first patrol base.

PLANNING CONSIDERATIONS

7-20. Leaders planning for a patrol base must consider the mission, and active and passive security measures. A patrol base must be located so it allows the unit to accomplish its mission:

- Observation posts and communication with observation posts.
- Patrol fire plan.
- Alert plan.
- Withdrawal plan from the patrol base to include withdrawal routes and a rally point or alternate patrol base.
- A security system that makes sure specific individuals are always awake.

- Enforcement of camouflage, noise, and light discipline and elimination of signals emissions to maximum extent possible.
- The conduct of required activities with minimum movement and noise.
- Priorities of work.

Security Measures

7-21. The following security measures should be considered:

- Select terrain the enemy probably would consider of little tactical value.
- Select terrain off main lines of drift.
- Select difficult terrain impeding foot movement, such as an area of dense vegetation, preferably bushes and trees spreading close to the ground.
- Select terrain near a source of water.
- Select terrain that is defendable for a short period and offers good cover and concealment.
- Avoid built up areas and known or suspected enemy positions.
- Avoid ridges and hilltops, except as needed for maintaining communications.
- Avoid small valleys.
- Avoid roads and trails.

Occupation

7-22. A patrol base is reconnoitered and occupied in the same manner as an ORP (see paragraph 7-11 and figure 7-1 on page 7-6); with the exception, the patrol will typically plan to enter at a 90-degree turn. The patrol leader leaves a two-Soldier observation post at the turn; the patrol covers tracks from the turn to the patrol base.

7-23. The patrol moves into the patrol base. Squad-sized patrols generally will occupy a cigar-shaped perimeter; platoon-sized patrols generally will occupy a triangle-shaped perimeter (see figure 7-3). The patrol leader inspects and adjusts the entire perimeter, as necessary.

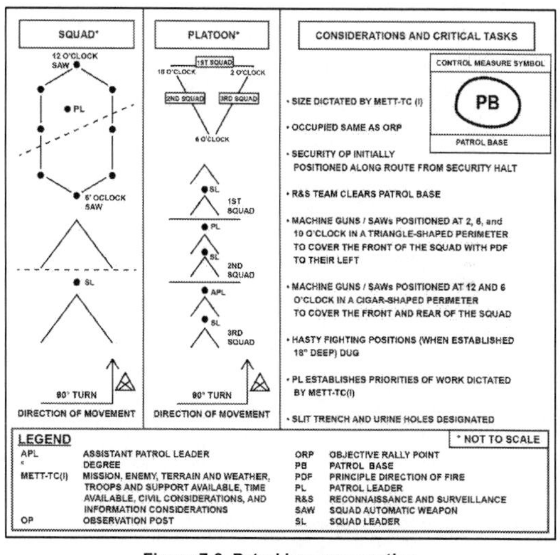

Figure 7-3. Patrol base occupation

7-24. After the patrol leader has checked each squad's portion of the perimeter, each squad leader sends a two-Soldier reconnaissance and surveillance team to the leader at the patrol's command post. The leader issues the three reconnaissance and surveillance teams a contingency plan, reconnaissance method, and detailed guidance on what to look for (enemy, water, built up areas or human areas, roads, trails, or possible rally points).

Note. Squad-sized patrols do not normally send out a reconnaissance and surveillance team at night.

7-25. The patrol leader directs the teams to use a particular reconnaissance method, such as a fan or box (see paragraphs 7-229 to 7-231), the route, and specific exit and reentry points:

- If the leader feels the patrol was tracked or followed, the leader may elect to wait in silence at 100 percent alert before sending out reconnaissance and surveillance teams.

- The patrol remains at 100 percent alert during this recon.
- The reconnaissance and surveillance teams conduct a security patrol, gathering the directed information. Following the patrol, they submit a sketch and detailed observations about the terrain surrounding the patrol base.
- Based on the reconnaissance and surveillance teams' reports, the patrol leader decides to either stay in the current patrol base and begin priorities of work or move the patrol base to an alternate location.

7-26. Key considerations when occupying a passive (clandestine) patrol base include:

- Squad or smaller-size patrol base established to rest Soldiers.
- Unit moves as a whole and occupies in force.
- Unit moves in at a 90-degree angle to the order of movement.
- A Claymore mine is emplaced on the route entering the patrol base.
- Alpha and bravo teams sit back-to-back facing outward.
- One individual (at least) on each team is alert and providing security.

PRIORITIES OF WORK

7-27. Once the leader is briefed by the reconnaissance and surveillance teams and determines the area is suitable for a patrol base, the leader establishes or modifies defensive work priorities to establish the defensive posture of the patrol base. Priorities of work are not a laundry list of tasks to be completed; priorities of work must consist of a task, a given time, and a measurable performance standard. For each priority of work, a clear standard must be issued to guide the element in the accomplishment of each task. It also must be designated whether the work will be controlled in a centralized or decentralized manner. Priorities of work are determined in accordance with the mission variables of METT-TC (I). Priorities of work may include the tasks in paragraphs 7-28 to 7-39 but are not limited to them.

Continuous Security

7-28. Prepare to use all active and passive measures to cover the entire perimeter all the time, regardless of the percentage of weapons used to cover all the terrain. Employ all elements, weapons, and personnel to meet conditions of the terrain, enemy, or situation. Additional considerations include (METT-TC [I] dependent):

- Emplacement of obstacles, including claymore mines.
- Construction of hasty individual fighting positions.
- Completion of standard range cards for each weapon system.
- Completion of teams, squads, and patrol sector sketches (before security is decreased from 100 percent).

Withdrawal Plan

7-29. The patrol leader designates the signal for withdrawal and order of withdrawal. The leader identifies a rally point(s) and alternate patrol base.

Continuous Communications

7-30. Communications must be maintained with higher headquarters, observation posts, and within the unit. Other Soldiers within the patrol may rotate duties with the patrol leader's RTO allowing accomplishment of continuous radio monitoring, radio maintenance, act as runners for the platoon leader, or conduct other priorities of work.

Mission Planning and Preparation

7-31. The patrol leader uses the patrol base to plan, issue orders, rehearse, and inspect. When time permits, the leader can use the patrol base to prepare for follow-on missions.

Weapons and Equipment Maintenance

7-32. The leader ensures medium machine guns, weapon systems, communications equipment, night vision devices and all other equipment are maintained. No more than one-third of any given type of system carried by the patrol should be disassembled for preventative checks and services at any one time. As a rule, weapons should not be disassembled for routine maintenance at night. If one of the patrol's medium machine guns is down for maintenance, then security levels for all remaining systems are raised.

Water Resupply

7-33. The APL (or platoon sergeant) organizes watering parties, as necessary. The watering party carries canteens in an empty rucksack or duffel bag and must have communications and contingency plans related to their making enemy contact en route to or returning from the water point or if the patrol base has to displace during their absence.

Mess Plan

7-34. At a minimum, security and weapons maintenance are performed prior to mess. Normally no more than half the patrol eats at one time. Soldiers typically eat 1 to 3 meters behind their fighting positions to avoid distracting those Soldiers providing security.

Rest and Sleep Plan Management

7-35. All leaders within the patrol must understand the problems associated with sleep deprivation and the consequences of not following the unit rest and sleep plan. The body needs regular rest to restore physical and mental vigor. Tired Soldiers are sluggish and react slower than normal; fatigue makes them more susceptible to sickness, and to making errors that could endanger both themselves and the entire patrol. For the best health, Soldiers should get 6 to 8 hours of uninterrupted sleep each day. As that is seldom possible in combat, use rest periods and off-duty time to rest or sleep.

7-36. The patrol leader must develop and enforce the unit sleep plan that provides Soldiers with a minimum of 4 hours of uninterrupted sleep in a 24-hour period. If sleep is interrupted, then 5 hours should be given. During continuous operations when uninterrupted sleep is not possible, blocks of sleep, which add up to 6 hours in a 24-hour period, are adequate for most people. Remember, 4 hours each 24-hour period is far from ideal. Do not go with only 4 hours of sleep each 24 hours for more than 2 weeks before paying back sleep debt. Recovery time should be approximately 8 to 10 hours of sleep each 24 hours over a 5- to 7-day period.

Alert Plan and Stand To

7-37. The patrol leader states the alert posture and stand to time. The leader develops the plan to ensure all positions are checked periodically, observation posts are relieved periodically, and at least one leader always is alert. The patrol typically conducts stand to at a time specified by unit tactical SOP such as 30 minutes before and after begin morning nautical twilight or end evening nautical twilight.

Resupply

7-38. Distribute or cross load ammunition, meals, equipment, and so on. Resupply operations, when required, are integrated within the scheme of maneuver, or executed as a standalone operation.

Sanitation and Personal Hygiene

7-39. The APL (or platoon sergeant) and combat medic ensure a straddle trench is prepared and marked. All Soldiers will brush their teeth, wash their face, shave, and wash their hands, armpits, groin, and feet. The patrol will not leave trash behind. The mission variables of METT-TC (I) are always taken into consideration prior to executing sanitation and personal hygiene activities. (See ATP 4-25.12 for additional information on basic field sanitation and hygiene.)

SECTION II – PLANNING AND PREPARATION

7-40. Planning and preparation are the processes by which the leader develops a specific course of action (COA) for execution focused on the expected results from the patrol. Planning and preparation help the patrol leader create and communicate a common vision and a shared understanding between subordinate leaders. The two processes result in an order that synchronizes the action of forces in time, space, and purpose to achieve objectives and accomplish missions. The leader relies on intuitive decision making and direct contact with subordinate leaders to integrate activities when circumstances are not suited for a more deliberate planning timeline.

Note. As with any mission assigned to the patrol, the patrol leader employs troop leading procedures (TLPs), see chapter 2, to develop the plan and prepare for the mission.

INITIAL COORDINATIONS AND ACTIVITIES

7-41. The patrol leader normally will receive the operation order (OPORD) in the battalion or company command post where communications are good and vital personnel are available for coordination. The leader must identify required actions on the objective, plan backward to the departure from friendly lines, then forward to the reentry of friendly lines. Because patrols act semi-independently, move beyond the supporting range of the parent unit, and often operate forward of friendly units, coordination and activities must be thorough and detailed. As the patrol leader plans, the following elements in paragraphs 7-42 to 7-56 are considered.

7-42. Environment, local situation, and possible threats. The patrol leader should coordinate an intelligence briefing covering the operational environment considerations, (specifically the enemy, terrain, weather, and civil considerations, and integrated into these variables, informational considerations) that might affect the patrol's mission, general and specific threats to the patrol, suspect persons, and vehicles and locations known to be in the patrol's area of operations (AO).

7-43. Mine and explosive device threat. The patrol leader should make a mine and explosive device risk assessment based on the latest information available. This will determine many of the actions of the patrol. Patrol members must be informed of the latest mine and explosive device threats and restrictions to the unit's tactical SOPs.

7-44. Operations update. The patrol leader should coordinate for an up-to-date briefing on the location and intentions of other friendly patrols and units in the patrol's AO. This briefing should include the existing control measures in effect, restricted terrain areas, special effects of the patrol's area, and all other operational issues affecting the patrol and its mission.

7-45. Mission and tasks. Every patrol leader should be given a specific task and purpose to accomplish with a patrol. Accordingly, each patrol member knows the mission and is aware of individual responsibilities.

7-46. Locations and route. The leader must brief the patrol on all pertinent locations and routes. Locations and routes may include drop-off points, pick-up points, planned routes; rally points, exit and re-entry points, and alternates for each should be covered in detail.

7-47. Posture. This is a vital consideration during a civil reconnaissance patrol. (See paragraph 7-209.) The patrol leader should not depart until the leader is sure the patrol completely understands what posture or attitude the leader wishes the patrol to present to the populace it encounters. The posture may be soft or hard depending on the situation and environment. The patrol posture may change several times during a patrol.

7-48. Biometric enrollments/biometric-enabled watchlist (BEWL). An additional consideration during civil reconnaissance may be the number of biometric enrollments accomplished as well as how many people were identified with organic biometric devices as Tier/Level 1 to 6 targets on the BEWL. Biometrics collections and its use prior to conducting essential tasks or activities enhance protection. Soldiers utilizing the BEWL loaded on handheld devices or other biometrics collect/match systems can identify individuals via prior biometric enrollments so that regardless of who they say they are their identities are known with certainty. (See FM 3-24.2 for additional information.)

7-49. Personnel recovery. The efforts taken to plan, prepare for, and effect the recovery and return to friendly command and control of U.S. military, DOD civilian and contractor personnel, or other personnel as directed by the President or Secretary of Defense, who are isolated in an operational environment that requires them to survive, communicate, organize, resist, escape, and evade. (See ATP 3-21.10 for additional information.)

7-50. Actions on contact and actions at the scene of an incident. These are likely to be part of the unit's tactical SOPs but should be covered especially if there are local variations or new patrol members.

7-51. ROE, rules of interaction, and rules for escalation of force. Each patrol member must know and understand these rules. (See ATP 3-21.10 for additional information.)

7-52. Communications plan/lost communications plan. Every patrol member should know the means in which the patrol plans to communicate, to whom, how, and when the patrol should report. The patrol leader must consider what actions the patrol will take in the event it loses communications. The unit may have established these actions in its tactical SOP, but all patrol members should be briefed on the communication plan and be given the appropriate frequencies, contact numbers, and passwords in effect.

7-53. Electromagnetic warfare (EW) countermeasures plan. This is especially important if the explosive device threat level is high. The patrol leader should clearly explain to all patrol members which EW devices are being employed and their significant characteristics. These issues may be covered by the unit's tactical SOP, but all patrol members should be briefed on the EW plan in effect during the patrol.

7-54. Standard and special uniforms and equipment. Equipment should be distributed evenly among the patrol members. The location of essential or unique equipment should be known by all members of the patrol. SOPs should be developed to stipulate what uniform is to be worn for various types of patrols. The dress state will be linked to threats proper preparations. All patrols must have a day and night capability regardless of the expected duration of the patrol.

7-55. Medical. Every Soldier should carry their own individual improved first-aid kit per unit tactical SOP. The leader should ensure that every patrol has a combat medic and one CLS qualified Soldier with a CLS bag. All patrol members must know who is responsible for carrying the bag and know how to use its contents.

7-56. Attachments. The patrol leader must ensure all personnel attached to the patrol are introduced to the other patrol members and briefed thoroughly on the tactical SOP; all patrol special orders; and existing chain of command. The following personnel may be attached to a unit going out on patrol:

- Machine gun, antiarmor, sniper, sapper, and air defense teams.
- Signal intelligence and human intelligence collection teams.
- Interpreters and host-nation security forces.
- Explosive ordnance disposal teams.

- Female Soldiers specifically designated and trained to search local women, when required.
- Military police or military working dog teams.

ESSENTIAL ELEMENTS AND SUPPORTING TASKS

7-57. The patrol leader ensures all essential elements and supporting tasks have been assigned to be performed on the objective, at rally points, at chance contact, at danger areas, at security or observation locations, along the routes, and at passage lanes. Example essential elements and supporting tasks are addressed in paragraphs 7-58 to 7-79.

KEY TRAVEL AND EXECUTION TIMES

7-58. The leader estimates time requirements for movement to the objective, this includes movement to an ORP, leader's reconnaissance of the objective, establishment of security and surveillance, and completion of all assigned tasks on the objective. The leader also estimates time from the objective back to the ORP and the movement to return through friendly lines.

PRIMARY AND ALTERNATE ROUTES

7-59. The leader selects primary and alternate routes to and from the objective. Return routes should differ from routes to the objective. (See figure 7-4.)

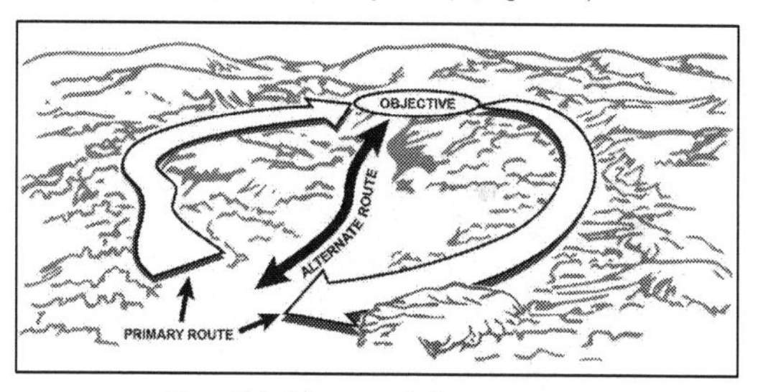

Figure 7-4. Primary and alternate routes

SIGNALS

7-60. The leader should consider the use of special signals. These include hand and arm signals, flares, pyrotechnics, voice, whistles, radios, and visible or nonvisible lasers. (See TC 3-21.60.) All signals are rehearsed to ensure all patrol members understand what they mean. The leader incorporates a PACE plan for all signals during the patrol.

The PACE plan is a communication plan that exists for a specific mission or task, not a specific unit, as the plan considers both intra- and inter-unit sharing of information. The PACE plan designates the order in which an element will move through available communications systems until contact can be established with the desired distant element.

CHALLENGE AND PASSWORD OUTSIDE OF FRIENDLY LINES

7-61. The challenge and password from the signal operating instructions must not be used when the patrol is outside friendly lines. The unit's tactical SOP should state the procedure for establishing a patrol challenge and password as well as other combat identification features and patrol markings. Two methods for establishing a challenge and password are the odd number system and running password.

Odd Number System

7-62. The leader specifies an odd number. The challenge can be any number less than the specified number. The password will be the number that must be added to it to equal the specified number, for example, the number is 9, the challenge is 4, and the password is 5.

Running Password

7-63. Signal operating instructions also may designate a running password. This code word alerts a unit that friendly are approaching in a less than organized manner and possibly under pressure. The number of friendly approaching follows the running password. For example, if the running password is "eagle," and seven friendlies are approaching, they would say "eagle seven."

LOCATIONS OF KEY LEADERS

7-64. The patrol leader considers where best to locate throughout each phase of the patrol, and where to locate the APL, and other essential leaders for each phase of the patrol. The APL normally is with the following elements for each type of patrol:

- On a raid or ambush, the APL can be with the patrol leader on the objective or control the support element from the support position.
- On an area reconnaissance, the APL can move with one of the area reconnaissance elements or supervise security in the ORP.
- On a zone reconnaissance, the APL can move with one of the zone reconnaissance elements or move with the reconnaissance element setting up the linkup point.

ACTIONS ON CHANCE CONTACT

7-65. The leader's plan must address actions on chance contact at each phase of the patrol. (See paragraphs 2-48 to 2-52 for additional information on actions on contact.) For the patrol's mission the plan must address—

- The handling of seriously wounded in action and killed in action personnel.
- The actions required to recover isolated Soldiers.
- The handling of prisoners captured because of chance contact who are not part of the planned mission.

Note. In both the offense and defense, the nine forms of contact are direct; indirect; nonhostile; obstacle; chemical, biological, radiological, and nuclear (CBRN); aerial; visual; electromagnetic; and influence.

7-66. Control measures help leaders anticipate chance contact (see paragraphs 7-7 to 7-11 for additional information on rally points). They include:

- ORP. An area designated for an arranged meeting from which to begin an action or phase of an operation or to return to after an action or operation.
- Rally point. A place designated by the leader where the patrol moves to reassemble and reorganize if it becomes dispersed.
- Release point. A location on a route where marching elements are released from centralized control. The release point also is used after departing the ORP.
- Linkup point. A point where two infiltrating elements in the same or different infiltration lanes are scheduled to consolidate before proceeding with their missions. (See chapter 6 section IV for additional information.)

ACTIONS AT DANGER AREAS

7-67. When analyzing the terrain through METT-TC (I) during TLP, the patrol leader (for example, the platoon leader) may identify danger areas. When planning the route, the leader marks the danger areas on the map overlay. The term danger area refers to areas on the route where the terrain could expose the patrol to enemy observation, fire, or both. If possible, the leader plans to avoid danger areas, but sometimes cannot. When the unit must cross a danger area, it does so as quickly and carefully as possible. During planning, the leader designates nearside and far-side rally points. If the patrol encounters an unexpected danger area, it uses the en route rally points closest to the danger area as far side and nearside rally points. Examples of danger areas include:

- Open areas. First, try boxing around if feasible (see figure 7-7 on page 7-23). If not, conceal the patrol on the near side and observe the area. Post security to give early warning. Send an element across to clear the far side. When cleared, cross the remainder of the patrol at the shortest exposed distance and as quickly as possible (see figure 7-6 on page 7-22).
- Roads and trails. Crossroads or trails at or near a bend, a narrow spot, or on low ground.
- Villages. Pass villages on the downwind side and far away from them. Avoid animals, especially dogs, which might reveal the patrol's presence.
- Enemy positions. Pass on the downwind side. (The enemy might have scout dogs.) Be alert for trip wires and warning devices.

- Minefields. Bypass minefields, if possible, even if it requires changing the route by a great distance. Clear a path through minefields only if necessary.
- Streams. Select a narrow spot in the stream offering concealment on both banks. Observe the far side carefully. Emplace nearside and far-side security for early warning. Clear the far side and cross rapidly but quietly.
- Wire obstacles. Avoid wire obstacles. (The enemy may cover obstacles with observation and fire.)

Crossing Danger Areas

7-68. Regardless of the type of danger area, when the patrol must cross one independently, or as the lead element of a larger force, it must perform the following:

- When the lead element signals "danger area" (relayed throughout the patrol), the patrol halts.
- The patrol leader moves forward, confirms the danger area, and determines what technique the patrol will use to cross. The APL (for example, the platoon sergeant) also moves forward to the patrol leader.
- The patrol leader informs all element leaders (for example, squad leaders) of the situation, the nearside and far-side rally points.
- The APL directs positioning of the nearside security (usually conducted by the trail element or squad). These two security elements may follow the APL forward when the patrol halts and a danger area signal is passed back.
- The patrol leader reconnoiters the danger area and selects the crossing point providing the best cover and concealment.
- Nearside security observes to the flanks and overwatches the crossing.
- When the nearside security is in place, the patrol leader directs the far-side security element (for example, a fire team) to cross the danger area.
- The far-side security element clears the far side. The team clears to a depth sufficient to allow the entire element to cross.
- The far-side security element leader establishes an observation post forward of the cleared area.
- The far-side security element signals to the element leader (for example, the squad leader) the area is clear. The element leader relays the message to the patrol leader.
- The patrol leader selects the method the patrol will use to cross the danger area.
- The patrol quickly and quietly crosses the danger area.
- Once across the danger area, the main body begins moving slowly on the required azimuth.
- The nearside security element, controlled by the APL, crosses the danger area where the patrol crossed. They may attempt to cover tracks left by the patrol.
- The APL ensures everyone crosses and sends up the report.

Note. Same principles stated above are used when crossing a smaller unit (such as a squad) across a danger area.

7-69. The patrol leader decides how the unit will cross based on the time the unit has, size of the unit, size of the danger area, fields of fire into the area, and amount of security available to post. Units may cross all at once, in buddy teams, or one Soldier at a time. A large unit normally crosses its elements one at a time. As each element crosses, it moves to an overwatch position or to the far-side rally point until told to continue movement.

Crossing of Linear Danger Areas

7-70. A linear danger area is an area where the patrol's flanks are exposed along a relatively narrow field of fire. Examples include streets, roads, trails, and streams. The patrol crosses a linear danger area in the formation and location specified by the patrol leader. (See figure 7-5.)

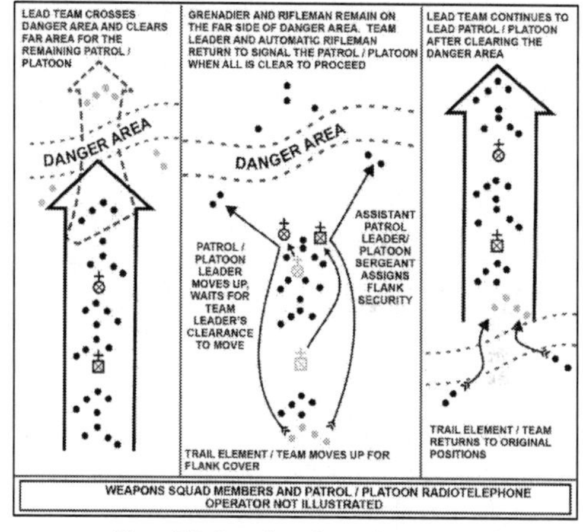

Figure 7-5. Crossing a linear danger area

Crossing of Large Open Areas

7-71. If the large open area is so large the patrol cannot bypass it due to the time needed to accomplish the mission, a combination of traveling overwatch and bounding overwatch is used to cross the large open area. (See figure 7-6.) The traveling overwatch technique is used to save time. The patrol (platoon or squad) moves using the bounding overwatch technique at any point in the open area where enemy contact may be expected. The technique also may be used once the patrol comes within range of enemy small-arms fire from the far side (about 250 meters). Once beyond the open area, the patrol re-forms and continues the mission. Depending on the security situation, level of threat, and desired tempo, the patrol leader may emplace machine guns on the near side to provide overwatch while crossing a large open area.

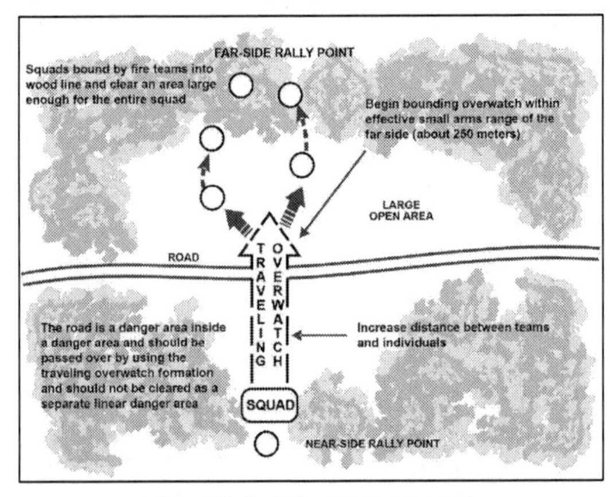

Figure 7-6. Crossing a large open area

Crossing of Small Open Areas

7-72. Small open areas are small enough to bypass in the time allowed for the mission. Two techniques can be used: contouring around the open area method or detour bypass method. (See figure 7-7.)

Contouring Around the Open Area Method

7-73. The leader designates a rally point on the far side with the movement azimuth. The leader then decides which side of the open area to contour around (after considering the enemy situation, distance, terrain, cover, and concealment), and moves around the open area. The leader uses the wood line and vegetation for cover and concealment. When the patrol arrives at the rally point on the far side, the leader reassumes the azimuth to the objective area and continues the mission.

Detour Bypass Method

7-74. The patrol turns 90 degrees to the right or left around the open area and moves in the direction of travel. Once the patrol has passed the danger area, the unit completes the box with another 90-degree turn and arrives at the far-side rally point, then continues the mission. The pace counts of the offset and return legs is not added to the distance of the planned route.

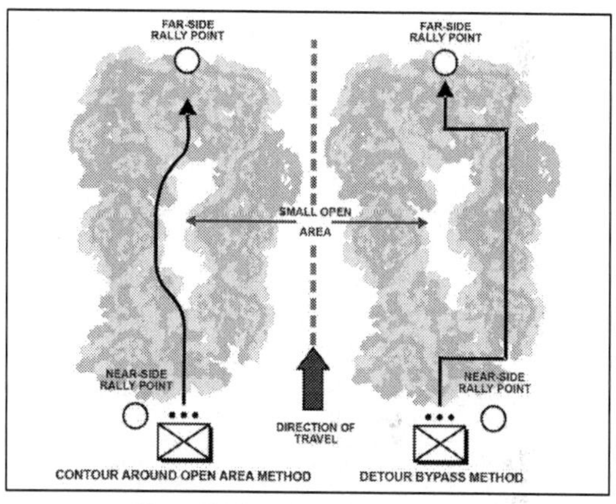

Figure 7-7. Crossing a small open area

Enemy Contact at Danger Areas

7-75. An increased awareness of the situation helps the patrol leader control the patrol when it makes contact with the enemy. If the patrol makes contact in or near the danger area, it moves to the designated rally points. Based on the direction of enemy contact, the leader designates the far- or nearside rally point. During limited visibility, the leader also can use laser systems to point out the rally points at a distance. If the patrol has a difficult time linking up at the rally point, dependent upon the enemy threat, the first element to arrive may position listening/observation post(s) away from the rally point along natural lines of drift and/or good observation points into the rally point. Also, it may mark the rally point (for limited periods) with an infrared light source to help direct the rest of the patrol to the location. During movement to the rally point, position updates allow separated elements to identify each other's locations. These updates help them linkup at the rally point by identifying friends and foes.

DEPARTURE FROM FRIENDLY LINES OR FIXED BASE

7-76. The departure from friendly lines, or from a fixed base, must be thoroughly planned and coordinated. The patrol leader coordinates with the unit through which the patrol conducts its forward and rearward passage of lines. The leader also coordinates patrol activities with its higher echelon and the leaders of other units patrolling in adjacent areas. Paragraphs 7-77 to 7-79 highlight additional leader considerations.

Higher Echelon Coordination

7-77. The patrol leader may routinely coordinate with the higher commander and/or elements of the battalion staff directly. Higher echelon units and the patrol leader should have well developed tactical SOPs with detailed checklists to preclude omitting items vital to mission accomplishment. Items coordinated between the patrol leader and higher commander and/or battalion staff include:

- Changes or updates in the enemy situation.
- Best use of terrain for routes, rally points, and patrol bases.
- Light and weather data.
- Changes in the friendly situation.
- The attachment of Soldiers with special skills or equipment (engineers, sniper teams, military working dog teams, FOs, or interpreters).
- Use and location of landing or pickup zones.
- Departure and reentry points of friendly lines.
- Direct and indirect fire support on the objective and along the planned routes, including alternate routes.
- Rehearsal areas and times. The terrain for rehearsal should be like the objective, to include buildings and fortifications if necessary. Coordination for rehearsals includes security of the area, and when appropriate, use of blanks, pyrotechnics, and live ammunition.
- Special equipment and ammunition requirements.
- Transportation support, including transportation to and from rehearsal sites.
- Signal plan, call signs frequencies, code words, pyrotechnics, and challenge and password.

Friendly Unit Coordination

7-78. The patrol leader must coordinate with the commander of the forward unit and leaders of other units patrolling in the same or adjacent areas. The coordination includes automated network control device information, signal plan, fire plan, running passwords, procedures for departure and reentry of lines, planned dismount points (when mounted during a part of the mission), initial rally points, and actions at departure and reentry points. Additional coordination includes:

- The patrol leader providing the forward unit leader with the patrol's identification, size, departure and return times, and AO.
- The forward unit commander providing the patrol leader with the following:
  - Additional information on terrain just outside the friendly unit lines.
  - Known or suspected enemy positions in the near vicinity.
  - Likely enemy ambush sites.
  - Latest enemy activity.
  - Detailed information on friendly positions, obstacles, and observation posts.
  - Friendly unit fire plan.
  - Support the forward unit can provide (direct and indirect fire support, litter teams, guides, communications, and quick reaction force).

Planning for Departure

7-79. In planning for departure of friendly lines, the patrol leader considers the following sequence of actions:

- Contacting friendly guides at the contact point.
- Moving to a coordinated initial rally point just inside friendly lines.
- Completing final coordination.
- Moving to and through the *passage point*—a designated place where passing units pass through the stationary unit (FM 3-90).
- Establishing a security-listening halt beyond the friendly unit's FPFs.

PRE- AND POST-DEPARTURE PREPARATIONS

7-80. Patrols should rehearse all specific tactical tasks and battle drills for any situation the patrol could encounter. Patrol preparation activities prior to departure and after departure are discussed in paragraphs 7-81 to 7-95.

COMMUNICATIONS CHECKS

7-81. Communications checks should be conducted with the patrol's own unit headquarters or command post before every patrol. Patrols should not leave the vicinity of the main body until all communication systems are operating correctly.

Patrol Manifest

7-82. When the situation allows, the patrol leader should submit a written patrol manifest to the commander or to command post personnel prior to departing the main body. Regardless of the situation, whenever the unit sends out a patrol there should be a specific list of the patrol members made before it departs. It should contain the following information:

- Patrol number or call sign designation.
- Unit designation of unit sending the patrol out.
- Patrol task and purpose (mission).
- Names and rank of patrol leader and all subordinate leaders.
- Estimated date-time-group out.
- Estimated date-time-group in.
- Brief description of the patrol's intended route.
- Complete names, rank, and unit of all members of the patrol, including attachments.
- Number, nomenclature, and serial number of all weapons with the patrol.
- Number, nomenclature, and serial number of all EW devices, radios, and other special or sensitive equipment with the patrol.
- Vehicle type and registration number, if appropriate.

7-83. The purpose of the manifest is to allow the higher headquarters to keep track of all the patrols that are out and those that have returned. If the patrol engages the enemy or fails to return on time without reporting, the headquarters has information on the size, capability, and intentions of the patrol that it may need. If the patrol suffers casualties or has a vehicle disabled, this manifest can be used to check that all personnel, weapons, and sensitive items were recovered.

Departure Report

7-84. The patrol leader should render a departure report just as the patrol departs the main body location or the base. Depending on the procedure established by the unit's tactical SOP, this might include a detailed listing of the patrol's composition. It also may simply state the patrol's call sign or patrol number and report its departure.

Weapons Status

7-85. Immediately upon leaving an established base or the main body position, the patrol leader, vehicle commanders (when vehicles attached), and team leaders should ensure all the patrol weapons are loaded and prepared for in accordance with ROE. EW equipment should be checked to ensure it is turned on if appropriate and all radio frequency settings should be confirmed.

7-86. When the patrol returns to the base, all Soldiers clear their weapon immediately after entering the protected area. The unit's tactical SOP normally will establish precise procedures for this clearing. The patrol leader (or APL) should ensure all individual and crew-served weapons are cleared.

EQUIPMENT

7-87. Equipment carried by the patrol will be environment- and task-specific. Types of equipment carried include—

- Radios. Radio equipment should be checked prior to every patrol ensuring it is serviceable and operates correctly. Batteries must be taken for expected duration of the patrol plus some extra for backup. Patrol members must be trained in the operation of all radio equipment. It is the patrol leader's responsibility to ensure radio equipment is switched on and working and communication checks are conducted prior to leaving the base location.
- Weapons. All weapons must be prepared for firing prior to departure from the larger unit. Slings should be used to ensure weapons do not become separated from Soldiers who became incapacitated. This also ensures a weapon cannot be snatched away from a distracted Soldier while they are speaking with locals and used against them.
- Ammunition. Sufficient ammunition, signal pyrotechnics, obscurants, and nonlethal munitions must be carried to enable the patrol to conduct its mission. The amount a patrol carries may be established by the unit's tactical SOP or the mission the patrol faces.
- Load-carrying equipment. Patrol members should carry sufficient team and personal equipment to enable them to accomplish other missions such as reassignment to a cordon position before returning to the larger unit for resupply. The unit's tactical SOP should establish the standard amount of equipment and supplies to be carried. The leader carefully considers the burden being placed on Soldiers going on a foot patrol, especially in extreme weather conditions or rugged terrain.
- Special purpose equipment (attached augmented teams and individuals). These equipment enablers include, in addition to special purpose equipment teams established internally, individuals with specific environmental skills and capabilities, and attachments such as civil affairs teams, military working dogs, public affairs teams, human intelligence collection teams, EW (protection and support) assets, and interpreters. The patrol leader leverages these enablers by effectively task organizing them within the formation. The leader considers direct fire control measures (see appendix A) and indirect fire coordination measures (see appendix B), and how they will affect supporting enablers, such as engineer support. All attachments, if possible, attend rehearsals to ensure they understand the patrol's purpose and their individual task(s) and purpose in the operation.
- Documentation. Team leaders are responsible to the patrol leader for ensuring appropriate documentation is carried by individuals for conducting the mission. Under normal circumstances, Soldiers should carry just their identification card and tags. The unit tactical SOP may prohibit or require the

carrying of other appropriate theater specific documentation such as cards with rules on escalation of force or ROE.

7-88. Several equipment checks (individual and precombat checks [PCCs]) and inspections (precombat inspections [PCIs]) are conducted prior to the patrol departing. (See paragraphs 2-120 to 2-123 for additional information on PCCs and PCIs.) Patrol checks/inspections emphasize the following areas:

- Individual equipment check. It is the responsibility of every patrol member to check their individual equipment. Soldiers should ensure all loose items of carried equipment are secured.
- Squad leader and team leader individual and equipment check. Leaders must ensure individual team members limit what they carry to which is required for the patrol. Equipment must be checked for serviceability.
- Patrol leader and APL inspections. Patrol leader and APL inspect individual and team equipment from each team prior to deploying, paying particular attention to the serviceability of mission specific equipment.

EXITING AND ENTERING A FIXED BASE

7-89. Exiting and entering a fixed operating base is a high-risk activity due to the way Soldiers are channeled through narrow entry or exit points. Enemies are known to monitor patrols leaving and entering base locations to identify and exploit patterns and areas of weakness. Patrols leaving and entering a base reduce the risks of attack by varying the points used to exit and enter the base, and routes used to transit the immediate area around the base. If this is not possible, extreme caution should be used in the vicinity of the exit and entry points. Patrol leaders ensure their patrols do not become complacent. Units should ensure close coordination between patrol leaders and guards at the entry point while the patrol is transiting the gate.

SECURITY CHECKS WHILE ON PATROL

7-90. Patrol members assist their patrol leader by consistently applying basic patrolling techniques. This gives the team leader more time to concentrate on assisting the patrol leader in the conduct of the patrol. Team members should concentrate on maintaining spacing, formation, alertness, conducting 5- and 25-meter checks and taking up fire positions without supervision.

FIVE- AND TWENTY-FIVE METER CHECKS

7-91. Every time a patrol stops, it should use a fundamental security technique known as the 5- and 25-meter check. The technique requires every patrol member to make detailed, focused examinations of the area immediately around the member, looking for anything out of the ordinary that might be dangerous or significant. Five-meter checks should be conducted every time a patrol member stops. Twenty-five-meter checks should be conducted when a patrol halts for more than a few minutes.

7-92. Soldiers should conduct a visual check using their unaided vision, and by using the optics on their weapons and binoculars. They should check for anything suspicious, and anything out of the ordinary. This might be as minor as bricks missing from walls, new string or wire run across a path, mounds of fresh soil, or other suspicious signs. Check the area at ground level through to above head height.

7-93. When the patrol makes a planned halt, the patrol leader identifies an area for occupation and stops 50 meters short of it. While the remainder of the patrol provides security, the patrol leader carries out a visual check using binoculars. After moving the patrol forward 20 meters from the position, the patrol leader conducts a visual check using optics on the weapon or with unaided vision.

7-94. Before occupying the position, each Soldier conducts a thorough visual and physical check for a radius of 5 meters. Each Soldier must be systematic, take time and show curiosity. Use touch and, at night, low-visibility lighting (infrared) or preferable thermal optics.

7-95. Obstacles must be physically checked for command wires. Fences, walls, wires, posts, and ground immediately underneath must be carefully felt by hand, without gloves.

SECTION III – COMBAT PATROLS

7-96. **A *combat patrol* is a patrol that provides security and harasses, destroys, or captures enemy troops, equipment, or installations.** When the commander assigns a combat patrol mission, the commander intends the patrol to make contact with the enemy and engage in close combat. A combat patrol always tries to remain undetected while moving, but when it discloses its location to the enemy, it is with a sudden and violent attack. For this reason, the patrol normally carries a significant number of weapons and ammunition. It may carry specialized munitions. A combat patrol collects and reports information gathered during the mission, whether related to the combat task or not. The three types of combat patrols are raid, ambush, and security.

ESSENTIAL ELEMENTS FOR A COMBAT PATROL

7-97. The three essential elements for a combat patrol are assault, support, and security. Assault elements accomplish the mission during actions on the objective. Support elements suppress or destroy enemy on the objective in support of the assault element. Security elements assist in isolating the objective by preventing enemy from entering and leaving the objective area as well as by ensuring the patrol's withdrawal route remains open. The size of each element is based on the situation and patrol leader's analysis of the mission variables of METT-TC (I).

ASSAULT ELEMENT

7-98. The assault element is the combat patrol's main effort. Its task is to conduct actions on the objective. In most cases, the assault element will accomplish the overall purpose. This element must be capable (through inherent capabilities or positioning relative to the enemy) of destroying or seizing the target of the combat patrol. Activities typically associated with the assault element include:

- Conduct of assault across the objective to destroy enemy equipment, capture or kill enemy, and clearing of key terrain and enemy positions.
- Maneuver close enough to the objective to conduct an immediate assault if detected.
- Being prepared to support itself if the support element cannot suppress the enemy.
- Provide support to a breach element (when established) in breaching of obstacles (see chapter 6 section I), if required.
- Plan detailed fire control and distribution.
- Conduct controlled withdrawal from the objective.

7-99. Analysis of the mission variables of METT-TC (I) may result in the requirement to organize special purpose teams and activities. Designated special purpose teams and activities may include:

- Surveillance teams. To establish and maintain covert observation of an objective for as long as it takes to complete the patrol's mission.
- Search teams. To find and collect documents, equipment, and information in the objective area.
- Detainee teams. To capture, secure, and document detainees (see paragraphs 4-48 to 4-53).
- Demolition teams. To plan and execute the destruction of obstacles and when necessary, enemy equipment.
- Breach team. To create lanes in protective obstacles (particularly for a raid) to assist the assault team in getting to the objective.
- Aid and litter teams. To identify, collect, render immediate aid, and coordinate casualty evacuation (CASEVAC).

SUPPORT ELEMENT

7-100. The support element suppresses the enemy on the objective using direct and indirect fires. The support element is a shaping effort setting conditions for the mission's main effort. This element must be capable of supporting the assault element (and breach team[s] when established). The support force can be divided into multiple elements.

7-101. The support element is organized to prevent a threat of enemy interference with the assault elements. The support force suppresses, fixes, or destroys elements on the objective. The support force's primary responsibility is to suppress enemy forces to prevent them repositioning against the main effort. The support force—

- Initiates fires and gains fire superiority with crew-served weapons and indirect fires.
- Controls rates and distribution of fires.
- Shifts/ceases fire on signal.
- Overwatches the withdrawal of the assault element.

SECURITY ELEMENT

7-102. The security element is a supporting effort having three roles. The first is to isolate the objective from enemy personnel and vehicles attempting to enter the objective area. These actions range from simply providing early warning, to blocking enemy movement. This element may require several different forces located in various positions. The patrol leader is careful to consider enemy reserves or response forces that will be alerted once the engagement begins. The second role is to prevent enemy from escaping the objective area. The third role is to secure the patrol's withdrawal route.

7-103. All elements of the patrol are responsible for their own local security. What distinguishes the security element is they are protecting the entire patrol. Their positions must be such they can, in accordance with their engagement criteria, provide early warning of approaching enemy.

7-104. The security element is organized to prevent a threat of enemy interference with the assault element. To facilitate the success of the assault element, the security element must fix, or block (or at a minimum screen) all enemy security or response forces located on parts of the battlefield away from the raid objective, or ambush site.

7-105. The security observation post graphic control measure symbol (see figure 7-8) identifies the security element position(s) during combat patrols. An ideal security observation post—

- Does not mask fires of the main body.
- Provides timely information to the main body. (Gives the leader enough time to act on information provided.)
- Can provide a support by fire position.

Note. The security observation post graphic control measure symbol is also used to identify a reconnaissance and security element position(s) during reconnaissance patrols.

Figure 7-8. Security observation post control measure symbol

CONDUCT OF THE RAID

7-106. A *raid* is a variation of attack to temporarily seize an objective with a planned withdrawal (FM 3-90). A raid is a variation of an attack, usually small scale, involving battalion size or smaller forces requiring detailed intelligence (precise, time-sensitive, all-source intelligence), planning, and preparation. At the platoon level, a raid is a surprise attack against a position or installation for a specific purpose other than seizing and holding the terrain. A raid patrol retains terrain just long enough to accomplish the intent of the raid. The raid ends with a planned withdrawal off the objective and a return to the main body. Characteristics of a raid include:

- Destruction of essential systems or facilities (command and control nodes, logistical areas, other high value areas).
- Collection of critical information.
- Capturing of hostages or prisoners.
- Confusing the enemy or disrupting their plans.
- Detailed information collection (significant collection assets committed during planning).
- Command and control from the higher headquarters to synchronize the operation.
- Taking advantage of a window of opportunity.

FUNDAMENTALS

7-107. The fundamentals of the raid include surprise and speed, coordinated fires, violence of action, and a planned exfiltration. *Exfiltrate* is a tactical mission task in which a unit removes Soldiers or units from areas under enemy control by stealth, deception, surprise, or clandestine means (FM 3-90). Surprise and speed are accomplished through infiltration and moving to the objective undetected. Coordinated and well synchronized direct and indirect fires seal off the objective. Violence of action overwhelms the enemy with fire and maneuver. The planned withdrawal allows friendly forces to move off of the objective in a well-organized manner while maintaining security. Raids are normally conducted in five phases (see paragraphs 7-110 to 7-122 for additional information)—

- Phase I – Insertion and/or infiltration.
- Phase II – Objective area sealed off (isolation).
- Phase III – Surprise attack.
- Phase IV – Objective seized, and task accomplished.
- Phase V – Withdrawal.

ORGANIZATION OF FORCES

7-108. The patrol leader considers the functions necessary to conduct a raid, and task organizes appropriately. The raid is normally organized as three elements; security element (see paragraphs 7-102 to 7-105), support element (see paragraphs 7-100 to 7-101), and the assault element (see paragraphs 7-98 to 7-99). Each sub-element is organized and equipped to do a specific part of the overall mission depending upon the specific mission, nature of the target, and enemy situation, and terrain.

PLANNING AND PREPARATION

7-109. Although the planning and preparation discussed in chapter 4 for the attack apply for the raid as well, there are some differences. Because a raid is normally conducted deep in enemy controlled territory and often conducted against an enemy of equal or greater strength, the plan must ensure the unit is not detected prior to initiating the assault. Detailed planning and preparation ensure the success of the raid as well as the survivability of the raid force during infiltration, actions on the objective, and withdrawal from the objective area.

Phases of the Raid

7-110. When planning the raid, the patrol leader considers the phases of the raid. As stated above, raids are normally conducted in five phases; insertion and/or infiltration, objective area sealed off (isolation), surprise attack, objective seized and task accomplished, and withdrawal. All phases should be rehearsed but with special emphasis on the objective seizure and task accomplishment phase, and the withdrawal phase.

7-111. Phase I – Insertion and/or infiltration. The raiding force inserts or infiltrates (see paragraph 4-16) into the objective area (see figure 7-9 on page 7-34). The force launching the raid plans if possible, launching the raid during an unexpected time or place by taking advantage of darkness and limited visibility and moving over terrain that the enemy may consider impassable. The raid force avoids detections through proper movement techniques and skillful camouflage and concealment to include taking advantage of natural terrain cover. During infiltration, the patrol moves to and occupies the ORP (see paragraph 7-11) according to the patrol SOP. The patrol then prepares for the leader's reconnaissance (see paragraphs 7-15 to 7-16) and the patrol leader:

- Leaves a five-point contingency plan with the APL.
- Establishes the release point; pinpoints the objective; contacts the APL (dependent upon the situation) to prepare Soldiers, weapons, and equipment.
- Emplaces the surveillance team (if used) to observe the objective; and verifies the intelligence assessment and reports any differences or changes to the intelligence assessment.
- Provides a five-point contingency plan to the surveillance team.
- Brings forward security teams on the leader's reconnaissance and emplaces them before the leader's reconnaissance leaves the release point.
- Verifies location of and routes to security, support, and assault positions.
- Conducts the reconnaissance without compromising the patrol.
- Conducts a reconnaissance of support by fire position first, then the assault position.
- Confirms, denies, or modifies the plan and issues instructions to squad leaders.

- Assigns positions and withdrawal routes to all elements.
- Designates control measures on the objective (element objectives, lanes, limit of advance (LOA), target reference points (TRPs), and assault line).

Note. Upon the return of the leader's reconnaissance, the patrol leader allows squad leaders time to disseminate information and confirm that their elements are ready.

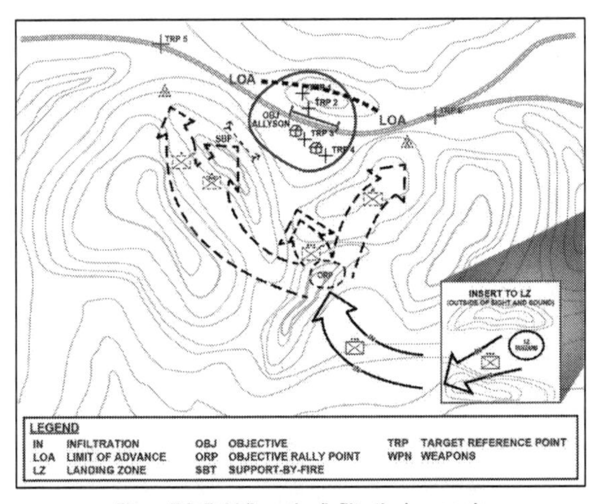

Figure 7-9. Raid (insertion/infiltration), example

7-112. Phase II – Objective area sealed off (isolation). The isolation of the objective area to seal off support or reinforcement from outside the objective, to include enemy air assets is key to the success of the operation. Raiding forces employ security elements, support elements, and both direct and indirect fires to successfully isolate the objective. (See figure 7-10.) Security elements occupy designated positions, moving undetected into positions that provide early warning and can seal off the objective from outside support or reinforcement. The support element leader moves the support element to designated positions then ensures the element can place well-aimed fire on the objective.

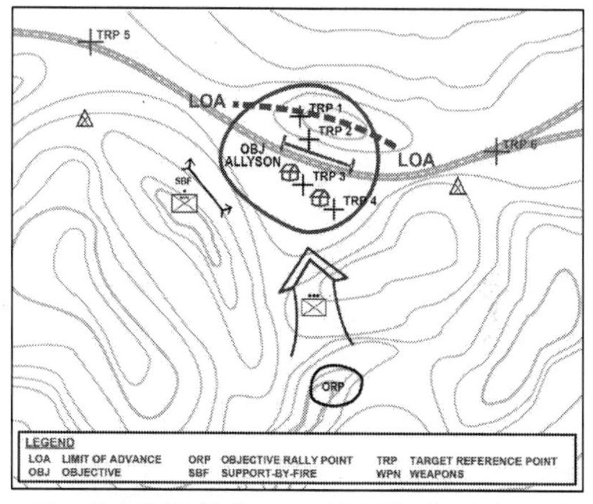

Figure 7-10. Raid (objective area sealed off/isolation), example

7-113. Phase III – Surprise attack. Any enemy forces at or near the objective are overcome in a violently executed surprise attack using all available firepower for shock effect. Assault forces move through the objective, while using the covering fire of support forces. A simple plan that is coordinated and rehearsed will allow for the assault element to conduct a rapid and precise assault into and through the objective. During phase three, the patrol leader moves with the assault element into the assault position. The assault position is normally the last covered and concealed position before reaching the objective. As it passes through the assault position, the assault element deploys into its assault formation. Its squads and fire teams deploy to place the bulk of their firepower to the front as they assault the objective. The assault element:

- Makes contact with the surveillance team to confirm any enemy activity on the objective.
- Ensures that the assault position is close enough for immediate assault if the element is detected early.
- Moves into position undetected and establish local security and fire control measures. Once in position:
  - Element leaders inform the patrol leader when their elements are in position and ready.
  - Patrol leader initiates the raid and directs the support element to fire.

- Patrol leader, upon gaining fire superiority, directs the assault element to move towards the objective.
- Assault element holds fire until engaged, or until ready to penetrate the objective.
- Patrol leader signals the support element to lift or shift fires.
- Support element lifts or shifts fires as directed, shifting fire forward of the assault element as it moves across the objective or to the flanks of targets or areas as directed in the order.

Note. When planning fires in support of a raid, the patrol leader chooses the at my command or time on target (expressed as a specific time, a not-later-than time, or a conclusive time period) as the method of engagement. This ensures the patrol leader has absolute control of fires on the target. For example, when rounds land on the target or if the fires do not happen the leader can cease-fire the mission before committing the assault. Due to the precise timing issues for a raid, when ready–method of engagement is rarely used. Finally, depending on when the patrol leader wants to incorporate fires, the leader needs to ensure the assault force is outside the danger close radius of the target.

7-114. Phase IV – Objective seized, and task accomplished. The assault element attacks and seizes the objective and accomplishes its assigned task quickly before any surviving enemy in the objective area can recover or be reinforced. It is imperative that assault element spend as little time as possible on the objective area while accomplishing its mission.

Note. The assault element may be required to breach a wire obstacle (see chapter 6 section I for additional information).

7-115. As the assault element moves onto the objective, it increases the volume and accuracy of fires (see figure 7-11). Squad leaders assign specific targets or objectives for their fire teams. Only when these direct fires keep the enemy suppressed can the rest of the unit maneuver. As the assault element gets closer to the enemy, there is more emphasis on suppression and less on maneuver. Ultimately, all but one fire team may be suppressing to allow that one fire team to break into the enemy position. Throughout the assault, Soldiers use proper individual movement techniques.

Note. During a deliberate attack, as addressed in chapter 4, when the enemy disposition is better understood the line formation is generally the preferred formation to use during the assault. Because the enemy situation on a raid objective may still be vague, assault elements generally retain their basic shallow wedge formation in order to maintain flexibility and security to their flanks in movement.

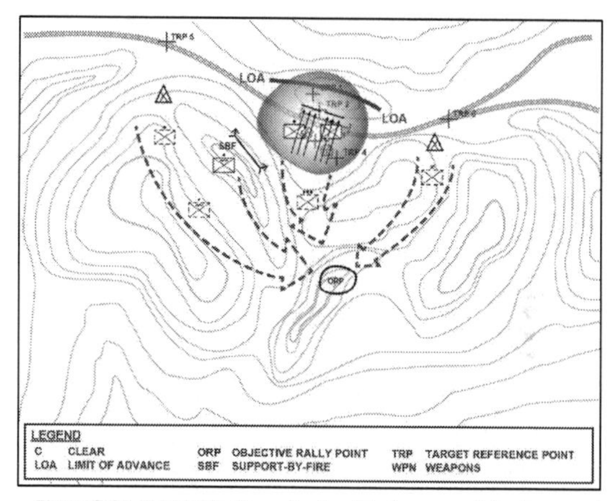

Figure 7-11. Raid (objective seized and task accomplished), example

7-116. Once the assault element assaults through the objective to the designated LOA, assault element leaders:

- Establish local security along the LOA and consolidate and reorganize as necessary.
- Provide liquid, ammunition, casualties, and equipment (known as LACE) reports to the patrol leader and APL.
- On order, special teams (see paragraph 1-31) accomplish all assigned tasks under the supervision of the patrol leader and/or APL, who are positioned to control the patrol.
- Special team leaders report to patrol leader or as directed to the APL when assigned tasks are complete.

7-117. The patrol leader or APL consolidates and reorganizes the patrol based on the contact:

- Establishes security, reorganizes key weapons, initiates the platoon casualty response plan (specifically tactical combat casualty care (TCCC), see chapter 8 section II), and prepares wounded Soldiers for CASEVAC or medical evacuation (MEDEVAC).
- Redistribute ammunition and supplies and relocate selected weapons to alternate positions if leaders believe that the enemy may have pinpointed them during the attack.

- Adjust (as directed by the patrol leader or APL) other positions for mutual support. Other position leaders provide LACE reports (see paragraph 8-13) to the patrol leader or APL.

7-118. Phase V – Withdrawal. The raiding force withdraws from the objective area to the ORP understanding that the patrol may move to the same ORP it moved to the objective from or may move to a different ORP, dependent on the situation or the extraction location. From the ORP the patrol can move rearward or to an extraction location. When using the same location for both insertion and extraction the patrol usually use a different route than what was used for movement to the ORP. The patrol leader generally considers a different location for extraction than that of the insertion location. Due to the nature of the raid normally being conducted in enemy held territory it is imperative that the routes, halts, linkup, and pickup locations be in areas where enemy contact is unlikely.

7-119. On order or signal of the patrol leader the assault element withdraws from the objective (see figure 7-12). Using prearranged signals, the assault line begins an organized withdrawal from the objective site, maintaining control and security throughout the withdrawal. The assault element bounds back near the original assault line and begins a single file withdrawal through any breached lane. All Soldiers move through the choke point for an accurate count. Once the assault element is a safe distance from the objective and the headcount is confirmed, the raid force can withdraw the support element. If the support elements were a part of the assault line, they withdraw together, and security is signaled to withdraw. Once the support is a safe distance off the objective, they notify the patrol leader, who contacts the security element and signals them to withdraw. All security teams link up at the release point and notify the patrol leader before moving to the ORP. Soldiers returning to the ORP immediately secure their equipment and establish all-around security. Once the security element returns, the raid force moves out of the objective area as soon as possible. Withdrawal actions include:

- Before withdrawing, the demolition team activates devices and charges.
- Support element or designated Soldier(s) in the assault element maintain local security during the withdrawal.
- Leaders report updated accountability and status to the patrol leader and APL.

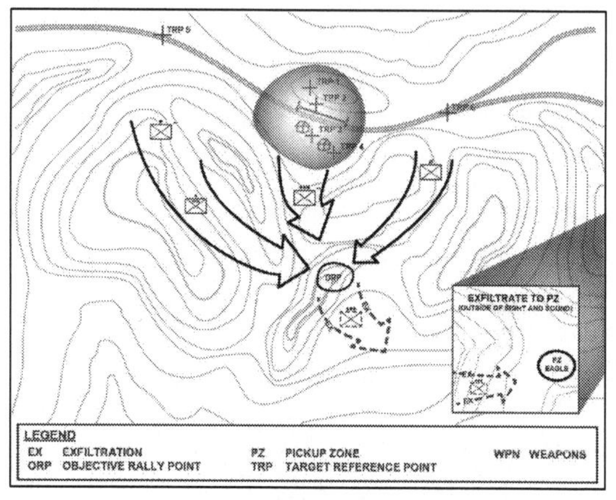

Figure 7-12. Raid (withdrawal/extraction), example

7-120. Elements withdraw from the objective in the order designated in the order (generally assault element first, support element second, and security element third) to the ORP. Then from the ORP, the patrol:

- Account for Soldiers and equipment.
- Disseminate information.
- Redistribute ammunition and equipment as required.

7-121. The patrol leader reports mission accomplishment to the higher headquarters from the ORP or during the withdrawal from the ORP rearward or to the extraction location. The report includes at a minimum:

- Raid assessment.
- Any information requirements and priority intelligence requirements (PIRs) gathered.

Movement to and Withdrawal from the Objective

7-122. The patrol leader considers how the patrol will arrive at the objective. The patrol leader assesses the need for additional lift capability to deliver the patrol closer to the objective, for example, rotary-/fixed-wing aircraft (see FM 3-99), or mounted ground movement (directly to the objective area or off set from the objective). It is important that the patrol leader ensures the insertion/infiltration is not compromised by the delivery platform(s). It is critical that the patrol leader conducts reverse planning, ensuring the development of the ground tactical plan prior to analyzing the delivery method for the patrol. If the delivery method requested is unable to support the ground tactical plan, the patrol leader re-assesses and makes necessary adjustments to allow for greater overall mission success. The patrol leader also considers site selection of landing zones, drop zones, ORPs, security observation posts, support locations, and assault positions. These specific sites must support the planned actions at the objective.

Additional Considerations

7-123. Additional planning and preparation considerations include but are not limited to—

- Security in all directions throughout the raid.
- Clear abort criteria for the raid based on the higher CCIRs. Criteria may include loss of personnel, equipment, or support assets, and changes in the enemy situation. Abort criteria is normally identified within each phase of the operation.
- Contingency plans for contact before and after actions on the objective.
- CASEVAC/MEDEVAC (when available) and raiding force extraction throughout the entire depth of the operation.
- Rally points for units to assemble during movement, to prepare for the attack, and to assemble after the mission is complete and when the force is ready to conduct their planned exfiltration.
- Contingency plans to prepare for and execute the recovery of isolated patrol members (see ATP 3-21.10 for additional information).

CONDUCT OF THE AMBUSH

7-124. An *ambush* is a variation of attack from concealed positions against a moving or temporarily halted enemy (FM 3-90). It can include an assault to close with and destroy the target or as an attack by fire. An ambush need not seize or hold ground. The general purpose of an ambush is to destroy or to harass enemy forces. The ambush combines the advantages of the defense with the advantages of the offense, allowing a smaller force with limited means the ability to destroy a much larger force. Ambushes are enemy-oriented. Terrain is held only long enough to conduct the ambush, and then the force withdraws.

7-125. Ambushes range from simple to complex and synchronized; short duration of minutes to long duration of hours; and within hand grenade range, to maximum standoff. Ambushes employ direct fire systems as well as other destructive means, such as antitank (AT) missiles, command-detonated mines and explosives, and indirect fires on the enemy force. The ambush may include an assault to close with and destroy the enemy or may just be a harassing attack by fire. Ambushes may be conducted as independent operations or as part of a larger operation.

7-126. The leader develops the ambush based on its purpose, form, time, and formation. The purpose of an ambush is either harassment or destruction. A harassing ambush is one in which attack is by fire only (meaning there is no assault element). A destruction ambush includes assault to close with and destroy the enemy.

7-127. A typical ambush is organized into three elements: assault, support, and security (see paragraphs 7-97 to 7-105). The assault element fires into the *kill zone*—the location where fires are concentrated in an ambush (FM 3-90). Its goal is to destroy the enemy force. When used, the assault force attacks into and clears the kill zone. It also may be assigned additional tasks, to include searching for items of intelligence value, capturing prisoners, photographing new types of equipment and when unable to take enemy equipment, completing the destruction of enemy equipment to avoid its immediate reuse. The support element supports the assault element by firing into and around the kill zone, and it provides the ambush's primary killing power. The support element attempts to destroy most of the enemy combat power before the assault element moves into the objective or kill zone. The security element isolates the kill zone, provides early warning of arrival of all enemy relief forces, and provides security for the assault and support elements. It secures the ORP and *blocks*—tactical mission tasks that deny the enemy access to an area or an avenue of approach (FM 3-90)—enemy avenues of approach into and out of the ambush site, which prevents the enemy from entering or leaving.

7-128. The three common forms of a small-unit ambush are point, area, and antiarmor ambushes. In a point ambush, Soldiers deploy to attack a single kill zone. In an area ambush, Soldiers deploy as two or more related point ambushes. These ambushes at separate sites are related by their purpose. (See figure 7-13 on page 7-42.) A unit smaller than a platoon normally does not conduct an area ambush. Antiarmor ambushes focus on moving or temporarily halted enemy armored vehicles. An antiarmor ambush can be either a point or area ambush.

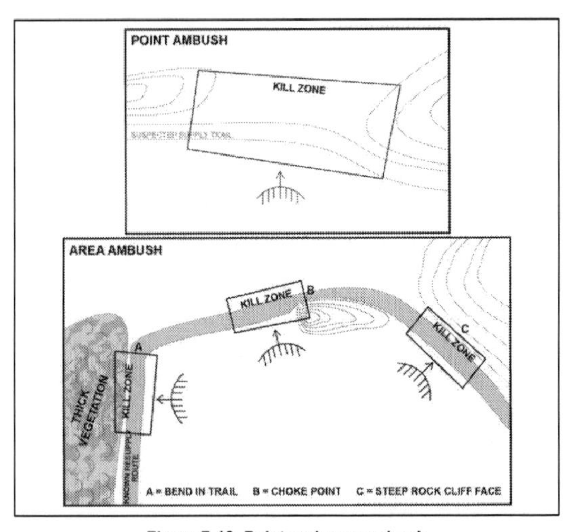

Figure 7-13. Point and area ambush

7-129. Based on the amount of time available to set an ambush, ambushes are hasty or deliberate. A hasty ambush is conducted based on an unanticipated opportunity. It is used when a patrol sees the enemy before the enemy sees it, and the patrol has time to act. The leader gives the prearranged signal to start the action and all Soldiers move to concealed firing positions, prepared to engage the enemy. Depending on the mission, the patrol may allow the enemy to pass if the enemy does not detect the patrol.

7-130. A deliberate ambush is conducted against a specific target at a location chosen based on current intelligence products and assessments. With a deliberate ambush, leaders plan and prepare based on detailed information allowing them to anticipate enemy actions and enemy locations. Detailed information includes type and size of target, organization or formation, routes and direction of movement, time the force will reach or pass certain points on its route, and weapons and equipment carried.

TERMINOLOGY

7-131. During terrain analysis, leaders identify at least four different locations: the ambush site, the kill zone, security observation posts (see paragraph 7-105 for additional information), and rally points (see paragraphs 7-7 to 7-11 for additional information on rally points). As much as possible, so-called "ideal" ambush sites should be avoided because alert enemies avoid them if possible and increase their vigilance and security when they must be entered. Therefore, surprise is difficult to achieve. Instead, unlikely sites should be chosen when possible. The following are characteristics of these four ideal positions.

Ambush Site

7-132. The ambush site is the terrain on which a point ambush is established. The ambush site consists of a support by fire position for the support element and an assault position for the assault element. An ideal ambush site—

- Has a good field of fire into the kill zone.
- Has good cover and concealment.
- Has a protective obstacle (natural or man-made).
- Has a covered and concealed withdrawal route.
- Makes it difficult for the enemy to conduct a flank attack.

Kill Zone

7-133. The kill zone is the part of an ambush site where fire is concentrated to isolate or destroy the enemy. An ideal kill zone has the following characteristics:

- Enemy forces are likely to enter it.
- It has natural tactical obstacles.
- Large enough to observe and engage the anticipated enemy force.

7-134. A near ambush is a point ambush with the assault element within reasonable assaulting distance of the kill zone (less than 50 meters). Close terrain, such as an urban area or heavy woods, may require this positioning. It also may be appropriate in open terrain in a "rise from the ground" ambush.

7-135. A far ambush is a point ambush with the assault element beyond reasonable assaulting distance of the kill zone (beyond 50 meters). This location may be appropriate in open terrain offering good fields of fire or when attack is by fire for a harassing ambush.

FORMATIONS

7-136. Many ambush formations exist. Paragraphs 7-137 to 7-143 discuss the linear, L-shaped, and V-shaped. All formations require leaders to exercise strict direct fire control. Leaders need to understand strengths and weaknesses of their units and plan accordingly. The formation selected is based on the following: terrain, visibility, Soldiers available, weapons and equipment, ease of control, and target to be attacked.

Linear Ambush

7-137. In an ambush using a linear formation, the assault and support elements parallel the target's route. This positions the assault and support elements on the long axis of the kill zone and subjects the target to flanking fire. (See figure 7-14.) Only a target that can be covered with a full volume of fire can be engaged in the kill zone. A dispersed target might be too large for the kill zone. This is the disadvantage of linear formations.

7-138. The linear formation is good in close terrain restricting the target's maneuver, and in open terrain where one flank is blocked by natural obstacles or can be blocked by other means such as claymores. Claymores or explosives can be placed between the assault and support elements and kill zone to protect the unit from counterambush actions.

7-139. When the ambushing unit deploys this way, it leaves access lanes through the obstacles so it can assault the target. An advantage of the linear formation is the relative ease by which it can be controlled under all visibility conditions.

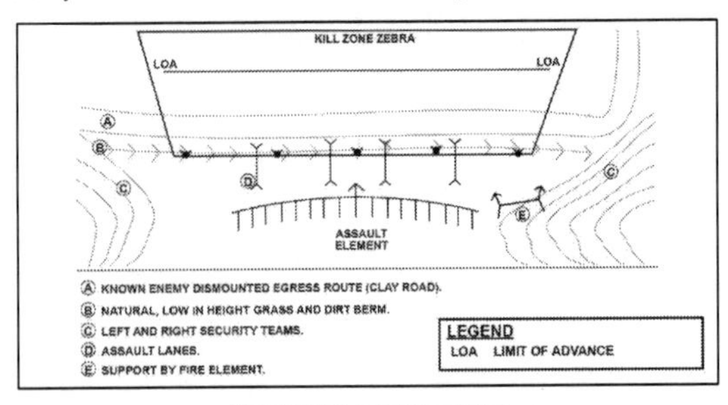

Figure 7-14. Linear ambush

L-Shaped Ambush

7-140. An ambush in the L-shaped formation (see figure 7-15) is a variation of the linear formation. The long leg of the L (assault element) is parallel to the kill zone. This leg provides flanking fire. The short leg (support element) is at the end of and at a right angle to the kill zone. This leg provides enfilade fire working with fire from the other leg. The L-shaped formation can be used at a sharp bend in a trail, road, or stream. Typically, crew-served weapons are positioned on the leg to maximize their effectiveness through enfilade fires.

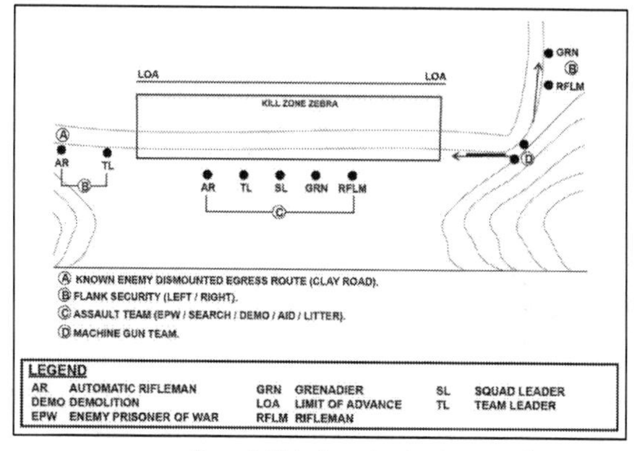

Figure 7-15. L-shaped ambush

V-Shaped Ambush

7-141. The V-shaped ambush assault elements (see figure 7-16 on page 7-46) are placed along both sides of the enemy route so they form a V. Take extreme care to ensure neither group fires into the other. This formation subjects the enemy to both enfilading and interlocking fire.

7-142. When performed in dense terrain, the legs of the V close in as the lead elements of the enemy force approach the point of the V. The legs then open fire from close range. Here, even more than in open terrain, all movement and fire is carefully coordinated and controlled to avoid fratricide.

7-143. A wider separation of the elements makes this formation difficult to control, and fewer sites favor its use. Its main advantage, it is difficult for the enemy to detect the ambush until well into the kill zone.

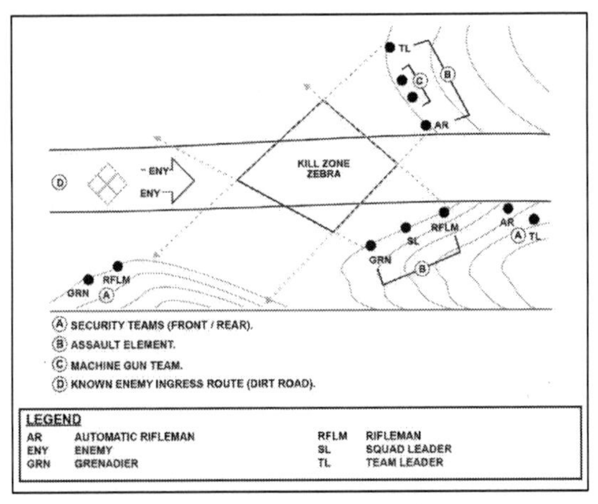

Figure 7-16. V-shaped ambush

FINAL PREPARATIONS

7-144. Final preparations begin with the unit occupying an ORP (see paragraph 7-11) and end with the main body prepared to depart for the ambush site. The unit halts at the ORP and establishes security. When ready, the leader conducts a leader's reconnaissance to confirm the plan and then returns to the ORP. The leader may position the flank ambush security elements during the leader's reconnaissance or upon returning to the ORP. If the security element returns to the ORP, the security elements leave the ORP first. Teams of the security element move to positions from which they can secure flanks of the ambush site and the ORP. (See figure 7-17.)

Note. Security elements should use a release point if there is a large distance between the ORP and objective.

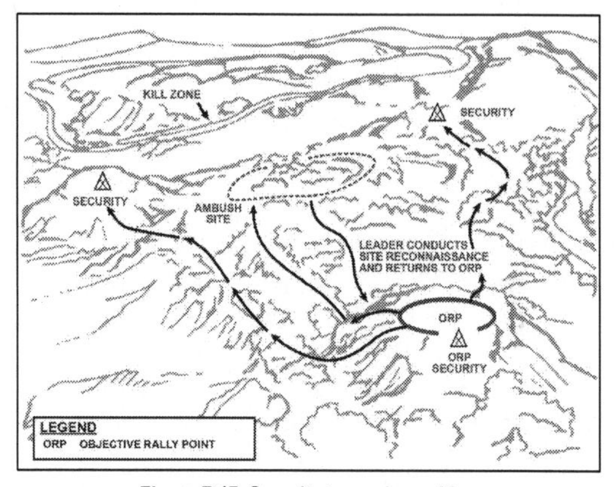

Figure 7-17. Security teams in position

OCCUPY THE SITE AND CONDUCT AMBUSH

7-145. Occupying the site and conducting the ambush begins with main body movement out of the ORP and ends when the leader initiates a withdrawal. Common control measures include:

- Kill zone.
- LOA.
- Attack by fire and/or support by fire position.
- Security and assault positions.
- TRPs.

Time of Occupation

7-146. As a rule, the ambush force occupies the ambush site at the latest possible time permitted by the tactical situation and amount of site preparation required. This reduces the risk of discovery and time Soldiers must remain still and quiet in position.

Occupying the Site

7-147. Security elements are positioned first to prevent surprise while the ambush is being established. When the security teams are in position, the support and assault elements leave the ORP and occupy their positions. If there is a suitable position, the support element can overwatch the assault element's move to the ambush site. If not, both elements leave the ORP at the same time. (See figure 7-18.)

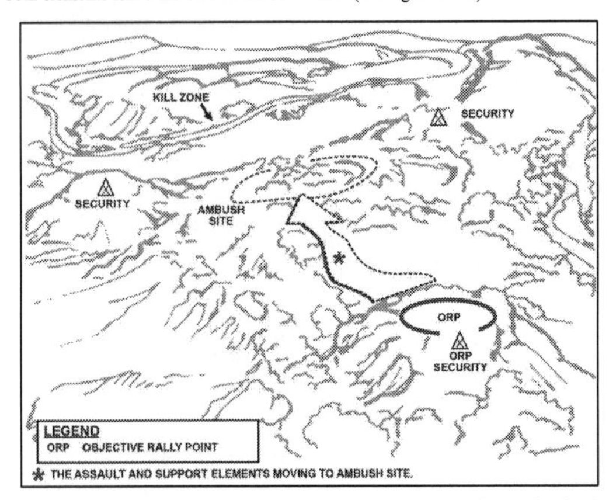

Figure 7-18. Assault and support elements moving to the ambush site

7-148. The main body moves into the ambush site from the rear. Ideally, the leader emplaces the most casualty-producing weapons first, ensuring they have line of sight along the entire kill zone. Once in place, the leader locates subordinate units to complement and reinforce the vital positions. The leader then selects a location where the leader can best initiate and control the action. Once on the objective, movement is kept to a minimum to enhance security measures.

Positions

7-149. Each Soldier must be hidden from the target and have line of sight into the kill zone. At the ambush site, positions are prepared with minimal change in the natural appearance of the site. Soldiers conceal debris resulting from preparation of positions. If sufficient water is available, Soldiers dampen the ground under their muzzles to decrease their firing signature to prevent detection once the ambush is triggered.

Confirming the Direct Fire Plan

7-150. Claymores, explosives, and grenade launchers may be used to cover dead space left by automatic weapons. All weapons are assigned sectors of fire to provide mutual support. The unit leader sets a time by which positions must be prepared.

Movement in the Kill Zone

7-151. The kill zone is not entered if entry can be avoided. When emplacing tactical obstacles, care is taken to remove tracks or signs that might alert the enemy and compromise the ambush. If claymores or explosives are placed on the far side, or if the appearance of the site might cause the enemy to check it, a wide detour around the kill zone should be made. Here, too, care is taken to remove all traces that might reveal the ambush. An alternate route from the ambush site also is planned.

Initiating the Ambush

7-152. Once all friendly elements are in position, the unit waits for enemy targets. When the target approaches, the security team spots it and alerts the ambush leader. The security team reports the target's direction of movement, size, and special weapons or equipment. Upon receipt of the report, the leader alerts the other elements.

7-153. When most of the enemy force is in the kill zone, the leader initiates the ambush with the most casualty-producing weapon, medium machine gun fire, or the detonation of mines or explosives. The blast of explosives may reduce accuracy of the initial volley of direct fires due to obscuration; however, the unit must immediately establish a high volume of fire and continue to fire until conditions are set to cease or shift fires. The assault element may conduct an assault through the kill zone to the LOA. If the assault element must assault the kill zone, the leader signals to cease or shift fire. This also signals the start of the assault. Besides destruction of the enemy force, other kill zone tasks can include searching for items of intelligence value, capturing prisoners, and completing the destruction of enemy equipment. When the assault element has finished its mission in the kill zone, the leader gives the signal to withdraw to the ORP.

7-154. Fire discipline is critical during an ambush. Soldiers do not fire until the signal is given. Then it must be delivered at once in the heaviest, most accurate volume possible. Well-trained gunners and well-aimed fire help achieve surprise and destruction of the target. If the enemy force is large, subordinate leaders designate individuals or teams to use various rates of fire (rapid, sustained, or cyclic) to prevent entire elements from conducting magazine changes at the same time and allow the unit to maintain a volume of fire in the kill zone. When the target is to be assaulted, the ceasing or shifting of fire also must be precise. If it is not, the assault is delayed, and the target has a chance to react. Sector stakes should be used if possible.

Withdrawal

7-155. The withdrawal begins once the assault element completes its actions on the objective and ends with consolidate/reorganization at a designated rally point. On signal, the unit withdraws to the ORP, reorganizes, and continues its mission. At a set terrain feature, the unit halts and disseminates information. If the ambush fails and enemy pursues, the unit withdraws by bounds. Units should use obscurants to help conceal the withdrawal. Obstacles already set along the withdrawal routes can help stop the pursuit.

CONDUCT A POINT AMBUSH

7-156. In a point ambush (see figures 7-14, 7-15, and 7-16 on pages 7-44, 7-45, and 7-46 respectively for examples of three different formations), Soldiers deploy to attack an enemy in a single kill zone. The patrol leader (or platoon leader) is the leader of the assault element. The APL (or platoon sergeant) probably will locate with the patrol leader in the assault element.

7-157. The security or surveillance teams should be positioned first. The support element should then be emplaced before the assault element moves forward. The support element must overwatch the movement of the assault element into position.

7-158. The patrol leader must check each Soldier once the Soldier emplaces. The patrol leader signals the surveillance team to rejoin the assault element if it is positioned away from the assault location. Actions of the assault element, support element, and security element are shown in table 7-3.

Table 7-3. Actions by ambush elements

Security Element	*Support Element*	*Assault Element*
• Identify sectors of fire for all weapons; emplace aiming stakes. • Emplace claymores and other protective obstacles. • Camouflage positions. • Secure the objective rally point. • Secure a route to the objective rally point, as required.	• Identify sectors of fire for all weapons, especially medium machine guns. • Emplace limiting stakes to prevent friendly fires from hitting the assault element in an L-shaped ambush. • Emplace claymores and other protective obstacles. • Camouflage positions.	• Identify individual sectors of fire assigned by the patrol leader; emplace aiming stakes. • Emplace claymores and other protective obstacles. • Emplace claymores, mines, or other explosives in dead space within the kill zone. • Camouflage positions. • Take weapons off safe when directed by the patrol leader.

7-159. The patrol leader instructs the security element (or teams) to notify the leader when the enemy approaches the kill zone using the size, activity, location, unit, time, and equipment (SALUTE) reporting format. The security element also must keep the patrol leader informed if additional enemy forces are following the lead enemy force. This will allow the leader to know if the enemy force meets the engagement criteria directed by the company commander. The patrol leader is prepared to let pass enemy forces which are too large or do not meet the engagement criteria. The leader must report to the higher commander enemy forces passing through the ambush unengaged.

7-160. The patrol leader initiates the ambush with the greatest casualty-producing weapon, typically a command-detonated claymore. The leader also must plan a back-up method, typically a medium machine gun, to initiate the ambush should the primary means fail. All Soldiers in the ambush must know the primary and back-up methods. The patrol should rehearse with both methods to avoid confusion and loss of surprise during execution of the ambush.

7-161. The patrol leader must include a plan for engaging the enemy during limited visibility. Based on the company commander's guidance, the leader should consider the use and mix of tracers and employment of illumination, night vision devices, and thermal weapon sights. For example, if Javelins are not used during the ambush, the leader still may employ the command launch unit (known as CLU) with its thermal sights in the security or support element to observe enemy forces. The M3 Multi-role, Antiarmor, Antipersonnel Weapon System (known as MAAWS), when replacing the Javelin, can provide the platoon with an antipersonnel as well as an antiarmor capability.

7-162. The patrol leader also may include the employment of indirect fire support (when available) in the plan. Based upon the company commander's guidance, the leader may employ indirect fires to cover flanks of the kill zone. This isolates an enemy force or assists the patrol's disengagement if the ambush is compromised, or the patrol departs the ambush site under pressure.

7-163. The patrol leader has a good plan (day and night) for signals that direct cease fire or shift fire and that signal the advance of the assault element into the kill zone to begin its search and collection activities. The leader should take into consideration the existing environmental factors. For example, obscurants may not be visible to the support element because of limited visibility or the lay of the terrain. Soldiers must know and practice relaying the signal during rehearsals to avoid the potential of fratricide.

7-164. The assault element must be prepared to move across the kill zone using individual movement techniques if there is return fire once they begin to search. Otherwise, the assault element moves across by bounding fire teams.

7-165. The assault element collects and secures all enemy prisoner of war and moves them out of the kill zone to an established location before searching dead enemy bodies. The enemy prisoner of war collection point should provide cover and should not be easily found by enemy forces following the ambush. The friendly assault element searches from the far side of the kill zone to the near side.

7-166. Once the bodies have been thoroughly searched, search teams continue in this manner until all enemy personnel in and near the kill zone have been searched. Enemy bodies should be marked once searched (for example, arms folded over the chest and legs crossed) to ensure thoroughness and speed and to avoid duplication of effort.

7-167. The patrol identifies and collects equipment to be carried back and prepares it for transport. Enemy weapon chambers are cleared and put on safe. The patrol also identifies and collects at a central point the enemy equipment to be destroyed. The demolition team prepares the fuze and awaits the signal to initiate. This is normally the last action performed before departing the ambush site. The flank security element returns to the ORP after the demolition team completes its task.

7-168. The flank security teams also may emplace antiarmor mines after the ambush has been initiated if the enemy is known to have armored vehicles that can quickly reinforce the ambushed enemy force. If a flank security team makes enemy contact, it fights as long as possible without becoming decisively engaged. It uses prearranged signals to inform the patrol leader it is breaking contact. The leader may direct a portion of the support element to assist the security element in breaking contact.

7-169. The patrol leader must plan the withdrawal of the patrol from the ambush site. The planning process should include the following:

- Elements normally are withdrawn in the reverse order they established their positions.
- Elements may return to the release point, then to the ORP, depending on the distance between the elements.
- The security element at the ORP must be alerted to assist the platoon's return.
- It maintains security of the ORP while the remainder of the platoon prepares to depart.

7-170. Actions back at the ORP include, but are not limited to, accounting for personnel and equipment, stowing captured equipment, and first aid (as necessary). Upon return personnel within the patrol reorganized as required, and ammunition and equipment are redistributed for movement out of the ORP.

CONDUCT AN AREA AMBUSH

7-171. In an area ambush, Soldiers deploy in two or more related point ambushes. The patrol may conduct an area ambush as part of a company offensive or defensive plan, or it may conduct a point ambush as part of a company area ambush (see ATP 3-21.10).

7-172. The platoon is usually the smallest unit to conduct an area ambush. Patrols conduct area ambushes (see figure 7-19) where enemy movement is largely restricted to trails or streams.

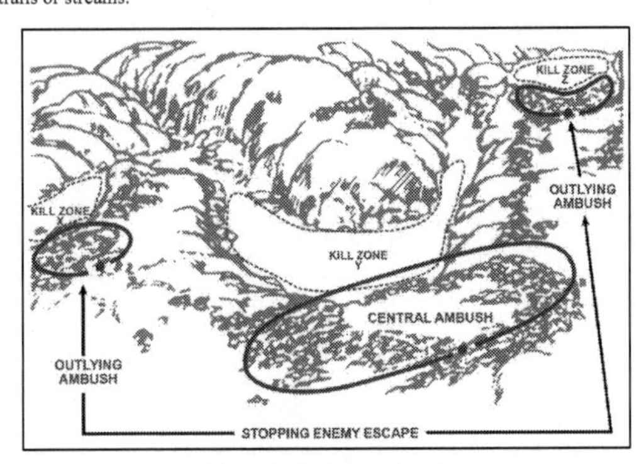

Figure 7-19. Area ambush

7-173. The patrol leader selects one principal ambush site around which to organize outlying ambushes. These secondary sites are located along the enemy's most likely avenue of approach and escape routes from the principal ambush site. Squad-size elements normally are responsible for each ambush site when conducting a platoon-size area ambush.

7-174. The patrol leader considers the mission variables of METT-TC (I) to determine the best employment of the weapons squad. The leader normally locates the medium machine guns with the support element in the principal ambush site.

7-175. Squad- or section-size elements responsible for outlying ambushes do not initiate their ambushes until the principal one has been initiated. They then engage to prevent enemy forces from escaping the principal ambush or reinforcing the ambushed force.

CONDUCT AN ANTIARMOR AMBUSH

7-176. Patrols conduct antiarmor ambushes (see figure 7-20) to destroy armored vehicles. The antiarmor ambush may be part of an area ambush. The antiarmor ambush consists of the assault element (armor-killer element) and support-security element.

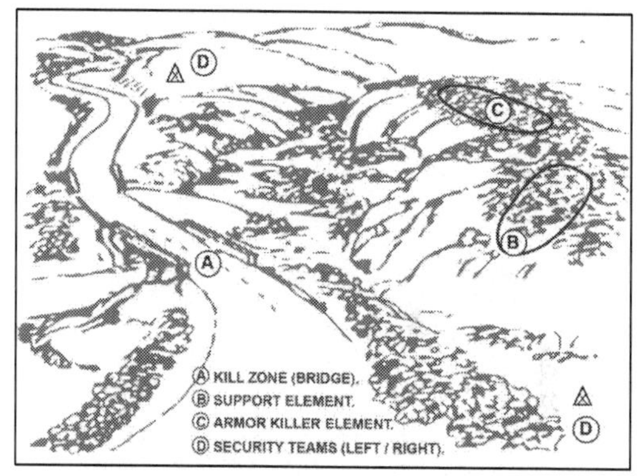

Figure 7-20. Antiarmor ambush

7-177. The armor-killer element is built around the close combat missile system (known as CCMS) Javelin (or M3 MAAWS). (See appendix D of this publication for information about employment of the Javelin [or M3 MAAWS].) Typically, antiarmor ambushes are executed at long range, and may only employ CCMS (or M3) fires, or those of the CCMS (or M3) and machine guns fires. The patrol leader should consider additional shoulder launched munitions (for example, AT4s, see appendix D) available to supplement the CCMS (or M3) fires. Depending on the terrain, these may mass fires in the kill zone, or simply be positioned to cover a planned withdrawal or simply fire in response to an assaulting enemy force. The leader considers the mission variables of METT-TC (I) to position all antiarmor weapons to take advantage of their best engagement aspect (rear, flank, or top). The remainder of the patrol must function as support-security elements in the same manner as the other forms of ambushes to cover dismounted enemy avenues of approach into the ambush site. However, to achieve stand-off and increase survivability, these elements may be out of range of the kill zone.

7-178. In a platoon-size antiarmor ambush, the company commander selects the general site of the ambush with the platoon leader finding a specific site restricting the movement of enemy armored vehicles out of the designated kill zone. The platoon leader should emplace weapons, so an obstacle is between the platoon and the kill zone. In a squad (with attachments from the weapons squad) antiarmor ambush, the platoon leader selects the general site of the ambush, and the squad leader then finds a site restricting the movement of enemy armored vehicles out of the kill zone.

7-179. The leader should consider the method for initiating the antiarmor ambush. The preferred method (common to all ambushes) should be a mass casualty-producing signal initiated by a reliable weapon system or the detonation of mines or explosives. The Javelin (M3 MAAWS) can be used to initiate the ambush, but even with its limited signature, it may be less desirable than an AT mine.

7-180. The armor-killer team destroys the first and last vehicle in the enemy formation, if possible. Designated weapons begin firing once the ambush has been initiated.

7-181. The leader must determine how the presence of dismounted enemy soldiers with armored vehicles will affect the success of the ambush. With appropriate terrain utilization, natural or reinforcing obstacles, and stand-off ranges, enemy dismounts should not be able to affect the ambush rapidly. The leader's choices include:

- Initiate the ambush as planned.
- Withdraw without initiating the ambush.
- Divert some medium machine gun fires from armored or lightly armored vehicles to engage enemy dismounts.
- Initiate the ambush with either CCMS or medium machine guns, upon completion conduct withdrawal.

7-182. Because of the speed, enemy armored forces can reinforce the ambushed enemy with, the leader should plan to keep the engagement short and have a quick withdrawal planned. The patrol, dependent on METT-TC (I), may not clear the kill zone as in other forms of ambushes.

SECURITY PATROLS

7-183. Security patrols prevent surprise of the main body by patrolling to the front, flank, and rear of the main body to detect and destroy enemy forces in the local area. These patrols detect and disrupt enemy forces conducting reconnaissance (counterreconnaissance) of the main body or massing to conduct an attack. They normally are away from the main body of the unit for limited periods, returning frequently to coordinate and rest. Security patrols do not operate beyond the range of communication and supporting fires from the main body, especially mortar fires, because they normally operate for limited periods, and are combat-oriented.

Note. Leaders orient their security effort by identifying a *security objective*—the most important entity to protect during that specific security effort (FM 3-90)—within a patrol's assigned area.

7-184. A security patrol is sent out from a unit location when the unit is stationary or during a halt to search the local area, detect enemy forces near the main body, and to engage and destroy the enemy within the capability of the patrol. This form of combat patrol normally is sent out by units operating in close terrain with limited fields of observation and fire. Although this form of combat patrol seeks to make direct enemy contact and to destroy enemy forces within its capability, it should try to avoid decisive engagement. When the main body is stationary, the security patrol prevents enemy infiltration, reconnaissance, surveillance, or attacks (for example, patrolling the area surrounding a battle position or the area between battle positions). When the main body is moving, the security patrol prevents the unit from being ambushed or coming into surprise chance contact (for example, a meeting engagement, see paragraph 4-72).

SECTION IV – RECONNAISSANCE PATROLS

7-185. A reconnaissance patrol employs many tactics, techniques, and procedures throughout the course of the patrol, one of which may include an extended period of surveillance. A reconnaissance patrol collects information and confirms or disproves the accuracy of information previously gained. The intent for this type of patrol is to avoid enemy contact and accomplish its mission without engaging in close combat. Reconnaissance patrols always try to accomplish their mission without being detected or observed. Because detection cannot always be avoided, a reconnaissance patrol carries the necessary arms and equipment to protect itself and break contact with the enemy. Platoons or squads generally perform three types of reconnaissance patrols: area reconnaissance, route reconnaissance, or zone reconnaissance.

Note. The main difference between the performance of reconnaissance operations and security operations is that reconnaissance operations orient on the enemy and terrain, while security operations (see chapter 6 section VII) orient on the friendly force, area, or facility. The reconnaissance fundamentals ensure continuous reconnaissance; do not keep reconnaissance assets in reserve; orient on the reconnaissance objective; report all required information rapidly and accurately; retain freedom of maneuver; gain and maintain enemy contact (find the enemy); and develop the situation rapidly (see FM 3-96 for additional information).

- *Area reconnaissance*—a form of reconnaissance operation that focuses on obtaining detailed information about the terrain or enemy activity within a prescribed area (FM 3-90).
- *Route reconnaissance*—a form of reconnaissance operation to obtain detailed information of a specified route and all terrain from which the enemy could influence movement along that route (FM 3-90).
- *Zone reconnaissance*—a form of reconnaissance operation that involves a directed effort to obtain detailed information on all routes, obstacles, terrain, and enemy forces within a zone defined by boundaries (FM 3-90).

Note. Leaders orient their reconnaissance effort by identifying a *reconnaissance objective*—the most important result desired from that specific reconnaissance effort (FM 3-90)—within a patrol's assigned area. The objective can be either threat or terrain based, depending on what type of operation its higher echelon is conducting. Every reconnaissance mission specifies a reconnaissance objective by priority, intent, and the information to obtain. For a platoon or squad, the reconnaissance objective is usually a small and narrowly defined area, such as a particular building, a bridge or suspected enemy position designated as a named area of interest (NAI).

ESSENTIAL ELEMENTS FOR A RECONNAISSANCE PATROL

7-186. A reconnaissance patrol normally travels light, with as few personnel and as little arms, ammunition, and equipment as possible. This increases stealth and cross-country mobility in close terrain. Regardless of how the patrol is armed and equipped, the patrol leader always plans for direct-fire contact with a hostile force. Leaders must anticipate where they may possibly be observed and control the hazard by planning their movement or employment to lessen their risk. If detected, or unanticipated opportunities arise, reconnaissance patrols must be able to rapidly transition to combat. A reconnaissance patrol is organized with a reconnaissance element and a security element.

RECONNAISSANCE ELEMENT

7-187. The reconnaissance element gathers information to answer PIRs which enable leaders to make tactical decisions. The primary means is reconnaissance with extended periods of surveillance enabled by tactical movement, and continuous and accurate reporting. Thorough and accurate reconnaissance and surveillance (see paragraph 2-3 and note on page 2-5) is important. However, avoiding detection is equally important. The following are activities normally associated with reconnaissance and extended periods of surveillance:

- Reconnoiter all terrain within the assigned area, route, or zone.

- Determine trafficability routes or potential avenues of approach (based on the personnel or vehicles to be used on the route):
  - Inspect and classify all bridges, overpasses, underpasses, and culverts on the route.
  - Locate fords or crossing sites near bridges on the route.
- Determine the time it takes to traverse the route.
- Reconnoiter to the limit of direct fire range—
  - Terrain influencing the area, route, or zone.
  - Built-up areas.
  - Lateral routes.
- Within capabilities, reconnoiter natural and man-made obstacles to ensure mobility along the route. Locate a bypass or reduce/breach, clear, and mark—
  - Lanes.
  - Defiles and other restrictive/severely restrictive terrain.
  - Minefields.
  - Contaminated areas.
  - Log obstacles such as abatis, log cribs, stumps, and posts.
  - AT ditches.
  - Wire entanglements.
  - Fills, such as a raised railroad track.
  - Other obstacles along the route.
- Zone reconnaissance and area reconnaissance accomplish the following activities within the unit's capability (unless otherwise ordered):
  - Find and report all enemy forces within the zone.
  - Based on engagement criteria, clear all enemy forces in the designated assigned area of the unit conducting reconnaissance.
  - Inspect and classify all bridges in the zone.
  - Inspect and classify all overpasses, underpasses, and culverts.
  - Locate fords, crossing sites, or obstacle bypasses in the zone.
  - Reconnoiter specific terrain in the zone, for example, an NAI.
  - Report reconnaissance information.
  - Confirm or deny commander's PIR.
- Extended periods of surveillance may include:
  - Decision points where decisions by the commander are required—for example, determining an enemy COA.
  - Information requirements to identify critical events linked to NAIs and target area of interests.
- Find all threat forces influencing movement along the area, route, or zone.
- Report information.

Note. Infantry rifle platoons and squads typically require engineer augmentation to complete a full technical inspection of bridges, roads, and culverts; however, platoons and squads do have the ability to conduct a general assessment.

SECURITY ELEMENT

7-188. The security element's primary purpose is to protect the reconnaissance element. The security element provides overwatch to the reconnaissance element from support by fire positions. This element must be able to observe the reconnaissance objective and avenues of approach as well as provide early warning. The security element is prepared to provide alternate communications to higher headquarters or relay for the reconnaissance element. Security elements must be able to engage the enemy with direct and indirect fire within and beyond the reconnaissance objective. If a reconnaissance element is compromised, the security element immediately provides overwhelming fires against enemy forces to enable the element in contact to withdraw. This worst-case scenario must be well-rehearsed and well thought out.

ACTIONS ON THE RECONNAISSANCE OBJECTIVE

7-189. The actual reconnaissance begins at the designated transition point and ends with a follow-on transition to tactical movement away from the reconnaissance objective. Leaders mark the follow-on transition point with a control measure like the first transition point, using a linkup point, an LOA, or a phase line. During this phase, leaders execute one of the three types of reconnaissance (area, zone, or route). These types of reconnaissance are distinguished by the scope of the reconnaissance objective. The general types of reconnaissance patrols Infantry units conduct are area, zone, and route. (See figure 7-21.)

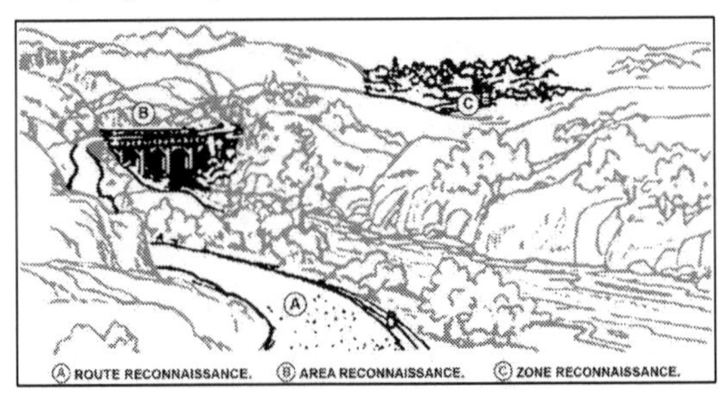

Figure 7-21. Types of reconnaissance patrols

ORGANIZATION-TASKS TO SUBORDINATE UNITS

7-190. Regardless of how reconnaissance (and/or surveillance) and security elements are organized, each element always maintains responsibility for its own local security. In a small reconnaissance patrol, the patrol headquarters may form a part of one of the subordinate elements rather than being a separate element. The number and size of the various teams and elements must be determined through the leader's mission analysis. There are three ways to organize reconnaissance and security elements.

7-191. The first technique is to organize the reconnaissance (and/or surveillance) elements separate from security elements. This technique is used when the security element can support the reconnaissance element from one location. This requires the reconnaissance objective to be defined clearly and the area to be fairly open.

7-192. The second technique is to organize the reconnaissance (and/or surveillance) elements and security elements together into reconnaissance and security teams. This technique is used when the reconnaissance objective is not defined clearly, or the teams are not mutually supporting, and each reconnaissance element potentially needs its own security force. Within the reconnaissance and security team, the reconnaissance can be done by one or two individuals while the rest of the element provides security. The number of Soldiers in a reconnaissance and security team varies depending on the mission. Usually a fire team (three to four Soldiers) is required for an adequate reconnaissance and still provide local security.

7-193. The third technique is to establish reconnaissance (and/or surveillance) and security teams with an additional separate security element. The separate security element also can act as a quick reaction force.

REVERSE-PLANNING PROCESS

7-194. The patrol leader uses a reverse-planning process to plan for a reconnaissance patrol. The leader first determines the reconnaissance objective, an information requirement corresponding to the terrain or enemy in a specific area, route, or zone; it may be designated by a control measure such as named area of interest (NAI), target area of interest, checkpoints, objective, route, phase lines, or boundaries. Once the patrol leader has clarified the reconnaissance objective, the leader determines the observation plan enabling the patrol to obtain the information required. After determining the observation plan, the leader determines the tactical movement necessary to position the elements of the patrol to achieve the plan.

INFORMATION REQUIREMENTS

7-195. Information requirements are the basis for the development of the CCIRs, the answers to which are needed to allow commanders to make tactical decisions. The controlling headquarters must clearly define the information requirements it wants the patrol to determine. The patrol leader must clarify these information requirements prior to conducting the mission. Table 7-4 on page 7-60 illustrates an example matrix used to capture the information requirements for the headquarters' collection plan.

Table 7-4. Infrared collection matrix, example

Information Requirement	*Location/Description*	*Time*	*Purpose*
Are enemy forces within small arms range of intersection?	NV 12349875 road intersection	From: 20 1700 Nov To: 21 0600 Nov	Facilitate the platoon's passage through the area
Legend: Nov–November			

7-196. Information requirements can be enemy-oriented, terrain-oriented, civil-oriented, or a combination. It is important the leader clarifies the requirement prior to conducting the reconnaissance. Knowing this orientation enables the leader to use disciplined initiative to meet the higher commander's information requirement.

7-197. Terrain-oriented information requirements focus on determining information on the terrain of a particular area, route, or zone. While the unit looks for enemy presence, the overall intent is to determine the terrain's usefulness for friendly purposes. For example, the company commander may send out a squad-sized reconnaissance patrol to identify a location, for example, the company's future TAA or battle position(s). The patrol leader may send out a squad or team-sized reconnaissance patrol to obtain information about a bridge on a proposed infiltration route.

7-198. Enemy-oriented information requirements focus on finding a particular enemy force. The purpose of enemy-oriented reconnaissance is to confirm or deny planning assumptions. While the unit may be given a terrain feature as a reference point, the overall intent is to find the enemy. This means if the enemy is not in the location referenced, the leader must demonstrate the initiative to find the enemy force within the mission parameters.

7-199. Civil-oriented information requirements focus on determining information on the human environment in a particular area, route, or zone. A civil-oriented information requirement is a larger, vaguer category requiring more clarification than the other two categories. Examples of civil-oriented information requirements are the physical infrastructures; service infrastructures such as sewer, water, electric, and trash; the political situation; demographics; and dislocated civilians.

Observation Plan

7-200. Once the patrol leader understands the information requirement, the leader develops an observation plan to obtain the information. The leader captures the observation plan as part of the patrol leader's COA sketch. This is done by asking two basic questions:

- What is the best location to obtain the information required?
- What is the best way to obtain the information without compromising the patrol?

7-201. The answer to the first question is: all vantage points and observation posts from which the patrol can best obtain the required information. A vantage point is a temporary position enabling observation of the enemy. It is meant to be occupied only until the enemy activity is confirmed or denied. The answer to the second question is: use the routes and number of teams necessary to occupy the vantage points and/or observation posts. An observation post is a position where military observations can be made, and fire can be directed and adjusted. Observation posts must possess appropriate communications. The observation post can be short-term (12 hours or less) or long-term, depending upon guidance from higher. Unlike a vantage point, the observation post normally is occupied, and surveillance is conducted for a specified period.

7-202. The patrol views the reconnaissance objective from as many perspectives as possible, using whatever combinations of observation posts and vantage points are necessary. The leader selects the tentative locations for patrol's vantage points, observation posts, and movement after analyzing METT-TC (I). These locations are proposed and are confirmed and adjusted as necessary by the actual leader on the ground. After analysis, the leader determines how many vantage points and observation posts must be established and where to position them. After deciding on these general locations, the leader designs the routes for necessary movement between these and other control measures (such as the release point[s] and linkup point[s]). Positions should have the following characteristics:

- Covered and concealed routes to and from each position.
- Unobstructed observation of the assigned area, route, or zone. Ideally, the fields of observation of adjacent positions overlap to ensure full coverage.
- Cover and concealment. Leaders select positions with cover and concealment to reduce their vulnerability on the battlefield. Leaders may need to pass up a position with favorable observation capability but no cover and concealment to select a position affording better survivability.
- A location not attracting attention. Positions should not be sited in such locations as a water tower, an isolated grove of trees, or a lone building or tree. These positions draw enemy attention and may be used as enemy artillery TRPs.
- A location not sky lining the observers. Avoid hilltops. Locate positions farther down the slope of the hill or on the side, provided there are covered and concealed routes into and out of the position.

7-203. The locations selected by the patrol leader are either long-range or short-range. Long-range positions must be far enough from the objective to be outside enemy's small-arms weapons, sensors, and other local security measures. Long-range positions are the most desirable method for executing a reconnaissance because the patrol does not come in close enough to be detected. If detected, the patrol can employ direct and indirect fires. Therefore, it is used whenever METT-TC (I) permits the required information to be gathered from a distance. Security must be maintained by—

- Selecting covered and concealed observation posts.
- Using covered and concealed routes in and around the AO.
- Deploying security elements, including sensors, to give early warning, and providing covering fire, if required.

7-204. Short-range positions are within the range of enemy local security measures and small arms fire. When information required cannot be obtained by a long-range position, reconnaissance elements move closer to the objective. The vantage points and routes used during short-range observation should be planned carefully and verified prior to using them. Doing so prevents detection by the enemy or friendly units from stumbling into one another or covering ground already passed over by another element.

7-205. Once a subordinate element establishes the observation post, the occupying element prepares a sector sketch for the position. Figure 7-22 is a similar sketch to an individual fighting position sketch but with some important differences. At a minimum, the sketch should include:

- A rough sketch of key and significant terrain.
- The location of the observation post and alternate observation post.
- The location of the hide position.
- Routes to the observation post and fighting positions.
- Sectors of observation.
- Direct and indirect fire control measures.

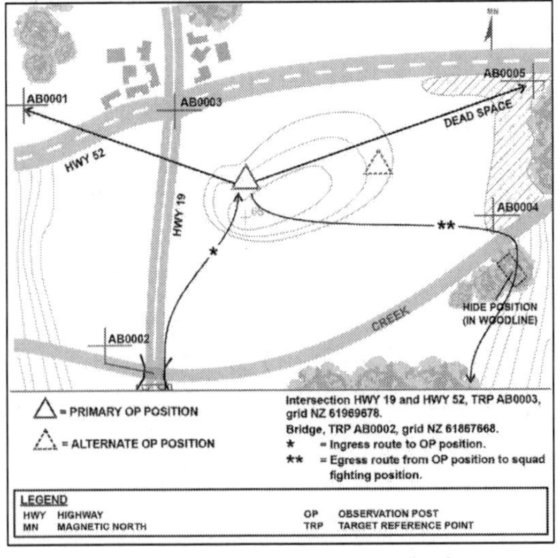

Figure 7-22. Observation post sector sketch

AREA RECONNAISSANCE PATROL

7-206. Area reconnaissance patrols focus on obtaining detailed information about the enemy activity, terrain, and/or specific civil considerations and integrated into these variables, informational considerations within a prescribed area. The area may be given as a grid coordinate, an objective, on an overlay. In an area reconnaissance, the patrol uses surveillance (vantage points or observation posts) around the objective to observe it and the surrounding area. The area may include a town, a neighborhood, a ridgeline, woods, an airhead, or any other feature critical to operations. The area may consist of a single point (such as a bridge or an installation). Areas are normally smaller than zones and not usually contiguous to other friendly areas targeted for reconnaissance. Because the area is smaller, units conduct an area reconnaissance quicker than a zone reconnaissance. Other unique techniques falling under area reconnaissance actions on the objective include point, contact, civil, and tracking reconnaissance patrols.

POINT RECONNAISSANCE

7-207. A point reconnaissance patrol goes straight to a specific location and determines the situation there. As soon as it does so, it either reports the information by radio or returns to the larger unit to report. This patrol can obtain, verify, confirm, or deny extremely specific information for the platoon leader or commander.

CONTACT RECONNAISSANCE

7-208. A contact reconnaissance patrol is a special type of reconnaissance patrol sent from one unit to physically contact and coordinate with another. Modern technology has reduced, but not eliminated, the need for contact patrols in some environment. In other environments, for example, against a peer threat, the need for contact patrols may be the best means of passing information safely. These type of patrols are often used today when a U.S. force must contact a non-U.S. coalition partner who lacks compatible communications or position-reporting equipment. Contact patrols may either go to the other unit's position or the units can link up at a designated contact point. The leader of a contact patrol provides the other unit with information about the location, situation, and intentions of the patrol, and obtains and reports the same information about the contacted unit back to the patrol. A contact patrol also observes and reports pertinent information about the area between the two units.

CIVIL RECONNAISSANCE

7-209. A civil reconnaissance patrol is a targeted, planned, and coordinated observation and evaluation of specific civil aspects of the environment. Civil reconnaissance focuses on the civil component, the elements which are best represented by areas, structures, capabilities, organizations, people, and events (ASCOPE). PIRs or information requirements focus on civil reconnaissance for purposes of collecting civil information to enhance situational understanding and facilitate decision making. Potential sources of civil information which a coordinated civil reconnaissance plan considers include:

- Ongoing ASCOPE assessments of the AO.
- Identified unknowns in civil information:
  - Gaps identified during collation and analysis.
  - Gaps remaining in the area study and area assessment.
- Civil affairs interaction, including but not limited to:
  - Host-nation government officials.
  - Religious leaders.
  - Tribal or clan leaders.
  - Dislocated civilian camp leaders.
  - Dislocated civilians on the move.
  - Infrastructure managers and workers.
  - Local industry personnel.
  - Medical and educational personnel.

7-210. The civil reconnaissance element enables information requirements for the purposes of facilitating tactical decision making. The following are activities normally associated with civil reconnaissance and extended periods of surveillance:

- Determine the size, location, and composition of society/human demographics.
- Identify essential infrastructure influencing military operations, including the following:
  - Political, government, and religious organizations and agencies.
  - Physical facilities and utilities (such as power generation, transportation, and communications networks).

Tracking Reconnaissance

7-211. A tracking reconnaissance patrol is normally a squad-size, possibly smaller, element. It is tasked to follow the trail of a specific enemy unit in order to determine its composition, final destination, and actions en route. Patrol members look for subtle signs left by the enemy as it moves. As the patrol tracks, the members of the patrol gather information about the enemy unit, the route it took, and surrounding terrain. Normally, a tracking reconnaissance patrol avoids direct fire contact with the tracked unit. Tracking reconnaissance patrols often use military working dog teams with scouting capability to help them detect enemy presence by scent, sound, and sight.

Actions in the Objective Area

7-212. Actions in the objective area for an area reconnaissance begin with the patrol in the ORP, followed by the execution of the observation plan, and end with a dissemination of information after a linkup of the patrol's subordinate elements. This dissemination of information can occur in the ORP, other contact point, or other designated location prior to the patrol's return to its higher unit.

Actions From the Objective Rally Point

7-213. The patrol occupies the ORP and conducts associated priorities of work. While the patrol establishes security and prepares for the mission, the patrol leader and selected personnel conduct a leader's reconnaissance (see paragraphs 7-15 and 7-16). The leader must accomplish three things during this reconnaissance: pinpoint the objective and establish surveillance, identify a release point and follow-on linkup point (if required), and confirm the observation plan.

7-214. Upon returning from the leader's reconnaissance, the patrol leader disseminates information and updates to the plan as required. Once ready, the patrol departs. The leader first establishes security. Once security is in position, the reconnaissance element moves along the specified routes to the observation posts and vantage points in accordance with the observation plan.

7-215. Upon nearing the objective, the patrol leader should establish a forward release point. It should be sited so it is well hidden, no closer than 200 meters from known enemy patrol routes, observation posts, or sentry positions. The forward release point provides the patrol leader with a temporary location close to the objective from which to operate. While the reconnaissance is in progress, the forward release point is manned by the patrol second in charge and RTO. Only vital transmissions are made while in the forward release point. The volume setting should be as low as possible on the radio, and if available, the operator should use an earphone.

7-216. The reconnaissance team makes its final preparation in the forward release point. Movement from the forward release point must be slow and deliberate. The patrol leader should allow sufficient time for the team to obtain the information. If time is limited, the team should only be required to obtain essential information. If the enemy position is large, or time is limited, the leader may employ more than one reconnaissance team. If this occurs, each team must have clearly defined routes for movement to and from the forward release point. They also must have clearly defined areas in which to conduct their reconnaissance in order to avoid contact with each other.

7-217. The reconnaissance team(s) normally consists of one to two observers and two security Soldiers. The security Soldiers should be close enough to provide protection to the observer, but far enough away, so their position is not compromised. When moving in areas near the enemy position, only one Soldier should move at one time. Accordingly, bounds should be short.

7-218. Once in position, the team(s) observes and listens to acquire the needed information. No eating, talking, or unnecessary movement occurs during this time. If a reconnaissance team cannot acquire the information needed from its initial position, it retraces the route and repeats the process. This method of reconnaissance is extremely risky. The reconnaissance team must remember the closer it moves to an objective, the greater the risk of being detected.

Multiple Reconnaissance and Surveillance Teams

7-219. When information cannot be gathered from just one observation post or vantage point, successive points may be used. Once determined, the patrol leader decides how the patrol will occupy them. The critical decision is determining the number of teams in the reconnaissance element. The advantages of a single team are the leader's ability to control the team, and a decreased probability of enemy detection. The challenges of a single team are the lack of redundancy, and the objective area is observed with just one team. The advantages of using multiple teams include providing the leader redundancy in accomplishing the mission, potentially finishing faster, and ability to look at the objective area from more than one perspective. Challenges include an increased likelihood of compromise and the challenge of controlling the teams.

7-220. The patrol leader may include a surveillance team in the reconnaissance of the objective from the ORP. The leader positions these surveillance teams while on the reconnaissance. The leader may move them on one route, posting them as they move, or may direct them to move on separate routes to their assigned locations.

ROUTE RECONNAISSANCE PATROL

7-221. A route reconnaissance patrol is conducted to obtain detailed information about one route and all its adjacent terrain, or to locate sites for emplacing obstacles. Route reconnaissance is oriented on a road, a narrow axis such as an infiltration lane, or on a general direction of attack. Patrols conducting route reconnaissance operations attempt to view the route from both the friendly and enemy perspective. Infantry rifle platoons and squads conducting the patrol require augmentation with technical expertise for a complete detailed route reconnaissance. However, platoons are capable of conducting hasty route reconnaissance or area reconnaissance of selected points along a route.

FOCUS AND CONDUCT

7-222. A route reconnaissance patrol focuses along a specific line of communications (such as a road, railway, or cross-country mobility corridor). It provides new or updated information on route conditions, for example, trafficability, determining radius of a curve, or degree of slope. Additional more detailed information may include obstacles and bridge classifications, and enemy and civilian activity along the route. This type of reconnaissance includes not only the route itself, but also all terrain along the route from which the enemy could influence the friendly force's movement. A route reconnaissance patrol is conducted to obtain and locate the following:

- Detailed information about trafficability on the route and all adjacent terrain.
- Detailed information about enemy activity or enemy force moving along a route.
- Sites for emplacing hasty obstacles to slow enemy movement.
- Obstacles, CBRN contamination, and so forth.

7-223. The patrol can be tasked to survey a route in a planned infiltration lane. After being briefed on the proposed infiltration, the patrol leader conducts a thorough map reconnaissance and plans a series of fans along the route. (See figure 7-23.) The coverage must reconnoiter all intersecting routes for a distance greater than the range at which enemy direct-fire weapons could influence the infiltrating forces.

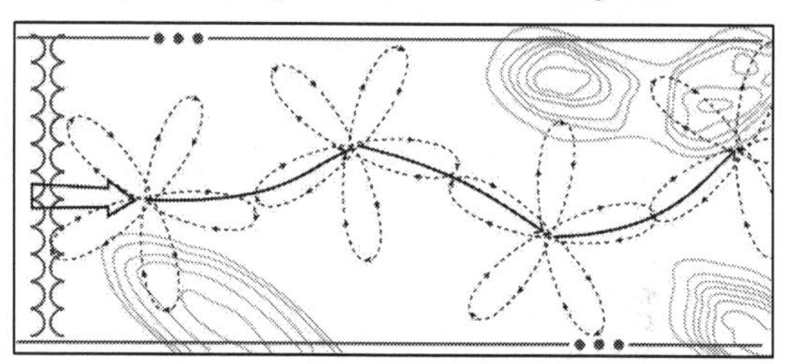

Figure 7-23. Route reconnaissance using fans

7-224. The patrol reports conditions likely to affect friendly movement. These conditions include:

- Presence of the enemy.
- Terrain information.
- Location and condition of bypasses, fords, and obstacles.
- Choke points.
- Route and bridge conditions.

7-225. If all or part of the proposed route is a road, the leader must treat the road as a danger area. The patrol moves parallel to the road, using a covered and concealed route. When required, reconnaissance (and/or surveillance) and security teams move close to the road to reconnoiter important areas. The patrol plans a different route for its return.

Control Measures

7-226. Control measures for a route reconnaissance create an AO for the unit conducting the reconnaissance. The leader should submit the patrol report in an overlay format including:

- Two grid references (required).
- Magnetic north arrow (required).
- Route drawn to scale (required).
- Title block (required).
- Route classification formula (required).
- Road curves with a radius of less than 45 degrees.
- Steep grades and maximum gradients.

- Road width of constrictions such as bridges and tunnels, with the widths and lengths of the traveled ways (in meters).
- Underpass limitations with limiting heights and widths.
- Bridge bypasses classified as easy, hard, or impossible.
- Civil or military road numbers or other designations.
- Locations of fords, ferries, and tunnels with limiting information.
- Causeways, snow sheds, or galleries if they are in the way. Data about clearance and load-carrying capacity should be included to permit an evaluation to decide whether to strengthen or remove them.

ZONE RECONNAISSANCE PATROL

7-227. A zone reconnaissance patrol is conducted to obtain information on enemy, terrain, and routes within a specified zone. Zone reconnaissance techniques include the use of moving elements, stationary teams, or multiple area reconnaissance actions.

MOVING ELEMENT TECHNIQUES

7-228. When moving elements are used, the elements (squads or fire teams) move along multiple routes to cover the whole zone. When the mission requires a patrol to saturate an area, the patrol uses one of the following techniques: the fan, the box, converging routes, or successive sectors.

Fan Method

7-229. When using the fan method, the leader first selects a series of ORPs throughout the zone from during the operation. The patrol establishes security at the first ORP. Upon confirming the ORP location, the leader confirms reconnaissance routes out from and back to the ORP. These routes form a fan-shaped pattern around the ORP. The routes must overlap to ensure the entire area is reconnoitered. Once the routes are confirmed, the leader sends out reconnaissance (and/or surveillance) and security teams along the routes. When all reconnaissance and security teams have returned to the ORP, the platoon collects and disseminates all information to every Soldier before moving on to the next ORP.

7-230. Each reconnaissance and security team moves from the ORP along a different fan-shaped route overlapping with others to ensure reconnaissance of the entire area. (See figure 7-24.) These routes should be adjacent to each other. Adjacent routes prevent the patrol from potentially making contact in two different directions. The leader maintains a reserve at the ORP.

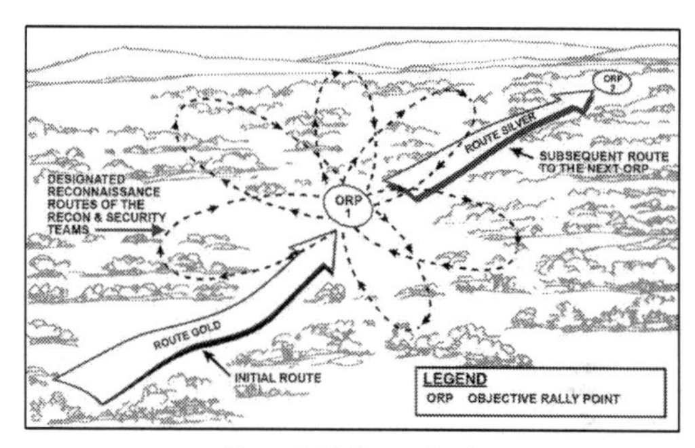

Figure 7-24. Fan method

Box Method

7-231. When using the box method, the leader sends reconnaissance (and/or surveillance) and security teams from the ORP along routes forming a boxed-in area. The leader sends other teams along routes through the area within the box. (See figure 7-25.) All teams meet at a linkup point at the far side of the box from the ORP.

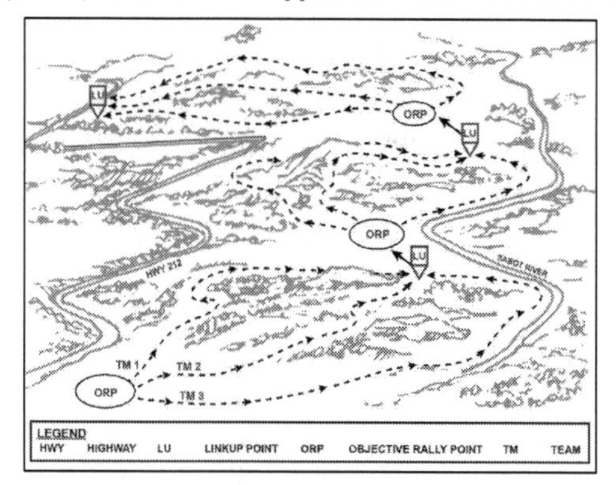

Figure 7-25. Box method

Converging Routes Methods

7-232. When using the converging routes method, the leader selects routes from the ORP through the zone to a rally point at the far side of the zone from the ORP. Each reconnaissance (and/or surveillance) and security team moves along a specified route and uses the fan method (see paragraph 7-229) to reconnoiter the area between routes. (See figure 7-26.) The leader designates a time for all teams to linkup. Once the unit arrives at the rally point, it halts and establishes security.

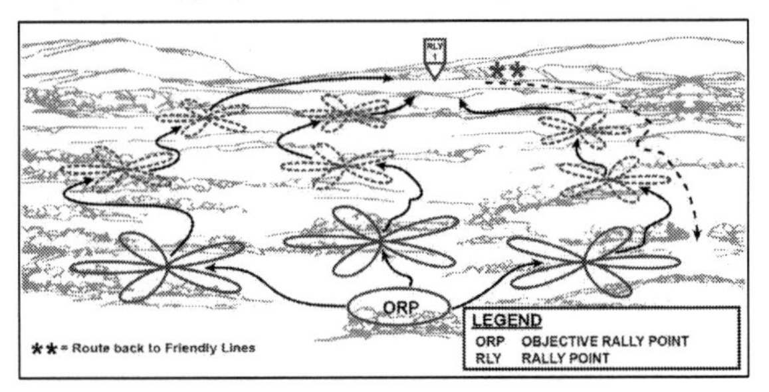

Figure 7-26. Converging routes method

Successive Sector Method

7-233. The successive sector method is a continuation of the converging routes method. (See figure 7-27.) The leader divides the zone into a series of sectors. The platoon uses the converging routes within each sector to reconnoiter to an intermediate linkup point where it collects and disseminates the information gathered to that point. It then reconnoiters to the next sector. Using this method, the leader selects an ORP, a series of reconnaissance routes, and linkup points. The actions from each ORP to each linkup point are the same as in the converging routes method. Each linkup point becomes the ORP for the next phase. Upon linkup at a linkup point, the leader again confirms or selects reconnaissance routes, a linkup time, and next linkup point. This action continues until the entire zone has been reconnoitered. Once the reconnaissance is completed, the unit returns to friendly lines.

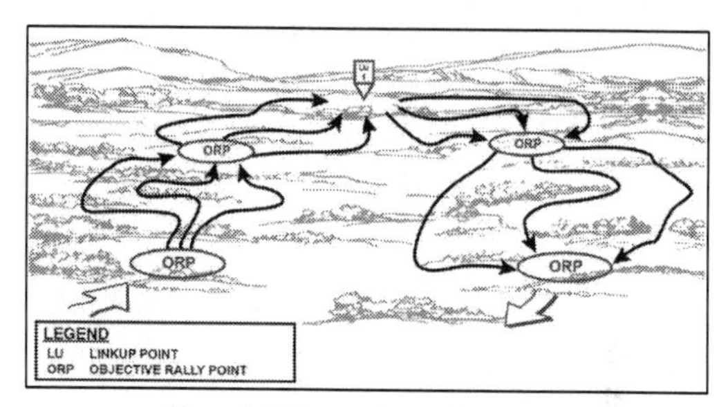

Figure 7-27. Successive sector method

STATIONARY ELEMENT TECHNIQUES

7-234. Using the stationary element technique, the leader positions reconnaissance (and/or surveillance) and security teams in locations where they can collectively observe the entire zone for long-term, continuous information gathering. (See figure 7-28.) The leader considers sustainment requirements when developing Soldiers' load plans.

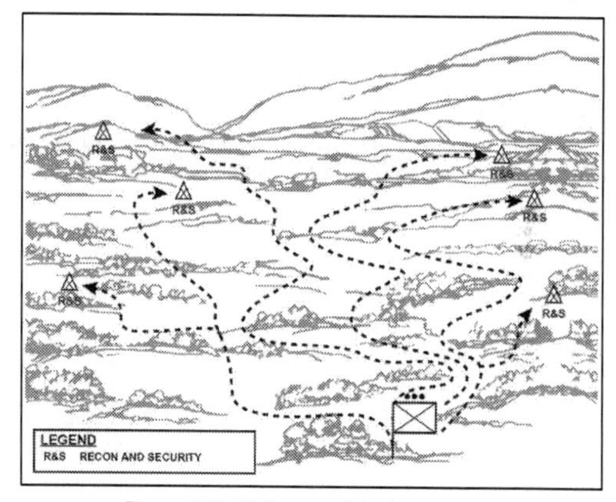

Figure 7-28. Stationary element technique

MULTIPLE AREA RECONNAISSANCE

7-235. When using multiple area reconnaissance, the leader tasks each of the subordinate units to conduct a series of area reconnaissance actions within the zone. (See figure 7-29.)

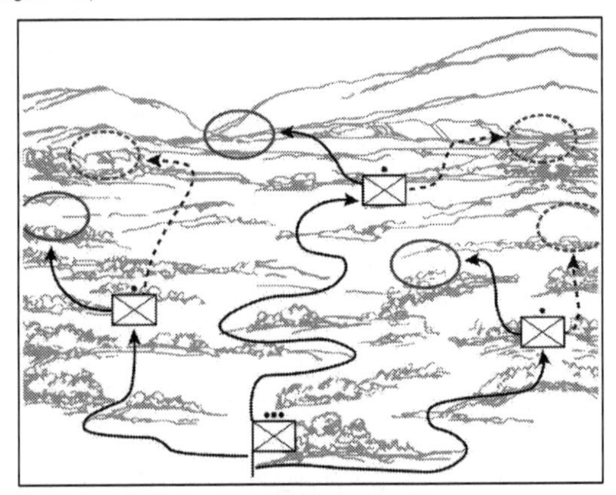

Figure 7-29. Multiple area reconnaissance

SECTION V – POST PATROL ACTIVITIES

7-236. Immediately on reentering a secure location or rejoining the unit, the patrol leader should positively verify all members of the patrol and attachments, and detainees are accounted for. The patrol leader should check in with the company or battalion command post as soon as possible after entering a secure location or rejoining the unit. Additional post patrol activities will include:

- Account for weapons and equipment.
- Debrief.
- Patrol report.

ACCOUNTING FOR WEAPONS AND EQUIPMENT

7-237. The patrol leader is responsible for verifying all the patrol's weapons, ammunition, munitions, and equipment are properly accounted for and reporting the status to the commander or the command post. Lost or missing equipment must be reported immediately. The patrol may be ordered to return to the area where it was lost, if it is assessed safe to do so, and try to find the item.

DEBRIEF

7-238. The patrol leader should conduct a "debrief" with the entire patrol as soon as possible after entering the secure location or rejoining the main body. This allows the leader to capture low-level information while the Soldiers' memories are fresh and information relevant. The leader should go over the notes taken by the patrol scribe chronologically to facilitate the discussion. Every patrol member should participate. If there was an interpreter or other attachments with the patrol, they should be debriefed as part of the patrol debriefing, allowing them to pass on information they obtained during the patrol.

7-239. Normally the debriefing is oral. Sometimes a written report is required. Information on the written report should include:

- Size and composition of the unit conducting the patrol.
- Mission of the platoon such as type of patrol, location, and purpose.
- Departure and return times.
- Routes use checkpoints, grid coordinates for each leg or include an overlay.
- Detailed description of terrain and enemy positions identified.
- Results of contact with the enemy.
- Unit status at the conclusion of the patrol mission, including the disposition of dead or wounded Soldiers.
- Number of isolated Soldiers' platoon/squad unable to recover during execution of the mission.
- Conclusions or recommendations.

PATROL REPORT

7-240. The patrol leader is responsible for the patrol report and may be assisted by the APL and specialist personnel attached to the patrol. Immediately after the debriefing, the patrol leader should render the patrol report to the commander. This report may be verbal or written, simple, or elaborate depending on the situation and commander's requirements.

7-241. The commander may have the patrol leader render the report to the battalion intelligence officer or to the duty officer at the battalion command post. The patrol report (see figure 7-30 on page 7-74) should include a description of the actual route taken by the patrol (as opposed to the planned route), including halt locations. If the unit uses digital command and control systems automatically tracking and displaying the patrol's route, the information is known already. If not, the patrol leader reports it.

7-242. When Global Positioning System (GPS) devices are used by the patrol, gathering route information is easier and faster. The actual route the patrol took is important for planning future patrol routes and actions. Enemy intelligence operations attempt to identify patterns set by U.S. and coalition patrols, including the locations of halts. This may result in attack against locations regularly used by security forces.

7-243. Additional information may include the number of biometric enrollments and identification on the BEWL; was anyone detained according to the instructions on the BEWL; and what is the status? (See FM 3-24.2 for additional information.)

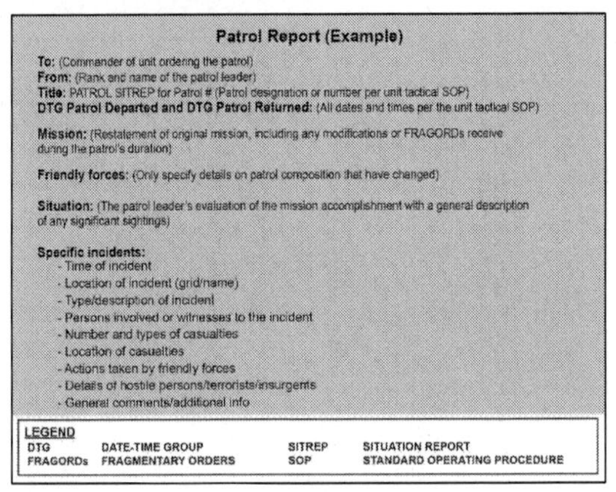

Patrol Report (Example)

To: (Commander of unit ordering the patrol)
From: (Rank and name of the patrol leader)
Title: PATROL SITREP for Patrol # (Patrol designation or number per unit tactical SOP)
DTG Patrol Departed and DTG Patrol Returned: (All dates and times per the unit tactical SOP)

Mission: (Restatement of original mission, including any modifications or FRAGORDs receive during the patrol's duration)

Friendly forces: (Only specify details on patrol composition that have changed)

Situation: (The patrol leader's evaluation of the mission accomplishment with a general description of any significant sightings)

Specific incidents:
- Time of incident
- Location of incident (grid/name)
- Type/description of incident
- Persons involved or witnesses to the incident
- Number and types of casualties
- Location of casualties
- Actions taken by friendly forces
- Details of hostile persons/terrorists/insurgents
- General comments/additional info

LEGEND

DTG	DATE-TIME GROUP	SITREP	SITUATION REPORT
FRAGORDs	FRAGMENTARY ORDERS	SOP	STANDARD OPERATING PROCEDURE

Figure 7-30. Patrol report, example

Chapter 8
Sustainment

Sustainment is the provision of logistics, financial management, personnel services, and health service support necessary to maintain operations until successful mission completion (ADP 4-0). In the Infantry rifle platoon, the platoon leader is responsible for planning sustainment; the platoon sergeant and squad leaders have the responsibility to execute the plan. The platoon sergeant is the platoon's primary sustainment operator and works closely with the company executive officer and first sergeant to ensure the platoon receives the required support for its assigned missions. The company normally forecasts supplies and "pushes" to the platoon rather than having the platoon "pull" them. This chapter describes sustainment in support of the Infantry rifle platoon. It concludes with a discussion of casualty response, specifically the three phases (care under fire, tactical field care, and tactical evacuation care) of tactical combat casualty care (TCCC). (See ATP 3-21.10 for additional information.)

Note. See ATP 3-21.18 for additional information on Soldier load.

SECTION I – SUSTAINING THE PLATOON

8-1. Sustaining the platoon is primarily a company- and battalion-level operation. While the company commander and executive officer plan the company's concept of support for the operation, the platoon leader and platoon sergeant are responsible for sustainment planning, preparation, execution, and assessment at platoon level. Sustainment is characterized by the following eight principles: integration, anticipation, responsiveness, simplicity, economy, survivability, continuity, and improvisation. (See FM 3-96 for additional information.) The platoon leader, with the assistance of the platoon sergeant, integrates these guiding principles to shape the sustainment concept of support for the platoon. This section describes sustainment operations in support of the platoon, specifically responsibilities, functions, tasks, and activities, and unit relationships throughout high operating tempo operations.

RESPONSIBILITIES

8-2. Sustainment responsibilities for the platoon include reporting and requesting support requirements through the company and ensuring sustainment elements execute operations properly when they arrive in the platoon's area. The platoon sergeant is normally in charge of these functions, with guidance and oversight provided by the platoon leader. Prior to and during operations, the platoon sergeant must submit accurate personnel and sustainment reports, along with other necessary information and requests to receive the needed support and services at the precise place and time needed. The following paragraphs (see paragraphs 8-3 to 8-5) address key leader sustainment specific responsibilities within the platoon and squad.

PLATOON LEADER

8-3. The platoon leader is responsible for developing the platoon's sustainment concept of support. The platoon leader (assisted by the platoon sergeant) develops the concept of support to sustain the platoon's scheme of maneuver. The platoon leader (assisted by the platoon sergeant) identifies support requirements for the platoon's mission to the company where the company headquarters consolidates it and passes it on to the battalion.

PLATOON SERGEANT

8-4. The platoon sergeant is the platoon's primary sustainment coordinator and operator, reporting directly to the platoon leader. The platoon sergeant executes the platoon's sustainment plan, relying heavily on platoon and company tactical standard operating procedures (SOPs). The platoon sergeant (assisted by squad leaders) supervises and controls the Soldiers and equipment within the platoon. During preparations for the mission, the platoon sergeant works closely with the platoon leader and squad leaders to determine specific support requirements of the tactical plan. The platoon sergeant then ensures proper arrangements are made to provide those support requirements. Additional sustainment responsibilities include:

- Determining the location of the platoon's resupply point (as required) based on data developed during initial planning and war-gaming processes.
- Compiling DA Form 2404 (*Equipment Inspection and Maintenance Worksheet*) or DA Form 5988-E (*Equipment Maintenance and Inspection Worksheet),* from the squad leaders and providing updates to the platoon leader as required.
- Ensuring the platoon executes sustainment operations according to the platoon and company sustainment plans.
- Leading the sustainment rehearsal in coordination with squad leaders to ensure integration into the platoon/company rehearsal.
- Assisting the platoon leader in developing sustainment priorities (to include load tailoring [see ATP 3-21.18]) and guidance according to the company's concept of support and enforcing them.

- Conducting close coordination with the company executive officer and first sergeant for planning and resourcing the platoon's mission.
- Coordinating and synchronizing human resources support with the company first sergeant. This includes personnel accountability reports, casualty reports, replacement operations, personnel readiness management, mail operations, essential personnel services, and other administrative or personnel requirements.
- Meeting the logistics package at the company logistics release point (known as LRP) or company trains; guiding it to the platoon resupply point and supervising platoon resupply operations (see paragraphs 8-16 to 8-25, or ATP 3-21.10 for additional information).
- Providing a platoon orientation for new personnel and, in consultation with the platoon leader, makes recommendations on replacements to leaders within the platoon.
- Directing and supervising evacuation of casualties, detainees, and damaged equipment.
- Directing and supervising the collection, initial identification, and evacuation of human remains to the mortuary affairs collection point.
- Maintaining the platoon-manning roster.
- Cross-leveling supplies and equipment throughout the platoon.
- Ensuring leaders within the platoon supervise the proper maintenance on all assigned equipment.
- Coordinating logistics/personnel requirements with attached or operational controlled units.

SQUAD LEADER

8-5. Squad leader (assisted by team leaders) sustainment duties include:

- Ensuring Soldiers perform proper maintenance on all assigned equipment.
- Ensuring Soldiers maintain personal hygiene.
- Compiling personnel and logistics reports for the platoon and submitting them to the platoon sergeant as directed or in accordance with unit SOP.
- Obtaining supplies, equipment (except Class VIII), and mail from the platoon sergeant and ensuring proper distribution.
- Cross-leveling supplies and equipment throughout the squad.
- Assisting the platoon sergeant in developing sustainment priorities (to include load tailoring [see ATP 3-21.18]).

Note. The squad leader ensures all squad members carry their own improved first-aid kit to sustain survivability. See paragraphs 1-43 and 1-46 to 1-48 for information on combat lifesaver (CLS) and the platoon's medic, respectively.

PLAN, PREPARE, EXECUTE, AND ASSESS

8-6. The platoon leader, with the assistance of the platoon sergeant, fully integrates and synchronizes the company's sustainment concept of support within the platoon's concept of operations. This concept of support describes how the unit will execute and receive sustainment during the operation. Planning during troop leading procedures (TLP) is continuous and concurrent with ongoing support preparation, execution, and assessment.

PLANNING

8-7. Company and platoon SOPs should be the basis for sustainment operations, with planning conducted to determine specific requirements and to prepare for contingencies. The platoon planning should address specific support matters of the mission. Deviations from the sustainment SOPs should be covered early in the planning process. In most situations, sustainment planning begins before receipt of the mission, as part of the ongoing process of refining the sustainment plan.

8-8. The platoon leader develops the platoon sustainment plan by first determining exactly what the platoon has on hand to accurately estimate the platoon's support requirements. It is critical for the company to know what the platoon has on hand, especially for designated critical supplies. This process is important not only in confirming the validity of the sustainment plan but also in ensuring the platoon's support requests are submitted as early as possible. The platoon leader formulates the sustainment plan and submits, through the platoon sergeant, support requests based on the results of the platoon's maneuver plan. The sustainment plan should provide answers based on the nature of the operation and specific tactical considerations, such as—

- What types of support will the platoon need?
- In what quantities will this support be required?
- Will contingency resupply (for example, Class I-water, Class V-ammunition) be required during the operation (see note on page 8-5)?
- Does this operation require pre-position supplies?
- What is the composition, disposition, and capabilities of the expected enemy threat? How will these affect sustainment during execution?
- Where and when will the expected contact occur?
- What are the platoon's expected casualties and equipment losses based on the nature and location of expected contact?
- What impact will the enemy's special weapons capabilities (such as chemical, biological, radiological, and nuclear [CBRN]) have on the battle and on expected sustainment requirements?
- How many enemy prisoners of war are expected, and where?
- How will terrain and weather affect the sustainment plan during the operation?
- What ground will provide the best security for maintenance collection points (when vehicles attached) and casualty collection points (CCPs)?

- What are the platoon's casualty evacuation (CASEVAC) and/or medical evacuation (MEDEVAC) routes?
- What are the company's dirty routes for evacuating contaminated personnel, vehicles, and equipment?
- When and where will the platoon need resupply?
- Based on the nature and location of expected contact, what are the best sites for the CCP?
- Where will the enemy prisoner of war collection points be located?
- What are the support requirements, by element and type of support?
- Which squad or team has priority for support and resupply?
- Will lulls in the operation permit support elements to conduct resupply operations in relative safety?
- Based on information developed during the sustainment planning, which resupply technique should the platoon use (see paragraphs 8-16 to 8-25, or ATP 3-21.10 for additional information)?

Note. Contingency resupply is the on-call delivery of prepackaged supplies during the execution phase of an operation. This type of on call delivery of a prepackaged resupply is generally used to support an operation of limited duration, such as an Airborne or air assault or other limited engagement of short duration. Contingency resupply operations for the platoon are identified during the TLP, normally during war gaming as the course of action (COA) is analyzed. Contingency resupply differs from a planned resupply (for example logistics package) or emergency (immediate) resupply, in that, prior to execution, triggers for delivery are developed to tie contingency resupply operations to the ground tactical plan. (See ATP 3-21.10 for additional information.)

PREPARATION

8-9. Preparation for sustainment consists of activities performed by the platoon to improve its ability to execute the operation. These preparations refine the sustainment concept of support. Preparations include but are not limited to plan refinement, rehearsals, coordination, inspections, and movements. Sustaining factors to consider, although not inclusive, include geography information and the availability of supplies and services, facilities, transportation, maintenance, and general skills (such as translators or laborers). Preparations may include building pre-packaged resupply drops; for instance, bags loaded with loaded rifle magazines, rations, water, batteries, and so forth.

8-10. Sustainment rehearsals help synchronize the platoon's concept of support within the company's overall operation. These rehearsals typically involve coordination and procedure drills for transportation support, resupply, maintenance and vehicle recovery (when attached), and MEDEVAC and CASEVAC. Throughout preparation, units rehearse battle drills and SOPs. Leaders place priority on those drills or actions they anticipate occurring during the operation. Sustainment rehearsals validate logistical synchronization with the concept of operations. Rehearsals focus on the supported and supporting unit with respect to sustainment operations across time and space as well as the method of support for specific actions (for example, contingency resupply) during the operation. If the platoon is using unfamiliar movement assets, rehearsals should include both evacuating injured Soldiers from those vehicles and loading casualties in them.

EXECUTION

8-11. The platoon leader and platoon sergeant plan and organize sustainment operations that can be executed rapidly, possibly with widely dispersed operations. Sustainment determines the depth and duration of the platoon operation and is essential to retaining and exploiting the initiative to provide the support necessary to maintain operations until the mission is completed. Failure to sustain operations could cause a pause or culmination of an operation resulting in the loss of the initiative. The platoon sergeant, reporting directly to the platoon leader, and company executive officer and first sergeant work closely prior to execution to synchronize all sustainment functions to allow the platoon maximum freedom of action.

ASSESSMENT

8-12. The platoon leader and platoon sergeant, through continued assessment of the current situation determine if the current operation is proceeding according to the company commander's intent and if planned future operations are supportable. For example, the platoon sergeant may assess the need to apply additional security measures to an LRP in the platoon's area of operations (AO). The mission of, threat to, and location of the LRP determines the degree of protection needed.

8-13. Additionally, knowing the status of liquid, ammunition, casualties, and equipment (known as LACE) through the reporting of subordinates provides information to the platoon leader and platoon sergeant on the impact of these areas on the mission. The status of these areas is compared to execution to assess whether they are shortages: now; will arise as shortages that can impact the mission; or are shortages that will not impact mission accomplishment. Examples include, knowing:

- The extent of casualties, the consumption of medical supplies.
- The status of ammunition and water, comparing consumed to expected continued consumption.
- The status of serviceable equipment and weapons systems.

SUPPLY AND FIELD SERVICES

8-14. Supply provides the materiel required to accomplish the mission. Supplies such as subsistence items, water, ammunition, and barrier materials are the most common. Field services maintain combat strength of the force by providing for its basic needs and promoting its health, welfare, morale, and endurance. Field services provide life support functions. Common services that may be available consist of shower and laundry, field feeding, sanitation services, mortuary affairs, and water purification.

8-15. The Infantry rifle platoon normally deploys with 72 hours of supplies specific to its assigned mission. Supply commodities are used in support of missions and are separated into ten supply classes:

- Class I – Perishable and semi-perishable subsistence items, water, and gratuitous health and comfort items, includes rations that are packaged as individual or group meals.
- Class II – Individual equipment and general supplies, items consist of common consumable items such as clothing, individual equipment, tentage, tool sets and kits, maps, administrative/housekeeping supplies, and CBRN protective equipment.
- Class III – Packaged fuel—petroleum, oils, and lubricants, consists of packaged petroleum, oils and lubricants that can be handled similarly to dry cargo.
- Class III – Bulk fuel—gasoline, diesel, and aviation fuel.

Note. Consumption rates of Class III packaged, and bulk are especially important when the Infantry rifle platoon has attached or in direct support vehicles and armored or Stryker vehicles.

- Class IV – Construction and barrier materials, consists of construction, fortifications, and barrier materials. Typically, a platoon will not have a requirement for Class IV materials unless in the defense. For example, gloves and picket pounders for handling of concertina wire and placement of pickets.
- Class V – Ammunition, unit *basic load*—the quantity of supplies required to be on hand within, and moved by a unit or formation, expressed according to the wartime organization of the unit or formation and maintained at the prescribed levels (JP 4-09)—is determined by the weapon density, number of Soldiers, and specific mission requirements over time.
- Class VI – Sundry personal demand items such as toiletry, hygiene, and small recreational items. In most cases, a Soldier deploys with a 30-day supply of health and comfort items. After the first 30 days, health and comfort packages are provided at 30-day intervals through Class I channels at the request of the unit commander.
- Class VII – Major end items, this class of supply is intensely managed and controlled through higher command channels. Detailed planning at all echelons includes recovery operations and battle loss/battle damage replacement operations. This is particularly import to the platoon when attached with mounted assets.
- Class VIII – Medical supplies.

Note. Typically, the medical platoon of the Infantry battalion will deploy with a 3-day supply of Class VIII to support the battalion. When forecasting Class VIII requirements, take the mission, location, projected casualty rates, and available medical assets into consideration.

- Class IX – Repair parts include individual repair parts and major assemblies such as engines, transmissions, and final drives, which are required to maintain equipment and operational readiness.

DISTRIBUTION AND RESUPPLY OPERATIONS

8-16. The platoon sergeant, in coordination with the platoon leader, is the primary coordinator and operator for synchronizing distribution and resupply operations for the platoon. Distribution encompasses the movement of personnel, materiel, and equipment in support of the platoon. Resupply operations cover all classes of supply, water, mail, and any other items usually requested. The company executive officer is responsible for requesting and coordinating support capabilities against the platoon's requirements. The platoon sergeant identifies these requirements through daily logistics status (known as LOGSTAT) reports. Whenever possible, the platoon conducts resupply on a regular basis, ideally during hours of limited visibility.

METHODS OF DISTRIBUTION

8-17. Distribution is the operational process of synchronizing all elements of the logistic system to deliver the "right things" to the "right place" at the "right time" to support the platoon's concept of operations. The elements of logistics include maintenance, transportation, supply, field services, distribution, operational contract support, and general engineering. The platoon sergeant requests and executes distribution operations based on supply requirements communicated by the platoon leader and squad leaders. The platoon sergeant communicates the platoon's requirements through LOGSTAT reports and other means, through the company executive officer or first sergeant, to the battalion.

8-18. Methods of distribution integrate and synchronize materiel management and transportation. Supporting sustainment units use the best distribution method dependent on the mission, the urgency of requirement, the threat, the supported unit's priority of support, time/distance, and other mission variables of METT-TC (I) mission, enemy, terrain and weather, troops and support available, time available, civil considerations, and informational considerations. The two methods of distribution are unit distribution and the three techniques (service station, tailgate, or in-position) of supply point distribution. (See ATP 3-21.10 for additional information on each method of distribution.)

METHODS OF RESUPPLY

8-19. Resupply operations require continuous and close coordination between the supporting and supported units. The two methods of resupply are planned resupply and emergency resupply. Planned resupply is the preferred method of resupply. At the platoon level, the company sustainment concept of support and LOGSTAT reports by the platoon sergeant, normally establish the requirement, timing, and frequency for planned resupply. A planned resupply may occur during an operation. Emergency resupply is the least preferred method of supply. While instances of emergency resupply may be required, especially when combat losses or a change in the enemy situation occurs, requests for it may indicate a breakdown in coordination and collaboration between supported and supporting units.

8-20. The platoon sergeant is responsible for synchronizing resupply operations for subordinate squads and other elements assigned or attached to the platoon. The platoon sergeant identifies requirements through daily LOGSTAT reports and sustainment planning conducted during TLP. (See ATP 3-21.10 for additional information on the methods of resupply to sustain the platoon during operations.)

Delivery Techniques of Resupply

8-21. Logisticians and supported units can use several techniques for resupply during planned and emergency resupply operations. See ATP 3-21.10 for additional information on the following delivery techniques for resupply:

- Logistics package. A grouping of multiple classes of supply and supply vehicles under the control of a single convoy commander, is a simple and efficient way to accomplish planned (routine) resupply.
- Contingency resupply. The on-call delivery of prepackaged supplies during the execution phase of an operation.
- LRP. The company most commonly executes supply point distribution by means of an LRP:
  - Any place on the ground where unit distribution vehicles take supplies met by the company representative (for example, the first sergeant).
  - Then takes the supplies forward to subordinate platoons for subsequent distribution.
- Pre-positioned supplies. The pre-positioning of supplies is a planned resupply technique that reduces the reliance on traditional convoy operations and other resupply operations.
- Cache. A pre-positioned and concealed supply point (different from standard pre-positioned supplies because the supported or supporting units conceal the supplies from the enemy whereas units may not conceal other pre-positioned supplies).
- Aerial delivery. By airland, airdrop, and sling-load operations that provides additional capability to resupply the platoon when the terrain or enemy situation limits access by ground transportation.

Echelon Support

8-22. How the Infantry battalion, including external and attached organizations and the brigade support battalion of the Infantry brigade combat team (IBCT), array in echelon varies widely based upon the mission variables of METT-TC (I). The forward support company, in support of the Infantry battalion's concept of support (prepared by the battalion logistics staff officer), plans and synchronizes echelon support—the method of supporting an organization arrayed within an area of operation (see ATP 4-90). Current mission, task organization, command and control, concept of support, capability and capacity, and terrain influence how support is echeloned. As the Infantry battalion's primary sustainment organization, the forward support company's organization facilitates echeloned support. Common echelon of support at the lowest level of sustainment is executed at the battalion and company echelons. (See ATP 3-21.20 and FM 3-96 for additional information.)

8-23. Company trains provide sustainment for a company during combat missions. Company trains (when established) usually include the first sergeant, supply sergeant, armorer, and combat medic. When pushed down to company level, a MEDEVAC team may collocate in the company trains. The supply sergeant can collocate in the battalion combat trains (when established) or battalion field trains if it facilitates logistics package operations. The first sergeant controls the company trains to ensure its operational responsiveness and survivability although the company commander may assign the responsibility to the company executive officer. The distance of the company trains from the company's combat mission is METT-TC (I) dependent. By placing at least one terrain feature between it and the enemy, the company trains will be concealed from the enemy's direct fire weapons.

8-24. Configuring the company trains with the assets described above gives the company virtually immediate access to essential logistical functions to support subordinate platoons while allowing the trains to remain in a covered and concealed position out of enemy contact. Because the security of support elements is critical to the success of any mission, battalion and company trains develop plans for continuous security. Where feasible, they plan and execute a perimeter defense. Trains, however, can lack the personnel and combat power to conduct this level of security effort. In such situations, they implement passive measures (see paragraph 5-143) to avoid detection by the from enemy during unit and supply point distribution operations.

8-25. METT-TC (I) ultimately dictates the actual distance at which the trains operate. In some instances, the company trains can be collocated with the battalion combat trains or even the battalion field trains. Figure 8-1 illustrates a notional battalion concept of support to the platoon echelon. (See ATP 3-21.10 for additional information.)

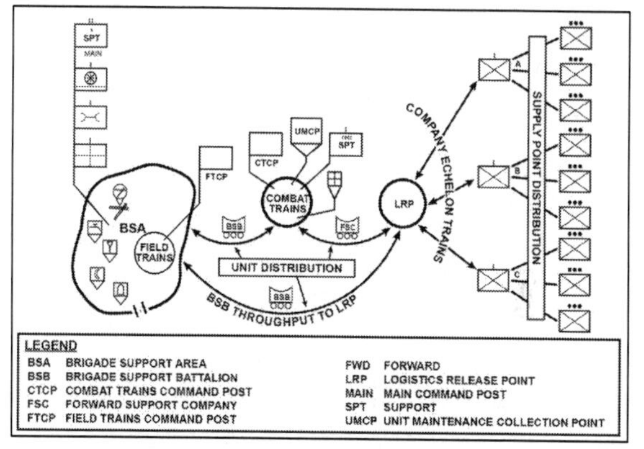

Figure 8-1. Sustainment concept of support (echelon support example)

MAINTENANCE

8-26. Maintenance is all actions taken to retain materiel in a serviceable condition. The maintenance of weapons and equipment is continuous. All Soldiers must know how to maintain their weapon and equipment according to the related technical manual. The platoon leader, platoon sergeant, and squad leaders must understand maintenance for every piece of equipment in the platoon.

8-27. Maintenance includes inspecting, testing, servicing, repairing, requisitioning, recovering, and evacuating vehicles and equipment. Maintenance at the platoon and squad level comprises thorough preventive maintenance checks and services, and accurate reporting of maintenance problems to the company.

8-28. Maintenance and the early identification of problems prevent equipment downtime and the reduction of combat effectiveness. The result of good preventive maintenance checks and services is a properly completed equipment inspection and maintenance forms. These forms, DA Form 2404 or DA Form 5988-E, are the primary means through which the platoon and squads obtain maintenance support or repair parts. The forms follow a pathway from crew level to the brigade support area (BSA) and back. Per unit SOP, the company executive officer or first sergeant supervises the flow of these critical maintenance documents and parts. The maintenance flow is as follows:

- Squad leaders and/or vehicle (when equipped) commanders collect the maintenance forms to give them to the platoon sergeant (or sends them via,

Joint Capabilities Release Logistics or Joint Battle Command-Platform) who then consolidates the forms for the platoon.

- The platoon sergeant gives a hard copy of the forms (or forwards an electronic version) to the executive officer or first sergeant, who reviews and verifies problems and deficiencies and requests parts needed for maintenance and repairs.
- The hard copy of the forms (or electronic versions) are consolidated at company level and then transmitted to the battalion and/or its supporting combat repair team.
- During the next logistics package operation, the completed hard copy forms are returned to the combat repair team to document completion of the repair.
- In the BSA, required repair parts are packaged for delivery during the next scheduled (planned) resupply or through emergency resupply means.
- When the repair or installation of the part requires higher skills and equipment than the operator, a combat repair team is dispatched to assess the repair and to install the part on site.
- The operator conducts initial maintenance, repair, and recovery actions on site. Once it is determined that the crew cannot repair or recover the equipment or vehicle (when attached), the platoon contacts the executive officer or first sergeant. If additional assistance is needed, the combat repair team assesses the damaged or broken equipment and makes necessary repairs to return the piece of equipment to a fully mission-capable or mission-capable status, if appropriate.

8-29. The unit SOP should detail when maintenance is performed (at least once a day in the field), to what standards, and who inspects it. The squad leader is often the one who inspects maintenance work whereas the platoon sergeant, platoon leader, first sergeant, executive officer, and commander conduct spot-checks. A technique is for the platoon leader or platoon sergeant to spot check equipment at the platoon's planned (routine) resupply time to ensure equipment is clean and the proper maintenance forms are complete before receiving Class I. Another technique is for each to spot-check a different squad whereas another is for each to check a single type of weapon or piece of equipment in all squads daily. Instructions for routine maintenance must be integrated into the SOP for patrol bases, tactical assembly areas (TAAs), defenses, and reorganization. They help ensure that Soldiers make a habit of maintenance and that they perform it without jeopardizing unit security. In addition to operator maintenance, selected Soldiers are trained to perform limited maintenance on damaged weapons and battle damage assessment and repair.

SECTION II – TACTICAL COMBAT CASUALTY CARE

8-30. TCCC is divided into the three phases—care under fire, tactical field care, and tactical evacuation care. TCCC occurs during a combat mission and is the military counterpart to prehospital emergency medical treatment. (See FM 4-02 for additional information.)

CARE UNDER FIRE

8-31. In the care under fire phase, combat medical personnel and their units are under effective hostile fire and are very limited in the care they can provide. In essence, only those lifesaving interventions that must be performed immediately are undertaken during this phase. Casualty care under fire has a positive impact on the morale of a unit. Casualties are cared for at the point of injury (or under nearby cover and concealment) and receive self- or buddy-aid, advanced first aid from the CLS, and/or emergency medical treatment from the platoon or company combat medics.

Note. Nonmedical personnel (specifically individuals performing self-aid and buddy-aid and CLSs [see paragraph 1-43]) within the platoon assist combat medics (see paragraphs 1-46 to 1-48) within platoons and the company senior combat medic in their duties. Individuals (self-aid and buddy-aid) and CLS administer appropriate TCCC. If needed, Soldiers are evacuated to the Role 1 medical treatment facility (MTF) (battalion aid station) in the battalion support area, or the Role 2 MTF (brigade support medical company of the brigade support battalion) in the BSA of the IBCT. (See ATP 3-21.20 and FM 3-96 for additional information.)

8-32. All platoon CLSs and the combat medic carry multiple blank versions and use (they complete all entries as fully as possible) DD Form 1380 (see figure 8-2 on page 8-14) to document pre-MTF care at the point of injury. Such care relates to both battle and nonbattle injuries. Once completed, DD Form 1380 is visibly attached to the patient when transferred to the CCP and/or to a Role 1 or Role 2 MTF. All entries on the DD Form 1380 will be made using a non-smearing pen or marker. All entries on the DD Form 1380 should be printed clearly, including the first responder's name.

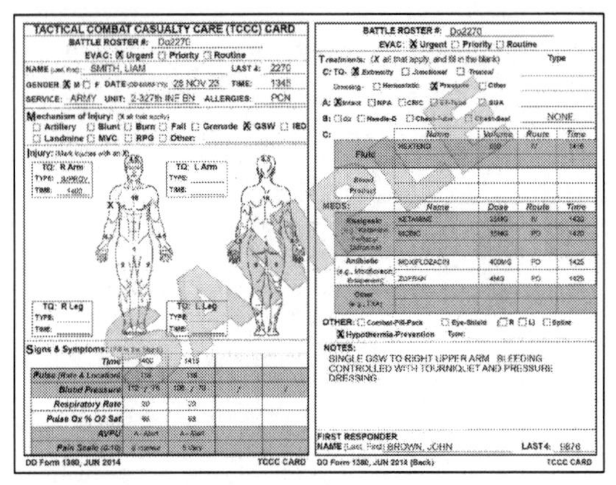

TACTICAL COMBAT CASUALTY CARE (TCCC) CARD

BATTLE ROSTER #: Do2270

EVAC: [X] Urgent [ ] Priority [ ] Routine

NAME (Last, First): SMITH, LIAM LAST 4: 2270

GENDER [X] M [ ] F DATE (DD-MMM-YY): 28 NOV 23 TIME: 1345

SERVICE: ARMY UNIT: 2-327th INF BN ALLERGIES: PCN

Mechanism of Injury: (X all that apply)

[ ] Artillery [ ] Blunt [ ] Burn [ ] Fall [ ] Grenade [X] GSW [ ] IED [ ] Landmine [ ] MVC [ ] RPG [ ] Other:

Injury: (Mark injuries with an X)

TQ: R Arm TYPE: IMPROV. TIME: 1400

TQ: L Arm TYPE: TIME:

TQ: R Leg TYPE: TIME:

TQ: L Leg TYPE: TIME:

Signs & Symptoms: (Fill in the blank)

Time	1400	1415		
Pulse (Rate & Location)	[illegible]	[illegible]		
Blood Pressure	[illegible] / [illegible]	[illegible] / [illegible]	/	/
Respiratory Rate	20	20		
Pulse Ox % O2 Sat	[illegible]	98		
AVPU	A - Alert	A - Alert		
Pain Scale (0-10)	[illegible]	[illegible]		

DD Form 1380, JUN 2014 TCCC CARD

BATTLE ROSTER #: Do2270

EVAC: [X] Urgent [ ] Priority [ ] Routine

Treatments: (X all that apply, and fill in the blank) Type

C: TQ- [X] Extremity [ ] Junctional [ ] Truncal

Dressing- [ ] Hemostatic [X] Pressure [ ] Other

A: [X] Intact [ ] NPA [ ] CRIC [ ] ET-Tube [ ] SGA

B: [ ] O2 [ ] Needle-D [ ] Chest-Tube [ ] Chest-Seal NONE

C:

	Name	Volume	Route	Time
Fluid	HEXTEND	[illegible]	IV	[illegible]
Blood Product				

MEDS:	Name	Dose	Route	Time
Analgesic (e.g., Ketamine, Fentanyl, Morphine)	KETAMINE	[illegible]	IV	[illegible]
	[illegible]	[illegible]	PO	[illegible]
Antibiotic (e.g., Moxifloxacin, Ertapenem)	MOXIFLOXACIN	400MG	PO	1425
	ZOFRAN	4MG	PO	1425
Other (e.g., TXA)				

OTHER: [ ] Combat-Pill-Pack [ ] Eye-Shield [ ] R [ ] L [ ] Splint

[X] Hypothermia-Prevention Type:

NOTES:

SINGLE GSW TO RIGHT UPPER ARM. BLEEDING CONTROLLED WITH TOURNIQUET AND PRESSURE DRESSING

FIRST RESPONDER

NAME (Last, First): BROWN, JOHN LAST 4: 9876

DD Form 1380, JUN 2014 (Back) TCCC CARD

Figure 8-2. DD Form 1380 (Tactical Combat Casualty Care [TCCC] Card)

8-33. During the fight, casualties should remain under cover where they received initial treatment (self- or buddy-aid). (See collective tasks–Evacuate Casualties 07-SQD-9033 and Evacuate Casualties 07-PLT-9033.) As soon as the situation allows, casualties are moved to the platoon CCP (when established). Once the casualties are collected, evaluated, and treated, they are prioritized for evacuation back to the company CCP. Once they arrive at the company CCP, the above process is repeated while awaiting their evacuation back to the battalion aid station or other facility in the battalion support area or BSA. Unit SOPs address these activities, to include the marking of casualties in limited visibility operations. Small, standard, or infrared chemical lights work well for this purpose.

8-34. An effective technique, particularly during an attack, is to task-organize a logistics team under the first sergeant. This team carries additional ammunition forward to the platoons and evacuates casualties to either the company or the battalion CCP. The commander determines the size of the team during mission analysis.

8-35. When the platoons are widely dispersed, casualties might be evacuated directly from the platoon CCP by nonmedical vehicles and personnel. However, casualties are usually moved to the company CCP before evacuation. If the capacity of the battalion's organic ambulances is exceeded, unit leaders may re-assign supply or other nonmedical vehicles to backhaul or otherwise transport non urgent casualties to the battalion aid station. In other cases, the platoon sergeant may direct platoon aid and litter teams to carry the casualties to the rear. Unless the threat environment is highly permissive, helicopter evacuation is unlikely to occur farther forward on the battlefield than the battalion's ambulance exchange points between the battalion aid station (Role 1) and the brigade's medical company (Role 2) in the BSA.

8-36. Leaders minimize the number of Soldiers required to evacuate casualties. Casualties with minor wounds can walk or even assist with carrying the more seriously wounded. Soldiers can make field-expedient litters by cutting small trees and putting the poles through the sleeves of buttoned uniform blouses. A travois, or skid, might be used for CASEVAC. Wounded are strapped on this type of litter, then one person can pull it. It can be made locally from durable, rollable plastic. Tie-down straps are fastened to it. In rough terrain, or on patrols, litter teams can evacuate casualties to the battalion aid station. Then, they are carried with the unit either until transportation can reach them or until they are left at a position for later pickup.

8-37. Unit SOP and operation order (OPORD) address casualty treatment and evacuation in detail. They cover the duties and responsibilities of key personnel, the evacuation of chemically contaminated casualties (on separate routes from noncontaminated casualties), and the priority for operating key weapons and positions. They specify preferred and alternate methods of evacuation and make provisions for retrieving and safeguarding the weapons, ammunition, and equipment of casualties. Slightly wounded personnel are treated and returned to duty by the lowest echelon possible. Platoon combat medics evaluate (under the control to the platoon sergeant) sick Soldiers and either treat or evacuate them as necessary. MEDEVAC and CASEVAC are rehearsed like any other critical part of an operation.

8-38. As casualties occur, the nearest observer informs the platoon sergeant who then informs the first sergeant via the most expedient method available (for example, radio voice). The first sergeant then submits a personnel status report to the battalion personnel staff officer section. This report documents duty status changes on all casualties. A casualty report is filled out when a casualty occurs, or as soon as the tactical situation permits. This usually is done by the Soldier's squad leader and turned in to the platoon sergeant, who forwards it to the first sergeant. A brief description of how the casualty occurred (including the place, time, and activity being performed) and who or what inflicted the wound is included. If the squad leader does not have personal knowledge of how the casualty occurred, the squad leader gets this information from Soldiers who have the knowledge.

8-39. DA Form 1156, (*Casualty Feeder Card*) (see figures 8-3a and 8-3b on page 8-16), are used to report those Soldiers who have been killed and recovered, and those who have been wounded. This form also is used to report captured or killed in action Soldiers who are missing or not recovered. The Soldier with the most knowledge of the incident should complete the witness statement. During lulls in the battle, the platoon forwards casualty information to the company headquarters. The first sergeant ensures a completed DA Form 1156 is forwarded to the Infantry battalion personnel staff officer, who then enters the data into the Defense Casualty Information Processing System.

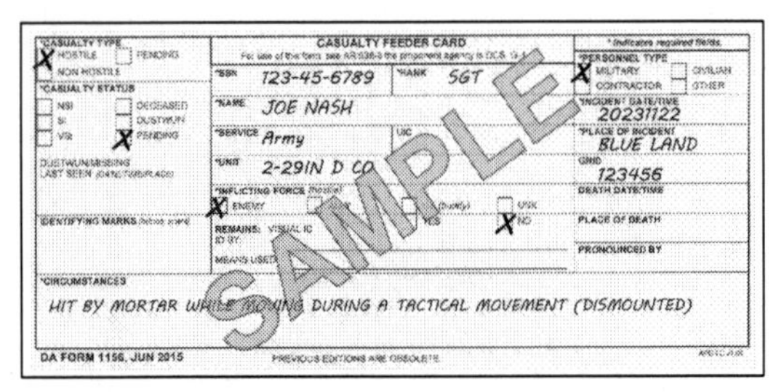

CASUALTY FEEDER CARD

*CASUALTY TYPE: X HOSTILE / NON HOSTILE / PENDING

*CASUALTY STATUS: NSI / SI / VSI / DECEASED / DUSTWUN / X PENDING

DUSTWUN/MISSING LAST SEEN

IDENTIFYING MARKS

*SSN: 123-45-6789 | *RANK: SGT

*NAME: JOE NASH

*SERVICE: Army | UIC

*UNIT: 2-29IN D CO

*INFLICTING FORCE: X ENEMY / UNK

REMAINS: VISUAL ID / ID BY / MEANS USED — X NO

*PERSONNEL TYPE: X MILITARY / CIVILIAN / CONTRACTOR / OTHER

*INCIDENT DATE/TIME: 20231122

*PLACE OF INCIDENT: BLUE LAND

GRID: 123456

DEATH DATE/TIME

PLACE OF DEATH

PRONOUNCED BY

*CIRCUMSTANCES: HIT BY MORTAR WHILE MOVING DURING A TACTICAL MOVEMENT (DISMOUNTED)

DA FORM 1156, JUN 2015 — PREVIOUS EDITIONS ARE OBSOLETE.

SAMPLE

Figure 8-3a. DA Form 1156 (Casualty Feeder Card) (front)

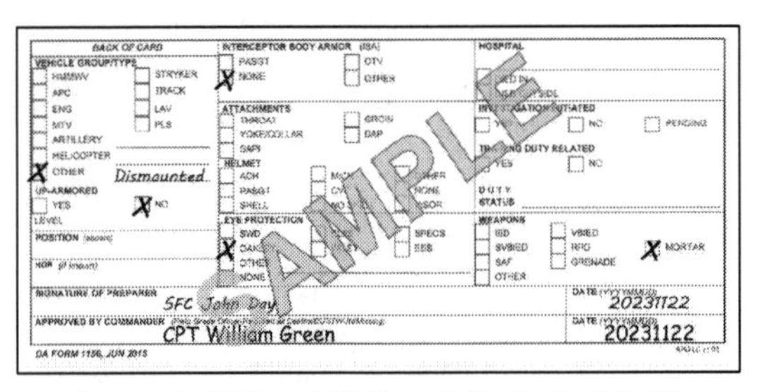

BACK OF CARD

VEHICLE GROUP/TYPE: HMMWV / STRYKER / APC / TRACK / ENG / LAV / MTV / PLS / ARTILLERY / HELICOPTER / X OTHER: Dismounted

UP-ARMORED: YES / X NO

LEVEL

POSITION

HOW

INTERCEPTOR BODY ARMOR (IBA): PASGT / OTV / X NONE / OTHER

ATTACHMENTS: THROAT / GROIN / YOKE/COLLAR / DAP / SAPI

HELMET: ACH / PASGT / SHELL / NONE / SENSOR

EYE PROTECTION: SWD / X OAKLEY / OTHER / NONE / SPECS / ESS

HOSPITAL

INVESTIGATION INITIATED: YES / NO / PENDING

TRAINING DUTY RELATED: YES / NO

DUTY STATUS

WEAPONS: IED / VBIED / SVBIED / RPG / SAF / GRENADE / OTHER / X MORTAR

SIGNATURE OF PREPARER: SFC John Day | DATE (YYYYMMDD): 20231122

APPROVED BY COMMANDER: CPT William Green | DATE (YYYYMMDD): 20231122

DA FORM 1156, JUN 2015

SAMPLE

Figure 8-3b. DA Form 1156 (Casualty Feeder Card) (back)

8-40. Before casualties are evacuated to the CCP or beyond, leaders remove all key operational or sensitive items and equipment, including communications security devices or signal operating instructions, maps, position location devices. Every unit should establish an SOP for handling the weapons and ammunition of its wounded or killed in action. Protective masks and other protective equipment must stay with the individual.

8-41. Casualties are taken to CCP for classification based on their medical condition, assigned evacuation precedence (urgent, priority, routine, and convenience), and availability of MEDEVAC platforms. Within a CCP, the combat medic conducts triage of all patients, takes the necessary steps to stabilize their conditions, and initiates the process of evacuating them to the rear for further treatment. The combat medic helps the first sergeant arrange evacuation via ground or air ambulance, or by nonstandard means.

See FM 4-02 and ATP 4-02.2 for additional information on evacuation precedence for Army operations at Roles 1 through 3 MTFs.

8-42. When possible, the battalion medical platoon ambulances provide evacuation and en route care from the Soldier's point of injury or the CCP to the battalion aid station. The ambulance team supporting the company or platoon works in coordination with the combat medic supporting the company or platoon. In mass casualty situations, nonmedical vehicles can be used to assist in CASEVAC as directed by the platoon leader or company commander. Plans for the use of nonmedical vehicles to perform CASEVAC should be included in the unit SOP. Ground ambulances from the brigade support medical company or other supporting ambulances evacuate patients from the battalion aid station back to the brigade support medical company MTF located in the BSA.

TACTICAL FIELD CARE

8-43. During the tactical field care phase, medical personnel and their patients are no longer under effective hostile fire and medical personnel can provide more extensive patient care. In this phase, interventions directed at other life-threatening conditions, as well as resuscitation and other measures to increase the comfort of the patient may be performed. The physician and physician assistant at the battalion aid station or during tailgate medicine support provide TCCC. *Tailgate medical support* is an economy of force device employed primarily to retain maximum mobility during movement halts or to avoid the time and effort required to set up a formal, operational treatment facility (for example, during rapid advance and retrograde operations) (FM 4-02). During tactical field care, personnel must be prepared to transition back to care under fire, or to prepare the casualty for tactical evacuation, as the tactical situation dictates. (See FM 4-02 and ATP 4-02.4 for additional information on tactical field care.)

Note. The Infantry battalion's organic medical resources within its headquarters and headquarters company include a medical platoon staffed with a field surgeon, physician assistant, and numerous combat medics. The mission of the battalion medical platoon is to provide Role 1 Army Health System support to the Soldiers of the Infantry battalion. Role 1 (also referred to as unit-level medical care) is the first medical care a Soldier receives. The medical platoon within the Infantry battalion is configured with a headquarters section, medical treatment squad, ambulance squad (ground), and combat medic section. The treatment squad consists of two teams (treatment team alpha and team bravo). The treatment squad operates the battalion aid station and provides Role 1 medical care and treatment (to include disease and nonbattle injury prevention, sick call, emergency medical treatment [including TCCC], and patient decontamination). Team alpha is clinically staffed with the physician assistant while team bravo is clinically staffed with the field surgeon. Medical platoon ambulances provide MEDEVAC and en route care from the Soldiers' point of injury, the CCP, or an ambulance exchange point to the battalion aid station. The ambulance squad is four teams of two ambulances composed of one emergency care sergeant and two ambulance aide/drivers assigned to each ambulance. (See ATP 4-02.4 for additional information on the medical platoon.)

TACTICAL EVACUATION

8-44. In the tactical evacuation phase, casualties are transported from the battlefield to MTFs. MTFs provide medical treatment and include the Role 1 facility (battalion aid station), Role 2 facility (brigade support medical company of the brigade support battalion), dispensaries, clinics, and hospital. Evacuation can be by either MEDEVAC (dedicated platforms [ground or air] manned with dedicated medical providers) or CASEVAC (ranging from nondedicated, but tasked, platforms [ground or air] augmented with medical equipment and providers to platforms of opportunity without medical equipment or providers).

Note. For the purposes of this discussion, CASEVAC will mean that which is done when moving casualties from the point of injury to the platoon CCP or company CCP. Ideally, casualties are transferred from a CCP to an MTF by a MEDEVAC asset. When this is not possible, the casualty is moved from the CCP, when required to move, aboard a nonmedical vehicle or aircraft to a MEDEVAC asset or an MTF.

8-45. *Casualty evacuation* is the movement of casualties aboard nonmedical vehicles or aircraft without en route medical care (FM 4-02). CASEVAC encompasses a wide spectrum of potential capability—depending on the mix of transport platform, medical equipment, and medical providers allocated to the mission. At the upper end of the spectrum, nondedicated platforms can be outfitted with the requisite medical equipment and MEDEVAC assets. At the lower end of the spectrum, CASEVAC can be no more than the transport of casualties using platforms of opportunity with no medical equipment or medical providers (in using such assets, the risk of not moving the casualty must outweigh the risk evacuating the casualty in such a manner). Effective CASEVAC complements MEDEVAC by providing additional evacuation capacity when number of casualties (workload) or reaction time exceeds the capabilities of MEDEVAC assets. CASEVAC requires detailed assessment and planning to achieve an effective integration of MEDEVAC and CASEVAC capabilities. (See ATP 4-02.13 for additional information on CASEVAC).

CAUTION

Casualties transported in a CASEVAC platform may not receive proper en route medical care or be transported to the appropriate MTF that can best address the casualty's medical needs. This may have an adverse impact on the casualty's prognosis, long-term disability or even death may result.

8-46. *Medical evacuation* is the timely and effective movement of the wounded, injured, or ill to and between medical treatment facilities on dedicated and properly marked medical platforms with en route care provided by medical personnel (ATP 4-02.2). MEDEVAC is the key factor to ensuring the continuity of care provided to Soldiers by providing en route medical care during evacuation and facilitating the transfer of patients between MTFs to receive the appropriate specialty care. This ensures that scarce medical resources (personnel, equipment, and supplies [to include blood]) can be rapidly transported to areas of critical need on the battlefield.

Note. The Army MEDEVAC system is comprised of dedicated, standardized MEDEVAC platforms (ground and air ambulances). These ambulances have been designed, staffed, and equipped to provide en route medical care to patients being evacuated and are used exclusively to support the medical mission, in accordance with the law of land warfare and the Geneva Conventions (see FM 6-27 for additional information).

Dedicated air MEDEVAC aircraft include specifically trained medical personnel to provide en route care. The 9-line MEDEVAC request (see GTA 08-01-004) is the standard method to request air ambulance MEDEVAC.

This page intentionally left blank.

Appendix A
Direct Fires

Suppressing or destroying the enemy with direct fires is fundamental to success in close combat. This appendix addresses direct fire planning and control in support of Infantry operations. Its primary focus is on the organization and employment of direct fires organic to the Infantry rifle platoon.

SECTION I – DIRECT FIRE PLANNING AND CONTROL

A-1. Direct fire is inherent in maneuver, as is close combat. Small-unit leaders of the Infantry rifle platoon must plan to focus, distribute, and shift the overwhelming mass of the platoon's direct fire capability at critical locations and times to succeed in combat. Efficient and effective fire control means the platoon acquires the enemy and masses the effects of direct fires to achieve decisive results in the close fight. This section covers the principles of direct fire, direct fire planning, and direct fire control. (See TC 3-20.31-4 for additional information.)

PRINCIPLES OF DIRECT FIRE

A-2. When executing direct fires, leaders must know how to apply several fundamental principles. The purpose of the principles of direct fire is not to restrict the actions of subordinates. Applied correctly, they help the platoon to accomplish its primary goal in direct fire engagements both to acquire first and shoot first. These principles give subordinates the freedom to act quickly upon acquisition of the enemy. This discussion focuses on the following principles:

- Destroy the greatest threat first.
- Mass the effects of direct fire.
- Employ the best weapon for the specific target.
- Avoid target overkill.
- Minimize exposure.
- Plan and implement control measures.
- Plan for limited visibility conditions.
- Plan for degraded capabilities.

DESTROY THE GREATEST THREAT FIRST

A-3. The order in which the platoon engages enemy forces is in direct relation to the danger they present. The threat posed by the enemy depends on the weapons, range, and positioning. Presented with multiple targets, a unit will, in almost all situations, initially concentrate fires to destroy the greatest threat first and then distribute fires over the remainder of the enemy force. What constitutes the greatest threat will vary with the situation. For instance, while enemy vehicles or crew-served weapons are normally the most dangerous, an enemy rifle squad in position to flank a defending platoon would become the most dangerous threat. Friendly direct fire planning must be able to shift, focus and distribute direct fires over the width and depth of the anticipated area of operations (AO).

Mass the Effects of Direct Fire

A-4. The platoon must mass its fires to achieve decisive results. Massing entails focusing fires at critical points and distributing the effects. Random application of fires is unlikely to have a decisive effect. For example, concentrating the platoon's fires at a single target may ensure its destruction or suppression; however, that fire control technique its unlikely to achieve a decisive effect on the enemy formation or position. Leaders must direct their formations' fires through careful planning and rehearsal of direct fire control measures and establish appropriate rates of fire to achieve mass.

Employ the Best Weapon for the Specific Target

A-5. Using the appropriate weapon increases the probability of rapid enemy destruction or suppression; at the same time, it saves ammunition. Leaders have many weapons with which to engage the enemy. Target type, range, and exposure are vital factors in determining the weapon and ammunition being employed, as are weapons and ammunition availability and desired targets effects. Additionally, leaders should consider individual crew capabilities when deciding on the employment of weapons. The platoon leader arrays forces based on the terrain, enemy, and desired effects of fires. As an example, when a platoon leader expects an enemy dismounted assault in restricted terrain, the platoon leader would employ subordinate squads, taking advantage of their ability to engage numerous fast-moving targets.

Avoid Target Overkill

A-6. Use only the amount of fire required to achieve necessary effects. Target overkill wastes ammunition and ties up weapons better employed acquiring and engaging other targets. The idea of having every weapon engage a different target, however, must be tempered by the requirement to destroy the greatest threats first. Two simple ways to minimize overkill are with sectors of fire, and by establishing fire patterns. Establishing primary and secondary sectors of fire ensures direct fires are planned throughout the width of the area. Leaders ensure that Soldiers first engage targets in their assigned primary sector and then engage targets in their secondary sector if their primary sector is clear. Soldiers should not engage outside of these sectors unless directed to. Utilizing one of the standard fire patterns (frontal, cross, or depth) acts similarly to an assigned sector, but also controls the sequence of targets to be engaged. For instance, in frontal fires, the right-most Soldier engages the right-most target first and works inwards. In prepared defensive positions with frontal cover, the default fire pattern is almost always a cross pattern which interlocks with adjacent fighting positions.

Minimize Friendly Exposure

A-7. Units increase their survivability by exposing themselves to the enemy only to the extent necessary to engage the enemy. Natural or man-made defilade provides the best cover from lethal direct fire munitions. Infantry Soldiers minimize their exposure by constantly seeking available cover, attempting to engage the enemy from the flank, remaining dispersed, firing from multiple positions, and limiting engagement times.

Plan And Implement Control Measures

A-8. The leader has numerous tools to assist in the planning and implementation of controlling direct fires. These tools include graphic control measures for friendly forces, engagement criteria, identification training for combat vehicles and aircraft, unit weapons safety posture, weapons control status, recognition markings, and a situational understanding to include standard range cards, area sketches, and rehearsals. Knowledge and employment of applicable control measures and rules of engagement are the primary means of preventing fratricide and noncombatant casualties.

Note. Because it is difficult to distinguish between friendly and enemy dismounted Soldiers, small-unit leaders must constantly monitor positions of friendly squads.

Plan for Limited Visibility Conditions

A-9. At night, limited visibility fire control equipment enables the platoon to engage enemy forces at nearly the same ranges that are applicable during the day. Obscurants such as dense fog, heavy obscuration, and blowing sand, however, can reduce the capabilities of thermal and infrared equipment. Leaders should develop contingency plans for such extreme limited visibility conditions. Although decreased acquisition capabilities have minimal effect on area fire, point target engagements likely will occur at decreased ranges. Typically, firing positions, whether offensive or defensive, must be adjusted closer to the area or point where the leader intends to focus fires. Another alternative is the use of visual or infrared illumination when there is insufficient ambient light for passive light intensification devices. While night vision capabilities can deliver high accuracy, their use can be hazardous. The use of tracers, lasers, infrared illuminators, or even the glow of unshielded optics can be detected by peer forces and then targeted for fires. Additionally, the use of optics tends to significantly limit peripheral vision by both individuals and units. Leaders must ensure that their forces deliberately widen their scan to reduce tunnel-vision and observe the full width of assigned sectors.

Note. Vehicles (when attached) equipped with thermal sights can assist squads in detecting and engaging enemy infantry forces in conditions such as heavy obscurants and low illumination.

PLAN FOR DEGRADED CAPABILITIES

A-10. Leaders initially develop plans based on their units' maximum capabilities; they make backup plans for implementation in the event of casualties or weapon damage or failure. While leaders cannot anticipate or plan for every situation, they should develop plans for what they view as the most probable occurrences. Building redundancy into these plans, such as having two systems observe the same AO, is an invaluable asset when the situation (and number of available systems) permits. Designating secondary sectors of fire provide a means of shifting fires if adjacent elements are knocked out of action.

DIRECT FIRE PLANNING

A-11. Leaders plan direct fires as part of the troop leading procedures (TLP). Determining where and how the platoon and subordinate units can and will mass fires is an essential step as leaders develop their concept of operations.

EMPLOYMENT OF DIRECT FIRES

A-12. A *field of fire* is the area that a weapon or group of weapons may cover effectively from a given position (FM 3-90). In selecting a position, a unit must balance how the field of fire will best gain an advantage while simultaneously providing cover and mitigating the effect of the enemy's weapons systems. (See figure A-1.)

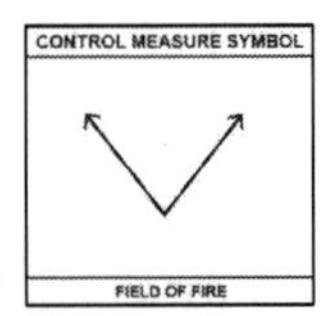

Figure A-1. Field of fire control measure symbol

ORIENT, FOCUS, DISTRIBUTE, AND SHIFT FIRES

A-13. Leaders plan direct fires to be able to command and control their fire. Determining where and how they can mass fires is an essential step in the process for planning direct fires in both the offense and defense. This planning must enable leaders to orient, focus, distribute, and shift their direct fires in execution.

A-14. Orienting fires allows the leader to effectively engage the enemy with direct fires. The plan must allow friendly forces to acquire enemy elements rapidly and accurately. Orienting friendly forces on probable enemy locations and on likely avenues of approach speeds target acquisition. Conversely, failure to orient subordinate elements results in slower acquisition, greatly increases the likelihood that enemy forces can engage first.

A-15. Leaders determine points or areas to focus combat power after identifying probable enemy locations. The leader's visualization of where and how the enemy may attack or defend enables the leader to determine the volume of fire necessary at points to have a decisive effect. Leaders establish weapons ready postures for the unit as well as triggers for initiating fires based on where and how they want to focus. Leaders evaluate the risk of fratricide and establish controls to prevent it (see paragraphs A-30 to A-31). Control measures, although not inclusive, include the designation of recognition markings, weapons safety posture (see paragraph A-57), and weapons control status (WCS) (see paragraph A-59).

A-16. Distribution of fires enables the firing unit to distribute the effects of fires throughout the width and depth of the enemy array. Shifting fires means the ability to shift fires from one area to another area rapidly. In the defense these are typically planned areas; however, in the offense, they may be ad hoc. Leaders typically shift the focus or distribution of fires from one area to another in response to an emerging threat or a changed enemy formation.

A-17. Leaders shift fires to refocus and redistribute the effects based on the evolving estimate of the situation as the engagement proceeds. Situational awareness becomes an essential part of the fire control process at this point. Leaders apply the same techniques and considerations, including fire control measures that they used earlier to focus fires.

Planning Techniques

A-18. Based on where and how they want to focus and distribute fires, leaders can establish the weapons ready postures of their elements as well as triggers for initiating fires. During mission preparation, leaders plan, and conduct rehearsals of direct fires (and of the fire control process) based on the mission variables of METT-TC (I) mission, enemy, terrain and weather, troops and support available, time available, civil considerations, and informational considerations.

A-19. The leaders plan direct fires in conjunction with development of their mission analysis and completion of the plan. Determining where and how leaders can and will mass fires are also essential steps as leaders develop their concept of operations.

A-20. After identifying probable enemy locations, leaders determine the points or areas on the ground where they can focus combat power against the enemy. Their visualization of where and how the enemy will attack or defend assists them in determining enemy locations and the volume of fire against the enemy to have a decisive effect. If they intend to mass the fires of more than one subordinate element, leaders must establish the means for distributing fires.

A-21. Based on where and how they want to focus and distribute fires, leaders can then establish the weapons ready postures for platoon elements as well as triggers for initiating fires. Additionally, they must evaluate the risk of fratricide and establish controls to prevent it; these measures include the designation of recognition markings, WCS, and weapons safety posture.

A-22. After determining where and how they will mass and distribute fires, leaders then must orient elements so they can rapidly and accurately acquire the enemy. They also can war-game the selected course of action (COA) or concept of operations to determine probable requirements for refocusing and redistributing fires and to establish other required controls. During mission preparation, leaders plan and conduct rehearsals of direct fires (and of the fire control process) based on their mission analysis.

A-23. Leaders must continue to apply planning procedures and considerations throughout execution. They must be able to adjust direct fires based on a continuously updated mission analysis, combining situational awareness with the latest available intelligence products and assessments. When necessary, they also must apply direct fire SOPs, which are covered in the following discussion.

STANDARD OPERATING PROCEDURES

A-24. A well-rehearsed direct fire SOP ensures quick, predictable actions by all members of the platoon. Leaders base the various elements of the SOP on the capabilities of their forces and on anticipated conditions and situations. SOP elements should include standing means for focusing fires, distributing their effects, orienting forces, and preventing fratricide. The platoon leader should adjust the direct fire SOP whenever changes to anticipated and actual METT-TC (I) become apparent.

A-25. If the platoon leader does not issue other instructions, the squads begin the engagement using the SOP. Subsequently, the platoon leader can use a fire command to refocus or redistribute fires. The following paragraphs discuss specific SOP provisions for orienting forces, orienting fires, focusing fires, distributing fires, shifting fires, and preventing fratricide.

Orienting Forces and Fires

A-26. A standard means of orienting friendly forces is to assign a principal direction of fire (known as PDF), using a target reference point (TRP), to orient each element on a probable enemy position or likely avenue of approach. To provide all-around security, the SOP can supplement the PDF with sectors using a friendly-based quadrant. The following example SOP elements illustrate the use of these techniques:

- The center (front) squad's PDF is TRP 2 (center) until otherwise specified; the squad is responsible for the front two quadrants.
- The left flank squad's PDF is TRP 1 (left) until otherwise specified; the squad is responsible for the left two friendly quadrants (overlapping with the center squad).
- The right flank squad's PDF is TRP 3 (right) until otherwise specified; the squad is responsible for the right two friendly quadrants (overlapping with the center squad).

Focusing Fires

A-27. TRPs are a common means of focusing fires. One technique is to establish a standard respective position for TRPs in relation to friendly elements and to number TRPs consistently, such as from left to right. This allows leaders to determine and communicate the location of a TRP quickly.

Distributing Fires

A-28. Three useful means of distributing the platoon's fires are TRPs, engagement priorities, and target array. While single TRPs (see paragraph A-41) allow leaders to focus fires, leaders can distribute fires by designating TRPs as the left and right limits of a squad's fires. For instance, 1st Squad covers from TRP 1 to TRP 2. Leaders can designate engagement priority (see paragraph A-61), by type of enemy vehicle or weapon, for each type of friendly weapons system. The target array technique (see paragraph A-76) can assist in distribution by assigning specific friendly elements to engage enemy elements of approximately similar capabilities.

Shifting Fires

A-29. Leaders establish standards to designate the shifting of fires. For example, the support element shifts fires as directed or in accordance with established standards, shifting fire forward of the assault element as it moves across the objective or to the flanks of targets or areas. Additionally, leaders establish triggers for shifting fires (or ceasing) based on battlefield events such as the movement of enemy or friendly forces. One standard is the use of a minimum safe line when a friendly element, such as a breach force, is moving toward an area of indirect fires. As the element approaches the minimum safe line, observers call for fires shift (or cease), allowing the friendly force to move safely in the danger area (see figure 4-9 on page 4-52). In the defense, designating secondary sectors of fire provide a means of shifting fires if adjacent elements are knocked out of action.

Preventing Fratricide

A-30. A primary means of minimizing fratricide risk is to establish a standing WCS (see paragraphs A-59 and A-60). The SOP also must dictate ways of identifying friendly rifle squads and other dismounted elements; techniques include using an infrared light source or detonating an obscuration grenade of a designated color at the appropriate time. Minimizing the risk of fratricide in the platoon can be accomplished through command and control systems (if equipped); however, this does not supplant the platoon leader's responsibility to plan for fratricide prevention.

A-31. Finally, the SOP must address the most critical requirement of fratricide prevention maintaining situational awareness. It must direct subordinate leaders to inform the platoon leader, adjacent elements, and subordinates whenever a friendly force is moving or preparing to move. Leaders must be especially cognizant of the heightened risk of fratricide when friendly elements enter or exit friendly lines, or when security elements, engineers, or other elements operate forwards of, or to the rear of, defensive positions. This information must always be disseminated to all levels.

DIRECT FIRE CONTROL

A-32. Small-unit leaders communicate to their subordinates the manner, method, and time to initiate, shift, and mass fires, and when to disengage by using direct fire control measures. Leaders control their unit's fires so they can direct the engagement of enemy systems to gain the greatest effect. Leaders use the intelligence preparation of the operational environment (known as IPOE) and information collection to determine the most advantageous way to use direct fire control measures to mass the effects on the enemy and reduce fratricide from direct fire systems. (See chapter 2 paragraphs 2-4 to 2-7 and 2-102 to 2-106 and ATP 2-01.3 and FM 3-55 [respectively] for additional information.)

Fire Control Process

A-33. To bring direct fires against an enemy force, leaders must continuously apply the steps of the fire control process. At the heart of this process are two critical actions: rapid, accurate target acquisition and massing of fire to achieve decisive effects on the target. Target acquisition is the detection, identification, and location of a target in sufficient detail to permit the employment of weapons. Massing entails focusing fires at critical points and distributing the fires for optimum effect. The following discussion examines target acquisition and massing of fires using the following basic steps of the fire control process:

- Identify probable enemy locations and determine the enemy scheme of maneuver.
- Determine where and how to mass fires.
- Orient forces to speed target acquisition.
- Shift fires to refocus or redistribute.

Identify Probable Enemy Locations and Determine the Enemy Scheme of Maneuver

A-34. Leaders plan and execute direct fires based on their mission analysis. An essential part of this process is the analysis of the terrain and enemy force, which aids the leader in visualizing how the enemy will attack or defend a particular piece of terrain. A defending enemy's defensive positions or an attacking enemy's support positions are normally driven by intervisibility. Typically, there are limited points on a piece of terrain providing both good fields of fire and adequate cover for a defender. Similarly, an attacking enemy will have only a limited selection of avenues of approach providing adequate cover and concealment.

A-35. Coupled with available terrain analysis products and intelligence products and assessments, an understanding of the effects of a specific piece of terrain on maneuver, figure A-2 will assist leaders in identifying probable enemy locations and likely avenues of approach both before and during the fight. Leaders may use all the following products or techniques in developing and updating the analysis:

- An enemy situation template (known as SITEMP) based on the analysis of terrain and enemy.
- A spot or contact report on enemy locations and activities.
- Information collection within the AO.

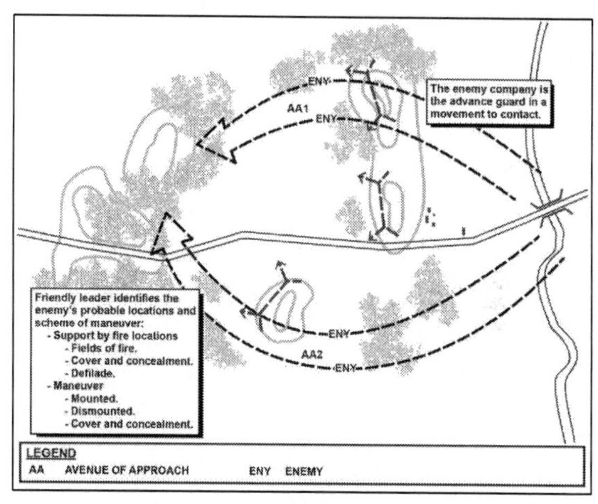

Figure A-2. Identifying probable enemy locations and determining enemy scheme of maneuver

Determine Where and How To Mass Fires

A-36. To achieve decisive effects, friendly forces must mass their fires. (See figure A-3 on page A-10.) Massing requires leaders to focus the fires of subordinate elements and to distribute the effects of the fires. Based on their mission analysis and their concept of operations, leaders identify points where they want to, or must, focus the unit's fires. Most often, these are locations they have identified as probable enemy positions or points along likely avenues of approach where the unit can mass fires. Because subordinate elements may not initially be oriented on the point where leaders want to mass fires, they may issue a fire command to focus the fires. At the same time, leaders must use direct fire control measures to distribute the fires of subordinate elements, which now are focused on the same point. In the offense, leaders should plan direct fire control measures to enable fires during the approach and to achieve mass at the objective. The ground scheme of maneuver is designed to enable the platoon to execute the direct fire plan.

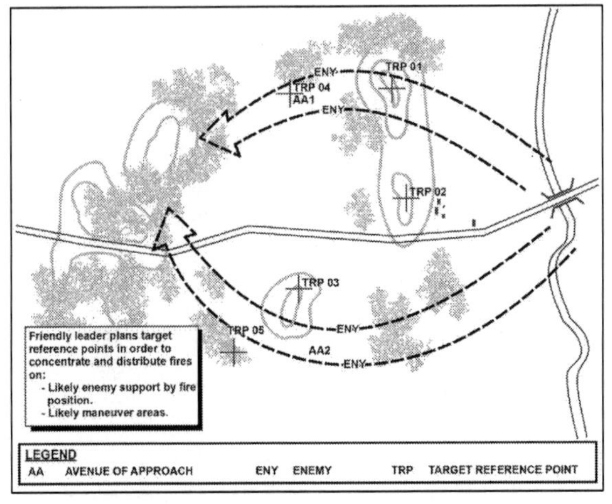

Figure A-3. Determining where and how to mass fires

Orient Forces to Speed Target Acquisition

A-37. To engage the enemy with direct fires, friendly forces must rapidly and accurately acquire enemy elements. (See figure A-4.) Orienting friendly forces on probable enemy locations and on likely avenues of approach will speed target acquisition. Conversely, failure to orient subordinate elements will result in slower acquisition; this greatly increases the likelihood enemy forces will be able to engage first. The clock direction orientation method, which is prescribed in most unit tactical standard operating procedures (SOPs), is good for achieving all-around security; however, it does not ensure friendly forces are most oriented to detect the enemy. To achieve this critical orientation, leaders typically designate TRPs on or near probable enemy locations and avenues of approach; they orient subordinate elements using directions of fire or sectors of fire. Normally, the gunners on crew-served weapons scan the designated direction, sector, or area while other crewmembers observe secondary sectors or areas to provide all-around security.

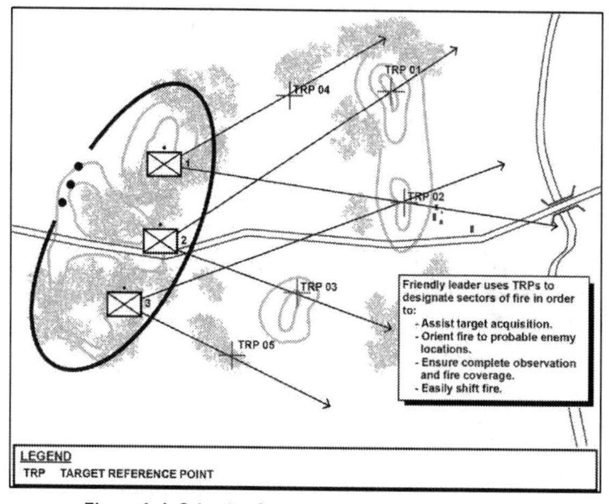

Figure A-4. Orienting forces to speed target acquisition

Shift Fires to Refocus and Redistribute

A-38. As the engagement proceeds, leaders must shift fires to refocus and redistribute the effects based on their evolving mission analysis. Situational awareness becomes an essential part of the fire control process at this point. Leaders apply the same techniques and considerations, including fire control measures they used earlier to focus and distribute fires (see figure A-5 on page A-12). A variety of situations will dictate shifting of fires, including the following:

- Appearance of an enemy force posing a greater threat than the one currently being engaged.
- Extensive attrition of the enemy force being engaged, creating the possibility of target overkill.
- Attrition of friendly elements engaging the enemy force.
- Change in the ammunition status of the friendly elements engaging the enemy force.
- Maneuver of enemy or friendly forces resulting in terrain masking.
- Increased fratricide risk as a maneuvering friendly element closes with the enemy force being engaged.

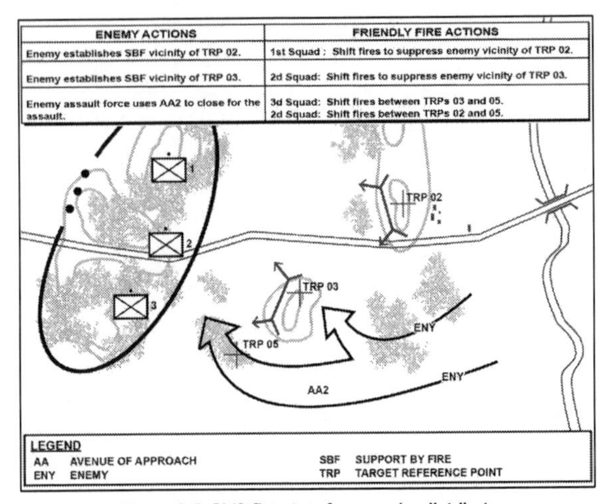

ENEMY ACTIONS	FRIENDLY FIRE ACTIONS
Enemy establishes SBF vicinity of TRP 02.	1st Squad : Shift fires to suppress enemy vicinity of TRP 02.
Enemy establishes SBF vicinity of TRP 03.	2d Squad: Shift fires to suppress enemy vicinity of TRP 03.
Enemy assault force uses AA2 to close for the assault.	3d Squad: Shift fires between TRPs 03 and 05. 2d Squad: Shift fires between TRPs 02 and 05.

Figure A-5. Shift fires to refocus and redistribute

Fire Control Measures

A-39. Leaders use fire control measures to control fires. Application of these concepts, procedures, and techniques assists the unit in acquiring the enemy, focusing fires on the enemy, distributing the effects of the fires, and preventing fratricide. At the same time, no single measure is sufficient to control fires. At the platoon level, fire control measures will be effective only if the entire unit has a common understanding of what they mean and how to employ them. The following discussion focuses on the various fire control measures employed by the Infantry platoon. Table A-1 lists terrain-based and threat-based control measures.

Table A-1. Common fire control measures

Terrain-Based Fire Control Measures	*Threat-Based Fire Control Measures*
Target reference point	Rules of engagement
Engagement area	Weapons ready posture
Sector of fire	Weapons safety posture
Direction of fire	Weapons control status
Terrain-based quadrant	Engagement priorities
Friendly-based quadrant	Trigger
Maximum engagement line; restrictive fire line; final protective line	Engagement techniques; fire patterns; target array

Terrain-Based Fire Control Measures

A-40. Leaders use terrain-based fire control measures to focus and control fires on a particular point, line, or area rather than on a specific enemy element. The following paragraphs describe the tactics, techniques, and procedures associated with this type of control measure.

Target Reference Point

A-41. A TRP (see paragraph 5-207) is a recognizable point on the ground leaders use to orient friendly forces, and to focus, distribute, shift, or otherwise control direct fires. In addition, when leaders designate TRPs as indirect fire targets, they can use the TRP when calling for and adjusting indirect fires. Leaders designate TRPs at probable enemy locations and along likely avenues of approach. These points can be natural or man-made. A TRP can be an established site. For example, a hill or a building, or an impromptu feature such as a burning enemy vehicle or obscurants generated by an artillery round can be designated as a TRP. Friendly units can construct markers to serve as a TRP. (See figure A-6 on page A-14.) Ideally, a TRP should be visible in three observation modes (unaided, passive-infrared, and thermal) so all forces can see them. Examples of TRPs include the following features and objects:

- Prominent hill mass.
- Distinctive building.
- Observable enemy position.
- Destroyed vehicle.
- Ground-burst illumination.
- Obscurant rounds for immediate engagements only; this is the least preferred method.

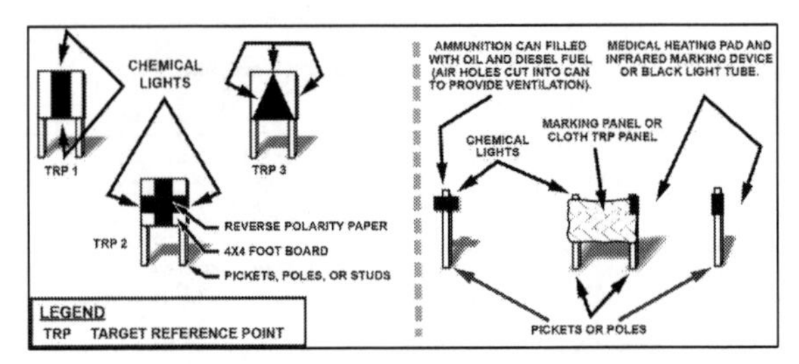

Figure A-6. Constructed target reference point markers, example

Engagement Area

A-42. This fire control measure is an area along an enemy avenue of approach where leaders intend to mass the fires of available weapons to destroy an enemy force. The size and shape of the engagement area is determined by the degree of relatively unobstructed intervisibility available to the unit's weapon systems in their firing positions and by the maximum range of those weapons. Typically, the platoon leader delineates responsibility within the engagement area by assigning each squad a sector of fire or direction of fire. (See chapter 5 section V for additional information.)

Sector of Fire

A-43. A sector of fire is a defined area being covered by direct fire. Leaders assign sectors of fire to subordinate elements, crew-served weapons, and individual Soldiers to ensure coverage of sectors; they also may limit the sector of fire of an element or weapon to prevent accidental engagement of an adjacent unit. In assigning sectors of fire, leaders consider the number and types of weapons available. Typically, leaders designate both primary and secondary sectors of fire with the primary scanned first and, if empty, the secondary is then scanned. In addition, leaders must consider acquisition system type and field of view in determining the width of a sector of fire. For example, while unaided vision has a wide field of view, its ability to detect and identify targets at range and in limited visibility conditions is restricted. Conversely, most fire control acquisitions systems have greater detection and identification ranges than the unaided eye, but their field of view is narrow. Means of designating sectors of fire include the following:

- TRPs.
- Clock direction.
- Terrain-based quadrants.
- Friendly-based quadrants.
- Azimuth or cardinal direction.

Direction of Fire

A-44. A direction of fire is an orientation or point used to assign responsibility for a particular area on the battlefield being covered by direct fire. Leaders designate directions of fire for purposes of acquisition or engagement by subordinate elements, crew-served weapons, or individual Soldiers. Direction of fire is most commonly employed when assigning sectors of fire would be difficult or impossible because of limited time or insufficient reference points, or when the purpose is to cover a narrow point on the ground such as a road junction, the mouth of a draw or similar features. Means of designating a direction of fire include the following:

- Closest TRP.
- Clock direction.
- Azimuth or cardinal direction.
- Tracer on target.
- Infrared laser pointer.
- Grenadier obscuration round.

Quadrants

A-45. Quadrants are subdivisions of an area created by superimposing an imaginary pair of perpendicular axes over the terrain to create four separate areas or sectors. Establish quadrants on the terrain, friendly forces, or on the enemy formation. The method of quadrant numbering is established in the unit SOP; however, care must be taken to avoid confusion when quadrants based on terrain, friendly forces, and enemy formations are used simultaneously.

Note. Techniques in which quadrants are based on enemy formations usually are referred to as target array; it is covered in the discussion of threat-based fire control measures.

Terrain-Based Quadrant

A-46. A terrain-based quadrant entails the use of a TRP, either existing or constructed, to designate the center point of the axes dividing the area into four quadrants. This technique can be employed in both the offense and defense. In the offense, platoon leaders designate the center of the quadrant using an existing feature or by creating a reference point. For example, using a ground burst illumination round, an obscurant marking round, or a fire ignited by incendiary or tracer rounds. The axes delineating the quadrants run parallel and perpendicular to the direction of movement. In the defense, platoon leaders designate the center of the quadrant using a planned TRP.

A-47. In examples shown in figure A-7 on page A-16, quadrants are marked using the letter "Q" and a number (Q1 to Q4); quadrant numbers are in the same relative positions as on military map sheets (from Q1 as the upper left-hand quadrant clockwise to Q4 as the lower left-hand quadrant).

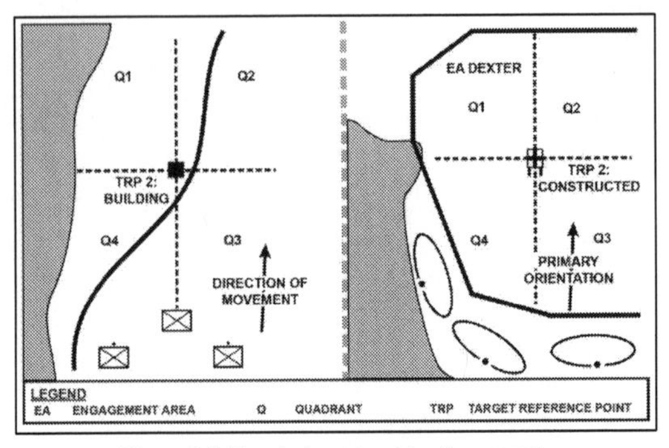

Figure A-7. Terrain-based quadrants, example

Friendly-Based Quadrant

A-48. The friendly-based quadrant technique entails superimposing quadrants over the unit's formation. The center point is based on the center of the formation, and axes run parallel and perpendicular to the general direction of travel, and the quadrant travel with the formation. For rapid orientation, the friendly quadrant technique may be better than the clock direction method; because different elements of a large formation rarely are oriented in the same exact direction and the relative dispersion of friendly forces causes parallax to the target. Figure A-8 illustrates use of friendly-based quadrants.

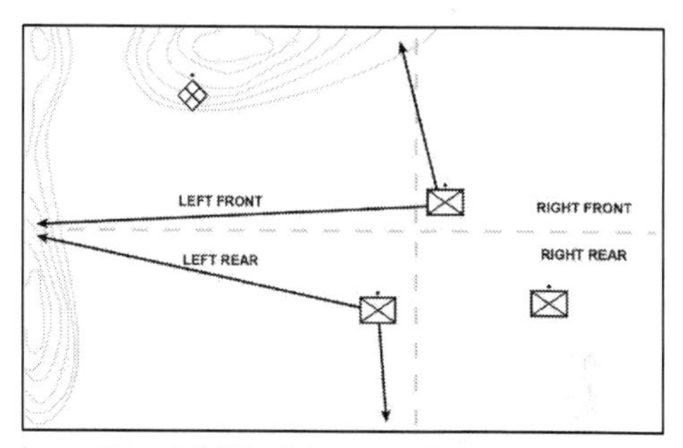

Figure A-8. Friendly-based quadrants, example

Maximum Engagement Line

A-49. Maximum engagement line is the linear depiction of the farthest limit of effective fire for a weapon or unit. This line is determined by the weapons or unit's maximum effective range and by the effects of terrain. For example, slope, vegetation, structures, and other features provide cover and concealment preventing the weapon from engaging to the maximum effective range. A maximum engagement line serves several purposes. Platoon leaders can use it to prevent crews from engaging beyond the maximum effective range, to define criteria of the establishment of triggers, and to delineate the maximum extent of sectors on the sector sketch.

Restrictive Fire Line

A-50. A restrictive fire line (RFL) is a linear fire control measure beyond which engagement is prohibited without coordination. In the offense, RFLs serve to deconflict maneuver elements during movement and planned maneuver. At the objective, platoon leaders can designate an RFL to prevent a base-of-fire element from firing into the area where an assaulting element is maneuvering. This technique is particularly important when armored vehicles support the maneuver of rifle squads. In the defense, platoon leaders may establish an RFL to prevent the unit from engaging a friendly rifle squad positioned in restricted terrain on the flank of an avenue of approach. Wherever used, RFLs are an essential element in fratricide reduction.

Final Protective Line

A-51. *Final protective line* (FPL) is a selected line of fire where an enemy assault is to be checked by interlocking fire of all available weapons and obstacles (FM 3-90). The unit reinforces this line with protective obstacles and with final protective fires (FPFs) whenever possible. Initiation of the FPFs is the signal for elements, crews, and individual Soldiers to shift fires to their assigned portion of the FPL. They spare no ammunition in repelling the enemy assault, a particular concern for medium machine guns and other automatic weapons. (See paragraphs B-20 to B-21 and paragraphs C-70 to C-71 for additional information on FPLs.)

Threat-Based Fire Control Measures

A-52. Platoon leaders use threat-based fire control measures to focus and control fires by directing their units to engage a specific enemy element rather than to fire on a point or area. The following paragraphs describe the tactics, techniques, and procedures associated with this type of control measure.

Rules of Engagement

A-53. Rules of engagement (ROE) specify the circumstances and limitations under which forces may engage; they include definitions of combatant and noncombatant elements and prescribe the treatment of noncombatants. Factors influencing ROE are national command policy, mission, commander's intent, the operational environment, and the law of war. ROE always recognize a Soldier's right of self-defense, but at the same time, they clearly define circumstances in which the Soldier may fire.

A-54. As an example, higher command directives, (ROE) have limited the use of antiarmor weapons in certain situations to prevent civilian casualties. Therefore, a subordinate commander establishes a WCS of WEAPONS TIGHT for antiarmor weapons.

Weapons Ready Posture

A-55. Weapons ready posture is the selected ammunition and range for individual and crew-served weapons in rifle squads. It is a means by which leaders use their mission analysis to specify the ammunition and range of the most probable anticipated engagement. Ammunition selection depends on the target type, but the leader may adjust it based on engagement priorities, desired effects, and effective range. Range selection depends on the anticipated engagement range; it is affected by terrain intervisibility, weather, and light conditions. Within the platoon, weapons ready posture affects the types and quantities of ammunition carried by rifle squads; consider the following:

A-56. Examples of weapons ready posture include:

- A grenadier, whose, most likely engagement is to cover dead space at 200 meters from the position, might load high explosive dual purpose (known as HEDP) ammunition and set 200 meters on the quadrant sight.
- To prepare for an engagement in a wooded area where engagement ranges are extremely short, an antiarmor specialist might plan to engage with an M136 AT4-series rocket instead of a Javelin.

Note. The M3 Multi-role, Antiarmor, Antipersonnel Weapon System (known as MAAWS), when replacing the Javelin, may provide the platoon with a better antiarmor and antipersonnel capability against enemy personnel, field fortifications, and enemy vehicles in a wooded areas where engagement ranges are decreased.

Weapons Safety Posture

A-57. Weapons safety posture is an ammunition handling instruction enabling platoon leaders to control the safety of their units' weapons precisely. Subordinate leaders' supervision of the weapons safety posture, as well as Soldiers' adherence to it, minimizes the risk of accidental discharge and fratricide. Table A-2 outlines procedures and considerations for the platoon when using the four weapons safety postures, listed in ascending order of restrictiveness—

- AMMUNITION LOADED.
- AMMUNITION LOCKED.
- AMMUNITION PREPARED.
- WEAPONS CLEARED.

A-58. When setting and adjusting the weapons safety posture, platoon leaders must weigh the desire to prevent accidental discharges against the requirement for immediate action based on the enemy threat. If the threat of direct contact is high, platoon leaders could establish the weapons safety posture as AMMUNITION LOADED. If the requirement for action is less immediate, they might lower the posture to AMMUNITION LOCKED or AMMUNITION PREPARED. Additionally, the platoon leader can designate different weapons safety postures for different elements of the unit.

Table A-2. Weapons safety posture levels

Weapons Safety Posture	*Infantry Squad Weapons and Ammunition*
Ammunition Loaded	Rifle rounds chambered. Machine gun and squad automatic weapon ammunition on feed tray; bolt locked to rear. Grenade launcher loaded. Weapons on manual safe.
Ammunition Locked	Magazines locked into rifles. Machine gun and squad automatic weapon ammunition on feed tray; bolt locked forward. Grenade launcher unloaded.
Ammunition Prepared	Magazines, ammunition boxes, launcher grenades, and hand grenades prepared but stowed in pouches/vests.
Weapons Cleared	Magazine, ammunition boxes, and launcher grenades removed; weapons cleared.

Weapons Control Status

A-59. The three levels of WCS outline the conditions, based on target identification criteria, under which friendly elements can engage. Platoon leaders set and adjust the WCS based on friendly and enemy disposition, and clarity of the situation. The higher the probability of fratricide, the more restrictive the WCS. The three levels, in descending order of restrictiveness, are—

- WEAPONS HOLD. Engage only if engaged or ordered to engage.
- WEAPONS TIGHT. Engage only targets positively identified as enemy.
- WEAPONS FREE. Engage targets not positively identified as friendly.

A-60. As an example, platoon leaders may establish the WCS as WEAPONS HOLD when friendly forces are conducting a passage of lines. By maintaining situational understanding of their own elements and adjacent friendly forces, however, they may be able to lower the WCS. In such a case, platoon leaders may be able to set a WEAPONS FREE status when they know there are no friendly elements in the vicinity of the engagement. This permits their elements to engage targets at extended ranges even though it is difficult to distinguish targets accurately at ranges beyond 2,000 meters under battlefield conditions. Another consideration is the WCS are extremely important for forces using combat identification systems. Establishing the WCS as WEAPONS FREE permits leaders to engage an unknown target when they fail to get a friendly response.

Engagement Priorities

A-61. Engagement priorities (see paragraph 5-206), which entail the sequential ordering of targets to be engaged, can serve one or more of the following critical fire control functions:

- Prioritize high-payoff targets (generally developed and identified during Step 3-Make a Tentative Plan [specifically, mission analysis] of TLP). In concert with the company's concept of operations, platoon leaders identify the high-payoff targets in their AO as a unit engagement priority. For example, destroying enemy engineer assets is the best way to prevent the enemy from reducing an obstacle.
- Employ the best weapons to the target. Establishing engagement priorities for specific friendly systems increases the effectiveness with which the unit employs its weapons. For example, the engagement priority of the Javelin could be enemy tanks first, then enemy personnel carriers; this would decrease the chance the platoon's lighter systems will have to engage enemy armored vehicles.
- Distribute the unit's fires. Establishing different priorities for similar friendly systems helps to prevent overkill and achieve distribution of fires. For example, platoon leaders may designate the enemy's tanks as the initial priority for the weapons squad, while making the enemy's personnel carriers the priority for one of the rifle squads.

Trigger

A-62. A trigger (see paragraph 5-208) is a specific set of conditions dictating initiation of fires. Often referred to as engagement criteria, a trigger specifies the circumstances in which subordinate elements should engage. The circumstances can be based on a friendly or enemy event. For example, the trigger for a friendly platoon to initiate engagement could be three or more enemy combat vehicles passing or crossing a given point or line. This line can be natural or man-made linear feature, such as a road, ridgeline, or stream. It also may be a line perpendicular to the unit's orientation, delineated by one or more reference points.

Engagement Techniques

A-63. Engagement techniques are effects-oriented fire distribution measures. The following engagement techniques are common in platoon operations:

- Point fire.
- Area fire.
- Simultaneous.
- Alternating fire.
- Observed fire.
- Sequential fire.
- Time of suppression.
- Reconnaissance by fire.

A-64. Point fire. Entails concentrating the effects of a unit's fire against a specific, identified target such as a vehicle, machine gun bunker, or antitank (AT) guided missile position. When leaders direct point fire, all unit weapons engage the target, firing until it is destroyed, or the required time of suppression has expired. Employing converging fires from dispersed positions makes point fire more effective because the target is engaged from multiple directions. The unit may initiate an engagement using point fire against the most dangerous threat, and revert to area fire against other, less threatening point targets.

A-65. Area fire. Involves distributing the effects of a unit's fire over an area in which enemy positions are numerous or are not obvious. If the area is large, leaders assign sectors of fire to subordinate elements using a terrain-based distribution method such as the quadrant technique. Typically, the primary purpose of the area fire is suppression; however, sustaining suppression requires judicious control of the rate of fire.

A-66. Simultaneous fires. To mass the effects of their fires rapidly or to gain fire superiority. For example, a unit may initiate a support by fire operation with simultaneous fire, then, revert to alternating or sequential fire to maintain suppression. Simultaneous fire also is employed to negate the low probability of the hit and kill of certain antiarmor weapons. For example, a rifle squad may employ simultaneous fire with its M136 AT4-series to ensure rapid destruction of an enemy armored fighting vehicle engaging a friendly position.

A-67. Alternating fire. Pairs of elements continuously engage the same point or area target one at a time. For example, an Infantry rifle company may alternate fires of two platoons; an Infantry rifle platoon may alternate the fires of its squads; or an Infantry rifle platoon may alternate the fires of a pair of medium machine guns. Alternating fire permits the unit to maintain suppression for a longer duration than does volley fire; it also forces the enemy to acquire and engage alternating points of fire.

A-68. Observed fire. Usually is used when the platoon is in protected defensive positions with engagement ranges more than 2,500 meters for stabilized systems (when attached) and 1,500 meters for unstabilized systems. It can be employed between elements of the platoon, such as the squad lasing and observing while the weapons squad engages. The platoon leader directs one squad to engage. The remaining squads observe fires and prepare to engage on order in case the engaging element consistently misses its targets, experiences a malfunction, or runs low on ammunition. Observed fire allows for mutual observation and assistance while protecting the location of the observing elements.

A-69. Sequential fire. Entails the subordinate elements of a unit engaging the same point or area target one after another in an arranged sequence. Sequential fire also can help to prevent the waste of ammunition, as when a platoon waits to see the effects of the first Javelin before firing another. Additionally, sequential fire permits elements already have fired to pass on information they have learned from the engagement. An example would be a Soldier who missed an enemy armored fighting vehicle with M136 AT4-series fires passing range and lead information to the next Soldier preparing to engage the enemy armored fighting vehicle with an M136 AT4-series.

A-70. Time of suppression. The period, specified by the platoon leader, during which an enemy position or force is required to be suppressed. Suppression time is typically dependent on the time it will take a supported element to maneuver. Usually, a unit suppresses an enemy position using the sustained rate of fire of its automatic weapons. In planning for sustained suppression, a leader must consider several factors: the estimated time of suppression, the size of the area being suppressed, the type of enemy force to be suppressed, range to the target, rates of fire, and available ammunition quantities.

A-71. Reconnaissance by fire. The process of engaging possible enemy locations to elicit a tactical response, such as return fire or movement. This response permits platoon leaders and their subordinate leaders to make target acquisition and to mass fires against the enemy element. Typically, platoon leaders direct a subordinate element to conduct the reconnaissance by fire (see paragraph 2-52). For example, they may direct an overwatching squad to conduct the reconnaissance by fire against a probable enemy position before initiating movement by a bounding element.

Fire Patterns

A-72. Fire patterns are a threat-based measure designed to distribute the fires of a unit simultaneously among multiple, similar targets. Platoons use those most often to distribute fires across an enemy formation. Leaders designate and adjust fire patterns based on terrain and anticipated enemy formation. The basic fire patterns, illustrated in figure A-9, include:

- Frontal fires.
- Cross fires.
- Depth fires.

A-73. Frontal fires. Leaders may initiate frontal fires when targets are arrayed in front of the unit in a lateral configuration. Weapon systems engage targets to their respective fronts. For example, the left flank weapon engages the left-most target; the right flank weapon engages the right-most target. As weapon systems destroy targets, weapons shift fires toward the center of the enemy formation from near to far.

A-74. Cross fires. Leaders initiate cross fires when targets are arrayed laterally across the unit's front in a manner permitting diagonal fires at the enemy's flank, or when obstructions prevent unit weapons from firing frontally. Right flank weapons engage the left-most targets; left flank weapons engage the right-most targets. Firing diagonally across an engagement area provides more flank shots, thus increasing the chance of kills; it also reduces the possibility of the enemy detecting friendly elements should the enemy continue to move forward. As friendly elements destroy targets, weapons shift fires toward the center of the enemy formation.

A-75. Depth fires. Leaders initiate depth fires when enemy targets disperse in-depth, perpendicular to the unit. Center weapons engage the closest targets; flank weapons engage deeper targets. As the unit destroys targets, weapons shift fires toward the center of the enemy formation.

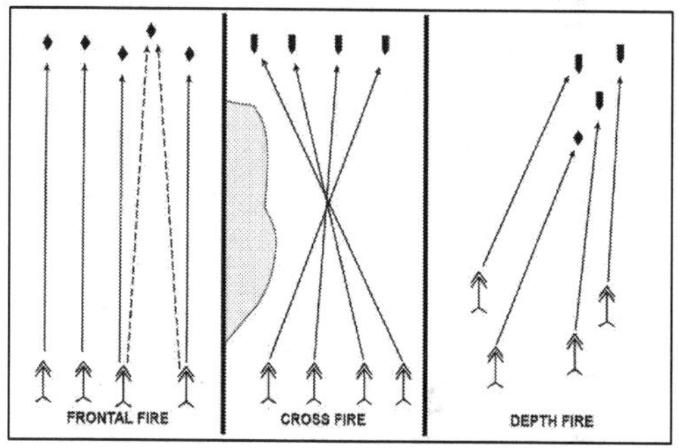

Figure A-9. Fire pattern, examples

Target Array

A-76. Target array enables leaders to distribute fires when an enemy force is concentrated, and terrain-based controls are inadequate. Forces create this threat-based distribution measure by superimposing a quadrant pattern on the enemy formation. Soldiers center the pattern on the enemy formation, with the axes running parallel and perpendicular to the enemy's direction of travel. The target array fire control measure is effective against an enemy with a well-structured organization and standardized doctrine. However, it may prove less effective against an enemy presenting few organized formations or those that do not follow strict prescribed tactics. Leaders describe quadrants using the quadrants' relative locations. The examples in figure A-10 on page A-24 illustrate the target array technique. Target arrays may be planned in both the offense or defense, or simply designated upon contact with an enemy force.

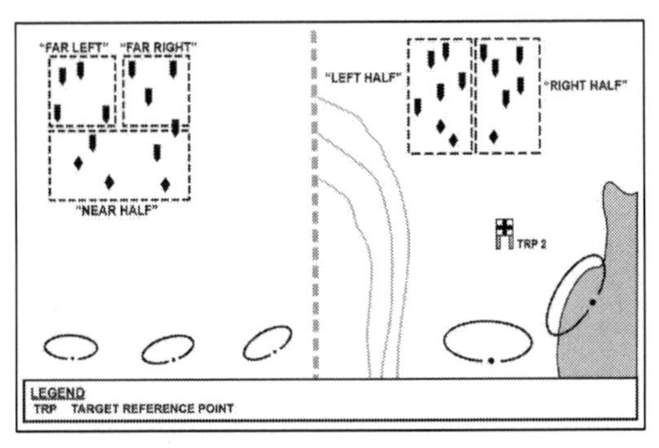

Figure A-10. Target array, examples

FIRE COMMANDS

A-77. Fire commands are oral orders issued by leaders to focus and distribute fires as required achieving decisive effects against the enemy. They allow leaders to articulate their firing instructions using a standard format rapidly and concisely. Fire commands (for Soldiers, leaders, crews, teams, and small units) include these elements, which are discussed in the following paragraphs:

- Alert.
- Weapon or ammunition (optional).
- Target description.
- Direction.
- Range (optional).
- Method.
- Control (optional).
- Execution.
- Termination.

Alert

A-78. The alert specifies the elements directed to fire. It does not require leaders initiating the command to identify the unit. Examples of the alert element (call signs and code words based on unit SOP) include the following:

- "GUIDONS" (all subordinate elements).
- "RED" (1st platoon only).

Weapon or Ammunition (Optional)

A-79. This element identifies the weapon and ammunition to be employed by the alerted elements. Leaders may designate the type and number of rounds to limit expenditure of ammunition. Examples of this element include the following:

- "JAVELIN."
- "TOP ATTACK" or "DIRECT ATTACK."

Target Description

A-80. Target description designates which enemy elements are to be engaged. Leaders may use the description to focus fires or achieve distribution. Examples of target description include the following:

- "TROOPS IN TRENCH."
- "BUNKER."
- "ARMORED PERSONNEL CARRIER."

Direction

A-81. This element identifies the location of the target. There are several ways to designate the location of target, including the following:

- Closest TRP. Example: "TRP 13."
- Clock direction. Example: "ONE O'CLOCK."
- Terrain quadrant. Example: "QUADRANT ONE."
- Friendly quadrant. Example: "LEFT FRONT."
- Target array. Example: "NEAR/FAR."
- Tracer on target. Example: "ON MY TRACER."
- Laser pointer. Example: "ON MY POINTER."

Range (Optional)

A-82. The range element identifies the distance to the target. Announcing range is not necessary for systems ranging finder-equipped or employing command-guided or self-guided munitions. For systems requiring manual range settings, leaders have a variety of means for determining range, including the following—

- Predetermined ranges to TRPs or phase lines.
- Handheld range finders.
- Range stadia.
- Mil reticle.

Method

A-83. Method describes to the firer the way or method the target(s) are engaged. Leaders use this element when presented with multiple targets to identify which target to engage first. For collective fire commands, this can also indicate the fire pattern used to engage the threats. Multiple methods may be used in one fire command.

Control (Optional)

A-84. The platoon leader may use this optional element to direct desired target effects, distribution methods, or engagement techniques. Subordinate leaders may include the control element to supplement the platoon leader's instructions and achieve distribution. Examples of information specified in the control element include the following:

- Target array. Example: "NEAR/FAR."
- Fire pattern. Example: "FRONTAL."
- Terrain quadrant. Example: "QUADRANT ONE."
- Engagement priorities. Example: "M320/SHOULDER-LAUNCHED MUNITIONS ENGAGE BUNKERS; MACHINE GUNS ENGAGE TROOPS."
- Engagement technique. Example: "VOLLEY."
- Target effect. Example: "AREA."

Execution

A-85. The execution element specifies when fires will be initiated. The platoon leader may wish to engage immediately, delay initiation, or delegate authority to engage. Examples of this element include the following:

- "FIRE."
- "AT MY COMMAND."
- "AT YOUR COMMAND."
- "AT PHASE LINE ORANGE."

Termination

A-86. Termination is the ninth element of the fire command. It informs the Soldiers to stop firing all weapons and systems in their control. All fire commands are terminated. This command may be given by any Soldier or crewmember for any reason.

A-87. The leader that issued the fire command is required to terminate the fire command at the completion of every engagement, regardless of another Soldier or crewmember announcing it. All fire commands, regardless of type or who issued them, are terminated by the announcement of CEASE FIRE.

SECTION II – STANDARD RANGE CARD AND SECTOR SKETCHES

A-88. Standard range cards are used to record firing data for individual or crew-served weapons. Sector sketches are used to record a unit's positioning of its weapons and direct fire control measures.

DA FORM 5517

A-89. A DA Form 5517 is a sketch of the assigned area for a direct fire weapon system on a given sector of fire. (See STP 7-11B1-SM-TG for additional information.) A standard range card aids in planning and controlling fires and aids the squad's gunners in acquiring targets during limited visibility. The individual Soldier or gunner prepares two copies of the standard range card. If alternate and supplementary firing positions are assigned, two copies are required for these as well.

TARGET AREAS AND TERRAIN FEATURES

A-90. Standard range cards show possible target areas and terrain features plotted with a firing position. The process of walking and sketching the terrain to create a standard range card allows the individual Soldier to become more familiar with their sector of fire. Soldiers should continually assess the area and, if necessary, update the standard range card. The standard range card is an aid for replacement personnel or platoons or squads to move into the position and orient on their sector of fire. To prepare a standard range card, the individual Soldier must know the following information:

- Sector of fire. A sector of fire is a piece of the battlefield for which a gunner is responsible.
- TRPs. Leaders designate natural or man-made features as reference points. A Soldier uses these reference points for target acquisition and range determination.
- Dead space. Dead space is an area that cannot be observed or covered by direct-fire systems within the sector of fire.
- Maximum engagement line. The maximum engagement line is the depth of the area and is normally limited to the maximum effective engagement range of the weapon systems. It can be less if objects prevent the Soldier or gunner from engaging targets at maximum effective ranges of an assigned weapon.
- Weapons reference point. The weapons reference point is an easily recognizable terrain feature on the map used to assist leaders in plotting the weapon position, squad, and (when attached) vehicle.

DATA SECTION

A-91. The gunner completes the position identification, date, weapon, and circle value according to STP 7-11B1-SM-TG. The table information is as follows:

- Number. Start with left and right limits, then list TRPs and reference points in numerical order.
- Direction and deflection. The direction is in degrees and taken from a lensatic compass.
- Elevation. Show the gun elevation reading (for example, for a medium machine gun) in tens or hundreds of mils. The smallest increment of measure on the elevation scale is tens of mils. Any number other than "0" is preceded

by a "plus" or "minus" symbol to show whether the gun needs to be elevated or depressed. Range must be indexed to have an accurate elevation reading.

- Range. This is the distance, in meters, from position to TRPs.
- Ammunition. List types of ammunition used.
- Description. List the name of the object.
- Remarks. Enter the weapons reference point data. As a minimum, weapons reference point data include a description of what the weapons reference point is, a six-digit or eight-digit grid coordinate of the weapons reference point, the magnetic azimuth, and the distance from the weapons reference point to the position.

SECTOR SKETCHES

A-92. As discussed earlier in this section, individual Soldiers in squads prepare standard range cards. Squad and platoon leaders prepare sector sketches. Section leaders may have to prepare sector sketches if they are assigned separate positions. Platoon leaders review their squads', and if applicable section's, sector sketches and ensure the sketches are accurate and meet their requirements. If they find gaps or other flaws, platoon leaders adjust weapons locations within the assigned area. Once platoon leaders approve the squad and section sector sketches, they prepare a consolidated report for the company commander and incorporate this into a consolidated platoon sector sketch. The platoon leader or platoon sergeant physically prepares the platoon sector sketch. The sector sketch can be on acetate taped to a map or it can be a hand drawn sketch. Accurate and detailed sketches aid in direct fire planning, and in direct fire control and distribution.

SQUAD SECTOR SKETCH

A-93. The squad leaders and section leaders make two copies of their sector sketches; one copy goes to the platoon leader; the other remains at the position. The squad leaders and section leaders draw sector sketches (see figure B-8 on page B-14 for an example sector sketch) as close to scale as possible, showing—

- Main terrain features in the assigned sector and the range to each.
- Each primary position.
- Engagement area or primary and secondary sectors of fire covering each position.
- Light and medium machine gun FPL or PDF.
- Type of weapon in each position.
- Reference points and TRPs in the assigned sector.
- Observation post locations.
- Dead space.
- Obstacles, including mines.
- Maximum engagement lines for all weapon systems.
- Indirect fire targets.

Platoon Sector Sketch

A-94. Squad leaders and section leaders prepare their sketches and submit them to the platoon leader. The platoon leader combines all sector sketches (and possibly separate standard range cards) to prepare a platoon sector sketch (see figure B-7 on page B-13 for an example sector sketch), which is drawn as close to scale as possible and includes a target list for direct and indirect fires. One copy is submitted to the company commander, one copy is given to the platoon sergeant, and one copy is kept by the platoon leader. At a minimum, the platoon sector sketch should show—

- Primary and secondary sectors of fire or engagement areas.
- Primary, alternate (when established), and supplementary (when established) squad positions.
- Remount points (when vehicles attached).
- Javelin and machine gun positions with a PDF.
- Machine gun (light and medium) FPLs.
- Maximum engagement lines for the medium machine gun.
- Observation posts.
- TRPs.
- Mines and other obstacles.
- Indirect fire target locations and FPF location (if applicable).
- Position and area of flanking unit (and when attached, vehicles).
- Priority engagement by weapon system and crew.

Coordination with Adjacent Units

A-95. Platoon leaders coordinate with adjacent platoons. Squad leaders coordinate with adjacent squads so that all positions and all platoons and squads are mutually supporting. The platoon leader must ensure that this coordination take place. Coordination is usually initiated from left to right. Gaps between positions are covered by fire at a minimum. Contact points are established to ensure friendly forces meet at some specific point on the ground to tie in their flanks. In many cases, the exchange of sector sketches will accomplish most of this. Typical information that is exchanged includes—

- Locations of primary, alternate, and supplementary positions; sectors of fire for medium machine guns, Javelins, and attached (if applicable) weapon systems.
- Location of dead space between platoons and how it is to be covered.
- Location of observation posts.
- Location and types of obstacles and how to cover them.
- Patrols (size, type, time of departure and return, and routes).

Note. When conditions require platoon, and/or squad positions to be arrayed in noncontiguous assigned area, security patrols (see chapter 7) between positions may be necessary.

This page intentionally left blank.

Appendix B
Fire Support Planning

Fire support planning is the continual process of selecting targets on which fires are prearranged to support a phase of the scheme of maneuver. Fire support planning is accomplished concurrently with maneuver planning at all echelons. Leaders conduct fire support planning to suppress, isolate, obscure, neutralize, destroy, deceive, or disrupt known, likely, or suspected targets, and to support the actions of the maneuver forces. Fires are planned for all phases of an operation. This appendix addresses the lethal fires coming from Infantry brigade combat team (IBCT) organic indirect fires, Army artillery and aviation assets, and joint and multinational artillery and aviation assets in support of the Infantry rifle platoon. (See ATP 3-21.10 and FM 3-09 for additional information.)

FIRE SUPPORT CONSIDERATIONS

B-1. Fire support planning starts as soon as the leader receives a mission. Once begun, fire support planning continues through the operation's completion. The primary aim of fire planning is to develop how fire is to be massed, distributed, and controlled to best support the leader's scheme of maneuver.

TARGETING CATEGORIES

B-2. The targeting process can be generally grouped into two categories: deliberate and dynamic. Deliberate targeting prosecutes planned targets. These targets are known to exist in the area of operations (AO) and have actions scheduled against them. Examples range from targets on target lists in the applicable order, targets detected in sufficient time to place in the joint air tasking cycle and get into the air tasking order, mission-type orders, or fire support plans. *Dynamic targeting* is targeting that prosecutes targets identified too late or not selected for action in time to be included in deliberate targeting (JP 3-60). Dynamic targeting (see ATP 3-60.1) prosecutes targets of opportunity and changes to planned targets or objectives. Targets of opportunity are targets identified too late, or not selected for action in time, to be included in deliberate targeting. Targets engaged as part of dynamic targeting are previously unanticipated, unplanned, or newly detected.

Target Types: Planned Targets and Targets of Opportunity

B-3. Deliberate targeting prosecutes planned targets. These targets are known to exist in the AO and have actions scheduled against them. The two types of planned targets are scheduled and on-call:

- Scheduled targets exist in the AO and are located in sufficient time so that fires or other actions upon them are identified for engagement at a specific, planned time.
- On-call targets have actions planned, but not for a specific delivery time. The commander expects to locate these targets in sufficient time to execute planned actions.

B-4. Dynamic targeting prosecutes targets of opportunity and changes to planned targets or objectives. Targets of opportunity are targets identified too late, or not selected for action in time, to be included in deliberate targeting. Targets engaged as part of dynamic targeting are previously unanticipated, unplanned, or newly detected. The two types of targets of opportunity are unplanned and unanticipated:

- Unplanned targets are known to exist in the AO, but no action has been planned against them. The target may not have been detected or located in sufficient time to meet planning deadlines. Alternatively, the target may have been located, but not previously considered of sufficient importance to engage.
- Unanticipated targets are unknown or not expected to exist in the AO.

Targeting Priorities

B-5. Targeting priorities must be addressed for each phase or critical event of an operation. Targeting priorities emphasize the identification of resources and activities the enemy can least afford to lose or that provide the greatest advantage (high-value target) and whose loss will significantly contribute to the success of the friendly COA (high-payoff target). Targeting links desired effects to activities and tasks involving specific fires. For example, a target area of interest is a point or area where the commander can acquire and engage high payoffs. The targeting decisions made are reflected in the following visual products:

- The high-payoff target list is a prioritized list of high-payoff targets by phase of the operation.
- A *high-payoff target* is a target whose loss to the enemy will significantly contribute to the success of the friendly course of action (JP 3-60).
- A *high-value target* is a target the enemy commander requires for the successful completion of the mission (JP 3-60).

Concept of Fires

B-6. Fire planning begins with the concept of fires. This essential component of the concept of operations complements the leader's scheme of maneuver detailing the leader's plan for direct and indirect preparatory and supporting fires. Fire planning requires a detailed knowledge of weapon characteristics and logistical capabilities of those providing the support. Although leaders may be augmented with personnel to assist in planning and controlling attached or supporting assets, the responsibility for planning and execution of fires lies with the leader. Leaders do not wait to receive the higher headquarters' plan to begin their own fire planning but begins as soon as possible to integrate fires into concept of operations and concept of operations of the higher headquarters.

Fire Support Teams

B-7. Company fire support team (FIST) headquarters personnel and platoon forward observers (FOs) provide support to subordinate platoons to plan and coordinate all available supporting fires, including mortars, field artillery, naval surface fire support, Army aviation attack, and close air support (CAS) integration. When attached, FISTs provide the platoon with fire support coordination, targeting, input for terminal attack control, and assessment capabilities. FOs are trained to adjust ground or naval gunfire and pass back battlefield information. Effective fires require qualified observers to call for and adjust fires on located targets. FOs, forward air controllers, naval gunfire spotter teams, joint fires observers, and joint terminal attack controllers (JTACs) train together and work effectively as a team to request, plan, coordinate, and place accurate fires on targets that create the desired effects. (See ATP 3-21.10 for additional information.)

B-8. Additional assets are allocated in either a command or support relationship. An example of a command relationship would be an attachment of a section from the weapons company (reference direct fire support, see note on page 3-39). The leader relies on the senior representative from the organization to provide expertise when planning. An example of a support relationship would be direct support from the artillery battalion or from an aviation attack company. When planning Army aviation attack fires (see paragraph B-64) or CAS from a supporting unit (see paragraph B-74), the leader normally receives someone from that organization to assist them. Attached FIST and JTACs are examples. Developing the concept of fire should be straightforward during deliberate operations because of the ability to conduct reconnaissance, planning, and preparation. However, during hasty operations the unit may have to rely on its internal SOPs and more hands-on control by the leader.

Tactical Uses of Planned Fires

B-9. Fires are used for many different tactical reasons. They include:

- Preparatory fire, delivered before an attack to weaken the enemy position. (See ADP 3-90 for additional information.)
- Supporting fires (covering fires). Supporting fires enable the friendly maneuver element to move by destroying, neutralizing, or suppressing enemy fires, positions, and observers.
- *Final protective fire* is an immediately available, prearranged barrier of fire designed to impede enemy movement across defensive lines or areas (JP 3-09.3).
- *Suppression*—temporary or transient degradation by an opposing force of the performance of a weapons system below the level needed to fulfill its mission

objectives (JP 3-01). *Suppression*—in the context of the computed effects of field artillery fires, renders a target ineffective for a short period of time producing suppression of enemy air defenses at least 3-percent casualties or materiel damage (FM 3-09).

- *Obscuration*—the employment of materials into the environment that degrade optical and/or electro-optical capabilities within select portions of the electromagnetic spectrum in order to deny acquisition by or deceive an enemy or adversary (ATP 3-11.50).
- Counterfire (indirect fires only). Counterfire is fire to destroy or neutralize enemy artillery/mortars. These missions normally are controlled at higher level headquarters. Counterfire radars are positioned to maintain radar coverage to ensure continuous coverage during rapid movement forward.
- Harassing fire is observed or predicted (unobserved) fire intended to disrupt enemy troop and vehicle movement, disturb their rest, and lower their morale.
- Illumination. While obscuration smoke hides friendly movements, illumination exposes enemy formations at night. Utilizing illumination rounds can be effective; however, friendly positions should remain in shadows, and enemy positions should be highlighted.

Echelonment of Fires

B-10. Echelonment of fires is the schedule of fire ranging from the highest caliber munitions to the lowest caliber munitions. The purpose of echeloning fires is to maintain constant fires on the enemy while using the optimum delivery system. As distances between friendly and enemy forces close, echeloning of fires is the principle of reducing the caliber of fires employed, reducing the associated risk. Leaders use risk estimate distance, surface danger zones, and minimum safe distance as per ATP 3-09.32, to manage associated risks. In the defense, triggers are tied to the progress of the enemy as it moves through the AO, enabling the leader to engage the enemy throughout the depth of the AO. In the offense, triggers are tied to the progress of the maneuver element as it moves toward the objective protecting the force and facilitating momentum up to the objective.

Defensive Echelonment

B-11. In defensive missions, echeloning fires are scheduled based on their optimum ranges to maintain continuous fires on the enemy, disrupting the formation and maneuver. Echelonment of fires in the defense places the enemy under increasing volumes of fire as they approach a defensive position. Aircraft and long-range indirect fire rockets and artillery deliver deep supporting fires. Close supporting fires, such as final protective fires (FPFs), are integrated closely with direct fire weapons such as Infantry weapons, tank support, and antiarmor weapon systems. Figure B-1 illustrates an example of defensive echelonment.

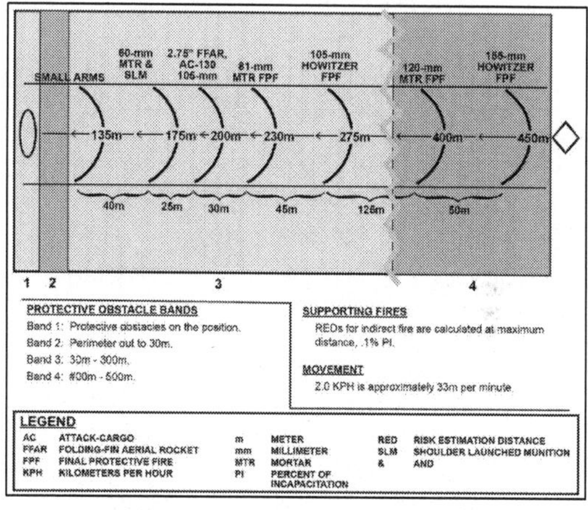

Figure B-1. Defensive echelonment of fires

Offensive Echelonment

B-12. In the offense, weapons are scheduled based on the point of a predetermined safe distance away from maneuvering friendly troops. When scheduled, fires provide protection for friendly forces as they move to and assault an objective. They also allow friendly forces to get in close with minimal casualties and prevent the defending enemy from observing and engaging the assault by forcing the enemy to take cover. The overall objective of offensive scheduled fires is to allow the friendly force to continue the advance unimpeded. (See figure B-2 on page B-6.)

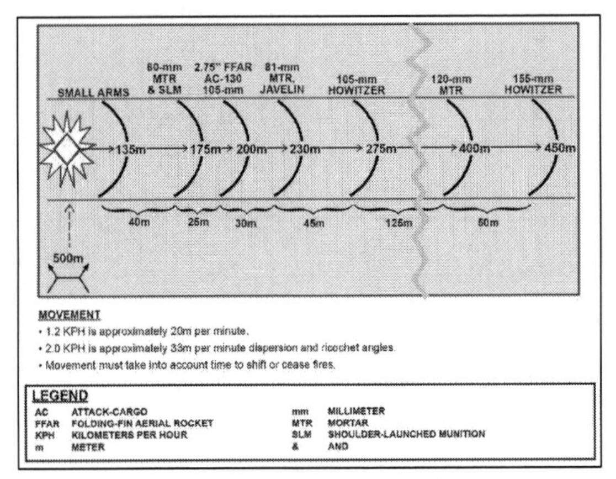

Figure B-2. Offensive echelonment of fires

B-13. As an example of echelonment of fires use during the conduct of a mission, consider an operation in which a platoon assaults an enemy position. (See figures B-3 through B-6 on pages B-7 through B-10.) As the lead elements of the unit approach the designated phase line en route to the objective, the leader orders the fire support officer to begin the preparation. Observers track friendly movement rates and confirm them. Other fire support officers in the chain of command may need to adjust the plan during execution based on unforeseen changes to anticipated friendly movement rates.

B-14. As the unit continues its movement toward the objective, the first weapon system engages its targets. It maintains fires on the targets until the unit crosses the next phase line corresponding to the risk estimate distance of the weapon system being fired.

B-15. To maintain constant fires on the targets, the next weapon system begins firing before the previous weapon system ceases or shifts. This ensures no break in fires, enabling the friendly forces' approach to continue unimpeded. However, if the unit rate of march changes, the indirect fire support system must remain flexible to the changes.

B-16. The fire support officer shifts and engages with each delivery system at the prescribed triggers, initiating the fires from the system with the largest risk estimate distance to the smallest. Once the maneuver element reaches the final phase line, the fire support officer ceases the final indirect fire system or shifts to targets beyond the objective to cease all fires on the objective. Direct fire assets in the form of supporting fires also are maintained until the final assault, then ceased or shifted to targets beyond the objective.

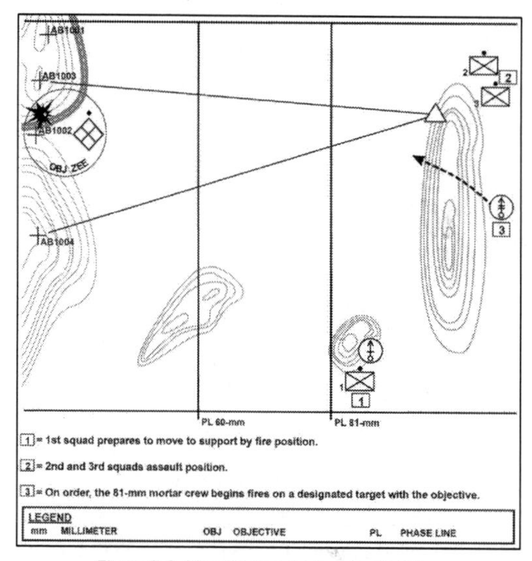

Figure B-3. 81-millimeter mortars begin firing

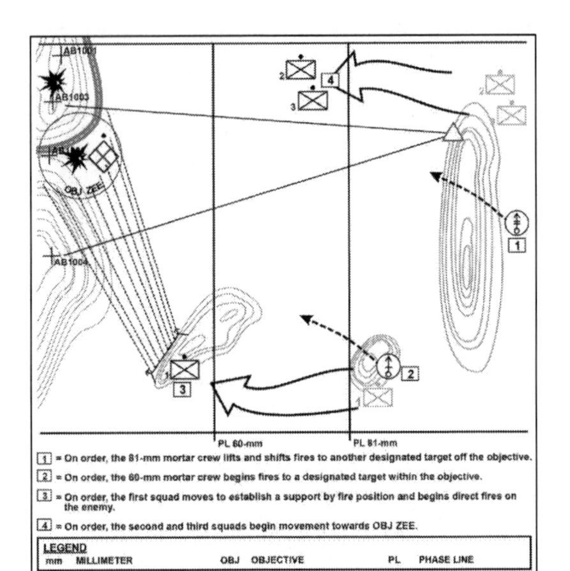

Figure B-4. 81-millimeter mortars shift, 60-millimeter mortars and supporting fires begin

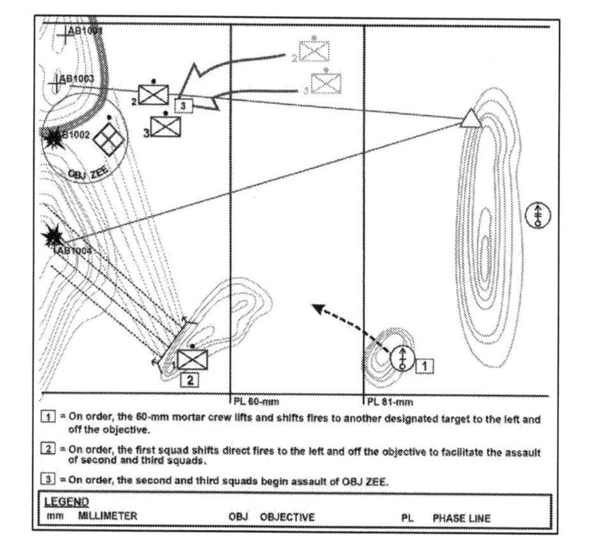

Figure B-5. 60-millimeter mortars shift

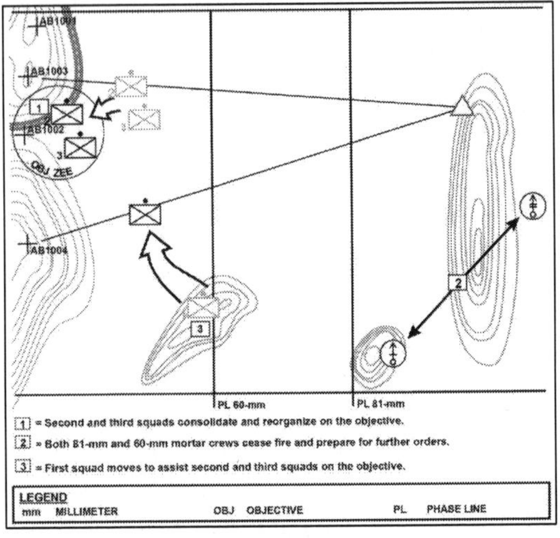

Figure B-6. Supporting fires cease

Fire Planning for the Defense

B-17. In the defense, fires are generally planned in three locations-in front of the unit positions, on the position's FPFs, and behind positions. To develop a defensive fire plan, the platoon leader within the company defense—

- Redefines designation of unit point or area targets and other control measures, such as TRPs, to coordinate the fire when more than one subordinate is firing into the same engagement area or AO.
- Masses fires on choke points and key terrain (for example, obstacles, water crossings, and dead space) to canalize, slow, and block the enemy's movements.
- Ensures fires are integrated into the obstacle plan.
- Considers the use of obscurants to support the obstacle plan.
- Refines and bases fire plans on the commander's guidance for fires and allocation of resources.

- Identifies requirements for positioning primary and alternate observers forward of friendly maneuver forces. Ensures that extraction guidelines are established and understood.
- Develops alternate plans in case these FOs are forced to withdraw prior to execution of fire support tasks.
- Determines the time needed for all fire support systems to be ready based on the scheme of maneuver. Ensure that these times are met.
- Determines how and recommend when to shift the priority of fires. Determine what will be the trigger to shift the priority of fires.
- Plans for the use of obscurants during periods of limited visibility to degrade enemy night vision capabilities.
- Receives target information from subordinates (normally provided on sector sketches or individual weapon standard range cards).
- Reviews target information to ensure fires are equally distributed across the entire unit's AO and sufficient control measures are established.
- Completes the unit's fire plan and gives sector sketch to the higher headquarters.

Final Protective Line and Final Protective Fire

B-18. Final protective line (FPL) is a line of fire selected where an enemy assault is to be checked by interlocking fire from all available weapons and obstacles. The defending unit reinforces this line (see paragraph A-51) with protective obstacles and with FPFs whenever possible. The FPL consists of all available measures, to include protective obstacles, direct fires, and indirect fires. The FPFs target the highest type of priority targets and takes precedence over all other fire targets. The FPFs differ from a standard priority target in that fire is conducted at the maximum rate until the mortars are ordered to stop, or until ammunition is depleted. If possible, the FPFs should be registered.

B-19. If Soldiers are in well-prepared defensive positions with overhead cover (known as OHC), FPFs can be adjusted close to the friendly positions, just beyond bursting range. If the threat situation warrants it (for example, in the event the friendly position is over run), the unit leader can call for artillery fires right on the unit's position using proximity or time fuzes for airbursts. Table B-1 on page B-12 shows indirect fire mortar weapon system characteristics being used when planning the FPF.

Table B-1. Normal final protective fires dimensions for each number of mortars

Weapon	*Number of Tubes*	*Width (meters)*	*Depth (meters)*	*Risk Estimated Distance, .1% PI*	*Risk Estimated Distance, 10% PI*
MORTARS					
120 mm	4	300	70	400 m	100 m
120 mm	2	150	70		
81 mm	4	150	40	230 m	80 m
81 mm	2	75	40		
60 mm	2	60	30	175 m	65 m
Legend: m–meters; mm–millimeters; PI–probability of incapacitation; %–percent					

Platoon Fire Planning

B-20. Squad leaders prepare their sketches and submit them to the platoon leader. The platoon leader combines all sectors of fire and sector sketches (and possibly separate standard range cards) to prepare a platoon sector sketch. A platoon sector sketch is drawn as close to scale as possible including a target list for direct and indirect fires. One copy is submitted to the company commander, one copy is given to the platoon sergeant, and one copy is maintained by the platoon leader. At a minimum, the platoon sector sketch should show the elements contained in figure B-7.

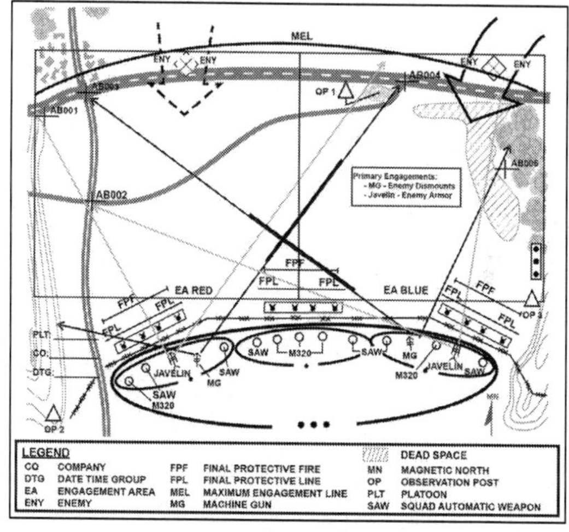

Figure B-7. Platoon sector sketch

Squad Fire Planning

B-21. Squad leaders make two copies of their sector sketches. One copy goes to the platoon leader; the other remains at the position. Squad leaders draw sector of fire sketches as close to scale as possible, showing the elements contained in figure B-8 on page B-14.

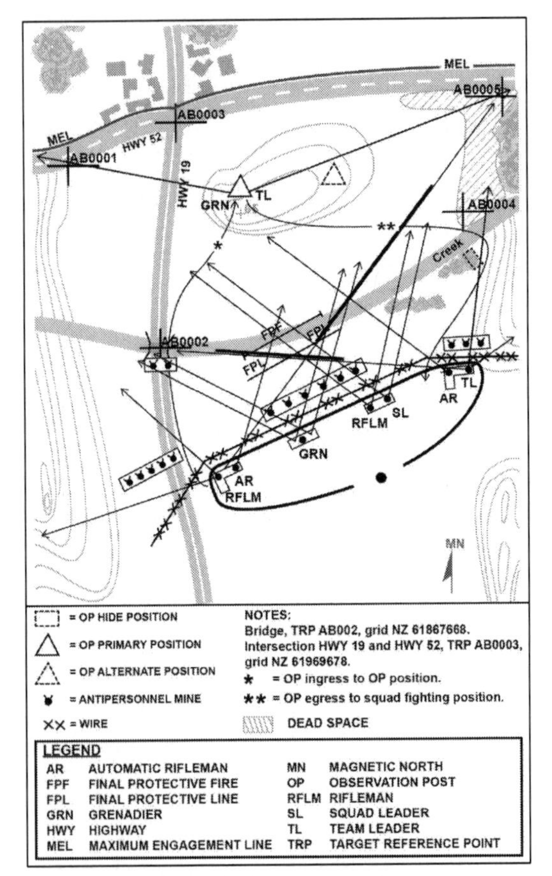

Figure B-8. Squad sector sketch

Fire Planning for the Offense

B-22. Offensive fire planning follows the same methodology as defensive fire planning within constraints of the situation. The presence of a friendly element requires a baseline of fire support planning and control throughout the entire mission (specifically movement and maneuver) to ensure indirect fire support is effective and efficient.

B-23. Leaders must plan how they will engage known or suspected enemy targets, where friendly suppressive fire may be needed, and how they will control their units' fires against both planned targets and targets of opportunity. Fire planning should include a thorough analysis of the type of threat expected. This will aid the supporting friendly element in tailoring the weapon and ammunition requirements to suit the situation.

B-24. Offensive fire planning supports four phases: planning and preparation, approach to the objective, actions on the objective, and follow through. The degree of completeness and centralization of offensive fire planning depends on the time available to prepare the offensive. Fires are planned in four locations on the battlefield short of the line of departure (LD)/line of contact (known as LC), LD/LC to the objective, on the objective, and beyond the objective. Table B-2 lists planning considerations for each of the four locations.

Table B-2. Planning considerations

PHASE	*PLAN FIRES TO:*
1) Planning and Preparation (Short of the LD/LC)	• Support unit in tactical assembly area. • Support unit's movement to the LD/LC. • Disrupt enemy reconnaissance forces. • Disrupt enemy defensive preparations. • Disrupt enemy spoiling attacks.
2) Approach to the Objective (LD/LC to the Objective)	• Begin echeloning fires for maneuver units. • Suppress and obscure for friendly breaching operations. • Suppress and obscure enemy security forces throughout movement. • Provide priority of fires to lead element. • Screen/guard exposed flanks.
3) Actions on the Objective (On the Objective)	• Fires to block enemy reinforcements. • Fires to suppress enemy direct fire weapons. • Suppress and obscure point of penetration. • Suppress and obscure enemy observation of friendly forces.
4) Follow Through (Beyond the Objective)	• Disrupt movement of enemy reinforcements during the assault. • Block avenues of enemy approach. • Disrupt enemy withdraw. • Screen friendly forces from enemy counterattacks during the assault. • Consolidate objective after the assault.
Legend: LC–line of contact; LD–line of departure	

B-25. Offensive fire planning is divided into two categories—preparatory and supporting fires. The concept of fires has artillery and mortars in support of an attack to gain and maintain fire superiority on the objective until the last possible moment. When this indirect fire ceases, the enemy should be stunned and ineffective for a few moments. Take full advantage of this period by executing any or all the following:

- Maintaining fire superiority. Small-arms fire from local and internal support by fire is continued as long as possible.
- Maneuver elements. Assaulting Soldiers must try to fire as they advance. Troops must observe fire discipline, as in many cases fire control orders will not be possible. They must arrive at the objective with ammunition.
- Audacity. Where the ground and vegetation do not prohibit movement, leading sections should move quickly over the last 30 or 40 meters to the enemy positions to minimize exposure.
- Combat vehicles. Vehicles used in the attack, or as direct fire support, continue to give close support.

B-26. When planning fires for the offense, leaders verify the fire element's task organization and ensure their plans coordinate measures for the attack, site exploitation, pursuit, and contingency plans. Leaders develop or confirm with the responsible level authority supporting systems are positioned and repositioned to ensure continuous fires throughout the operation. Mutual support of fire systems promotes responsive support and provides the commanders of maneuver units freedom of action during each critical event of the engagement or battle.

B-27. There exists a diverse variety of munitions and weapon systems, direct and indirect, to support close offensive missions. To integrate direct and indirect fire support, the leader must understand the mission, commander's intent, concept of operations, and critical tasks to be accomplished. Leaders plan fires to focus on enemy capabilities and systems being neutralized. Critical tasks include:

- Softening enemy defenses by delivering preparatory fires.
- Suppressing and obscuring enemy weapon systems to reduce enemy standoff capabilities.
- Scheduling indirect fires with direct fires to provide protection for friendly forces as they move to and assault an objective.
- Deconflicting fires during the approach and assault through the use of restrictive fire lines (RFLs), restrictive fire area, or no fire area.
- Continuous in-depth fire support (accomplished by proper positioning of systems), including support by fire and attack by fire positions.
- Interdicting enemy counterattack forces, isolating the defending force, and preventing its reinforcement and resupply.

MORTAR AND FIELD ARTILLERY FIRES

B-28. The majority of fire support to an Infantry rifle platoon is provided by indirect fire support systems of the Infantry rifle company. Indirect fire support systems include mortars and field artillery cannon and rocket fires. (See ATP 3-09.32 for a detailed listing of indirect fire system capabilities and characteristics.) Indirect fire support systems may be under direct command of the maneuver battalion/company or may be in a supporting role. Indirect fire targets during movement are planned on probable locations of enemy attempts to attack the movement. (See ATP 3-21.90 and ATP 3-09.42 for additional information.)

CALL FOR FIRE

B-29. Call for fire (see table B-3) is the request for fire containing data necessary for obtaining the required mortar and artillery fire on a target. The ability for mortars and artillery to engage targets from reverse-slopes and areas of defilade is a tremendous advantage, especially in adverse terrain. As with other operations, employing indirect fires in adverse terrain and climate does have its challenges. (See ATP 3-21.10, ATP 3-09.32, and FM 3-09 for additional information.) Unique challenges include—

- Unpredictable weather conditions affecting accuracy of rounds.
- Targets located on peaks and steep terrain, making adjustments difficult.
- Intervening crests requiring placement of observers on dominating heights for overwatch.
- Limited terrain suitable for firing positions to cover a particular movement.
- Mortar and artillery locations ideal for range and coverage unsuitable due to intervening adverse terrain features.
- Locations tactically positioned but in an area with difficult or limited access.
- Shifting mortar and artillery assets to alternate locations requiring significant time and engineering and logistical efforts.

Table B-3. Artillery and mortar call for fire

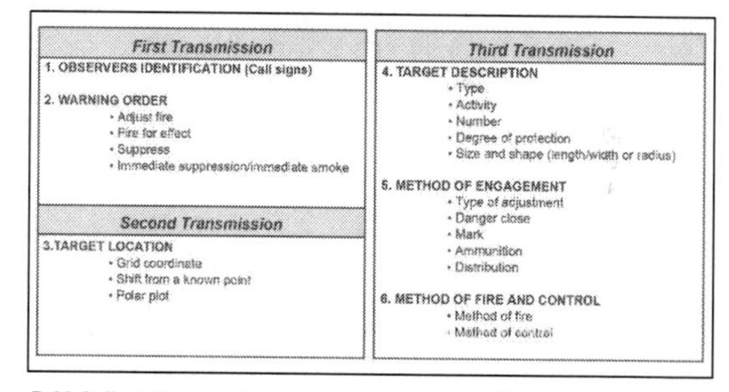

First Transmission	*Third Transmission*
1. OBSERVERS IDENTIFICATION (Call signs) 2. WARNING ORDER • Adjust fire • Fire for effect • Suppress • Immediate suppression/immediate smoke	4. TARGET DESCRIPTION • Type • Activity • Number • Degree of protection • Size and shape (length/width or radius) 5. METHOD OF ENGAGEMENT • Type of adjustment • Danger close • Mark • Ammunition • Distribution 6. METHOD OF FIRE AND CONTROL • Method of fire • Method of control
Second Transmission	
3. TARGET LOCATION • Grid coordinate • Shift from a known point • Polar plot	

B-30. Indirect fires must be integrated and synchronized in time, space, and purpose over the entire concept of operations. Integration means all available assets are planned and used throughout an operation. Synchronization means these assets are sequenced in time, space, and purpose in an optimal manner, producing complementary and reinforcing effects of the maneuver element.

TARGET EFFECTS PLANNING

B-31. Not only must indirect fire support planners determine what enemy targets to hit, and when, but also must decide how to attack each enemy target. Leaders should consider all the aspects of target effects when planning fires. Although this section is specific to mortars, the following concepts generally apply to most indirect fires. (See ATP 3-09.32 for additional information.)

High Explosive Ammunition

B-32. When mortar rounds impact, they throw fragments in a pattern never truly circular, and may even travel irregular, based on the round's angle of fall, the slope of the terrain, and type soil. However, for planning purposes, each mortar high explosive (HE) round is considered to have a circular lethal bursting area. Figure B-9 shows a scale representation of the lethal bursting areas of mortar rounds.

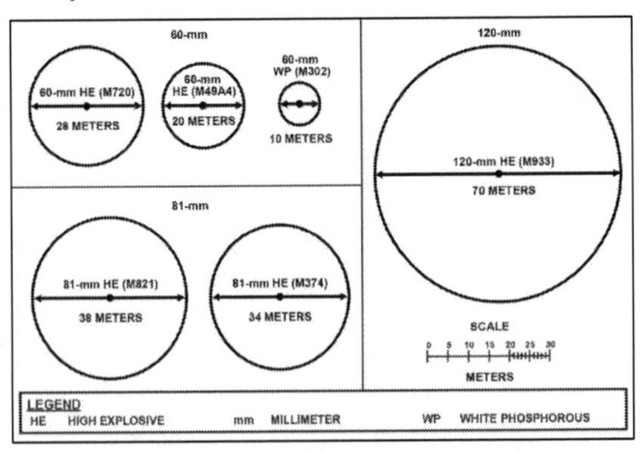

Figure B-9. Comparison of lethal bursting area for U.S. mortar rounds

---

Note. The bursting radius is the standard method to compare the effectiveness of shells. It is defined as the distance from the bursting point within 50 percent of standing targets will become casualties. The bursting diameter is therefore twice the bursting radius. The above mortar-bursting areas (diameters) are estimations since the type of round, fuze, range, and target surface all affect the mortar's bursting diameter.

Fuze Setting

B-33. The decision concerning what fuze setting to use depends on the position of the enemy. Exposed enemy troops standing up are best engaged with impact or near surface burst fuze settings. The round explodes on, or near, the ground. Shell fragments travel outward perpendicular to the long axis of the standing target. (See figure B-10.)

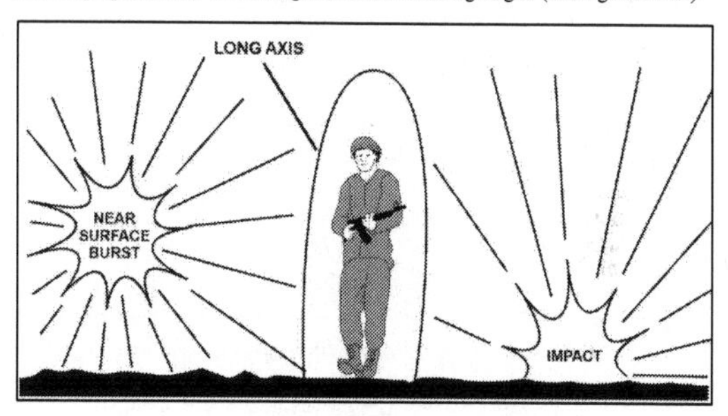

Figure B-10. Standing targets

B-34. If exposed enemy troops are lying prone, the proximity fuze setting is most effective. The rounds explode high above the ground, and fragments coming downward are traveling once again perpendicular to the long axis of the targets. (See figure B-11.)

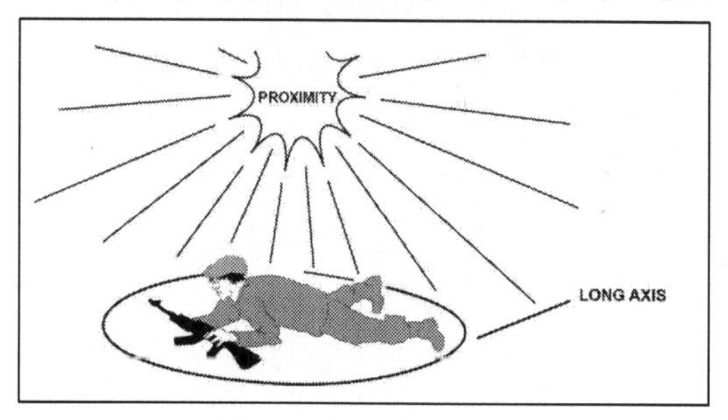

Figure B-11. Prone targets

B-35. The proximity setting is the most effective if the enemy is in open fighting positions without OHC. Even proximity settings will not always produce effects if the positions are deep. (See figure B-12.)

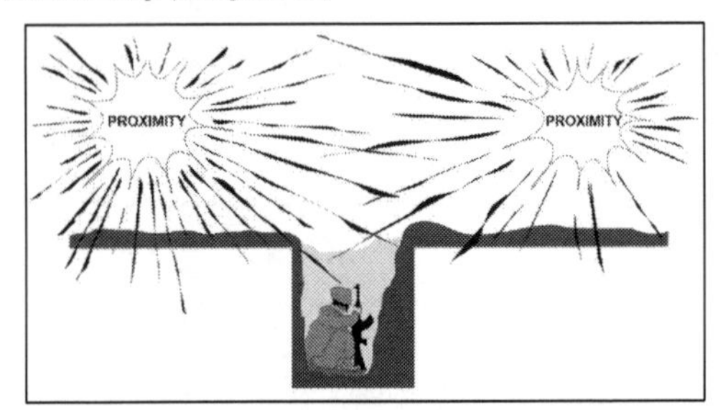

Figure B-12. Targets in open fighting positions

B-36. The delay fuze setting is most effective when the enemy is below triple canopy jungle or in fighting positions with OHC. Light mortars will have little effect against OHC. Even medium mortars have limited effect. Heavy mortars can destroy a bunker, or enemy troops beneath jungle canopy with a hit or near miss. (See figure B-13.)

Figure B-13. Targets beneath triple canopy jungle

Effects of Cover on High Explosive Rounds

B-37. Enemy forces normally will be either standing or prone. They maybe in the open or protected by varying degrees of cover. Each of these changes the target effects of mortar fire.

B-38. Surprise mortar fire is always more effective than fire against an enemy who has been warned and is seeking cover. Recent studies have shown a high casualty rate can be achieved with only two rounds against an enemy platoon standing in the open. The same studies required 10 to 15 rounds to duplicate the casualty rate when the platoon was warned by adjusting rounds and sought cover. If the enemy soldiers merely lay prone, they significantly reduce the effects of mortar fire. Mortar fire against standing enemy forces is almost twice as effective as fire against prone targets.

B-39. Proximity fire is usually more effective than surface-burst rounds against targets in the open. The effectiveness of mortar fire against a prone enemy is increased by about 40 percent by firing proximity-fuzed rounds rather than surface-burst rounds.

B-40. If the enemy is in open fighting positions without OHC, proximity-fuzed mortar rounds are about five times as effective as impact-fuzed rounds. When fired against troops in open fighting positions, proximity-fuzed rounds are only 10 percent as effective as they would be against an enemy in the open. The greatest effectiveness against troops in open fighting positions, the charge with the lowest angle of fall should be chosen. It produces almost two times as much effect as the same round falling with the steepest angle.

B-41. If the enemy has prepared fighting positions with OHC, only impact-fuzed and delay-fuzed rounds will have much effect. Proximity-fuzed rounds can restrict the enemy's ability to move from position to position, but they will cause few, if any, casualties. Impact-fuzed rounds cause some blast and suppressive effect. Delay-fuzed rounds can penetrate and destroy a position but must achieve a direct hit. Only the 120-mm mortar with a delay-fuze setting can damage a fighting position with 18 inches of OHC (see chapter 5 section VI). This type of fighting position cannot be destroyed by light or medium mortar rounds. Consider use of white phosphorous (WP) to drive enemy personnel out of fighting positions.

Suppressive Effects of High Explosive Mortar Rounds

B-42. Suppression from mortar is not as easy to measure as the target effect. It is the psychological effect produced in the mind of the enemy preventing the enemy from returning fire or carrying on their duties. Inexperienced or surprised Soldiers are more easily suppressed than experienced, warned Soldiers. Soldiers in the open are much more easily suppressed than those with OHC. Suppression is most effective when mortar fires first fall; as they continue, their suppressive effects lessen. HE rounds, when used, are the most suppressive and have a great psychological effect on the enemy. WP is used to ignite and suppress enemy materiel. (See figure B-14 on page B-22.) Examples include if a—

- 60-mm mortar round lands within 20 meters of a target, the target probably will be suppressed, if not hit.

- 60-mm mortar round lands within 35 meters of a target, there is a 50 percent chance it will be suppressed. Beyond 50 meters, little suppression takes place.
- 81-mm mortar round lands within 30 meters of a target, the target probably will be suppressed, if not hit.
- 81-mm mortar round lands within 75 meters of a target, there is a 50 percent chance the target will be suppressed. Beyond 125 meters, little suppression takes place.
- 120-mm mortar round (proximity-fuzed) lands within 65 meters of target, the target probably will be suppressed, if not hit. Within 125 meters of a target, there is a 50 percent chance the target will be suppressed. Beyond 200 meters, little suppression takes place.

Note. Because the effects of suppression end almost immediately after the enemy realizes the suppression mission has stopped, friendly forces must be prepared to act (for instance assault) immediately upon achieving suppression of enemy forces.

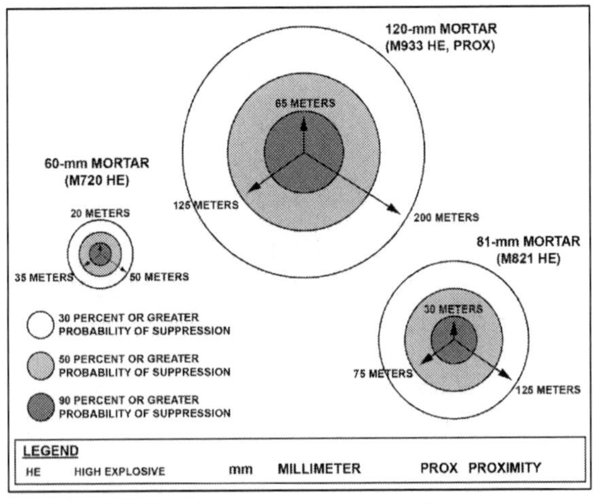

Figure B-14. Suppressive effects of common U.S. mortar rounds

B-43. Obscuration fires use smoke and WP ammunition to degrade the enemy by obscuring their view of the battlefield. (High explosive ammunition may also obscure their view with dust and fires, but the unit should not rely on it as the primary means.) Mortars only have WP or red phosphorous while artillery has smoke, WP, and red phosphorous. Obscurants are subject to changes in wind direction and terrain contours; therefore, its use must be coordinated with other friendly units affected by the mission. Used properly, obscuration fires can—

- Slow enemy vehicles to blackout speeds.
- Obscure the vision of enemy direct fire weapon crews.
- Reduce accuracy of enemy-observed fires by obscuring observation posts and command posts.
- Cause confusion and apprehension among enemy Soldiers.
- Limit the effectiveness of the enemy's visual command and control signals.

B-44. Screening fires are closely related to obscuration fires; they also involve the use of smoke and WP. However, screening fires mask friendly maneuver elements to disguise the nature of their missions. Screening fires may assist in consolidate by placing smoke in areas beyond the objective. They may also be used to deceive the enemy to believe that a unit is maneuvering when it is not. Screening fires require the same precautions as obscuration fires.

B-45. The WP round is used mainly to produce immediate, close point obscuration. It can be used to screen the enemy's field of fire for short periods, which allows the unit to maneuver against the enemy. The 60-mm WP round is not sufficient to produce a long-lasting, wide-area obscurants screen, but the much larger WP round from the heavy mortar is.

B-46. WP rounds generally should not be used solely to produce casualties due to the law of war principle of humanity, which forbids the infliction of suffering, injury, or destruction unnecessary to accomplish a legitimate military purpose. Unnecessary suffering would be implicated because of the persistent burning WP causes in the wounds it produces and the availability of a superior round for producing casualties, the HE round. While the bursting WP round can produce casualties among exposed enemy troops the casualty producing radius of the WP round is much less than that of the HE round. Generally, more casualties can be produced by firing HE ammunition than by firing WP. A few WP rounds mixed into a fire mission of HE rounds for a valid purpose (that is target marking) may increase the suppressive effect of the fire because of the significant psychological effect a WP burst may have on exposed troops.

B-47. As with other weapons, it is prohibited to make the civilian population as such, individual civilians, or civilian objects, the object of attack by WP. Although bursting WP rounds are not designed to cause casualties, the fragments of the shell casing and bits of burning WP can cause injuries. Inhalation from WP is also a primary hazard once the smoke screen is going in. Feasible precautions to reduce the risk of harm to civilians must be taken. (See FM 6-27.)

B-48. The WP rounds can be used to mark targets, especially for attack by aircraft. Base-ejecting obscurants' rounds, such as the 81-mm M819 red phosphorous round, produce a dispersed obscurant cloud, normally too indistinct for marking targets.

B-49. The effects of atmospheric stability can determine whether mortar obscurants are effective at all or, if effective, how much ammunition will be needed:

- During unstable conditions, mortar obscurants and WP rounds are almost ineffective the obscurant does not spread but often climbs straight up and quickly dissipates.
- Under moderately unstable atmospheric conditions, base-ejecting obscurants' rounds are more effective than bursting WP rounds. The M819 red phosphorous round of the M252 mortar screens for over 2 and a half minutes.
- Under stable conditions, both red phosphorous and WP rounds are effective.
- The higher the humidity, the better the screening effects of mortar rounds.

B-50. The M819 red phosphorous round loses up to 35 percent of its screening ability if the ground in the target area is covered with water or deep snow. During extremely cold and dry conditions over snow, up to four times the number of obscurant's rounds than expected may be needed to create an adequate screen. The higher the wind velocity, the more bursting WP rounds are needed, and less effective burning obscurant's rounds become.

B-51. If the terrain in the target area is swampy, rain-soaked, or snow-covered, then burning obscurant's rounds may not be effective. These rounds produce obscurants by ejecting felt wedges soaked in red phosphorous. These wedges then burn on the ground, producing a dense, long-lasting cloud. If the wedges fall into mud, water, or snow, they can be extinguished. Shallow water can reduce the obscurants produced by these rounds by as much as 50 percent. Bursting WP rounds are affected little by the terrain in the target area, except deep snow and cold temperatures which can reduce the obscurant cloud by about 25 percent.

Illumination

B-52. Illumination missions are important functions for mortars. Atmospheric stability, wind velocity, and wind direction are the most important factors when planning target effects for illumination mortar rounds. The terrain in the target area also affects illumination rounds. Illumination rounds can be used to disclose enemy formations, to signal, or to mark targets. There are illumination rounds available for all mortars.

B-53. The 60-mm illumination round available now is the standard cartridge, illuminating M83A3. This round has a fixed time of delay between firing and start of the illumination. The illumination lasts for about 25 seconds, providing moderate light over a square kilometer.

B-54. The 60-mm illumination round does not provide the same degree of illumination as do the rounds of the heavier mortars and field artillery. However, it is sufficient for local, point illumination. The small size of the round can be an advantage where illumination is desired in an area, but adjacent friendly forces do not want to be seen.

The 60-mm illumination round can be used without degrading the night vision devices of adjacent units.

B-55. The medium and heavy mortars can provide excellent illumination over wide areas. The 120-mm mortar illumination round provides 1 million candlepower for 60 seconds.

B-56. The M320 40-mm grenades, as well as all mortars have the capability to deliver infrared illumination rounds in addition to the more common white light.

Special Illumination Techniques

B-57. Illumination is always planned for attacks to be conducted in limited visibility. The following are three mortar special illumination techniques.

B-58. An illumination round fired extremely high over a general area will not always alert an enemy force that it is being observed. However, it will provide enough illumination to optimize the use of image intensification (starlight) scopes such as night vision devices.

B-59. An illumination round fired to burn on the ground will prevent observation beyond the flare into the shadow. This is one method of countering enemy use of image intensification devices. A friendly force could move behind the flare with greater security.

B-60. An illumination round fired to burn on the ground can be used to mark targets during day or night. Illumination rounds have an advantage over WP as target markers during high winds. The obscurant cloud from a WP round will be blown quickly downwind. The obscurants from the burning illumination round will continue to originate from the same point, regardless of the wind. Multiple illumination rounds are used to spread the illumination over the width and depth of the illuminated area.

Considerations When Using Thermal Sights

B-61. Although illumination rounds may aid target acquisition when friendly forces are using image intensification devices (such as night vision devices), this is not so when thermal sights are used. As the illumination flares burn out and land on the ground, they remain as a distinct hot spot seen through thermal sights for several minutes. This may cause confusion, especially if the flare canisters are between the enemy and friendly forces. WP rounds also can cause these hot spots which can make target identification difficult for gunners using thermal sights (for example, Javelin).

AIR-GROUND OPERATIONS

B-62. *Air-ground operations* are the simultaneous or synchronized employment of ground forces with aviation maneuver and fires to seize, retain, and exploit the initiative (FM 3-04). Employing the combined and complementary effects of air and ground maneuver and fires through air-ground operations presents the enemy with multiple dilemmas and ensures that aviation assets are positioned to support ground maneuver. Air-ground operations increase the overall combat power, mission effectiveness, agility, flexibility, and survivability of the entire combined arms team. Air-ground operations ensure that all members of the combined arms team, whether on the ground or in the air, work toward common and mutually supporting objectives to meet the higher commander's intent.

Army Aviation

B-63. Army aviation attack and reconnaissance units use maneuver to concentrate and sustain combat power at critical times and places to find, fix, and destroy enemy forces. During the planning process, Army aviation attack and reconnaissance units are integrated into the Infantry unit's scheme of maneuver to ensure responsiveness, synergy and agility during actions on the objective or upon contact with the enemy. Pre-mission development of control measures provides a foundation for the successful integration of Army aviation into Infantry operations. Among these control measures are engagement criteria; the triggers and conditions for execution; target control measures, such as TRPs or point targets; fire support coordination measures, such as no-fire area or restrictive fire area; engagement areas; and airspace coordinating measures, such as aerial ingress and egress routes and restricted operations zones.

Army Aviation Attack Call for Fire

B-64. *Call for fire* is a standardized request for fire containing data necessary for obtaining the required fire on a target (FM 3-09). Army aviation attack targets are planned on probable enemy locations. Army aviation attack call for fire is a coordinated attack by Army aviation attack against enemy forces in close proximity to friendly units. Army aviation attack call for fire (see figure B-15) is not synonymous with CAS flown by joint and multinational aircraft. Terminal control from ground units or controllers is not required due to aircraft capabilities and enhanced situational understanding of the aircrew. Depending on the enemy situation, Army aviation attack can be on station during times when contact is most likely to occur. Air-ground integration ensures frequencies are known and markings are standardized to prevent fratricide.

B-65. Coordination between ground maneuver units and aviation attack units maximizes the capabilities of the combined arms team while minimizing the risk of fratricide and friendly fire. To ensure adequate air-ground integration, the following major problem areas should be addressed:

- Ensure aircrews understand ground tactical plan and unit commander's intent.
- Ensure adequate common control measures are used to allow both air and ground unit's maximum freedom of maneuver.
- Ensure aircrews and ground forces understand methods of differentiating between enemy and friendly forces on the ground.
- Ensure aircrews understand ground force methods of designating or marking targets.

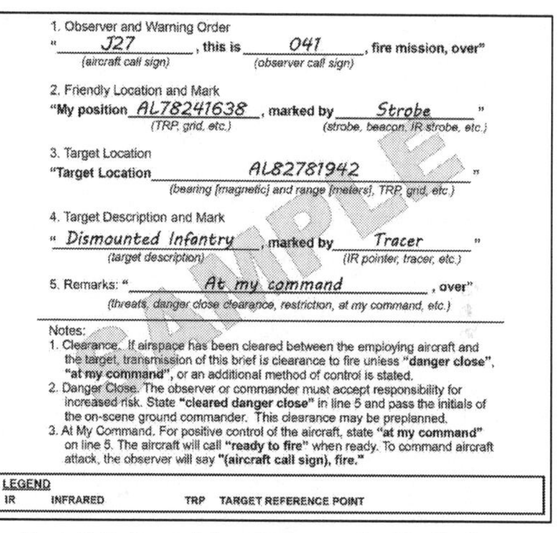

1. Observer and Warning Order
" J27 (aircraft call sign), this is 041 (observer call sign), fire mission, over"

2. Friendly Location and Mark
"My position AL78241638 (TRP, grid, etc.), marked by Strobe (strobe, beacon, IR strobe, etc.)"

3. Target Location
"Target Location AL82781942 (bearing [magnetic] and range [meters], TRP, grid, etc.)"

4. Target Description and Mark
" Dismounted Infantry (target description), marked by Tracer (IR pointer, tracer, etc.)"

5. Remarks: " At my command (threats, danger close clearance, restriction, at my command, etc.), over"

Notes:
1. Clearance. If airspace has been cleared between the employing aircraft and the target, transmission of this brief is clearance to fire unless **"danger close"**, **"at my command"**, or an additional method of control is stated.
2. Danger Close. The observer or commander must accept responsibility for increased risk. State **"cleared danger close"** in line 5 and pass the initials of the on-scene ground commander. This clearance may be preplanned.
3. At My Command. For positive control of the aircraft, state **"at my command"** on line 5. The aircraft will call **"ready to fire"** when ready. To command aircraft attack, the observer will say **"(aircraft call sign), fire."**

LEGEND
IR INFRARED
TRP TARGET REFERENCE POINT

Figure B-15. Army aviation attack request call for fire format

Employment Considerations

B-66. Mission success in Army aviation attack call for fire employment depends on leaders conducting detailed planning and coordination between the aerial attack team and ground unit already engaged in close combat. Once execution begins, there must be integration of the fires and movement of both maneuver and aerial elements.

B-67. During call for fire, the flight lead must have direct communication with the on-scene ground force commander to provide direct fire support. After receiving the call for fire brief from ground forces, pilots must be able to positively identify friendly locations before engagement. Once the crew has identified both enemy and friendly locations, flight leads formulate an attack plan and brief the supported commander and their other attack team members. (See ATP 3-09.32 for Army aviation attack call for fire capabilities.)

B-68. Planning for attack reconnaissance helicopter support usually begins at the battalion level or above. The aviation battalion provides the aviation brigade and Infantry battalion with information on locations, routes, and communications before the attack team's departure from its assembly area (AA). As part of this effort, Infantry companies and platoons usually provide information for aviation attack call for fire employment. All platoon Soldiers should familiarize themselves with the procedures used to call for attack reconnaissance helicopter support. If attack reconnaissance helicopter assets are working for their battalion, the platoon and company provide suppressive fires on any known or suspected threat air defense artillery locations.

B-69. Critical elements of the planning process are the procedures and resources used in marking and identifying targets and friendly positions. Leaders consider these factors thoroughly, regardless of the time available to the ground and air commanders.

B-70. The aerial attack team coordinates directly with the lowest-level unit in contact on the Infantry platoon and company radio command nets. Whenever practicable, before the attack team launches the aviation attack call for fire operation, the leader or commander conducts final coordination with the attack reconnaissance helicopters in a concealed position known as the aerial holding area. The holding area is a point in space within the supported unit's AO oriented toward the threat; it allows the attack team to receive requests for immediate aviation attack call for fire and expedite the attack. The aerial holding area could be an alternate battle position positioned out of range of the threat's direct fire and indirect fire weapons ranges.

B-71. Final coordination between the ground and helicopter units must include agreement on methods of identifying and marking friendly and threat positions. This should take advantage of the equipment and capabilities of the attack team, including the forward-looking infrared system (when equipped), the thermal imaging system, and night vision devices. Final coordination should identify limitations of the attack team as well. For example, ground units must understand that without a forward-looking infrared system attack helicopters will not be able to identify infrared light sources. In this case the ground unit may mark friendly positions with smoke or an enemy position with tracer fire.

B-72. Coordination also should cover the battle position, and assault by fire, or support by fire positions used by the attack reconnaissance helicopters. The leader should offset these positions from the ground maneuver unit and the gun-target line of supporting fires to maximize the effects of the attack team's weapons and to minimize the risk of fratricide. To prevent indirect fires within the AO or zone from posing a danger to the helicopters, the commander informs direct support artillery and organic mortars of the aerial positions. (See ATP 3-09.32 for additional information.)

CLOSE AIR SUPPORT

B-73. Infantry battalions' allocated CAS sorties may allocate assets to individual rifle companies. *Close air support* is air action by aircraft against hostile targets that are in close proximity to friendly forces and that require detailed integration of each air mission with the fire and movement of those forces (JP 3-09.3). CAS can be employed to blunt an enemy attack; to support the momentum of the ground attack; to help set conditions for Infantry operations as part of the overall counterfire fight; to disrupt, delay and destroy enemy second echelon forces and reserves; and to provide cover for friendly movements. The effectiveness of CAS is related directly to the degree of local air superiority attained. Until air superiority is achieved, competing demands between CAS and counterair operations may limit sorties apportioned for the CAS role. CAS is the primary support given to committed battalions and IBCT by Air Force, Navy, and Marine aircraft.

Mission

B-74. The IBCT normally plans and controls CAS. However, this does not preclude the battalion from requesting CAS, receiving immediate CAS to support a company-level operation, or accepting execution responsibility for a planned CAS mission. CAS is another means of fire support available to the battalion. In planning CAS missions, the commander must understand the capabilities and limitations of CAS and synchronize CAS missions with both the battalion fire plan and scheme of maneuver. CAS capabilities and limitations such as windows for use, targets, observers, and airspace coordination present some unique challenges, but the commander must plan CAS with maneuver the same way indirect artillery and mortar fires are planned. When executing a CAS mission, the battalion must have a plan that synchronizes CAS with maneuver and the scheme of fires of maneuver companies.

Preplanned Close Air Support

B-75. Battalion planners, in coordination with maneuver companies, forward CAS requests as soon as they can be forecasted. These requests for CAS normally do not include detailed timing information because of the lead time involved. Preplanned CAS requests involve any information about planned subordinate schemes of maneuver, even general information, which can be used in the apportionment, allocation, and distribution cycle.

Immediate Close Air Support

B-76. Immediate requests are used for air support mission requirements identified too late to be included in the current air tasking order. Those requests initiated below battalion level are forwarded to the battalion main command post by the most rapid means available. At battalion level, the commander, fire support officer, air liaison officer, and battalion operations staff officer consider each request. Approved immediate CAS requests are transmitted by the tactical air control party over the Joint Air Request Net directly to the air support operations center who works with the division Joint Air Ground Integration Center to execute the air tasking order, integrate airspace use and users, and control CAS and other air support missions in the division AO.

Executing Close Air Support

B-77. Units having a reasonable expectation of utilizing CAS will need to have a qualified JTAC, forward air controller (airborne) (FAC[A]), or JFC available. In the extreme circumstance that CAS is needed and a JTAC, (FAC[A]) or joint fires observer is not available, then aircrews executing CAS under these circumstances must be in contact with the ground commander (or the commander's designated representative). Aircrews bear increased responsibility for the detailed integration required to minimize fratricide normally done by a JTAC/FAC (A)/joint fires observer. In these rare circumstances, the aircrews providing CAS assist the ground movement commander to the greatest extent possible to bring fires to bear.

B-78. The flow and prosecution of CAS targets normally begins with a check-in briefing between the aircrew and the JTAC or the FAC(A). The JTAC will give a ground situation update followed by a game plan and a CAS 9-line brief (see figure B-16). (See ATP 3-09.32 for information on check-in brief, situation update briefing.) A game plan is a concise situational awareness-enhancing tool to inform all players of the flow of the CAS mission. At a minimum, the game plan will contain the type of control and method of attack. The type of *terminal attack control*—the authority to control the maneuver of and grant weapons release clearance to attacking aircraft (JP 3-09.3)—and method of attack are separate and independent constructs. The method of attack is broken down into two categories: bomb on target and bomb on coordinate. The method of attack conveys the JTAC's/FAC(A)'s intent for the aircraft prosecution of the target; either the aircraft will be required to acquire the target (bomb on target) or not (bomb on coordinate). These two categories define how the aircraft will acquire the target or mark the target. Any type of control can be utilized with either method of attack, and no type of control is attached to one particular method of attack. (See ATP 3-21.10 appendix D for additional information.)

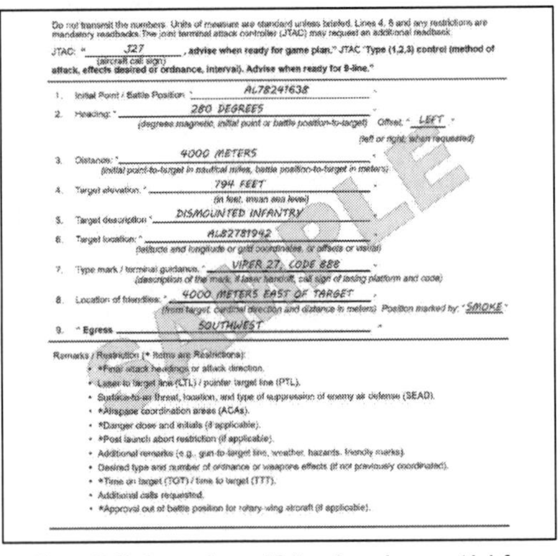

Do not transmit the numbers. Units of measure are standard unless briefed. Lines 4, 8 and any restrictions are mandatory readbacks. The joint terminal attack controller (JTAC) may request an additional readback.

JTAC: "J27 (aircraft call sign), **advise when ready for game plan.**" JTAC "**Type (1,2,3) control (method of attack, effects desired or ordnance, interval). Advise when ready for 9-line.**"

1. Initial Point / Battle Position: "AL78241638"
2. Heading: "280 DEGREES" (degrees magnetic, initial point or battle position-to-target) Offset: "LEFT" (left or right, when requested)
3. Distance: "4000 METERS" (initial point-to-target in nautical miles, battle position-to-target in meters)
4. Target elevation: "794 FEET" (in feet, mean sea level)
5. Target description: "DISMOUNTED INFANTRY"
6. Target location: "AL82781942" (latitude and longitude or grid coordinates, or offsets or visual)
7. Type mark / terminal guidance: "VIPER 27, CODE 888" (description of the mark, if laser handoff, call sign of lasing platform and code)
8. Location of friendlies: "4000 METERS EAST OF TARGET" (from target, cardinal direction and distance in meters) Position marked by: "SMOKE"
9. "**Egress** SOUTHWEST"

Remarks / Restriction (* items are Restrictions):

- *Final attack headings or attack direction.
- Laser to target line (LTL) / pointer target line (PTL).
- Surface-to-air threat, location, and type of suppression of enemy air defense (SEAD).
- *Airspace coordination areas (ACAs).
- *Danger close and initials (if applicable).
- *Post launch abort restriction (if applicable).
- Additional remarks (e.g., gun-to-target line, weather, hazards, friendly marks).
- Desired type and number of ordnance or weapons effects (if not previously coordinated).
- *Time on target (TOT) / time to target (TTT).
- Additional calls requested.
- *Approval out of battle position for rotary-wing aircraft (if applicable).

Figure B-16. Game plan and 9-line close air support brief

Appendix C

Machine Gun Employment and Theory

Whether organic to the unit or attached, machine guns provide the heavy volume of close and continuous fire needed to achieve fire superiority. They are the Infantry platoon's most effective weapons against a dismounted enemy force. These formidable weapons can engage enemy targets beyond the capability of individual weapons with controlled and accurate fire. This appendix addresses the M249 light machine gun and M240 medium machine gun organic to the Infantry rifle platoon.

TECHNICAL DATA

C-1. Leaders must know the technical data of their assigned weapon systems and associated ammunition to maximize their killing and suppressive fires while minimizing the risk to friendly forces. This section discusses machine gun technical data of the Infantry platoon and squad.

C-2. Machine gun fire has different effects on enemy targets depending on the type of system, ammunition used, range to target, and nature of the target. It is important gunners and leaders understand technical aspects of each system and different ammunition available to ensure the machine guns are employed in accordance with their capabilities. Machine guns use several different types of standard military ammunition. Soldiers should use only authorized ammunition manufactured to U.S. and North Atlantic Treaty Organization specifications.

C-3. The following paragraphs discuss the technical data for the tactical employment of M249 and M240-series machine guns. See the specific publication of the machine guns listed in table C-1 on page C-2 for complete information regarding their technical data.

Table C-1. Machine gun technical data

Weapon	*M249*	*M240-Series*
Field Manual	TC 3-22.249	TC 3-22.240
Description	5.56-mm gas-operated automatic weapon	7.62-mm gas-operated medium machine gun
Weight	16.41 lbs. (gun with barrel) 16 lbs. (tripod)	27.6 lbs. (gun with barrel) 20 lbs. (tripod)
Length	104 cm	110.5 cm
Sustained Rate of Fire Rounds/Burst Interval Minutes to Barrel Change	50 RPM 6 to 9 rounds 4 to 5 seconds 10 minutes	100 RPM 6 to 9 rounds 4 to 5 seconds 10 minutes
Rapid Rate of Fire Rounds/Burst Interval Minutes to Barrel Change	100 RPM 6 to 9 rounds 2 to 3 seconds 2 minutes	200 RPM 10 to 13 rounds 2 to 3 seconds 2 minutes
Cyclic Rate of Fire	850 RPM in continuous burst barrel change every 1 minute	650 to 950 RPM in continuous burst barrel change every 1 minute
Maximum Effective Ranges	Bipod/point: 600 m Bipod/area: 800 m Tripod/area: 1,000 m Grazing: 600 m	Bipod/point: 600 m Tripod/point: 800 m Bipod/area: 800 m Tripod/area: 1,100 m Suppression: 1,800 m Grazing: 600 m
Maximum Range	3,600 m	3,725 m
Legend: cm–centimeter; lbs.–pounds; m–meters; mm–millimeter; RPM–rounds per minute; TC–training circular		

M249 Light Machine Gun

C-4. The M249 light machine gun is organic to the Infantry rifle squad. It provides squads with a light automatic weapon for employment during assault. (See figure C-1.) The M249 also can be used in the medium machine gun role in defensive missions or support by fire position. The M249 fires from the bipod, the shoulder, the hip, or from the underarm position. The hip and underarm positions normally are used for close in fire during an assault when the M249 gunner is on the move and does not have time to set the gun in the bipod position. It is best used when a high rate of fire is needed immediately. Accuracy of fire is decreased when firing from either the hip or shoulder.

Figure C-1. M249 light machine gun, bipod mode

C-5. Available M249 ammunition is classified as follows (see table C-2):

- M855 5.56-mm ball. For use against light materiel, utility/unarmored vehicles, and personnel, but not armored vehicles.
- M856 5.56-mm tracer. Generally used for adjustments after observation, incendiary effects, and signaling. When tracer rounds are fired, they normally are mixed with ball ammunition in a ratio of four ball rounds to one tracer round.

Table C-2. M249 light machine gun ballistic data

Available M249 Cartridges	Maximum Range (Meters)	Tracer Burnout (Meters)	Uses
Ball, M855	3,600	—	Light materials, personnel
Tracer, M856	3,600	900	Observation and adjustment of fire, incendiary effects, signaling

M240B Medium Machine Gun

C-6. Two medium machine guns and crews are found in the weapons squad. (See figure C-2 on page C-4.) The M240B can be fired in the assault mode in emergencies, but normally is fired from the bipod or tripod platform. It also can be vehicle mounted. Platoon leaders (through their weapons squad leader) employ the M240B medium machine guns with a rifle squad to provide long range, accurate, sustained fires against dismounted Infantry, apertures in fortifications, buildings, and lightly armored vehicles. The M240B also provides a high volume of short-range fire in self-defense against aircraft. Machine gunners use point, traversing, searching, or searching and traversing fire to kill or suppress targets.

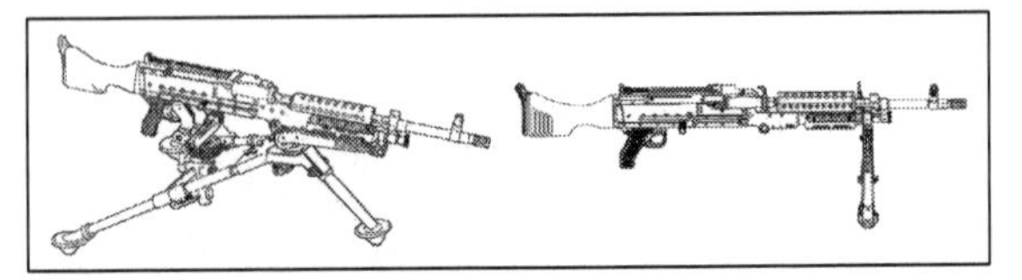

Figure C-2. M240B medium machine gun, bipod- and tripod-mounted

C-7. Available M240 medium machine gun ammunition is classified as follows (see table C-3):

- M80 7.62-mm ball. For use against light materiel and personnel.
- M61 7.62-mm armor piercing. For use against lightly armored targets.
- M62 7.62-mm tracer. For observation of fire, incendiary effects, signaling, and for training. When tracer rounds are fired, they normally are mixed with ball ammunition in a ratio of four ball rounds to one tracer round.

Table C-3. M240 medium machine gun ammunition

Available M240 Cartridges	*Maximum Range (Meters)*	*Tracer Burnout (Meters)*	*Uses*
Ball, M80	3,725	—	Light materials, personnel
Armor Piercing, M61	3,725	—	Lightly armored targets
Tracer, M62	3,725	900	Observation and adjustment of fire, incendiary effects, signaling

M240L Medium Machine Gun

C-8. The M240L short barrel reduces the length of the standard barrel by 4 inches and the weight by .5 pounds, while maintaining accurate fire at extended ranges. The shorter barrel improves mobility in military operations in urban terrain environments. The M240L is 5.1 pounds lighter than the M240B and is 5.6 pounds lighter with the short barrel installed. (See figure C-3.)

Note. The M240L incorporates titanium construction and alternative manufacturing methods for fabricating major M240B components to achieve significant weight savings. These improvements reduce the Soldier's combat load while allowing easier handling and movement of the weapon.

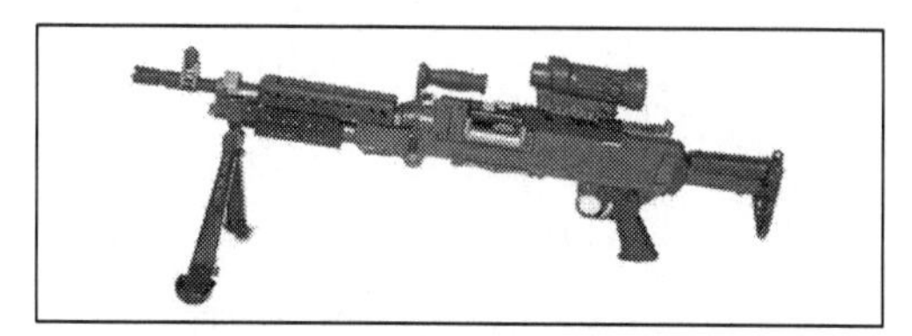

Figure C-3. M240L Medium machine gun, bipod mode

COMBAT TECHNIQUES OF FIRE

C-9. Combat techniques of fire illustrates the characteristics of machine gun fire, the types of enemy targets engaged, and how to apply machine gun fire on those enemy targets. Read the appropriate publications as shown in table C-1 on page C-2 for more weapon-specific information on engaging enemy targets with a particular machine gun.

CHARACTERISTICS OF FIRE

C-10. The gunner's or leader's knowledge of the machine gun is not complete until the gunner or leader learns about the action and effect of the projectiles when fired. The following definitions will help the leader, gunner, and assistant gunner understand the characteristics of fire of the weapons squad's machine guns.

Line of Sight

C-11. Line of sight is an imaginary line drawn from the firer's eye through the sights to the point of aim. Example line of sight restrictions can be obstacles, terrain, and vegetation driven.

Burst of Fire

C-12. A burst of fire is a number of successive rounds fired with the same elevation and point of aim when the trigger is held to the rear. The number of rounds in a burst can vary depending on the type of fire employed.

Trajectory

C-13. Trajectory is the curved path of the projectile in its flight from the muzzle of the weapon to its impact. The major factors influencing trajectory are the velocity of the round, gravity, rotation of the round, and air resistance. As the range to the target increases, so does the curve of trajectory. (See figure C-4 on page C-6.)

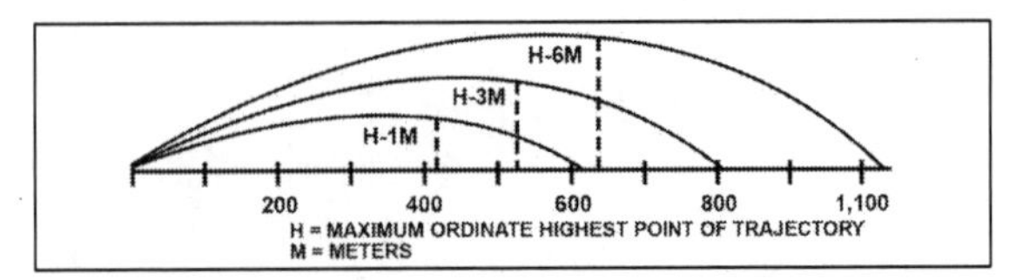

Figure C-4. Trajectory and maximum ordinate

Cone of Fire

C-14. The cone of fire is the pattern formed by the different trajectories in each burst as they travel downrange. Vibration of the weapon and variations in ammunition and atmospheric conditions all contribute to the trajectories making up the cone of fire. (See figure C-5.)

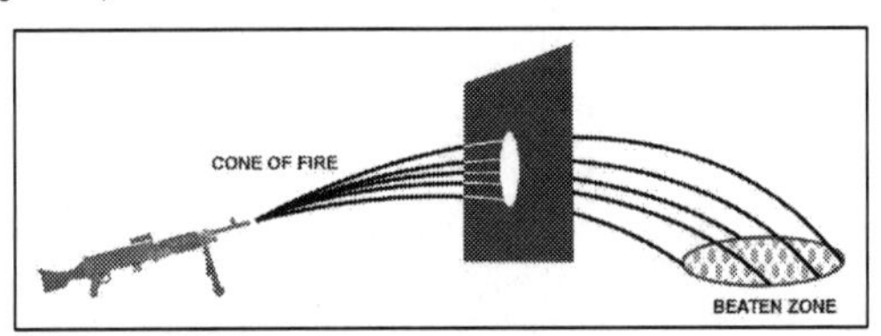

Figure C-5. Cone of fire and beaten zone

Beaten Zone

C-15. The beaten zone is the elliptical pattern formed when the rounds within the cone of fire strike the ground or target. (See figure C-5.) The size and shape of the beaten zone change as a function of the range to and slope of the target but is normally oval or cigar shaped, and density of rounds decreases toward the edges. Gunners and automatic riflemen should engage targets to take maximum effect of the beaten zone. The simplest way to do this is to aim at the center base of the target. Most rounds will not fall over (beyond) the target and falling short creates ricochets into the target.

C-16. Effective beaten zone. Leaders and gunners strive to position their gun teams to achieve an effective beaten zone. Because of dispersion, only part of the beaten zone in which 85 percent of the rounds fall is considered the effective beaten zone.

C-17. Effect of range on the beaten zone. As the range to the target increases, the beaten zone becomes shorter and wider. Conversely, as the range to the target decreases, the beaten zone becomes longer and narrower. (See table C-4.)

C-18. Effect of slope on the beaten zone. The length of the beaten zone for given ranges varies according to the slope of the ground. On rising ground, the beaten zone becomes shorter but remains the same width. On ground sloping away from the gun, the beaten zone becomes longer but remains the same width.

Table C-4. Beaten zones of the M240

M240-Series Machine Gun
Range: 500 meters (1 meter wide by 110 meters long)
Range: 1,000 meters (2 meters wide by 75 meters long)
Range: 1,500 meters (3 meters wide by 55 meters long)
Range: 2,000 meters (4 meters wide by 50 meters long)

Danger Space

C-19. This is the space between the muzzle of the weapon and target where trajectory does not rise above 1.8 meters (the average height of a standing Soldier) including the beaten zone. When firing over level or uniformly sloping terrain danger space would include the M240 and M249 attaining grazing fires (see paragraph C-23) to a maximum of 600 meters. Gunners should consider the danger space of weapons when planning overhead fires. Plunging fire (see paragraph C-24) occurs when there is little or no danger space from the muzzle of the weapon to the beaten zone.

Surface Danger Zone

C-20. Surface danger zones were developed for each weapon and are defined as the area in front, back, or side of the muzzle of the weapon providing a danger to friendly forces when the weapon is fired. The surface danger zone is not just the area comprising the cone of fire as it moves downrange. It also involves the possible impact area on both sides of the gun target line and possible dispersion of materiel caused by the strike of the rounds, the possible ricochet area, and areas to the rear adversely affected by the effects of firing the weapon. (See figure C-6 on page C-8.)

C-21. Surface danger zones were developed primarily for ranges and must be complied with when training, but they also should be complied with in combat, when possible, to minimize risk to friendly forces. (See DA Pam 385-63 for additional information.)

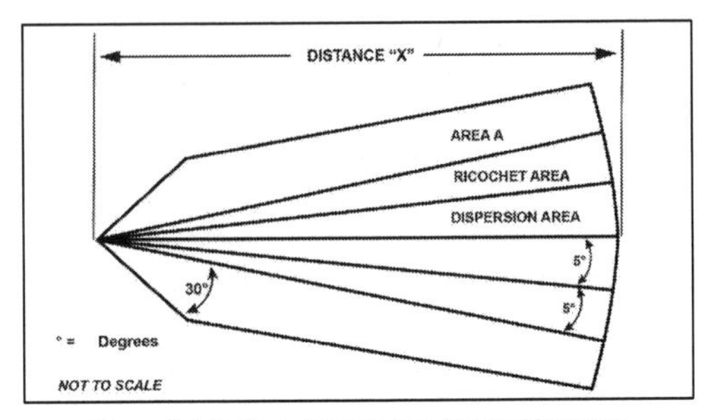

Figure C-6. Surface danger zones for machine guns

CLASSIFICATIONS OF AUTOMATIC WEAPONS FIRE

C-22. The U.S. Army classifies automatic weapons fires with respect to the ground, target, and weapon. Fires with respect to the ground include grazing and plunging fire.

Classification of Fires with Respect to the Ground

C-23. Grazing fires. Automatic weapons achieve grazing fire when the center of the cone of fire does not rise more than 1 meter above the ground. Grazing fire is employed in the final protective line (FPL) in the defense and is only possible when the terrain is level or sloping uniformly. When firing over level or uniformly sloping terrain, the machine gun M240-series and M249 can attain a maximum of 600 meters of grazing fire. Typically, an FPL is planned where the gunner can achieve grazing fire. The unit plans indirect fires, including M320s, to cover dead space along the FPL.

C-24. Plunging fires. Plunging fire occurs when there is little or no danger space from the muzzle of the weapon to the beaten zone. It occurs when weapons fire at long range, when firing from high ground to low ground, when firing into abruptly rising ground, or when firing across uneven terrain, resulting in a loss of grazing fire at points along the trajectory. (See figure C-7.)

Note. Folds or depressions in the ground preventing a target from being engaged from a fixed position are termed dead space. Paragraph C-72 discusses methods of determining dead space.

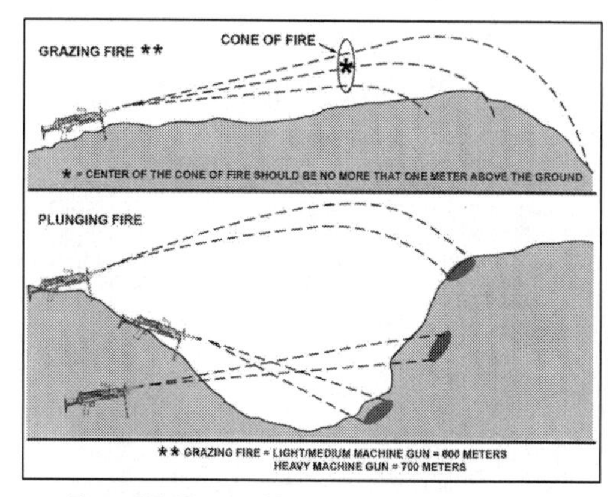

Figure C-7. Classes of fires with respect to the ground

Classification of Fires with Respect to the Target

C-25. Fires with respect to the target include enfilade, frontal, flanking, and oblique fire. (See figure C-8 on page C-10, figures C-9 and C-10 on page C-11.) These targets normally are presented to gun teams by the enemy and must be engaged as they are presented. For example, if the enemy presents its flank to the gun crew as it moves past their position from the left or right, the gun crew will have no choice but to employ flanking fire on the enemy.

C-26. Leaders and gunners strive to position their gun teams where they can best take advantage of the machine gun's beaten zone with respect to an enemy target. Channeling the enemy by use of terrain or obstacles so they approach a friendly machine gun position from the front in a column formation is one example. In this situation, the machine gun would employ enfilade fire on the enemy column, and effects of the machine gun's beaten zone would be much greater than if it engaged enemy column from the flank.

C-27. Enfilade fire. Fires that occur when the long axis of the beaten zone coincides or nearly coincides with the long axis of the target. It can be frontal fire on an enemy column formation or flanking fire on an enemy line formation. This is the most desirable class of fire with respect to the target because it makes maximum use of the beaten zone. Leaders and gunners always should strive to position the guns to the extent possible engaging enemy targets with enfilade fire.

C-28. Frontal fire. Fires that occur when the long axis of the beaten zone is at a right angle to the front of the target. This type of fire is highly desirable when engaging a column formation. It then becomes enfilade fire as the beaten zone coincides with the long axis of the target. Frontal fire is not as desirable when engaging a line formation because most of the beaten zone normally falls below or after the enemy target.

C-29. Flanking fire. Fires that are delivered directly against the flank of the target. Flanking fire is highly desirable when engaging an enemy line formation. It then becomes enfilade fire as the beaten zone will coincide with the long axis of the target. Flanking fire against an enemy column formation is least desirable because most of the beaten zone normally falls before or after the enemy target.

C-30. Oblique fire. Gunners and automatic riflemen achieve oblique fire when the long axis of the beaten zone is at an angle other than a right angle to the front of the target. Oblique fire is less desirable than enfilade fire with respect to the target because it does not maximize the use of the beaten zone.

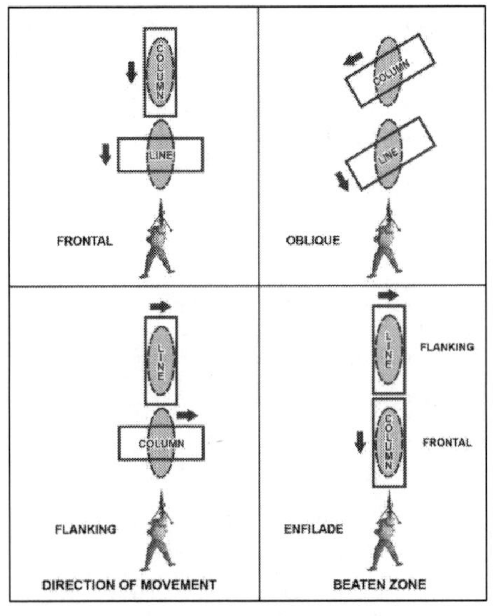

Figure C-8. Classes of fire with respect to the target

Figure C-9. Frontal fire and flanking fire

Figure C-10. Oblique fire and enfilade fire

Classification of Fires with Respect to the Machine Gun

C-31. Fires with respect to the weapon include fixed, traversing, searching, traversing and searching, swinging traverse, and free gun fires. (See figure C-11 on page C-13.)

C-32. Fixed fire. Fire is delivered against a stationary point target when the depth and width of the beaten zone covers the target with little or no manipulation needed. After the initial burst, the gunners follow changes or movement of the target without command.

C-33. Traversing fire. Traversing disperses fires in width by successive changes in direction, but not elevation. It is delivered against a wide target with minimal depth.

When engaging a wide target requiring traversing fire, the gunner selects successive aiming points throughout the target area. These aiming points should be close enough together to ensure adequate target coverage. However, they do not need to be so close as to waste ammunition by concentrating a heavy volume of fire into a small area.

C-34. Searching fire. Searching distributes fires in-depth by successive changes in elevation. It is employed against a deep target or a target having depth and minimal width, requiring changes in only the elevation of the gun. The amount of elevation change depends upon the range and slope of the ground.

C-35. Traversing and searching fire. This class of fire is a combination in which successive changes in direction and elevation result in the distribution of fires both in width and depth. It is employed against a target whose long axis is oblique to the direction of fire.

C-36. Swinging traverse. Swinging traverse fire is employed against targets requiring major changes in direction but little or no change in elevation. Targets may be dense, wide, in close formations moving slowly toward or away from the gun, or vehicles or mounted troops moving across the front. If tripod mounted, the traversing slide lock lever is loosened enough to permit the gunner to swing the gun laterally. When firing swinging traverse, the weapon normally is fired at the cyclic rate of fire. Swinging traverse consumes a lot of ammunition and the weapon does not have a beaten zone because the rapid traverse distributes the impact of the rounds laterally across a wide area.

C-37. Free gun. Fire delivered against moving targets rapidly engaging with fast changes in both direction and elevation. Examples are aerial targets, vehicles, mounted troops, or Infantry in relatively close formations moving rapidly toward or away from the gun position. When firing free gun, the weapon normally is fired at the cyclic rate of fire. Free gunfire consumes a lot of ammunition and does not have a beaten zone because each round seeks its own area of impact.

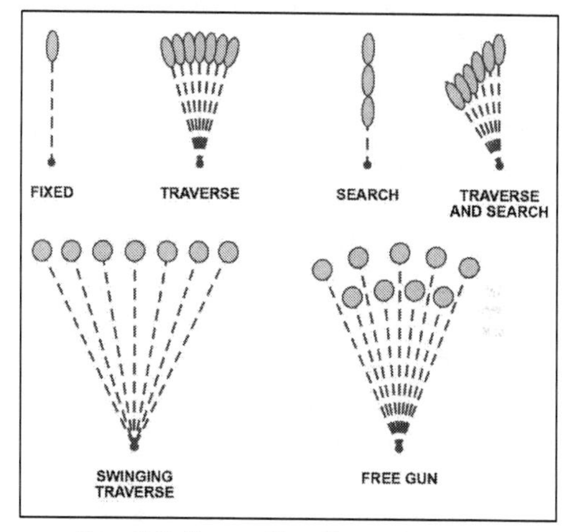

Figure C-11. Classes of fire with respect to the gun

APPLICATION OF FIRE

C-38. Application of fire consists of the methods the gunner uses to cover an enemy target area. Training these methods of applying fire can be accomplished only after the weapons squad leader and gunners have learned how to recognize the different types of targets they may find in combat. They also must know how to distribute and concentrate their fire, and how to maintain the proper rate of fire. Normally, the gunner is exposed to two types of targets in the squad or platoon area of operations (AO): enemy soldiers and supporting automatic weapons. Leaders must ensure targets have priority and are engaged immediately.

C-39. Machine gun fire must be distributed over the entire target area. Improper distribution of fire results in gaps allowing the enemy to escape or use its weapons against friendly positions without opposition.

C-40. The method of applying fire to a target is generally the same for either a single gun or a pair of guns. Direct lay is pointing the gun for direction and elevation, so the sights are aligned directly on the target. Fire is delivered in width, depth, or in a combination of the two. To distribute fire properly, gunners must know where to aim, how to adjust their fire, and direction to manipulate the gun. The gunner must aim, fire, and adjust on a certain point of the target. Binoculars may be used by the leader to facilitate fire adjustment.

Sight Picture

C-41. A correct sight picture has the target, front sight post, and rear sight aligned. The sight picture has sight alignment and placement of the aiming point on the target. The gunner aligns the front sight post in the center of the rear sight and aligns the sights with the target. The top of the front sight post is aligned on the center base of the target.

Beaten Zone

C-42. The gunner ensures throughout firing the center of the beaten zone is maintained at the center base of the target for maximum effect from each burst of fire. When this is done, projectiles in the upper half of the cone of fire will pass through the target if it has height, and projectiles in the lower half of the beaten zone may ricochet into the target. (See figure C-12.)

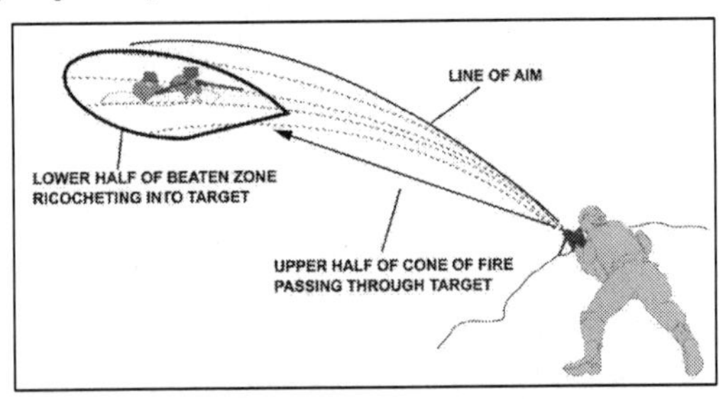

Figure C-12. Line of aim and placement of center of beaten zone on target

C-43. The gunner must move the beaten zone in a certain direction over the target. The direction depends on the type of target and whether the target is engaged with a pair of guns or a single gun. When engaging targets other than point targets with a pair of guns, the targets are divided so fire is distributed evenly throughout the target area. Fire delivered on point target, or a specific area of other target configurations is called concentrated fire.

Target Engagements

C-44. Gunners engage targets throughout their respective sectors. They must know how to engage all types of targets, either individually or with other gunners. When a single gunner is assigned targets, the gunner is responsible for covering the entire target. When a pair of gunners engage an enemy target, each gunner normally is responsible for covering one half of the target. The gunners must be prepared to engage the entire target should the other gun go down.

Enemy Troops and Light Vehicles

C-45. Gunners' targets in combat are normally enemy troops and light vehicles in various formations or displacements, which require distribution and concentration of fire. These targets often have both width, depth, and application of machine gun fire is designed to cover the area in which the enemy is known or suspected to be completely. These targets may be easy to see or may be indistinct and difficult to locate. The size of the target, stated in terms of the number of aiming points required to engage it completely, determines its type.

Self-Defense Against Hostile Low-Flying, Low-Performance Aircrafts

C-46. The machine gun can provide units with a self-defense capability against hostile low-flying, low-performance aircraft. These guns are employed in the air defense role as part of the unit's local defense. The machine guns are not components of an integrated and coordinated air defense system. Unless otherwise directed, hostile aircraft within range of the gun (about 800 meters maximum effective range) should be engaged. The decision will be made by the commander or leader. Typical targets are surveillance, reconnaissance, and liaison aircraft; troop carriers; helicopters; and drones.

C-47. Employment of machine guns used for air defense is guided by the following defensive design factors:

- Defensive design should produce an equally balanced defense in all directions unless a forced route of ingress/egress exists.
- Machine guns should be positioned so the maximum number of targets can be engaged, continuous fire can be delivered, and likely routes of approach are covered.
- Machine guns used to defend march columns should be interspersed in the convoy, with emphasis on the lead and rear elements.

Selection and Engagement Control

C-48. These actions depend upon visual means. The sites selected for guns must provide maximum observation and unobstructed sectors of fire. Units furnished machine guns in sufficient numbers should position them within mutually supportive distances of 90 to 360 meters. Each gun is assigned a primary and secondary sector of fire. Weapon crews maintain constant vigilance in their primary sectors of fire, regardless of the sector in which the guns are engaged.

Distributed and Concentrated Fire

C-49. Distributed fire is delivered in width and depth such as at an enemy formation. Concentrated fire is delivered at a point target such as an automatic weapon or an enemy fighting position.

Rate of Fire

C-50. The size and nature of the enemy target determines how machine gun fire is applied. Automatic weapons fire in one of three rates: rapid, sustained, or cyclic. The rates of fire for each machine gun are shown in table C-1 on page C-2. The situation normally dictates the rate used, but the availability of ammunition and need for barrel changes play important roles as well. The rate of fire must be controlled to cover the target adequately, but not waste ammunition or destroy the barrel.

Rapid Fire

C-51. Rapid rate of fire places an exceptionally high volume of fire on an enemy position. Machine gunners normally engage targets at the rapid rate to suppress the enemy quickly. Rapid fire requires more ammunition than sustained fire and requires frequent barrel changes.

Sustained Fire

C-52. Once the enemy has been suppressed, machine gunners fire at the sustained rate. Sustained fire conserves ammunition and requires only infrequent barrel changes, but it might not be enough volume of fire to suppress or destroy.

Cyclic Rate of Fire

C-53. To fire the cyclic rate, the gunner holds the trigger to the rear while the assistant gunner feeds ammunition into the weapon. This normally is used only to engage aerial targets in self-defense or to fire the FPLs in the defense to protect the perimeter. This produces the highest volume of fire the machine gun can fire but can permanently damage the machine gun and barrel and should be used only in case of emergency.

Limited Visibility

C-54. Gunners have difficulty detecting and identifying targets during limited visibility. The leader's ability to control the fires of the gunners' weapons also is reduced; therefore, the leader may instruct the gunners to fire without command when targets present themselves.

C-55. Gunners should engage targets only when they can identify the targets, unless ordered to do otherwise. For example, if one gunner detects a target and engages it, the other gunner observes the area fired upon and adds fire only if the gunner can identify the target or if ordered to fire.

C-56. Often referred to as engagement criteria, a trigger specifies the circumstances in which subordinate elements should engage. The circumstances can be based on a friendly or enemy event. For example, the trigger for a friendly platoon to initiate engagement could be three or more enemy combat vehicles passing or crossing a given point or line. This line can be any natural or man-made linear feature, such as a road, ridgeline, or stream. It may also be a line perpendicular to the unit's orientation, delineated by one or more reference points.

C-57. Tracer ammunition helps a gunner engage targets during limited visibility and should be used if possible. It is important to note in certain circumstances the enemy will have an easy time identifying the machine gun's position if the gunner uses tracer ammunition. The need to engage targets must be balanced with the need to keep the guns safe before deciding to employ tracers. If firing unaided or during limited visibility, gunners must be trained to fire low at first and adjust upward. This overcomes the tendency to fire high.

C-58. When two or more gunners are engaging linear targets, linear targets with depth, or deep targets, they do not engage these targets as they would when visibility is good. With limited visibility, the center and flanks of these targets may not be defined clearly. Therefore, each gunner observes the tracers and covers what each gunner believes to be the entire target.

Techniques

C-59. Techniques of fire include assault fire, overhead fire, and fire from a defilade position. Generally, only automatic rifles use assault fire (see paragraphs C-126 to C-128 when the medium machine gun is used in the assault element).

Assault Fire

C-60. Automatic riflemen use assault fire when in close combat. Assault fire involves firing without the aid of sights using the hip, shoulder, and underarm positions. The underarm position is best when rapid movement is required. In all three positions, automatic riflemen adjust their fire by observing the tracer and impact of the bullets on the target area. Additional considerations for automatic riflemen using assault fire include:

- Maintaining alignment with the rest of the assault element.
- Reloading rapidly.
- Aiming low and adjusting the aim upward toward the target.
- Distributing fires across the objective when not engaging enemy automatic weapons.
- Automatic riflemen can control rates of fire by firing short bursts as they advance, for instance firing every other left step, every left step, or every step.

Overhead Fire

C-61. Gunners can use overhead fire when there is sufficient low ground between the machine gun and target area of the maneuver friendly forces. A machine gun on a tripod is capable of delivering this type of fire because of the small and uniform dispersion of the cone of fire. Gunners must accurately estimate range to the target and establish a safety limit imaginary line parallel to the target where fire would cause casualties to friendly Soldiers. Gun crews and leaders must be aware of this safety limit. Leaders must designate signals for lifting or shifting fires. Gunners should not attempt overhead fires if the terrain is level or slopes uniformly, if the barrel is badly worn, or if visibility is poor.

C-62. Gunner's rule. The gunner's rule can be applied when the friendly troops are at least 350 meters in front of the gun position and range to the target is 850 meters or less. (See figure C-13.) The tripod mounted machine gun rule follows:

- Lay the gun on the target with the correct sight setting to hit the target.
- Without disturbing the lay of the gun, set the rear sight at a range of 1,600 meters.
- Look through the sights and notice where the new line of aim strikes the ground. This is the limit of troop safety. When the feet of the friendly troops reach this point, fire must be lifted or shifted.

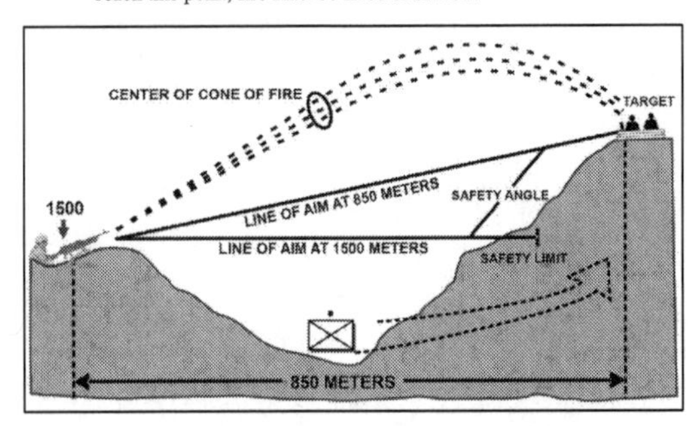

Figure C-13. Application of gunner's rule

C-63. Leader's rule. When the range to the target is greater than 850 meters, overhead fire should be delivered only in an emergency. Even then, fire should extend only to a range at which the tracers or strike of the bullets can be seen by the gunner. In this situation, the leader's rule applies. (See figure C-14.) The platoon or section leader uses the leader's rule only when the target is greater than 850 meters. The rule follows:

- Select a point on the ground where it is believed friendly troops can advance with safety.
- Determine the range to this point by the most accurate means available.
- Lay the gun on the target with the correct sight setting to hit the target.
- Without disturbing the lay of the gun, set the rear sight to 1,600 meters or the range to the target plus 500 meters, whichever is the greater of the two ranges. Under no conditions should the sight setting be less than 1,500 meters.
- Note the point where the new line of aim strikes the ground:
  - If it strikes at the selected point, that point marks the limit of safety.
  - If it strikes short of the selected point, it is safe for troops to advance to the point where the line of aim strikes the ground and to an unknown point beyond. If fire is called for after friendly troops advance farther than the

point where the line of aim strikes the ground, this farther point is determined by testing new selected points until the line of aim and selected point coincide.

- If it clears the selected point, it is safe for troops to advance to the selected point and to an unknown point beyond. If it is advantageous to have troops advance beyond the selected point, this farther point must be determined by testing new selected points until the line of aim and selected point coincide. This point marks the line of safety.

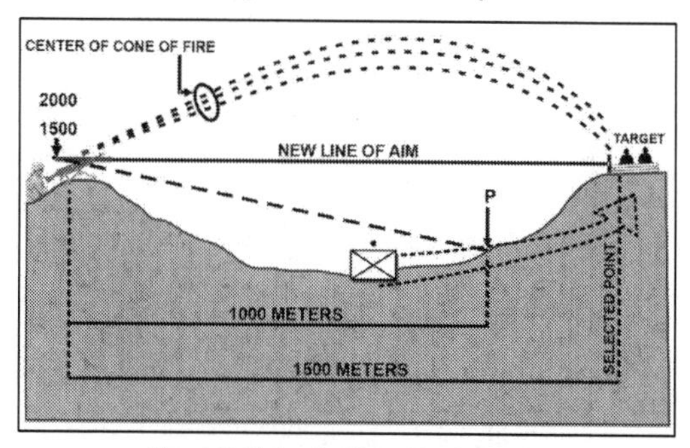

Figure C-14. Application of leader's rule

Fire from a Defilade Position

C-64. Defilade positions protect gunners from frontal or enfilading fires. (See figure C-15 on page C-20.) Cover and concealment may not provide the gunner a view of some or all of the target area. In this instance, some other member of the platoon or squad must observe the impact of the rounds and communicate adjustments to the gunner. (See figure C-16 on page C-20.) Gunners and leaders must consider the complexity of laying on the target. They also must consider the gunner's inability to make rapid adjustments to engage moving targets, which includes the ease with which targets are masked and the difficulty in achieving grazing fires for an FPL.

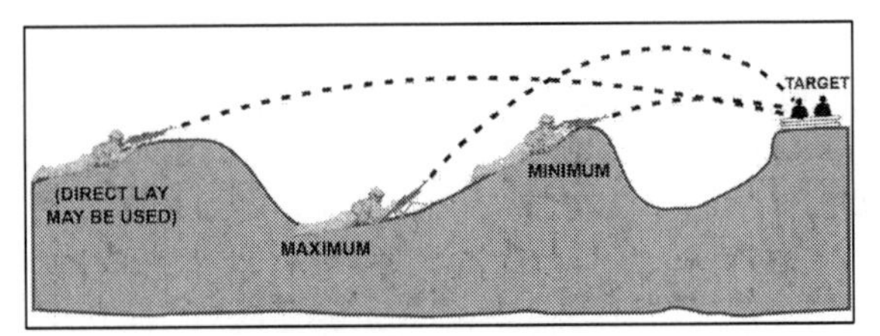

Figure C-15. Defilade positions

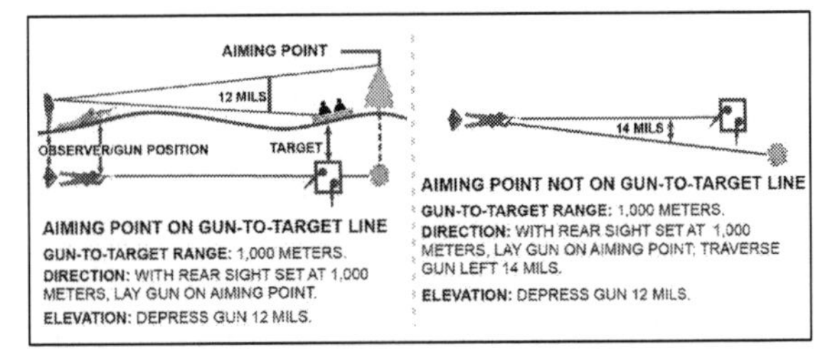

Figure C-16. Observer adjusting fire

PREDETERMINED FIRES

C-65. Predetermined fires organize the battlefield for gunners. They allow the leader and gunner to select potential targets or target areas most likely to be engaged or have tactical significance. This includes dismounted enemy avenues of approach, likely positions for automatic weapons, and probable enemy assault positions.

TERMINOLOGY

C-66. The terminology used by gunners for predetermined fires maximizes the effectiveness of the machine gun during good as well as limited visibility. It enhances fire control by reducing the time required to identify targets, determine range, and manipulate the weapon onto the target. Abbreviated fire commands and previously recorded data enable the gunner to aim or adjust fire on the target quickly and accurately. Selected targets should be fired on in daylight whenever practical to confirm data. The standard range card (GTA 07-10-001 [*Machine Gunner's Card*] outlines the process in preparing a machine gun range card for the weapon's primary position) identifies the targets and provides a record of firing data. DA Form 5517 provides a record of firing data and aids defensive fire planning. If time permits, the gunner also prepares standard range cards for the weapon's alternate and supplementary positions. Gunners need to know several terms associated with predetermined fire.

Sector of Fire

C-67. A *sector of fire* is that area assigned to a unit or weapon system in which it will engage the enemy according to the established engagement priorities (FM 3-90). Individual Soldiers and gunners normally are assigned a primary and a secondary sector of fire.

Planned Targets

C-68. The platoon leader or weapons squad leader will designate planned targets within the primary and secondary sectors of fire. The gunner will number them, determine firing data, and can then fire them as directed whether visible, on command, or in periods of limited visibility.

Final Protective Line

C-69. An FPL is a predetermined line along which grazing fire is placed to stop an enemy assault. If an FPL is assigned, (see figure C-17) the machine gun is sighted along it except when other targets are being engaged. An FPL becomes the machine gun's part of the unit's FPFs. An FPL is fixed in direction and elevation. However, a small shift for search must be employed to prevent the enemy from crawling under the FPL and to compensate for irregularities in the terrain or the sinking of the tripod legs into soft soil during firing. Fire must be delivered during all conditions of visibility. The FPL is always the left or right limit of an assigned primary sector of fire.

C-70. A good FPL covers the maximum area with grazing fire. Grazing fire can be obtained over various types of terrain out to a maximum of 600 meters. To obtain the maximum extent of grazing fire over level or uniformly sloping terrain, the gunner sets the rear sight at 600 meters. The gunner then selects a point on the ground the gunner estimates to be 600 meters from the machine gun, and the gunner aims, fires, and adjusts on that point. To prevent enemy soldiers from crawling under grazing fire, the gunner searches (downward) by lowering the muzzle of the weapon.

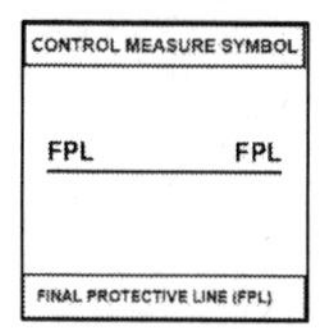

Figure C-17. Final protective line control measure symbol

Principal Direction of Fire

C-71. Principal direction of fire (known as PDF) is assigned to a gunner to cover an area having good fields of fire or has a likely dismounted avenue of approach. It also provides mutual support to an adjacent unit. Machine guns are sited using the PDF if an FPL has not been assigned. If a PDF is assigned and other targets are not being engaged, machine guns remain on the PDF. A PDF has the following characteristics:

- It is used only if an FPL is not assigned; it then becomes the machine gun's part of the unit's FPFs.
- When the target has width, direction is determined by aiming on one edge of the target area and noting the amount of traverse necessary to cover the entire target.
- The gunner is responsible for the entire wedge-shaped area from the muzzle of the weapon to the target, but elevation may be fixed for a priority portion of the target.

Note. Figure C-18 shows the sector of fire of an automatic rifle or machine gun. In this example, the PDF is along the left side of the sector. The field A shows the equipment symbol for an automatic rifle or machine gun.

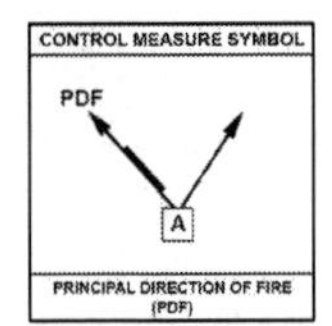

Figure C-18. Principal direction of fire control measure symbol

Dead Space and Grazing Fire

C-72. The extent of grazing fire and dead space may be determined in two ways. In the preferred method, the machine gun is adjusted for elevation and direction. A squad member then walks along the FPL while the gunner aims through the sights. In places where the Soldier's waist (midsection) falls below the gunner's point of aim, dead space exists. Hand and arm signals (see TC 3-21.60) must be used to control the Soldier who is walking and to obtain an accurate account of the dead space and its location. Another method is to observe the flight of tracer ammunition from a position behind and to the flank of the weapon.

Primary Sector of Fire

C-73. The primary sector of fire is assigned to the gun team to cover the most likely avenue of enemy approach from all types of defensive positions. Secondary sectors of fire, when there are no targets in the primary sector or when the leader needs to cover the movement of another friendly element, correspond to another element's primary sector of fire to obtain mutual support. When used, the tripod will be set to fire into the primary sector of fire. If assigned, the FPL will generally be the left or right limit of the primary sector of fire.

Secondary Sector of Fire

C-74. The secondary sector of fire is assigned to the gun team to cover the second most likely avenue of enemy approach. It is fired from the same gun position as the primary sector of fire; however, the gun will probably have to be dismounted from the tripod to fire within the secondary sector of fire.

Field Expedient Techniques

C-75. When laying the machine gun for predetermined targets, the gunner can use field expedients as a means of engaging targets when other sources are not available. These can include following:

Base Stake Technique

C-76. A base stake is used to define sector limits and may provide the lay of the FPL or predetermined targets along a primary or secondary sector limit. This technique is effective in all visibility conditions. The gunner uses the following steps:

- Defines the sector limits by laying the gun for direction along one sector limit and by emplacing a stake along the outer edge of the folded bipod legs. Rotates the legs slightly on the receiver, so the gunner takes up the "play." Uses the same procedure for placing a stake along the opposite sector limit.
- Lays the machine gun along the FPL by moving the muzzle of the machine gun to a sector limit. Adjusts for elevation by driving a stake into the ground so the top of the stake is under the gas cylinder extension. This allows a few mils of depression to cover irregularities in the terrain.
- Lays the machine gun to engage other targets within a sector limit. Done in a primary sector by using the procedure described previously, except the gunner keeps the elevation fixed.

Notched-Stake or Tree-Crotch Technique

C-77. The gunner uses the notched-stake or tree-crotch technique with the bipod mount to engage predetermined targets within a sector or to define sector limits. This technique is effective during all conditions of visibility and requires little additional materiel. The gunner uses the following steps:

- Drives either a notched stake or tree crotch into the ground where selected targets are anticipated. Places the stock of the machine gun in the nest of the stake or crotch and adjusts the weapon to hit the selected targets and to define the sector limits.

- Digs shallow, curved trenches or grooves for the bipod feet. (These trenches allow for rotation of the bipod feet as the gunner moves the stock from one crotch or stake to another.)

Horizontal Log or Board Technique

C-78. This technique is used with the bipod or tripod mount to mark sector limits and engage wide targets. It is good for all visibility conditions and is best suited for flat, level terrain. The gunner uses the following steps:

Bipod-Mounted Machine Gun

C-79. Using a bipod-mounted machine gun, the gunner places a log or board beneath the stock of the weapon so the stock can slide across it freely. The gunner digs shallow, curved trenches or grooves for the bipod feet to allow rotation of the feet as the gunner moves the stock along the log or board. (The gunner may mark the sector limits by notching or placing stops on the log or board. The gunner uses the bipod firing position and grip.)

Tripod-Mounted Machine Gun

C-80. Using a tripod-mounted machine gun, the gunner places a log or boards beneath the barrel, positioning it so the barrel, when resting on the log or board, is at the proper elevation to obtain grazing fire. When appropriate, the gunner marks the sector limits as described of the bipod in the preceding paragraph. (This technique is used only if a traversing and elevating mechanism is not available.)

FIRE CONTROL

C-81. Fire control includes all actions of leaders and Soldiers in planning, preparing, and applying fire on a target. Leaders select and designate targets. They also designate the midpoint and flanks or ends of a target unless they are obvious to the gunner. Gunners fire at the instant desired. They then adjust fire, regulates the rate of fire, shift from one target to another, and cease fire. When firing, gunners should continue to fire until the target is neutralized or until signaled to do otherwise by their leader.

C-82. Predetermined targets, including the FPL or PDF, are engaged on order or by SOP. The signal for calling these fires normally is stated in the defensive order. Control these predetermined targets by using hand and arm signals, voice commands, or pyrotechnic devices. Gunners fire the FPL or PDF at the sustained rate of fire unless the situation calls for a higher rate. When engaging other predetermined targets, the sustained rate of fire also is used unless a different rate is ordered.

METHODS OF FIRE CONTROL

C-83. The noise and confusion of battle may limit the use of some of the following fire control methods. Therefore, leaders must select a method or combination of methods to accomplish the mission.

Oral

C-84. The oral fire control method can be effective, but sometimes leaders may be too far away from the gunner, or the noise of the battle may make it impossible for them to hear. The primary means of the oral fire control method is the issuance of a fire command (see paragraph C-90).

Hand and Arm Signals

C-85. Hand and arm signals are an effective fire control method when the gunner can see the leader. All gunners must know the standard hand and arm signals. The leader gets the gunner's attention and points to the target. When the gunner returns the READY signal, the leader commands FIRE. (See TC 3-21.60 for hand and arm signals.)

Prearranged Signals

C-86. Prearranged signals are either visual or sound signals such as casualty-producing devices (rifle or claymore), pyrotechnics, whistle blasts, or tracers. These signals should be included in SOPs. If leaders want to shift fire at a certain time, they give a prearranged signal such as obscurants or pyrotechnics. Upon seeing the signal, the gunner shifts fire to a prearranged point.

Personal Contact

C-87. In many situations, leaders must issue orders directly to individual Soldiers. Personal contact is used more than other methods by Infantry rifle leaders. Leaders must use maximum cover and concealment to keep from disclosing the position or themselves.

Standard Range Cards

C-88. When using the standard range card method of fire control, leaders must ensure all standard range cards are current and accurate. Once this is accomplished, leaders may designate certain targets for certain weapons with the use of limiting stakes or with fire commands. They also should designate no-fire zones or restricted fire areas to others. The vital factor in this method of fire control is gunners must be well disciplined and pay attention to detail.

Standard Operating Procedures

C-89. SOPs are actions to be executed without command developed during the training of the squads. Their use eliminates many commands and simplifies the leader's fire control. SOPs for certain actions and commands can be developed to make gunners effective. Some examples follow:

- Observation. The gunners continuously observe their sectors.
- Fire. Gunners open fire without command on appropriate targets appearing within their sectors.

- Check. While firing, the gunners periodically check with the leader for instructions.
- Return fire. The gunners return enemy fire without order, concentrating on enemy automatic weapons.
- Shift fire. Gunners shift their fires without command when more dangerous targets appear.
- Rate of fire. When gunners engage a target, they initially fire at the rate necessary to gain and maintain fire superiority.
- Mutual support. When two or more gunners are engaging the same target and one stops firing, the other increases the rate of fire and covers the entire target. When only one gunner is required to engage a target and the leader has alerted two or more, the gunner not firing aims on the target and follows the movements of the target. The gunner does this to fire instantly in case the other machine gun malfunctions or ceases fire before the target has been eliminated.

FIRE COMMANDS

C-90. A fire command is given to deliver fire on a target quickly and without confusion. When leaders decide to engage a target not obvious to the squad, they must provide it with the information needed to engage the target. Leaders must alert the Soldiers; give a target direction, description, and range; name the method of fire; and give the command to fire. There are initial fire commands and subsequent fire commands.

C-91. It is essential the commands delivered by the weapons squad leader, are understood, and echoed by the assistant gunner or gun team leader and gunner. Table C-5 provides an example of the weapons squad fire commands and actions taken by the weapons squad leader, assistant gunner, gun team leader, and gunner.

Table C-5. Weapons squad fire commands and actions

Action	*WSL Commands*	*AG/GTL Command and Action*	*Gunner Action*	*Gunner Responses*
WSL or GTL identifies TGT within gun team's sector.	LIGHT-SKINNED TRUCK, 3 O'CLOCK, 400 METERS, ON MY LASER.	LIGHT-SKINNED TRUCK, 3 O'CLOCK, 400 METERS, ON MY LASER, ONCE ON TARGET, ENGAGE.	Gunner looks for laser and identifies TGT. Gunner traverses and gets on TGT. Gunner engages TGT with correct rate of fire.	TARGET IDENTIFIED. TARGET ACQUIRED.
Gun team (or weapons squad) go to bipod.	GUN 1-BIPOD.	Repeats GUN 1-BIPOD and identifies location for gun.	Gets down beside AG/GTL un-collapsing bipod legs.	GUN 1 UP once ready to fire.
Gun team go to tripod.	GUN 1-TRIPOD.	Repeats GUN 1-TRIPOD and lays down tripod (if not done) and prepares to lock gun on tripod.	Gunner picks up gun and places into tripod. Gunner gets AG/GTL to lock it in. Once locked in, the AG/GTL collapses bipod legs.	GUN 1 UP once ready to fire.
Barrel change.	N/A	GUN 1 PREPARE FOR BARREL CHANGE. GUN 1 BARREL CHANGE.	Fires one more burst. Waits for barrel change.	Repeats AG/GTL command. Once done, GUN 1 UP.
Displace gun.	GUN 1 OUT OF ACTION, PREPARE TO MOVE.	GUN 1 OUT OF ACTION, PREPARE TO MOVE. Breaks down barrel bag, prepares to move.	Gunner takes gun off tripod, continues to orient towards target on bipod, and prepares to move.	GUN 1, READY TO MOVE.
WSL identifies sector of fire for gun teams. **Day-marks with tracer.** **Night-marks with PEQ-series night vision device/tracer.**	GUN 1, LEFT, CENTER, RIGHT SECTORS ON MY MARK. DO YOU IDENTIFY? (Always marks left to right.)	Using binoculars identifies sectors and states, GUN 1 IDENTIFIES. Adjusts gunner onto target.	Gunner makes necessary adjustments, tells AG/GTL whether gunner identifies or not. Engages or makes further adjustments.	SECTOR IDENTIFIED to AG/GTL once identifies.

Legend: AG–assistant gunner; GTL–gun team leader; N/A–not applicable; TGT–target; WSL–weapons squad leader

Initial Fire Commands

C-92. Initial fire commands are given to adjust onto the target, change the rate of fire after a fire mission is in progress, interrupt fire, or terminate the alert. Commands for all direct-fire weapons follow a pattern including similar elements. There are six elements in the fire command of the machine gun: alert; direction; description; range; method of fire; and command to open fire. The gunners repeat each element of fire command as it is given.

Alert

C-93. This element prepares the gunners for more instructions. Platoon leaders may alert both gunners in the squad and may have only one fire, depending upon the situation. To alert and have both gunners fire, leaders announce FIRE MISSION. If they desire to alert both gunners but have only one fire, they announce GUN NUMBER ONE, FIRE MISSION. In all cases, upon receiving the alert, the gunners load their machine guns and place them on fire.

Description

C-94. The target description creates a picture of the target in the gunners' minds. To properly apply their fire, the Soldiers must know the type of target they are to engage. Leaders should describe it briefly. If the target is obvious, no description is necessary.

Direction

C-95. This element indicates the general direction to the target. It may be given in one or a combination of the following methods.

C-96. Oral. The leader orally gives the direction to the target in relation to the position of the gunner. For example, FRONT, LEFT FRONT, or RIGHT FRONT.

C-97. Pointing. Leaders designate a small or obscure target by pointing with their finger or aiming with a weapon. When they point with their finger, a Soldier standing behind them should be able to look over their shoulder and sight along their arm and index finger to locate the target. When aiming the weapon at a target, a Soldier looking through the sights should be able to see the target. Leaders also may use lasers in conjunction with night vision devices to designate a target to the gunner.

C-98. Tracer ammunition. Tracer ammunition is a quick and sure method of designating a target not clearly visible. When using this method, leaders first should give the general direction to direct the gunner's attention to the target area. To prevent the loss of surprise when using tracer ammunition, leaders do not fire until they have given all elements of the fire command except the command to fire. Leaders may fire their individual weapon. The firing of the tracers then becomes the last element of the fire command, and it is the signal to open fire.

CAUTION

Soldiers must be aware night vision devices, temporary blindness ("white out") may occur when firing tracer ammunition at night or when exposed to other external light sources. Lens covers may reduce this effect.

C-99. Reference points. Another way to designate obscure targets is to use easy-to-recognize reference points. All leaders and gunners must know terrain features and terminology used to describe them. (See TC 3-25.26 for additional information.) When using a reference point, the word "reference" precedes its description. This is done to avoid confusion. The general direction to the reference point should be given.

Range

C-100. Leaders always announce the estimated range to the target. The range is given, so the gunner knows how far to look for the target and what range setting to put on the rear sight. Range is announced in meters. However, since the meter is the standard unit of range measurement, the word "meters" is not used. With machine guns, the range is determined and announced to the nearest hundred or thousand example, THREE HUNDRED, or ONE THOUSAND.

Method of Fire

C-101. This element includes manipulation and rate of fire. Manipulation dictates the class of fire with respect to the weapon. It is announced as FIXED, TRAVERSE, SEARCH, or TRAVERSE AND SEARCH. (See paragraphs C-31 to C-37.) Rate controls the volume of fire (sustained, rapid, and cyclic). Normally, the gunner uses the sustained rate of fire. When a sustain rate of fire is to be used, rate of fire is omitted from the fire command. The method of fire of the machine gun is usually 3- to 5-round bursts (M249) and 6- to 9-round bursts (M240-series).

Command to Open Fire

C-102. When fire is to be withheld so surprise fire can be delivered on a target or to ensure both gunners open fire at the same time, leaders may preface the command to commence firing with AT MY COMMAND or AT MY SIGNAL. When the gunners are ready to engage the target, they report READY to the leader. The leader then gives the command FIRE at the specific time desired. If immediate fire is required, the command FIRE is given without pause and gunners fire as soon as they are ready.

Subsequent Fire Commands

C-103. Subsequent fire commands are used to make adjustments in direction and elevation, to change rates of fire after a fire mission is in progress, to interrupt fires, or to terminate the alert. If the gunner fails to engage a target properly, the leader must correct the gunner promptly by announcing or signaling the desired changes. When these changes are given, the gunner makes the corrections and resumes firing without further command.

C-104. Adjustments in direction and elevation with the machine gun always are given in meters one finger is used to indicate 1 meter and so on. Adjustment for direction is given first. Example: RIGHT ONE ZERO METERS or LEFT FIVE METERS. Adjustment for elevation is given next. Example: ADD FIVE METERS or DROP ONE FIVE METERS. These changes may be given orally or with hand and arm signals:

- Changes in the rate of fire are given orally or by hand and arm signals.
- To interrupt firing, the leader announces CEASE FIRE, or the leader signals to cease fire. The gunners remain on the alert. They resume firing when given the command FIRE.
- To terminate the alert, the leader announces CEASE FIRE, END OF MISSION.

Doubtful Elements and Corrections

C-105. When gunners are in doubt about elements of fire commands, they reply, SAY AGAIN RANGE, TARGET. Leaders then announces THE COMMAND WAS, repeats the element in question, and continues with the fire command.

C-106. When leaders make an error in the initial fire command, they correct it by announcing CORRECTION, and gives the corrected element. When leaders make an error in the subsequent fire command, they may correct it by announcing CORRECTION. They then repeat the entire subsequent fire command.

Abbreviated Fire Commands

C-107. Fire commands do not need be complete to be effective. In combat, leaders give only the elements necessary to place fire on a target quickly and without confusion. During training, however, they should use all the elements to get gunners in the habit of thinking and reacting properly when a target is to be engaged. After the gunner's initial training in fire commands, the gunner should be taught to react to abbreviated fire commands, using one of the following methods.

Oral

C-108. When leaders want to place the fire of one machine gun on an enemy machine gun, they may want to use oral commands. For example, leaders can quickly tell the gunner to fire on an enemy gun.

Hand and Arm Signals

C-109. Battlefield noise and distance between the gunners and leaders often make it necessary to use hand and arm signals to control fire (see figure C-19 on page C-32). When an action or movement is to be executed by only one of the gunners, a preliminary signal is given to the gunner only. The following are commonly used signals for fire control:

- Ready. Gunners indicate they are ready to fire by yelling UP or having the assistant gunner raise the hand above the head toward the leader.
- Commence firing or change rate of firing. Leaders bring their hand (palm down) to the front of their body about waist level and moves it horizontally in front of the body. To signal an increase in the rate of fire, they increase the speed of the hand movement. To signal slower fire, they decrease the speed of the hand movement.
- Change direction or elevation. Leaders extend their arm and hand in the new direction and indicates the amount of change necessary by the number of fingers extended. The fingers must be spread so gunners can easily see the number of fingers extended. Each finger indicates 1 meter of change of the weapon. If the desired change is more than 5 meters, leaders extend their hand the number of times necessary to indicate the total amount of change. For example, right nine would be indicated by extending the hand once with five fingers showing and a second time with four fingers showing for a total of nine fingers.
- Interrupt or cease firing. Leaders raise their hand and arm (palm outward) in front of their forehead and brings it downward sharply.
- Other signals. Leaders can devise other signals to control their weapons. A detailed description of hand and arm signals is given in TC 3-21.60.

Figure C-19. Hand and arm signals

MACHINE GUN USE

C-110. Despite their post-Civil War development, modern machine guns did not exhibit their full potential in battle until World War I. Although the machine gun has changed, the role of the machine gun and machine gunner has not. The mission of machine guns in battle is to deliver fires when and where leaders want them in both the offense and defense. Machine guns rarely, if ever, have independent missions. Instead, they provide their unit with accurate, heavy fires to accomplish the mission.

TACTICAL ORGANIZATION OF THE MACHINE GUN

C-111. The accomplishment of the platoon's or squad's mission demands efficient and effective machine gun crews. Leaders consider the mission and organize machine guns to deliver firepower and direct fire support to area(s) or point(s) needed to accomplish the assigned mission.

C-112. Infantry rifle platoons have an organic weapons squad consisting of a weapons squad leader, two-Javelin teams (see appendix D), and two-medium machine gun teams.

(See chapter 1 for additional information on weapons squad.) Depending on the platoon's or squad's mission, there could be additional machine gun teams attached to the platoon or squad.

C-113. The medium machine gun team has a gunner, assistant gunner, and ammunition bearer. In some units the senior member of the gun team is the gunner. In other units the assistant gunner is the senior gun team member who also serves as the gun team leader.

C-114. Table C-6 illustrates equipment carried by the weapons squad. Table C-7 on page C-34 illustrates the duty positions within the weapons squad and gives possible duty descriptions and responsibilities. The table shows possible position and equipment use only. Individual unit SOPs and available equipment dictate the exact role each weapons squad member plays within the squad.

Note. See appendix D for information on the Javelin (and when replaced with the M3 Multi-role, Antiarmor, Antipersonnel Weapon System [known as MAAWS]) teams in the weapons squad.

Table C-6. Weapons squad equipment by position, example

	WSL	*AG/GTL*	*Gunner*	*Ammunition Bearer*
Weapon	M4-series (w/ 7 magazines*)	M4-series (w/ 7 magazines*)	M240-series (50 to 100 rounds)	M4-series (w/ 7 magazines)
Day Optic	M68 or M150 Optics	M68 or M150 Optics	M145 Machine Gun Optic	M68 Optics
Laser	PEQ-15	PEQ-15	PEQ-15	PEQ-15
Additional Equipment	3x magnifier**	3x magnifier** Spare barrels***	3x magnifier**	Tripod traversing and elevating mechanism
M240-Series Ammunition	100 rounds	300 rounds	100 rounds	300 rounds
Miscellaneous	Whistle pen gun flare** Other shift signals** VS-17 panel Binoculars****	M-17 pistol cleaning kit Binoculars****	M-17 pistol cleaning kit CLP for 72 hours*****	N/A
*WSL and AG/GTL load tracer rounds (4:1 mix) in magazines for marking targets.				
**3x magnifier, flares, and shift signals are readily accessible at all times.				
***Spare barrels marked by relative age with 1/4 pieces of green tape on carrying handle. Oldest barrel=2 parallel strips, second newest barrel=1 strip, and newest barrel=no tape				
****Binoculars carried in the assault pack or in suitable pouch on vest (mission dependent).				
*****Gunners always carry enough CLP for 72 hours of operations.				
Legend: AG–assistant gunner; CLP–cleaner, lubricant, preservative; GTL–gun team leader; N/A–not applicable; w/–with; WSL–weapons squad leader				

Table C-7. Weapons squad duty positions and responsibilities

Weapons Squad Responsibilities	
WSL	Senior squad leader within the platoon. Responsible for all training and employment of the medium machine guns. The WSL's knowledge, experience, and tactical proficiency influence the effectiveness of the squad. The WSL controls/adjusts rates of fire, assesses effectiveness of fires, monitors ammunition expenditures, cross-levels ammunition between gun teams, and knows and leads gun teams through crew drills.
AG/GTL	-AG/GTL is a team leader with the responsibilities of a fire team leader. -If second in the gun team's chain of command, the AG is always fully capable of taking the GTL position. -GTL is responsible for the team members and all the gun equipment. -GTL redistributes crew duties if short the ammunition bearer and prepares and submits DA Form 5517 (*Standard Range Card).* -GTL knows and lead gun team through crew drills. -TL and the team will be tactically proficient and knowledgeable on this ATP and applicable TCs and TMs applying to the medium machine gun. -GTL assists the WSL on the best way to employ the M240-series. GTL enforces field discipline while the gun team is employed. -GTL leads by example in all areas. Sets the example in all things. -GTL assists the WSL in all areas. Advises the WSL of problems either tactical or administrative. -AG is responsible for all action concerning the gun. -AG/GTL calls the ammunition bearer if ammunition is needed or actively seeks it out if the ammunition bearer is not available. Constantly updates the WSL on the round count and serviceability of the M240-series. -When the gun is firing, AG/GTL spots rounds and makes corrections to the gunner's fire. Also watches for friendly troops to the flanks of the target area or between the gun and target. -If the gunner is hit by fire, AG/GTL immediately assumes the role of the gunner. -AG/GTL is always prepared to change the gun's barrel (spare barrel is always out when the gun is firing). Ensures the hot barrel is not placed on live ammunition or directly on the ground when it comes out of the gun.
Machine Gunner	-Primary responsibility is to the gun. Focused on its cleanliness and proper function. Immediately reports abnormalities to the GTL or WSL. -If necessary for gunner to carry M240-series ammunition, carries it in on the back so the AG/GTL can access it without stopping the fire of the gun. -Always carries the necessary tools of the gun to be properly cleaned, along with a sufficient amount of oil for the gun's proper function.
Ammunition Bearer	-The ammunition bearer is the RFLM/equipment bearer of the gun team. -Normally the newest member of the gun team. Ammunition bearers must quickly learn everything they can, exert maximum effort at all times, and attempt to outdo the gun team members in every situation. -Follows the gunner without hesitation. During movement moves to the right side of the gunner and no more than one 3- to 5-meter rush away from the gun. -During firing, pulls rear security and if the gunner comes under enemy fire, provides immediate suppression while the gun moves into new position. -Responsible for the tripod and traversing and elevating mechanism. -They always must be clean and ready for combat. Responsible for replacing them, if necessary.
Legend: AG–assistant gunner; ATP–Army techniques publication; DA–Department of Army; GTL–gun team leader; TC–training circular; TL–team leader; TM–training manual; WSL–weapons squad leader	

SECURITY

C-115. Security includes all command measures to protect against surprise, observation, and annoyance by the enemy. The gun team is responsible for its immediate local security, specifically provided by the assistant gunner and/or ammunition bearer for close in local security to the gunner, who is fixated on deeper targets. Though the principal unit security measures against ground forces include employment of observation posts, security patrols, and detachments covering the front flanks and rear of the unit's most valuable weapons systems and vulnerable areas. The composition and strength of these detachments depends on the size of the main body, its mission, and nature of the opposition expected. The presence of machine guns with security detachments augments their firepower to delay, attack, and defend, by virtue of inherent firepower.

C-116. The potential of air and ground attacks on the unit demands every possible precaution for maximum security while on the move. Where this situation exists, the machine gun crew must be thoroughly trained in the hasty delivery of antiaircraft fire and of counterfire against enemy ground forces. The distribution of the medium machine guns in the formation is critical. The medium machine gun crew is constantly on the alert, particularly at halts, ready to deliver fire as soon as possible. If leaders expect a halt to exceed a brief period, they carefully choose medium machine gun positions to avoid unduly tiring the medium machine gun crew. If they expect the halt to extend for a long period, they can have the medium machine gun crew take up positions in support of the unit. The crew covers the direction from which they expect enemy activity as well as the direction from which the unit came. Leaders select positions permitting the delivery of fire in the most probable direction of enemy attack, such as valleys, draws, ridges, and spurs. They choose positions offering obstructed fire from potential enemy locations.

EMPLOYMENT OF FIRE AND MOVEMENT

C-117. The employment of fire and movement is essential and greatly depends upon the other during maneuver. Without the support of covering fires, maneuvering in the presence of enemy fire can result in disastrous losses. Covering fires, especially providing fire superiority, allow maneuvering in the offense. However, fire superiority alone rarely wins battles. The primary objective of the offense is to advance, occupy, and hold the enemy position.

Machine Gun as a Base of Fire

C-118. Machine gun fire from a support by fire position must be the minimum possible to keep the enemy from returning fire. Ammunition must be conserved so the guns do not run out of ammunition.

C-119. The weapons squad leader positions and controls the fires of all medium machine guns in the element. Machine gun targets include essential enemy weapons or groups of enemy targets either on the objective or attempting to reinforce or counterattack. In terms of engagement ranges, medium machine guns in the base-of-fire element may find themselves firing at targets within a range of 800 meters. The nature of the terrain, desire to achieve some standoff, and the mission variables of METT-TC (I) mission, enemy, terrain and weather, troops and support available, time available, civil considerations, and informational considerations prompt the leader to the correct tactical positioning of the base-of-fire element.

C-120. The medium machine gun quickly delivers an accurate, high-volume rate of lethal fire on large areas. When accurately placed on the enemy position, medium machine gun fires can achieve fire superiority for duration of the firing. Troops advancing in the attack should take full advantage of this period to maneuver to a favorable position from where they can facilitate the last push against the enemy. In addition to creating enemy casualties, medium machine gun fire destroys the enemy's confidence and neutralizes their ability to engage the friendly maneuver element.

C-121. All vocal commands from the leaders to change the rates of fire are accompanied simultaneously by hand and arm signals. There are distinct phases of rates of fire employed by the base-of-fire element:

- Initial heavy volume (rapid rate) to gain fire superiority.
- Slower rate to conserve ammunition (sustained rate) while still preventing return fire as the assault moves forward.
- Increased rate as the assault nears the objective.
- Shift fires (commonly called walking fires) forward of the assault line.
- Lift and/or shift to other targets or targets of opportunity.

C-122. Machine guns in the support by fire role should be set in and assigned a primary and secondary sector of fire as well as a primary and alternate position. They are suppressive fire weapons used to suppress known and suspected enemy positions. Therefore, gunners cannot be allowed to empty all their ammunition into one bunker simply because it is all they can identify at the time.

C-123. The support by fire position, not the assault element, is responsible for ensuring there is no masking of fires. The assault element might have to mask the support by fire line because it has no choice on how to move. It is the support by fire gunner's job to shift fires continually or move gun teams or the weapons squad to support the assault and prevent masking.

C-124. Shift and shut down the weapons squad gun teams one at a time, not all at once. M320 and mortar or other indirect fire can be used to suppress while the medium machine guns are moved to where they can fire.

C-125. Leaders must consider the surface danger zones of the machine guns when planning and executing the lift and/or shift of the support by fire guns. The effectiveness of the enemy on the objective will play a large role in how much risk should be taken with respect to the lifting or shifting of fires. Once the support by fire line is masked by the assault element, fires are shifted and/or lifted to prevent enemy withdrawal or reinforcement.

Machine Gun with the Maneuver Element

C-126. The medium machine guns seldom accompany the maneuver element. The gun's primary mission is to provide covering fire. The medium machine guns only are employed with the maneuver element when the AO assigned to the assault, platoon, squad, or company is too narrow to permit proper control of the guns. The medium machine guns then are moved with the unit and readied to employ on order from the leader and in the direction needing the supporting fire.

C-127. When medium machine guns move with the element undertaking the assault, the maneuver element brings the medium machine guns to provide additional firepower. These weapons are fired from a bipod, in an assault mode, from the hip, or from the underarm position. They target enemy automatic weapons anywhere on the unit's objective. Once the enemy's automatic weapons have been destroyed (if any), the gunners distribute their fire over their assigned sector. In terms of engagement ranges, the medium machine gun in the assault engages within 300 meters of its target and frequently at point-blank ranges.

C-128. Where the AO is too wide to allow proper coverage by the platoon's organic medium machine guns, the platoon or squads can be assigned additional medium machine guns or personnel from within the company. This may permit the platoon or squads to accomplish its assigned mission. The medium machine guns are assigned a sector to cover and move with the maneuver element.

Machine Guns in the Offense

C-129. In the offense, the platoon leaders have the option to establish their base-of-fire element with one or two medium machine guns, the M249 light machine gun, the M320 weapon system, or a combination of the weapons. The platoon sergeant or weapons squad leader may position this element and control its fires when the platoon scheme of maneuver is to conduct the assault with the Infantry squads.

M249 Light Machine Gun in the Offense

C-130. In the offense, M249s target enemy-supporting weapons and can be fired from fixed positions anywhere on the squad's objective. When the enemy's supporting weapons have been destroyed, or if there are none, the machine gunners distribute their fire over the portion of the objective corresponding to their team's position.

M240-Series Medium Machine Guns in the Offense

C-131. In the offense, the M240-series machine gun, when placed on a tripod, provides stability and accuracy at greater ranges than the bipod, but it takes more time to maneuver the machine gun should the need arise. The machine gunners target essential enemy weapons until the assault element masks their fires. They also can be used to suppress the enemy's ability to return accurate fire, or to hamper the maneuver of the enemy's assault element. They fix the enemy in position and isolate by cutting off their avenues of reinforcement. They then shift their fires to the flank opposite the one being assaulted and continue to target automatic weapons providing enemy support and engage enemy counterattack. M240-series fires also can be used to cover the gap created between the forward element of the friendly assaulting force and terrain covered by indirect fires when the indirect fires are lifted and shifted. On signal, the machine gunners and base-of-fire element displace to join the assault element on the objective.

MACHINE GUNS IN THE DEFENSE

C-132. The platoon's defense centers on its machine guns. The medium machine gun is the base for the defense, they are planned first, around a PDF or FPL, and then other planned targets within the primary sector of fire. They engage targets in secondary sector of fire when not firing the primary sector, and, when necessary, transition to the FPL. Machine gun teams are also prepared to fire in support of the unit's counterattack. Platoon leaders site the rifle squad to protect the machine guns against the assault of a dismounted enemy formation. The machine gun provides the necessary range and volume of fire to cover the squad's front in the defense.

C-133. The primary requirement of a suitable machine gun position in the defense is its effectiveness in accomplishing specific missions. The position should be accessible and afford cover and concealment. Machine guns are sited to protect the front, flanks, and rear of occupied portions of defensive positions, and to be mutually supporting. Attacking troops usually seek easily traveled ground providing cover from fire. Every machine gun should have three positions: primary, alternate, and supplementary. Each of these positions should be chosen by leaders to ensure their sector is covered and machine guns are protected on their flanks.

C-134. Leaders site the machine gun to cover the entire sector or to overlap sectors with the other machine guns. The engagement range may extend from more than 1,000 meters where the enemy begins the assault to point-blank range. Machine gun targets include enemy automatic weapons and command and control elements.

C-135. Machine gun fire is distributed in width and depth in a defensive position. Leader can use machine guns to subject the enemy to increasingly devastating fire from the initial phases of the attack, and to neutralize partial successes the enemy might attain by delivering intense fires in support of counterattacks. The machine gun's tremendous firepower enables the unit to hold ground. This is what makes it the backbone or framework of the defense.

M249 Light Machine Gun in the Defense

C-136. In the defense, the M249 adds increased firepower without the addition of manpower. Characteristically, M249s are light, fire rapidly, and have more ammunition than the rifles in the squad they support. Under certain circumstances, the platoon leader may designate the M249 machine gun as a platoon crew-served weapon.

M240-Series Medium Machine Guns in the Defense

C-137. In the defense, the medium machine gun provides sustained direct fires covering the most likely or most dangerous enemy dismounted avenues of approach. It protects friendly units against the enemy's dismounted close assault. Platoon leaders position their machine guns to concentrate fires in locations where they want to inflict the most damage to the enemy. Leaders also place them where they can take advantage of grazing enfilade fires, stand-off or maximum engagement range, and best observation of the target area. Machine guns provide overlapping and interlocking fires with adjacent units and cover tactical and protective obstacles with traversing or searching fires. When FPFs are called for, machine guns (aided by M249 fires) place a barrier of fixed, direct fire across the platoon or squad front. Leaders position machine guns to—

- Concentrate fires where they want to kill the enemy.
- Fire across the platoon and squad front.
- Cover obstacles by direct fire.
- Tie in with adjacent units.

AMMUNITION PLANNING

C-138. Leaders must carefully plan the rates of fire to be employed by machine guns as they relate to the mission and amount of ammunition available. The weapons squad leader must understand fully the mission the amount of available ammunition and application of machine gun fire needed to support fully all vital events of the mission. Planning ensures the guns do not run out of ammunition.

C-139. A mounted platoon or squad might have access to enough machine gun ammunition to support the guns throughout its operation. A dismounted platoon or squad with limited resupply capabilities must plan for only the basic load to be available. In either case, leaders must consider vital events the guns must support during the mission. They must plan the rate of machine gun fire needed to support the vital events, and amount of ammunition needed for scheduled rates of fire.

C-140. Leaders must make an estimate of the total amount of ammunition needed to support all the machine guns. They then must adjust the amount of ammunition used for each event to ensure enough ammunition is available for all phases of the operation. Because medium machine guns may see different parts of the battlefield, their expenditure rates will vary. The weapons squad leader must monitor ammunition consumption to ensure rates are not too high, and cross-level ammunition between gun teams as necessary. Examples of planning rates of fire and ammunition requirements for a platoon or weapons squad's machine guns in the attack follow.

Know Rates of Fire

C-141. Leaders and gunners must know how much ammunition is required to support the different rates of fire each platoon or weapons squad machine gun and assault weapon will require. Coupling this knowledge with an accurate estimate of the length of time and rates of fire their guns are scheduled to fire will ensure enough ammunition resources to cover the entire mission. As part of an example of the planning needed to use M240-series in support by fire roles, the rates of fire of the M240-series are listed in table C-8 on page C-40.

Table C-8. M240-series rates of fire

Rates of Fire	
Sustained	• 100 rounds per minute • Fired in 6- to 9-round bursts • 4 to 5 seconds between bursts (barrel change every 10 minutes)
Rapid	• 200 rounds per minute • Fired in 10- to 12-round bursts • 2 to 3 seconds between bursts (barrel change every 2 minutes)
Cyclic	• 650 to 950 rounds per minute • Continuous burst (barrel change every minute)

Ammunition Requirement

C-142. Leaders must calculate the number of rounds needed to support every machine gun throughout all phases of the operation. Ammunition must be allocated for each vital event and to support movement with suppressive fires. Additional amounts of ammunition, when required, are carried/distributed (for example, every Soldier may be required to carry an extra 100 rounds for the guns) across the platoon or pre-stocked in the defense.

Appendix D

Shoulder-Launched Munitions and Close Combat Missile System

The Infantry rifle squad and weapons squad employ shoulder-launched munitions (known as SLMs), the Javelin close combat missile system (known as CCMS), and the M3 Multi-role, Antiarmor, Antipersonnel Weapon System (known as MAAWS) to destroy enemy personnel, field fortifications, and to disable enemy vehicles at ranges from 15 to 2,000 meters. They can engage targets in the assault, support by fire, and defensive roles, and are the Infantry platoon's highest casualty producing organic weapons when used against armored enemy vehicles.

Note. Leaders must employ SLMs (see TM 3-23.25 for safety considerations), the Javelin CCMS, medium (see TC 3-22.37 for safety considerations), and the M3 MAAWS (see TC 3-22.84 for safety considerations) to minimize danger to friendly Soldiers caused by the surface danger zones or backblast danger zones. They must weigh the risk of firing the missile in close proximity to friendly assault forces against the need to suppress or destroy enemy fortifications or vehicles from support by fire or attack by fire positions.

SHOULDER-LAUNCHED MUNITIONS

D-1. SLMs are used against field fortifications, enemy vehicles, or other similar enemy targets. Infantry Soldiers are issued SLMs as rounds of ammunition in addition to their assigned weapons. SLMs include the M136A1 AT4-CS, M136 AT4; the M72A2/A3 light antitank weapon (known as LAW), improved M72A4/5/6/7 LAW; and M141 bunker defeat munitions (known as BDMs). The M141 BDM, also referred to as the shoulder-launched multipurpose assault weapon disposable. Table D-1 on page D-2 lists select SLM specifications. (See TM 3-23.25 for additional information.)

D-2. All SLMs are lightweight, self-contained, single-shot, disposable weapons consisting of unguided free flight, fin-stabilized, rocket-type cartridges packed in expendable, telescoping launchers (except the M136 AT4/AT4-CS which does not telescope) also serving as storage containers. The only requirement for their care is a visual inspection. SLMs can withstand extreme weather and environmental conditions, including arctic, tropical, and desert climates.

D-3. SLMs increase the lethality and survivability of the Infantryman and provide a direct fire capability to defeat enemy personnel within armored platforms. BDM provides the Soldier a direct fire capability to defeat enemy personnel located within field fortifications, bunkers, caves, masonry structures, and lightly armed vehicles and to suppress enemy personnel in lightly armored vehicles.

D-4. The individual Soldier will use SLM to engage threat combatants at close ranges, across the street or from one building to another. The Soldier may employ SLM as a member of a support by fire element to incapacitate threat forces threatening the assault element. When the assault element clears a building, the leader may reposition the SLM gunner inside to engage a potential counterattack force.

Note. Several numbers in table D-1 are rounded off and might not represent exact numbers.

Table D-1. Shoulder-launched munitions

Shoulder-Launched Munition	*M136 AT4 DODIC C995*	*M136A1 (AT4-CS) DODIC HA35*	*M72A2/A3 LAW DODIC H557*
Technical Manual	TM 3-23.25	TM 3-23.25	TM 3-23.25
Carry Weight	15.0 lbs.	17.0 lbs.	5.0 lbs.
Length: **Carry Extended:**	40 inches N/A	41 inches N/A	25 inches 35 inches
Caliber	84-mm	84-mm	66-mm
Muzzle Velocity	290 m/s 950 f/s	225 m/s 738 f/s	144.8 m/s 475 f/s
Operating Temperature	-40° to 60° C -40° to 140° F	-40° to 60° C -40° to 140° F	-40° to 60° C -40° to 140° F
Maximum Effective Range	300 m	300 m	Stationary – 200 m Moving – 165 m
Maximum Range	2,100 m	2,100 m	1,000 m
Minimum Arming Range	10 m	10 m	10 m
Legend: °–degree; AT–antitank; C–Celsius; CS–confined space; DODIC–Department of Defense identification code; F–Fahrenheit; f/s–feet per second; LAW–light antitank weapon; lbs.–pounds; N/A–not applicable; m–meters; mm–millimeters; m/s–meters per second; TM–technical manual			

Table D-1. Shoulder-launched munitions (continued)

Shoulder- Launched Munition	*M72A4/5/6/7 Improved LAW DODIC HA29*	*M141 BDM DODIC HA08*
Technical Manual	TM 3-23.25	TM 3-23.25
Carry Weight	8.0 lbs.	16.0 lbs.
Length: Carry Extended:	31 inches 39 inches	32 inches 55 inches
Caliber	60-mm	83-mm
Muzzle Velocity	200 m/s 656 f/s	217 m/s 712 f/s
Operating Temperature	-40° to 60° C -40° to 140° F	-32° to 49° C -20° to 120° F
Maximum Effective Range	200 m	500 m
Maximum Range	1,400 m	2,000 m
Minimum Arming Range	25 m	15 m

Legend: °–degree; BDM–bunker defeat munitions; C–Celsius; DODIC–Department of Defense identification code; F–Fahrenheit; f/s–feet per second, LAW–light antitank weapon; lbs.–pounds; m–meters; mm–millimeter; m/s–meters per second; TM–technical manual

M136 AT4/M136A1 AT4-CS

D-5. The M136 AT4 is a lightweight, self-contained, SLM designed for use against the improved armor of light-armored vehicles. It provides lethal fire against light-armored vehicles and has some effect on most enemy field fortifications.

D-6. The M136A1 AT4-CS is similar to the M136 AT4 but uses a different propulsion system. This system enables the M136A1 AT4-CS to be fired from an enclosure.

D-7. The M136 AT4 and M136A1 AT4-CS is a round of ammunition with an integral, rocket-type cartridge. The cartridge consists of a fin assembly with tracer element, a point detonating fuze, and a high explosive antitank (known as HEAT) warhead. (See figures D-1 and D-2 on page D-4.)

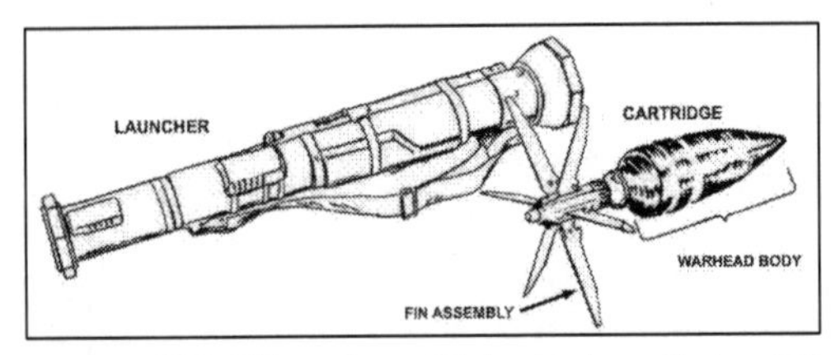

Figure D-1. M136 AT4 launcher and high explosive antitank cartridge

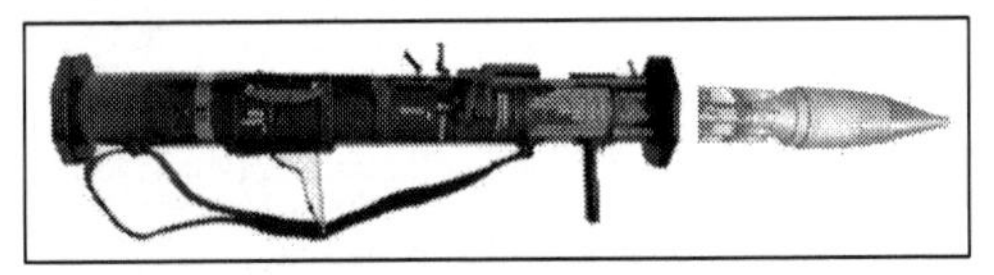

Figure D-2. M136A1 AT4-CS launcher and high explosive antitank cartridge

M72-Series Light Antitank Weapon

D-8. The M72-series (M72A3 and M72A7) are lightweight and self-contained SLMs consisting of a rocket packed in a launcher. They are man-portable and may be fired from either shoulder. The launcher, which consists of two tubes, one inside the other, serves as a watertight packing container of the rocket and houses a percussion-type firing mechanism activating the rocket.

D-9. The M72A3 LAW contains a nonadjustable propelling charge and a 66-mm rocket. Every M72A3 has an integral HEAT warhead in the rocket's head (or body) section. (See figures D-3 and D-4.) Although the M72A3 mainly is employed as an antiarmor weapon, it may be used with limited success against secondary targets such as gun emplacements, pillboxes, buildings, or light vehicles.

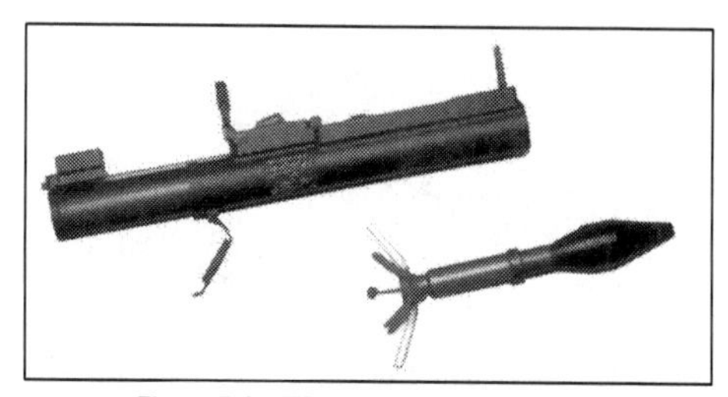

Figure D-3. M72A3 light antitank weapon

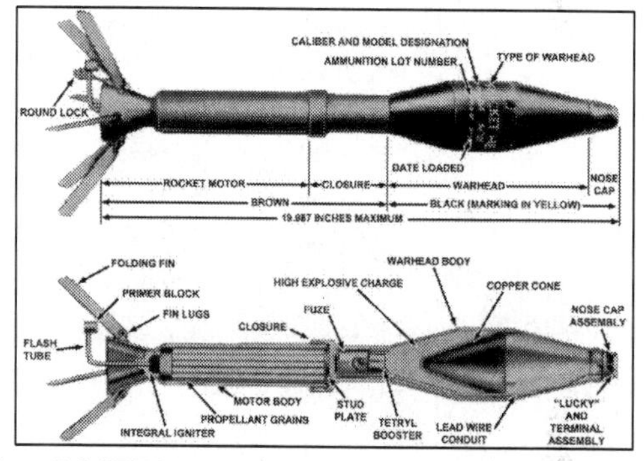

Figure D-4. M72A3 light antitank weapon 66-millimeter high explosive antiarmor rocket

D-10. The M72A7 is the improved LAW currently employed by the platoon. It is a compact, lightweight, single-shot, disposable weapon optimized to defeat lightly armored vehicles at close combat ranges. (See figure D-5 on page D-6.) The M72A7 offers enhanced capabilities beyond the original M72-series. The improved M72 consists of a 66-mm unguided rocket prepackaged at the factory in a telescoping, throwaway launcher. The system performance improvements include a higher velocity rocket motor extending the weapon effective range, increased lethality warhead, lower and more consistent trigger release force, rifle-type sight system, and better overall system reliability and safety. The weapon contains a 66-mm rocket and an integral

HEAT warhead. The warhead is designed to penetrate 150-mm (5.9 inches) of homogenous armor and is optimized for maximum fragmentation behind light armor, Infantry fighting vehicles, and urban walls.

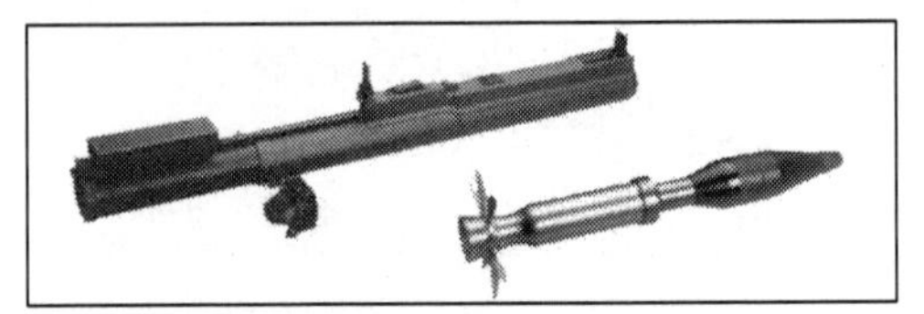

Figure D-5. Improved M72A7 light antitank weapon with rocket

M141 Bunker Defeat Munitions

D-11. The M141 BDM was developed to defeat enemy bunkers and field fortifications. (See figure D-6.) The M141 BDM is a disposable, lightweight, self-contained, man-portable, shoulder fired, HE multipurpose munition.

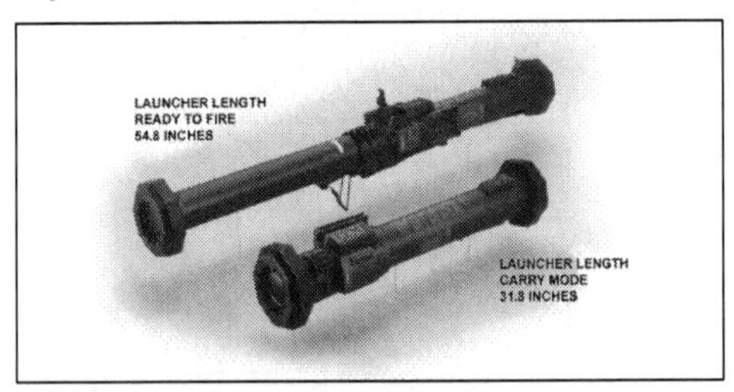

Figure D-6. M141 bunker defeat munitions

D-12. The M141 BDM utilizes the 83-mm HEDP assault rocket. (See figure D-7.) The 83-mm HEDP assault rocket warhead consists of a dual mode fuze and 2.38 pounds of A-3 explosive.

D-13. Warhead function, in quick or delay mode, is determined automatically by the fuze when the rocket impacts a target. The M141 BDM is fired at hard or soft targets without selection steps required by the gunner. This automatic feature assures the kill mechanism is employed. Warhead detonation is instantaneous when impacting a hard target, such as a brick or concrete wall or an armored vehicle. Impact with a softer target, such as a sandbagged bunker, results in a fuze time delay permitting the rocket to penetrate the target before warhead detonation.

D-14. The M141 BDM can destroy bunkers but is not optimized to kill the enemy soldiers within masonry structures in urban terrain or armored vehicles. The M141 BDM can penetrate masonry walls, but multiple rounds may be necessary to deliver sufficient lethality against enemy personnel behind the walls. The M141 BDM has been used with great success in destroying personnel and equipment in enemy bunkers, field fortifications, and caves in recent operations.

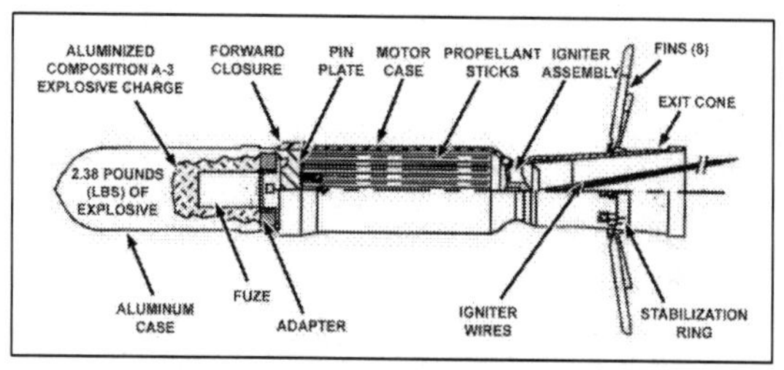

Figure D-7. M141 bunker defeat munition high explosive dual purpose assault rocket

JAVELIN-CLOSE COMBAT MISSILE SYSTEM

D-15. The Javelin is used primarily to defeat main battle tanks and other armored combat vehicles. (In the current force, the CCMS category of weapons includes the Javelin and tube launched, optically tracked, wire guided. The tube launched, optically tracked, wire guided CCMS, heavy is organic to the Infantry weapons company of the Infantry battalion [see ATP 3-21.20 for information on employment].) The Javelin provides overmatch antitank (AT) fires during the assault and provide extended range capability for engaging armor during both offensive and defensive missions. The system has a moderate capability against bunkers, buildings, and other fortified targets commonly found during combat in urban areas.

FIRE-AND-FORGET CAPABILITY

D-16. The Javelin is a fire-and-forget, shoulder-fired man-portable CCMS consisting of a reusable M98A1 (Block 0) and the improved M98A2 (Block 1), command launch unit (known as CLU) and a round. (See figure D-8 on page D-8.) The CLU houses the daysight, night vision sight, controls, and indicators. The round consists of the missile, the launch tube assembly, and battery coolant unit. The launch tube assembly serves as the launch platform and carrying container of the missile. (See TC 3-22.37 for additional information.)

D-17. The Javelin CCMS's primary role is to destroy enemy armored vehicles out to 2,000 meters with the M98A1 and 2,500 meters with the M98A2. The Javelin can be employed in a secondary role of providing an attack by fire capability against point targets such as bunkers and crew-served weapons positions. In addition, the Javelin CLU can be used alone as an aided vision device for reconnaissance, security operations, and surveillance. The Javelin gunner should be able to engage up to three targets in 2 minutes.

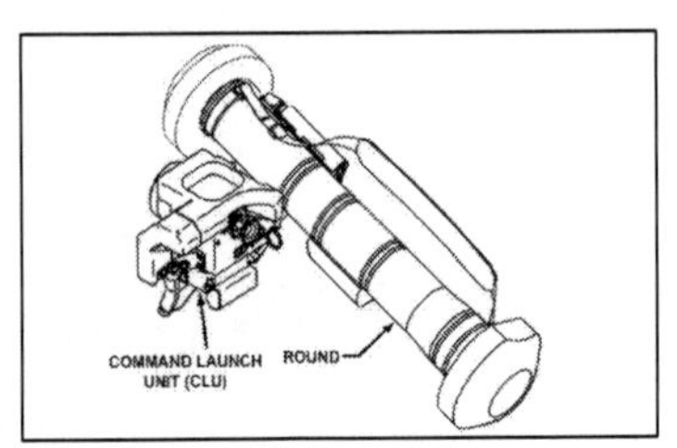

Figure D-8. Javelin close combat missile system

COMMAND LAUNCH UNIT

D-18. The M98A1 and M98A2 CLU is the reusable portion of the Javelin system. It contains the controls and indicators. The CLU provides increased utility to the Infantry platoon by allowing accurate surveillance out to 2.5 plus kilometers in both day and night. CLUs have been used to spot and destroy enemy snipers in hidden positions more than 1,000 meters away.

D-19. Tables D-2 through D-4 (on pages D-9 to D-10) list the Javelin's capabilities. In addition, it features the physical characteristics of the CLU, and physical characteristics of the round.

Table D-2. Javelin capabilities and features

Javelin Missile System	
Type of System	Surface attack guided missile and M98A1 and M98A2 CLU fire and forget
Crew	One- to three-Soldier teams based on TO&E
Missile modes	Top attack (default)
	Direct attack
Ranges	Top attack mode minimum effective engagement: 150 meters
	Direct attack mode minimum effective engagement range: 65 meters
	Maximum effective engagement range (direct attack and top attack modes): 2,000 meters for the M98A1 and 2,500+ meters for the M98A2 with the fire-and-forget guided missile (FGM)-148 round
Flight Time	About 14 seconds at 2,000 meters with the M98A1 and 2,500+ meters with the M98A2
Backblast Area	Primary danger zone extends out 25 meters at a 60-degree (cone-shaped) angle
	Caution zone extends the cone-shaped area out to 100 meters
Firing from Inside Enclosures	Minimum room length: 15 feet
	Minimum room width: 12 feet
	Minimum room height: 7 feet
Legend: CLU–command launch unit; TO&E–table of organization and equipment	

Table D-3. Physical characteristics of the command launch unit

Command Launch Unit		
M98A1/M98A2 Command Launch Unit	With battery, carrying bag, and cleaning kit	
	Weight: 14.16 lbs. (6.42 kg)/14.99 lbs. (6.80 kg)	
	Length: 13.71 in (34.82 cm)/19.29 in (49.00 cm)	
	Height: 13.34 in (33.88 cm)/13.00 in (33.02 cm)	
	Width: 19.65 in (49.91 cm)/1,650 in (41.91 cm)	
Sights	Day sight	
	Magnification: 4X	
	Field-of-view (FOV): 4.80° x 6.40° /6.4°x 4.8°	
	Night Vision Sight	
	Wide field-of-view (WFOV) magnification: 4.2X	
	WFOV: 4.58° x 6.11°	
	Narrow field-of-view (NFOV) magnification: 9.2X/12X	
	NFOV: 2.00° x 3.00° /2° x 1.5° (approximately)	
Battery Type	Lithium Sulfur Dioxide (LiSO2) BA-5590/U (Nonchargeable)	
	Nickel metal hydride battery, BB-390A/U rechargeable (training use only)	
	Number required: 1	
	NSN: 6135-01-036-3495	
	Weight: 2.2 lbs. (1.00 kg)	
	Life:	4.0 hrs 120°F (49°C)
		3.0 hrs between 50°F to 120°F (10°C to 49°C)
		1.0 hrs between -20°F to 50°F (-49°C to 10°C)
		0.5 hrs above 120°F (49°C)
Legend: °–degree; C–Celsius; cm–centimeters; F–Fahrenheit; hrs–hours; lbs.–pounds; kg–kilograms; in–inches; NSN–national stock number		

Table D-4. Physical characteristics of the round

Javelin Missile Round	
Complete Round (Launch tube assembly with missile and BCU)	Weight: 35.14 pounds (lbs.) (15.97 kilograms [kg])
	Length: 47.60 inches (in) (120.90 centimeters [cm])
	Diameter with end caps: 11.75 in (29.85 cm)
	Inside diameter: 5.52 in (14.00 cm)
Battery Coolant Unit (BCU)	Weight: 2.91 lbs. (1.32 kg)
	Length: 8.16 in (20.73 cm)
	Width: 4.63 in (11.75 cm)
	Battery type: lithium, nonrechargeable
	Battery life: 4 minutes of BCU time
	Battery coolant gas: argon

CAPABILITIES AND LIMITATIONS

D-20. The Javelin provides the platoon leader with a unique antiarmor capability. However, the leader must also understand antiarmor limitations in order to employ the system effectively. (See table D-5.)

Table D-5. Javelin capabilities and limitations

	CAPABILITIES	*LIMITATIONS*
FIREPOWER	• Maximum effective range is 2,000 meters. • Fire-and-forget capability. Missile I2R system gives missile ability to guide itself to the target when launched by the gunner. • Two missile flight paths: • Top attack – impacts on top of target. • Direct attack – impacts on front, rear, or flank of target. • Gunner can fire up to three missiles within 2 minutes. • Dual-shaped charge warhead can defeat known enemy armor. • NVS sees little degradation of target image. • Countermeasures used by enemy are countered by the NVS filter.	• Command launch unit sight cannot discriminate targets past 2,000 meters. • NVS cool-down time is from 2.5 to 3.5 minutes. • Seeker's cool-down time is about 10 seconds. • BCU life, once activated, is only about 4 minutes. • FOV can be rendered useless during limited visibility conditions (rain, snow, sleet, fog, haze, obscurants, dust, and night). Visibility is limited by the following: • Day FOV relies on daylight to provide the gunner a suitable target image; limited visibility conditions may block sun. • NVS uses the infrared naturally emitted from objects. Infrared crossover is the time at both dawn and dusk that terrain and targets are close enough in temperature to cause targets to blend in with their surroundings. • Natural clutter occurs when the sun heats objects to a temperature close enough to surrounding terrain that it causes a target to blend in with terrain. • Artificial clutter occurs when there are man-made objects that emit large amounts of infrared (for example, burning vehicles). • Heavy fog reduces the capability of the gunner to detect and engage targets. • Flight path of missile is restricted in wooded, mountainous, and urban terrain. • Gunner must have LOS of the seeker to lock onto a target.

Table D-5. Javelin capabilities and limitations (continued)

	CAPABILITIES	*LIMITATIONS*
MOBILITY	• Man-portable. • Fire-and-forget capability allows gunner to shoot and move before missile impact. • Soft launch capability allows it to be fired from inside buildings and bunkers. • Maneuverable over short distances of the gunners.	• Weight of Javelin makes maneuvering slow over long distances. • The Javelin round is bulky and restricts movement in heavily-wooded or vegetative terrain.
PROTECTION	• Passive infrared targeting system used to acquire lock-on cannot be detected. • Launch motor produces a small signature. • Fire-and-forget feature allows gunner to take cover immediately after missile is launched.	• Gunners must partially expose themselves to engage the enemy. • CLU requires an LOS to acquire targets.
Legend: BCU–battery coolant unit; CLU–command launch unit; FOV–field of view; I2R–missile imaging infrared system; LOS–line of sight; NVS–night vision sight		

M3 MULTI-ROLE, ANTIARMOR, ANTIPERSONNEL WEAPON SYSTEM

D-21. The M3 MAAWS is an 84-mm breach loaded and laterally, percussion fired recoilless rifle. The barrel has an internal steel liner. The liner is made with a laminate of epoxy and carbon fiber. The weapon is recoilless as part of the propellant gases escape rearwards through the venturi, equalizing the recoil force. The normal sight of the weapon is the telescopic sight; with the open sights being an auxiliary means of aiming.

D-22. The M3 MAAWS consists of components, assemblies, subassemblies, and individual parts. Figure D-9 on page D-12 illustrates the major components of the M3 MAAWS. Soldiers must be familiar with the components, assemblies, subassemblies, and individual parts, considerations, and how they interact during operation:

- Components are a uniquely identifiable group of fitted parts, pieces, assemblies, or subassemblies that are required and necessary to perform a distinctive function in the operation of the weapon.
- Components are usually removable as one piece and are considered indivisible for a particular purpose or use.
- Assemblies are a group of subassemblies and parts that are fitted to perform a specific set of functions during operation and cannot be used independently for any other purpose.
- Subassemblies are a group of fitted parts that perform a specific set of functions during operation.
- Subassemblies are compartmentalized to complete a specific task.

- Subassemblies may be grouped with other assemblies, subassemblies, and parts to create a component.
- Parts are the individual items that perform a function when attached to a subassembly, assembly, or component that serves a specific purpose.

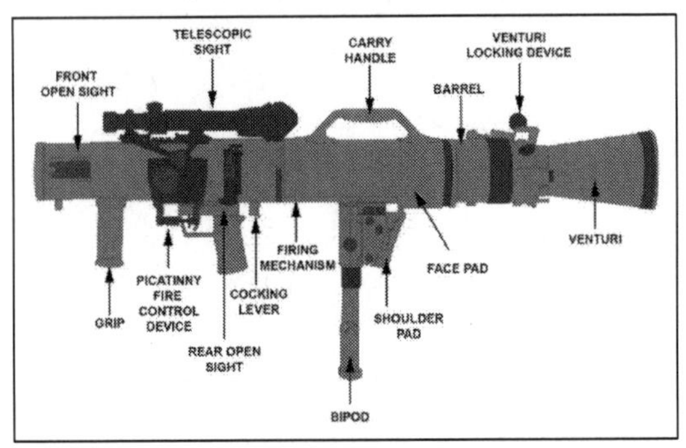

Figure D-9. Major components

D-23. The design of the M3 MAAWS allows it to engage lightly armored targets at ranges up to 700 meters, and soft-skinned vehicles and similar targets at ranges up to 1,300 meters. Gunners can fire the M3 MAAWS from standing, kneeling, sitting, or prone positions. A subcaliber adapter 553B is available for realistic live fire training with the M3 MAAWS. Table D-6 list the technical data of the M3 MAAWS.

Table D-6. Equipment technical data

M3 Multi-role, Antiarmor, Antipersonnel Weapon System	
Caliber	84 mm
Weight	20 lbs. (approximately 9 kgs)
Weight With Bipod and Telescopic Sight	22.6 lbs. (approximately 10 kgs)
Weight With ITWS and Storm	23.8 lbs. (approximately 11 kgs)
Weight of Telescopic Sight	2.0 lbs. (0.09 kgs)
Weight of Bipod	0.60 lbs. (0.25 kgs)
Length	41.93 lbs. (1065 kgs)
Weight of Transport Package with Weapon and Accessories	64 lbs. (29 kgs)
Dimensions of Transport Package	26.28 x 17.83 x 10.55 inches (1173 x 453 x 268 mm)
Practical Rate of Fire	Approximately 6 Rounds Per Minute
Legend: ITWS–integrated thermal weapon sight; kgs–kilograms; lbs.–pounds; mm–millimeter; STORM–small, tactical, optical rifle—mounted	

D-24. The primary role of the M3 MAAWS gunner is to engage a target with a well-aimed shot. Consistently hitting a target with precision is a complex interaction of factors immediately before, during, and after firing the round. The interactions include maintaining postural steadiness, establishing and maintaining the proper aim on the target, stabilizing the weapon while pressing the trigger, and adjusting for environmental and battlefield conditions. (See TC 3-22.84 for additional information.)

EMPLOYMENT CONSIDERATIONS

D-25. The platoon's objective is to concentrate combat power at the right time and place, by massing fires rather than by massing forces, and by presenting the enemy with multiple threats. A lethal mix of the SLMs and Javelin CCMS (or the M3 MAAWS) provide the Infantry rifle platoon with the flexibility to employ multiple systems designed to deliver maximum direct fire lethality and destroy enemy formations in close combat. At close combat ranges (15 to 300 meters), SLMs provide Soldiers with the ability to deliver direct fire lethality at close proximity to the enemy. At close combat extended ranges (maximum effective out to 2,000 meters), a mix of SLMs and Javelin provide the platoon with overwhelming combat overmatch. When the Javelin is replaced with the M3 MAAWS (dependent on the ammunition round selected) the platoon engages enemy personnel, lightly armored targets at ranges up to 700 meters, and soft-skinned vehicles and similar targets at ranges up to 1,300 meters. These weapons serve as vital components by applying overlapping and interlocking fires to achieve synergy and mutual support for maneuver.

URBAN OPERATIONS AND FIELD FORTIFICATIONS

D-26. Operations in complex terrain and urban environments alter the basic nature of close combat. History tells us engagements are more frequent and occur more rapidly when engagement ranges are close. Studies and historical analyses have shown only 5 percent of all targets are more than 100 meters away. About 90 percent of all targets are located 50 meters or less from the identifying Soldier. Few personnel targets will be visible beyond 50 meters. Engagements usually occur at 35 meters or less.

D-27. Soldiers employ SLM in the short, direct fire, close-quarter engagement range of close combat. Their use is preferable in urban areas where other direct fire (M1-series Abrams tank and M2-series Bradley fighting vehicle), indirect fire (artillery, mortars), Army aviation attack, and close air support (CAS) are incapable of operating due to risks of fratricide and collateral damage. In close combat, Soldiers employ SLM against a wide variety of targets. These include personnel armed with individual and crew-served weapons fighting from armored platforms (T-72); light armored personnel carriers and Infantry fighting vehicles (BMPs [Boyevaya Mashina Pekhotys], BRDMs [Boyevaya Razvedyvatelnaya Dozornaya Mashina], and BTRs [bronyetransportyor]); modified personnel/Infantry vehicles; lightly armed vehicles; and enemy in fortified positions, behind walls, inside caves and masonry buildings, and within earthen bunkers.

> ***Note.*** Firing the M136A1 AT4-CS from a confined area requires a room that is a minimum of 12 feet wide and 15 feet long and has a ceiling height of at least 7 feet. (See TM 3-23.25 for additional information on firing from a confined area.)

D-28. Javelin CCMS (or M3 MAAWS) teams provide overwatching AT fires during the attack of a built-up area. They are employed best in areas along major thoroughfares and in upper floors (Javelin only) of buildings or roofs to attain long-range fields of fire. Because the minimum engagement distance limits firing opportunities in the confines of densely built-up areas, the Javelin (or the M3 MAAWS) may not be the weapon of choice in the urban environment. Urban area hazards include fires caused by both friendly and enemy forces may cause target acquisition and lock-on problems, clutter on the battlefield may cause lock-on problems, and line of sight communications limited by structures. The Javelin's (or the M3 MAAWS's) unique flight path forces the gunner to think in three dimensions. Other urban environment hazards include overhead obstacles such as street signs, light poles, and wires, which could impede the missile's flight path.

> ***Notes.*** The Javelin has a soft launch system, allowing it to be fired safely from inside of a building, provided that the room is at least 7 feet high, 12 feet wide, and 15 feet deep. (See TC 3-22.37 for additional information of firing from enclosures.)
>
> Firing from fighting positions or enclosures is prohibited for the M3 MAAWS. (See TC 3-22.84 for additional information on ammunition firing restrictions.)

Shoulder-Launched Munitions Employment Against Field Fortifications and Buildings

D-29. The current inventory of the M136 AT4 and M141 BDM SLMs provide the platoon with the capability to incapacitate personnel within earth and timber bunkers, masonry buildings, and light armored vehicles. However, neither system (the M136 AT4 or M141 BDM) is fully capable of fire from enclosure. The M136A1 AT4-CS SLM can be fired safely from an enclosure to incapacitate personnel within earth and timber bunkers, masonry buildings, and light armored vehicles.

D-30. The M141 BDM was specifically designed to better enhance the destruction of field fortifications and buildings. (See table D-7.) The M141 BDM contains an HEDP round with a dual-mode fuze that automatically adjusts for the type of target on impact. For soft targets, such as sandbagged bunkers, the M141 BDM warhead automatically adjusts to delayed mode and hits the target with high kinetic energy. This energy propels the warhead through the barrier and into the fortification or building, where the fuze detonates the warhead, causing greater damage.

PLD	probable line of deployment
PPEP	personal protective equipment posture
RFL	restrictive fire line
RM	risk management
ROE	rules of engagement
RS	reduced sensitivity
RTO	radiotelephone operator
S-2	battalion or brigade intelligence staff officer
SALUTE	size, activity, location, unit, time, and equipment
SDM	squad-designated marksman
SITEMP	situation template
SLM	shoulder-launched munition
SOP	standard operating procedure
STP	Soldier training publication
TAA	tactical assembly area
TC	training circular
TCCC	tactical combat casualty care
TLP	troop leading procedures
TM	technical manual
TRP	target reference point
U.S.	United States
WARNORD	warning order
WCS	weapons control status
WP	white phosphorous

SECTION II – TERMS

actions on contact

A process to help leaders understand what is happening and to take action. (FM 3-90)

air-ground operations

The simultaneous or synchronized employment of ground forces with aviation maneuver and fires to seize, retain, and exploit the initiative. (FM 3-04)

air movement

Air transport of units, personnel, supplies, and equipment including airdrops and air landings. (JP 3-36)

JP	joint publication
JSLIST	joint service lightweight integrated suit technology
JTAC	joint terminal attack controller
LACE	liquid, ammunition, casualties, and equipment
LAW	light antitank weapon
LC	line of contact
LD	line of departure
LOA	limit of advance
LOGSTAT	logistics status
LRP	logistics release point
MAAWS	Multi-role Antiarmor, Antipersonnel Weapon System
MCOO	modified combined obstacle overlay
MCRP	Marine Corps reference publication
MCTP	Marine Corps training publication
MCWP	Marine Corps warfighting publication
MEDEVAC	medical evacuation
METT-TC (I)	mission, enemy, terrain and weather, troops and support available, time available, civil considerations, and informational considerations
mm	millimeter
MOPP	mission-oriented protective posture
MTF	medical treatment facility
NAI	named area of interest
NTTP	Navy tactics, techniques, and procedures
OAKOC	observation and fields of fire, avenues of approach, key terrain, obstacles, and cover and concealment
OHC	overhead cover
OPORD	operation order
ORP	objective rally point
PACE	primary, alternate, contingency, and emergency
PCC	precombat check
PCI	precombat inspection
PD	point of departure
PDF	principal direction of fire
PIR	priority intelligence requirement

withdraw

To disengage from an enemy force and move in a direction away from the enemy. (ADP 3-90)

zone

An operational area assigned to a unit in the offense that only has rear and lateral boundaries. (FM 3-0)

zone reconnaissance

A form of reconnaissance operation that involves a directed effort to obtain detailed information on all routes, obstacles, terrain, and enemy forces within a zone defined by boundaries. (FM 3-90)

This page intentionally left blank.

References

All websites accessed on 15 December 2023.

REQUIRED PUBLICATIONS

These documents must be available to intended users of this publication.

DOD Dictionary of Military and Associated Terms. September 2023.

FM 1-02.1. *Operational Terms*. 9 March 2021.

FM 1-02.2. *Military Symbols*. 18 May 2022.

RELATED PUBLICATIONS

These documents are cited in this publication.

JOINT PUBLICATIONS

Most joint publications are available online: https://www.jcs.mil/Doctrine.

Most Department of Defense directives are available online: https://www.esd.whs.mil/DD.

DODD 2310.01E. *The DOD Detainee Program*. 15 March 2022.

DODD 2311.01. *DOD Law of War Program*. 2 July 2020.

DODD 3115.09. *DOD Intelligence Interrogations, Detainee Debriefings, and Tactical Questioning*. 11 October 2012.

JP 1, Volume 1. *Joint Warfighting*. 27 August 2023.

JP 1, Volume 2. *The Joint Force*. 19 June 2020.

JP 2-0. *Joint Intelligence*. 26 May 2022.

JP 3-0. *Joint Campaigns and Operations*. 18 June 2022.

JP 3-01. *Countering Air and Missile Threats*. 6 February 2023.

JP 3-04. *Information in Joint Operations*. 14 September 2022.

JP 3-06. *Joint Urban Operations*. 20 November 2013.

JP 3-07.3. *Peace Operations*. 1 March 2018.

JP 3-09. *Joint Fire Support*. 10 April 2019.

JP 3-09.3. *Close Air Support*. 10 June 2019.

JP 3-10. *Joint Security Operations in Theater*. 25 July 2019.

JP 3-13.3. *Operations Security*. 6 January 2016.

JP 3-13.4. *Military Deception*. 14 February 2017.

JP 3-15. *Barriers, Obstacles, and Mines in Joint Operations*. 26 May 2022.

JP 3-18. *Joint Forcible Entry Operations*. 27 June 2018.

JP 3-31. *Joint Land Operations*. 3 October 2019.

JP 3-35. *Joint Deployment and Redeployment Operations*. 31 March 2022.

JP 3-36. *Joint Air Mobility and Sealift Operations*. 4 January 2021.

JP 3-50. *Personnel Recovery*. 14 August 2023.

JP 3-60. *Joint Targeting*. 28 September 2018.

JP 3-85. *Joint Electromagnetic Spectrum Operations*. 22 May 2020.

JP 4-09. *Distribution Operations*. 14 March 2019.

JP 5-0. *Joint Planning*. 1 December 2020.

ARMY PUBLICATIONS

Most Army doctrinal publications are available online: https://armypubs.army.mil. Collective tasks and drills are available online: https://rdl.train.army.mil.

ADP 1-01. *Doctrine Primer*. 31 July 2019.

ADP 3-0. *Operations*. 31 July 2019.

ADP 3-28. *Defense Support of Civil Authorities*. 31 July 2019.

ADP 3-37. *Protection*. 31 July 2019.

ADP 3-90. *Offense and Defense*. 31 July 2019.

ADP 4-0. *Sustainment*. 31 July 2019.

ADP 5-0. *The Operations Process*. 31 July 2019.

ADP 6-0. *Mission Command: Command and Control of Army Forces*. 31 July 2019.

ADP 6-22. *Army Leadership and the Profession*. 31 July 2019.

ATP 2-01.3. *Intelligence Preparation of the Battlefield*. 1 March 2019.

ATP 3-01.8. *Techniques for Combined Arms for Air Defense*. 29 July 2016.

ATP 3-01.60. *Counter-Rocket, Artillery, and Mortar Operations*. 10 May 2013.

ATP 3-01.81. *Counter-Unmanned Aircraft System (C-UAS)*. 11 August 2023.

ATP 3-06/MCTP 12-10B. *Urban Operations*. 21 July 2022.

ATP 3-06.1/MCRP 3-20.4/NTTP 3-01.04/AFTTP 3-2.29. *Multi-Service Tactics, Techniques, and Procedures for Aviation Urban Operations*. 22 February 2022.

ATP 3-09.32/MCRP 3-31.6/NTTP 3-09.2/AFTTP 3-2.6. *JFIRE. Multi-Service Tactics, Techniques, and Procedures for Joint Application of Firepower*. 29 November 2023.

ATP 3-09.42. *Fire Support for the Brigade Combat Team*. 1 March 2016.

ATP 3-11.32/MCWP 10-10E.8/NTTP 3-11.37/AFTTP 3-2.46. *Multi-Service Tactics, Techniques, and Procedures for Chemical, Biological, Radiological, and Nuclear Passive Defense*. 13 May 2016.

ATP 3-11.50. *Battlefield Obscuration*. 15 May 2014.

ATP 3-12.3. *Electromagnetic Warfare Techniques*. 30 January 2023.

ATP 3-21.10. *Infantry Rifle Company*. 14 May 2018.

ATP 3-21.18. *Foot Marches*. 13 April 2022.

ATP 3-21.20. *Infantry Battalion*. 28 December 2017.

ATP 3-21.50. *Infantry Small-Unit Mountain and Cold Weather Operations*. 27 August 2020.

ATP 3-21.51. *Subterranean Operations*. 1 November 2019.

ATP 3-21.90/MCTP 3-01D. *Tactical Employment of Mortars*. 9 October 2019.

ATP 3-28.1/MCRP 3-30.6/NTTP 3-57.2/AFTTP 3-2.67/CGTTP 3-57.1. *DSCA. Multi-Service Tactics, Techniques, and Procedures for Defense Support of Civil Authorities (DSCA)*. 11 February 2021.

ATP 3-34.80. *Geospatial Engineering*. 22 February 2017.

ATP 3-34.81/MCWP 3-17.4. *Engineer Reconnaissance*. 1 March 2016.

ATP 3-37.34/MCTP 3-34C. *Survivability Operations*. 16 April 2018.

ATP 3-60.1/MCRP 31.5/NTTP 3-60.1/AFTTP 3-2.3. *Multi-Service Tactics, Techniques, and Procedures for Dynamic Targeting*. 5 January 2022.

ATP 3-90.4/MCTP 3-34A (MCWP 3-17.8). *Combined Arms Mobility*. 10 June 2022.

ATP 3-90.8/MCTP 3-34B. *Combined Arms Countermobility*. 30 November 2021.

ATP 3-90.97. *Mountain Warfare and Cold Weather Operations*. 29 April 2016.

ATP 3-90.98/MCTP 12-10C. *Jungle Operations*. 24 September 2020.

ATP 3-90.99/MCTP 12-10D. *Desert Operations*. 7 April 2021.

ATP 3-94.4. *Reconstitution Operations*. 5 May 2021.

ATP 4-02.2. *Medical Evacuation*.12 July 2019.

ATP 4-02.4. *Medical Platoon*. 12 May 2021.

ATP 4-02.13. *Casualty Operations*. 30 June 2021.

ATP 4-14. *Expeditionary Railway Center Operations*. 22 June 2022.

ATP 4-25.12. *Unit Field Sanitation Teams*. 30 April 2014.

ATP 4-90. *Brigade Support Battalion*. 18 June 2020.

ATP 5-19. *Risk Management*. 9 November 2021.

ATP 6-02.53. *Techniques for Tactical Radio Operations*. 13 February 2020.

ATTP 3-06.11. *Combined Arms Operations in Urban Terrain*. 10 June 2011.

Collective Task 07-SQD-9033. *Evacuate Casualties*. 11 April 2022.

Collective Task 07-PLT-9033. *Evacuate Casualties*. 11 April 2022.

DA Pam 385-63. *Range Safety*. 16 April 2014.

Drill 03-PLT-D0071. *React to a Chemical Agent Attack – Platoon*. 28 January 2022.

Drill 07-PLT-D8015. *React to Aircraft While Dismounted – Platoon.* 23 July 2023.

Drill 07-PLT-D9412. *Breach of a Mined Wire Obstacle – Platoon.* 30 August 2022.

Drill 07-PLT-D9483. *React to Nuclear Attack – Platoon.* 11 March 2021.

Drill 07-PLT-D9501. *React to Direct Fire Contact While Dismounted – Platoon.* 26 July 2022.

Drill 07-PLT-D9502. *React to Ambush (Dismounted) – Platoon.* 26 July 2022.

Drill 07-PLT-D9504. *React to Indirect Fire While Dismounted – Platoon.* 26 July 2022.

Drill 07-PLT-D9505. *Break Contact – Platoon.* 26 July 2022.

Drill 07-PLT-D9508. *Establish Security at the Halt – Platoon.* 11 July 2022.

Drill 07-PLT-D9510. *Enter a Trench to Secure a Foothold – Platoon.* 30 August 2022.

Drill 07-PLT-D9511. *Conduct a Hasty Attack – Platoon.* 11 July 2022.

Drill 07-PLT-D9514. *Conduct a Platoon Assault.* 30 August 2022.

Drill 07-SQD-D9406. *Knock Out a Bunker – Squad.* 12 August 2022.

Drill 07-SQD-D9501. *React to Direct Fire Contact While Dismounted – Squad.* 26 July 2022.

Drill 07-SQD-D9502. *React to Ambush (Dismounted) – Squad.* 26 July 2022.

Drill 07-SQD-D9504. *React to Indirect Fire While Dismounted – Squad.* 26 July 2022.

Drill 07-SQD-D9505. *Break Contact – Squad.* 26 July 2022.

Drill 07-SQD-D9506. *Dismount a Vehicle Under Direct Fire – Squad.* 12 August 2022.

Drill 07-SQD-D9509. *Enter and Clear a Room – Squad.* 11 July 2022.

Drill 07-SQD-D9515. *Conduct a Squad Assault.* 11 July 2022.

FM 3-0. *Operations.* 1 October 2022.

FM 3-04. *Army Aviation.* 6 April 2020.

FM 3-09. *Fire Support and Field Artillery Operations.* 30 April 2020.

FM 3-11. *Chemical, Biological, Radiological, and Nuclear Operations.* 23 May 2019.

FM 3-13. *Information Operations.* 6 December 2016.

FM 3-24.2. *Tactics in Counterinsurgency.* 21 April 2009.

FM 3-55. *Information Collection.* 3 May 2013.

FM 3-63. *Detainee Operations.* 2 January 2020.

FM 3-90. *Tactics.* 1 May 2023.

FM 3-96. *Brigade Combat Team.* 19 January 2021.

FM 3-98. *Reconnaissance and Security Operations.* 10 January 2023.

FM 3-99. *Airborne and Air Assault Operations*. 6 March 2015.

FM 4-02. *Army Health System*. 17 November 2020.

FM 4-40. *Quartermaster Operations*. 22 October 2013.

FM 5-0. *Planning and Orders Production*. 16 May 2022.

FM 6-0. *Commander and Staff Organization and Operations*. 16 May 2022.

FM 6-27/MCTP 11-10C. *The Commander's Handbook on the Law of Land Warfare*. 7 August 2019.

FM 7-0. *Training*. 14 June 2021.

GTA 07-04-008. *Infantry Reference Card for Linkup Operations*. 1 November 2018.

GTA 07-10-001. *Machine Gunner's Card*. 1 June 2002.

GTA 07-10-003. *Infantry Reference Card for Small Unit Leaders (Troop Leading Procedures)*. 1 November 2021.

GTA 08-01-004. *MEDEVAC Request Card*. 17 August 2016.

Individual Task 071-317-0000. *Prepare an Antiarmor Range Card*. 29 September 2020.

STP 7-11B1-SM-TG. *Soldier's Manual and Trainer's Guide, MOS 11B, Infantry, Skill Level 1*. 21 September 2020.

TC 3-20.31-4. *Direct Fire Engagement Process (DIDEA)*. 23 July 2015.

TC 3-21.60. *Visual Signals*. 17 March 2017.

TC 3-21.75. *The Warrior Ethos and Soldier Combat Skills*. 13 August 2013.

TC 3-21.76. *Ranger Handbook*. 26 April 2017.

TC 3-22.9. *Rifle and Carbine*. 13 May 2016.

TC 3-22.23. *M18A1 Claymore Munition*. 15 November 2013.

TC 3-22.37. *Javelin-Close Combat Missile System, Medium*. 13 August 2013.

TC 3-22.84. *M3 Multi-Role, Anti-Armor Anti-Personnel Weapon System*. 18 July 2019.

TC 3-22.240. *Medium Machine Gun*. 28 April 2017.

TC 3-22.249. *Light Machine Gun M249 Series*. 16 May 2017.

TC 3-25.26. *Map Reading and Land Navigation*. 15 November 2013.

TM 3-23.25. *Shoulder-Launched Munitions*. 15 September 2010.

TM 9-1015-262-10. *Operator Manual for Rifle, 84 MM Recoilless, M3 NSN 1015-01-314-1770 (EIC: 7RR)*. 15 February 2019.

WEBSITES

Army Training Network at https://atn.army.mil

Central Army Registry at https://atiam.train.army.mil/catalog/dashboard

Geneva Convention Relative to the Treatment of Prisoners of War of August 12, 1949. https://www.loc.gov/law/help/us-treaties/index.php.

PRESCRIBED FORMS

This section contains no entries.

REFERENCED FORMS

Unless otherwise indicated, DA forms are available online: https://armypubs.army.mil.

DD forms are available online: https://www.esd.whs.mil/Directives/forms.

DA Form 1156. *Casualty Feeder Card.*

DA Form 2028. *Recommended Changes to Publications and Blank Forms.*

DA Form 2404. *Equipment Inspection and Maintenance Worksheet.*

DA Form 4137. *Evidence/Property Custody Document.*

DA Form 5517. *Standard Range Card.*

DA Form 5988-E. *Equipment Maintenance and Inspection Worksheet.* (Available from Global Combat Support System-Army [GCSS-A].)

DD Form 1380. *Tactical Combat Casualty Care (TCCC) Card.* (Available through normal publications supply channels.)

DD Form 2745. *Enemy Prisoner of War (EPW) Capture Tag.* (Available through normal publications supply channels.)

Index

Entries are by paragraph number.

Entries are by paragraph number.

Entries are by paragraph number.

ATP 3-21.8
11 JANUARY 2024

By Order of the Secretary of the Army:

RANDY A. GEORGE
General, United States Army
Chief of Staff

Official:

MARK F. AVERILL
Administrative Assistant
to the Secretary of the Army
2400202

DISTRIBUTION:
Active Army, Army National Guard, and United States Army Reserve. To be distributed in accordance with the initial distribution number (IDN) 110782, requirements for ATP 3-21.8.

alternate position

A defensive position that the commander assigns to a unit or weapon system for occupation when the primary position becomes untenable or unsuitable for carrying out the assigned task. (FM 3-90)

ambush

A variation of attack from concealed positions against a moving or temporarily halted enemy. (FM 3-90)

approach march

The advance of a combat unit when direct contact with the enemy is intended. (FM 3-90)

area defense

A type of defensive operation that concentrates on denying enemy forces access to designated terrain for a specific time rather than destroying the enemy outright. (ADP 3-90)

area of influence

An area inclusive of and extending beyond an operational area wherein a commander is capable of direct influence by maneuver, fire support, and information normally under the commander's command or control. (JP 3-0)

area of interest

That area of concern to the commander, including the area of influence, areas adjacent to it, and extending into enemy territory. (JP 3-0)

area reconnaissance

A form of reconnaissance operation that focuses on obtaining detailed information about the terrain or enemy activity within a prescribed area. (FM 3-90)

area security

A type of security operation conducted to protect friendly forces, lines of communications, installation routes and actions within a specific area. (FM 3-90)

assailable flank

A flank exposed to attack or envelopment. (ADP 3-90)

assault

(Army) A short and violent well-ordered attack against a local objective. (FM 3-90)

assault position

A covered and concealed position short of the objective from which final preparations are made to assault the objective. (ADP 3-90)

assembly area

An area a unit occupies to prepare for an operation. (FM 3-90)

attack

A type of offensive operation that defeats enemy forces, seizes terrain, or secures terrain. (FM 3-90)

attack by fire

A tactical mission task using direct and indirect fires to engage an enemy from a distance. (FM 3-90)

attack by fire position

The general position from which a unit performs the tactical task of attack by fire. (ADP 3-90)

attack position

(Army) The last position an attacking force occupies or passes through before crossing the line of departure. (ADP 3-90)

avenue of approach

A path used by an attacking force leading to its objective or to key terrain. Avenues of approach exist in all domains. (ADP 3-90)

basic load

The quantity of supplies required to be on hand within, and moved by a unit or formation, expressed according to the wartime organization of the unit or formation and maintained at the prescribed levels. (JP 4-09)

battle drill

Rehearsed and well understood actions made in response to common battlefield occurrences. (ADP 3-90)

battle handover

A coordinated mission between two units that transfers responsibility for fighting an enemy force from one unit to another. (FM 3-90)

battle handover line

A designated phase line where responsibility transitions from the stationary force to the moving force and vice versa. (ADP 3-90)

battle position

A defensive location oriented on a likely enemy avenue of approach. (ADP 3-90)

block

1. A tactical mission task that denies the enemy access to an area or an avenue of approach. (FM 3-90) 2. An obstacle effect that integrates fire planning and obstacle effort to stop an attacker along a specific avenue of approach or prevent the attacking force from passing through an engagement area. (FM 3-90)

bounding overwatch

A movement technique used when contact with enemy forces is expected. (FM 3-90)

breach

1. A tactical mission task in which a unit breaks through or establishes a passage through an enemy obstacle. (FM 3-90) 2. A synchronized combined arms activity under the control of the maneuver commander conducted to allow maneuver through an obstacle. (ATP 3-90.4)

breach area

A defined area where a breach occurs. (ATP 3-90.4)

bypass

A tactical mission task in which a unit deliberately avoids contact with an obstacle or enemy force. (FM 3-90)

bypass criteria

Measures established by higher echelon headquarters that specify the conditions and size under which enemy units and contact may be avoided. (FM 3-90)

call for fire

A standardized request for fire containing data necessary for obtaining the required fire on a target. (FM 3-09)

canalize

(Army) A tactical mission task in which a unit restricts enemy movement to a narrow zone. (FM 3-90)

casualty evacuation

(Army) The movement of casualties aboard nonmedical vehicles or aircraft without en route medical care. (FM 4-02)

checkpoint

A predetermined point on the ground used to control movement, tactical maneuver, and orientation. (FM 3-90)

clear

A tactical mission task in which a unit eliminates all enemy forces within an assigned area. (FM 3-90)

close air support

Air action by aircraft against hostile targets that are in close proximity to friendly forces and that require detailed integration of each air mission with the fire and movement of those forces. (JP 3-09.3)

close combat

Warfare carried out on land in a direct-fire fight, supported by direct and indirect fires and other assets. (ADP 3-0)

combat load

The minimum mission-essential equipment and supplies as determined by the commander responsible for carrying out the mission, required for Soldiers to fight and survive immediate combat operations. (FM 4-40)

combat outpost

A reinforced observation post capable of conducting limited combat operations. (FM 3-90)

***combat patrol**

A patrol that provides security and harasses, destroys, or captures enemy troops, equipment, or installations.

combat power

(Joint) The total means of destructive and disruptive force that a military unit/formation can apply against an enemy at a given time. (JP 3-0)

combined arms

The synchronized and simultaneous application of arms to achieve an effect greater than if each element was used separately or sequentially. (ADP 3-0)

command and control

The exercise of authority and direction by a properly designated commander over assigned and attached forces in the accomplishment of the mission. (JP 1, Vol 2)

command and control warfighting function

The related tasks and a system that enable commanders to synchronize and converge all elements of combat power. (ADP 3-0)

commander's intent

A clear and concise expression of the purpose of an operation and the desired objectives and military end state. (JP 3-0)

concealment

Protection from observation or surveillance. (FM 3-96)

concept of operations

(Army) A statement that directs the manner in which subordinate units cooperate to accomplish the mission and establishes the sequence of actions the force will use to achieve the end state. (ADP 5-0)

confirmation brief

A brief subordinate leaders give to the higher commander immediately after the operation order is given to confirm understanding. (ADP 5-0)

consolidate

To organize and strengthen a captured position to use it against the enemy. (FM 3-90)

contact point

In land warfare, a point on the terrain, easily identifiable, where two or more units are required to make contact. (JP 3-50)

contain

A tactical mission task in which a unit stops, holds, or surrounds an enemy force. (FM 3-90)

control

(Army) A tactical mission task in which a unit maintains physical influence over an assigned area. (FM 3-90)

conventional warfare

A violent struggle for domination between nation-states or coalitions of nation-states. (FM 3-0)

coordination point

A point that indicates a specific location for the coordination of tactical actions between adjacent units. (FM 3-90)

cordon and search

A variation of movement to contact where a friendly force isolates and searches a target area. (FM 3-90)

counterattack

A variation of attack by a defending force against an attacking enemy force. (FM 3-90)

countermobility

A set of combined arms activities that use or enhance the effects of natural and man-made obstacles to prevent the enemy freedom of movement and maneuver. (ATP 3-90.8)

counterreconnaissance

A tactical mission task that encompasses all measures taken by a unit to counter enemy reconnaissance and surveillance efforts. (FM 3-90)

cover

(Army) 1. Protection from the effects of fires. (FM 3-96) 2. A type of security operation done independent of the main body to protect them by fighting to gain time while preventing enemy ground observation of and direct fire against the main body. (ADP 3-90)

cyberspace operations

The employment of cyberspace capabilities where the primary purpose is to achieve objectives in or through cyberspace. (JP 3-0)

deception means

Methods, resources, and techniques that can be used to convey information to the deception target. (JP 3-13.4)

decisive point

Key terrain, key event, critical factor, or function that, when acted upon, enables commanders to gain a marked advantage over an enemy or contribute materially to achieving success. (JP 5-0)

decisive terrain

Key terrain whose seizure and retention is mandatory for successful mission accomplishment. (ADP 3-90)

decisively engaged

A fully committed force or unit that cannot maneuver or extricate itself. (FM 3-90)

defeat

To render a force incapable of achieving its objectives. (ADP 3-0)

defeat in detail

Concentrating overwhelming combat power against separate parts of a force rather than defeating the entire force at once. (ADP 3-90)

defensive operation

An operation to defeat an enemy attack, gain time, economize forces, and develop conditions favorable for offensive or stability operations. (ADP 3-0)

delay

When a force under pressure trades space for time by slowing down the enemy's momentum and inflicting maximum damage on enemy forces without becoming decisively engaged. (ADP 3-90)

delay line

A phase line over which an enemy is not allowed to cross before a specific date and time or enemy condition. (FM 3-90)

deliberate operation

An operation in which the tactical situation allows the development and coordination of detailed plans, including multiple branches and sequels. (ADP 3-90)

demonstration

(Army) A variation of tactical deception used as a show of force in an area where a unit does not seek a decision and attempts to mislead an adversary. (FM 3-90)

deny

A task to hinder or prevent the enemy from using terrain, space, personnel, supplies, or facilities. (ATP 3-21.20)

destroy

A tactical mission task that physically renders an enemy force combat-ineffective until reconstituted. (FM 3-90)

detachment left in contact

An element left in contact as part of the previously designated security force while the main body conducts its withdrawal. (FM 3-90)

directed obstacle

An obstacle directed by a higher commander as a specified task to a subordinate unit. (ATP 3-90.8)

disengage

A tactical mission task in which a unit breaks contact with an enemy to conduct another mission or to avoid becoming decisively engaged. (FM 3-90)

disengagement criteria

Protocols that specify those circumstances where a friendly force must break contact with the direct fire and observed indirect fire to avoid becoming decisively engaged or to preserve friendly combat power. (FM 3-90)

disengagement line

A phase line located on identifiable terrain that, when crossed by the enemy, signals to defending elements that it is time to displace to their next position. (ADP 3-90)

dismounted movement

A movement of troops and equipment mainly by foot, with limited support by vehicles. (FM 3-90)

disrupt

1. A tactical mission task in which a unit upsets an enemy's formation or tempo and causes the enemy force to attack prematurely or in a piecemeal fashion. (FM 3-90) 2. An obstacle effect that focuses fire planning and obstacle effort to cause the enemy to break up its formation and tempo, interrupt its timetable, commit breaching assets prematurely, and attack in a piecemeal effort. (FM 3-90)

domain

A physically defined portion of an operational environment requiring a unique set of warfighting capabilities and skills. (FM 3-0)

double envelopment

A variation of envelopment where forces simultaneously attack along both flanks of an enemy force. (FM 3-90)

dynamic targeting

Targeting that prosecutes targets identified too late or not selected for action in time to be included in deliberate targeting. (JP 3-60)

electromagnetic protection

(Joint) Division of electromagnetic warfare involving actions taken to protect personnel, facilities, and equipment from any effects of friendly or enemy use of the electromagnetic spectrum that degrade, neutralize, or destroy friendly combat capability. (JP 3-85)

electromagnetic warfare

Military action involving the use of electromagnetic and directed energy to control the electromagnetic spectrum or to attack the enemy. (JP 3-85)

emission control

The selective and controlled use of electromagnetic, acoustic, or other emitters to optimize command and control capabilities while minimizing, for operations security: a. detection by enemy sensors, b. mutual interference among friendly systems, and/or c. enemy interference with the ability to execute a military deception plan. (JP 3-85)

enabling operation

An operation that sets the friendly conditions required for mission accomplishment. (FM 3-90)

encirclement

Where one force loses its freedom of maneuver because an opposing force is able to isolate it by controlling all ground lines of communications and reinforcement. (FM 3-90)

end state

The set of required conditions that defines achievement of the commander's objectives. (JP 3-0)

enemy

A party identified as hostile against which the use of force is authorized. (ADP 3-0)

engagement area

An area where the commander masses effects to contain and destroy an enemy force. (FM 3-90)

engagement criteria

Protocols that specify those circumstances for initiating engagement with an enemy force. (FM 3-90)

engagement priority

Identifies the order in which the unit engages enemy systems or functions. (FM 3-90)

envelopment

A form of maneuver in which an attacking force avoids an enemy's principal defense by attacking along an assailable flank. (FM 3-90)

exfiltrate

A tactical mission task in which a unit removes Soldiers or units from areas under enemy control by stealth, deception, surprise, or clandestine means. (FM 3-90)

exploitation

(Army) A type of offensive operation following a successful attack to disorganize the enemy in depth. (FM 3-90)

feint

A variation of tactical deception that makes contact solely to deceive the adversary as to the location, time of attack or both. (FM 3-90)

field of fire

The area that a weapon or group of weapons may cover effectively from a given position. (FM 3-90)

final coordination line

A phase line close to the enemy position used to coordinate the lifting or shifting of supporting fires with the final deployment of maneuver elements. (ADP 3-90)

final protective fire

An immediately available, prearranged barrier of fire designed to impede enemy movement across defensive lines or areas. (JP 3-09.3)

final protective line

A selected line of fire where an enemy assault is to be checked by interlocking fire from all available weapons and obstacles. (FM 3-90)

fire and movement

The concept of applying fires from all sources to suppress, neutralize, or destroy the enemy, and the tactical movement of combat forces in relation to the enemy (as components of maneuver, applicable at all echelons). At the squad level, fire and maneuver entails a team placing suppressive fire on the enemy as another team moves against or around the enemy. (FM 3-96)

fire superiority

The dominating fires of one force over another force that permits that force to maneuver at a given time and place without prohibitive interference by the other. (FM 3-90)

fix

1. A tactical mission task in which a unit prevents the enemy from moving from a specific location for a specific period. (FM 3-90) 2. An obstacle effect that focuses fire planning and obstacle effort to slow an attacker's movement within a specified area, normally an engagement area. (FM 3-90)

fixing force

A force designated to supplement the striking force by preventing the enemy from moving from a specific area for a specific time. (ADP 3-90)

follow and assume

A tactical mission task in which a committed force follows and supports a lead force conducting an offensive operation and continues mission if lead force cannot continue. (FM 3-90)

follow and support

A tactical mission task in which a committed force follows and supports a lead force conducting an offensive operation. (FM 3-90)

forced march

A march longer or faster than usual or in adverse conditions. (FM 3-90)

forms of maneuver

Distinct tactical combinations of fire and movement with a unique set of doctrinal characteristics that differ primarily in the relationship between the maneuvering force and the enemy. (ADP 3-90)

forward boundary

A boundary that delineates the forward edge of a unit's area of operation. (FM 3-90)

forward edge of the battle area

The foremost limits of a series of areas in which ground combat units are deployed to coordinate fire support, the positioning of forces, or the maneuver of units, excluding areas in which covering or screening forces are operating. (JP 3-09.3)

forward line of own troops

A line that indicates the most forward positions of friendly forces in any kind of military operation at a specific time. (FM 3-90)

forward passage of lines

Occurs when a unit passes through another unit's positions while moving toward the enemy. (ADP 3-90)

fragmentary order

An abbreviated operation order issued as needed to change or modify an order or to execute a branch or sequel. (JP 5-0)

fratricide

The unintentional killing or wounding of friendly or neutral personnel by friendly firepower. (ADP 3-37)

frontal attack

A form of maneuver in which an attacking force seeks to destroy a weaker enemy force or fix a larger enemy force in place over a broad front. (FM 3-90)

gap

1. An area free of obstacles that enables forces to maneuver in a tactical formation. (FM 3-90) 2. A ravine, mountain pass, river, or other terrain feature that presents an obstacle that may be bridged. (ATP 3-90.4)

guard

A type of security operation done to protect the main body by fighting to gain time while preventing enemy ground observation of and direct fire against the main body. (ADP 3-90)

hasty operation

An operation in which a commander directs immediately available forces, using fragmentary orders, to perform tasks with minimal preparation, trading planning and preparation time for speed of execution. (ADP 3-90)

high-payoff target

A target whose loss to the enemy will significantly contribute to the success of the friendly course of action. (JP 3-60)

high-value target

A target the enemy commander requires for the successful completion of the mission. (JP 3-60)

human dimension

Encompasses people and the interaction between individuals and groups, how they understand information and events, make decisions, generate will, and act within an operational environment. (FM 3-0)

infiltration

A form of maneuver in which an attacking force conducts undetected movement through or into an area occupied by enemy forces. (FM 3-90)

infiltration lane

A control measure that coordinates forward and lateral movement of infiltrating units and fixes fire planning responsibilities. (FM 3-90)

information dimension

The content and data that individuals, groups, and information systems communicate and exchange, as well as the analytics and technical processes used to exchange information within an operational environment. (FM 3-0)

information environment

The aggregate of social, cultural, linguistic, psychological, technical, and physical factors that affect how humans and automated systems derive meaning from, act upon, and are impacted by information, including the individuals, organizations, and systems that collect, process, disseminate, or use information. (JP 3-04)

informational considerations

Those aspects of the human, information, and physical dimensions that affect how humans and automated systems derive meaning from, use, act upon, and are impacted by information. (FM 3-0)

interdict

A tactical mission task in which a unit prevents, disrupts, or delays the enemy's use of an area or route in any domain. (FM 3-90)

isolate

A tactical mission task in which a unit seals off an enemy, physically and psychologically, from sources of support and denies it freedom of movement. (FM 3-90)

key terrain

(Army) An identifiable characteristic whose seizure or retention affords a marked advantage to either combatant. (ADP 3-90)

kill zone

The location where fires are concentrated in an ambush. (FM 3-90)

knowledge management

The process of enabling knowledge flow to enhance shared understanding, learning, and decision making. (ADP 6-0)

lateral boundary

A boundary defining the left or right limit of a unit's assigned area. (FM 3-90)

leadership

The activity of influencing people by providing purpose, direction, and motivation to accomplish the mission and improve the organization. (ADP 6-22)

limit of advance

A phase line used to control forward progress of the attack. (ADP 3-90)

line of contact

A general trace delineating the locations where friendly and enemy forces are engaged. (FM 3-90)

line of departure

In land warfare, a line designated to coordinate the departure of attack elements. (JP 3-31)

linkup

A type of enabling operation that involves the meeting of friendly ground forces, which occurs in a variety of circumstances. (FM 3-90)

linkup point

A designated place where two forces are scheduled to meet. (FM 3-90)

local security

The low-level security activities conducted near a unit to prevent surprise by the enemy. (ADP 3-90)

main battle area

The area where the commander intends to deploy the bulk of their unit to defeat an attacking enemy. (FM 3-90)

main effort

A designated subordinate unit whose mission at a given point in time is most critical to overall mission success. (ADP 3-0)

maneuver

(Army) Movement in conjunction with fires. (ADP 3-90)

march column

All march serials using the same route for a single movement under control of a single commander. (FM 3-90)

march serial

A subdivision of a march column organized under one commander. (FM 3-90)

march unit

A subdivision of a march serial. (FM 3-90)

medical evacuation

The timely and effective movement of the wounded, injured, or ill to and between medical treatment facilities on dedicated and properly marked medical platforms with en route care provided by medical personnel. (ATP 4-02.2)

meeting engagement

A combat action that occurs when a moving force engages an enemy at an unexpected time and place. (FM 3-90)

mission

The essential task or tasks, together with the purpose, that clearly indicates the action to be taken and the reason for the action. (JP 3-0)

mission command

(Army) The Army's approach to command and control that empowers subordinate decision making and decentralized execution appropriate to the situation. (ADP 6-0)

mission orders

Directives that emphasize to subordinates the results to be attained, not how they are to achieve them. (ADP 6-0)

mission statement

A short sentence or paragraph that describes the organization's essential task(s), purpose, and action containing the elements of who, what, when, where, and why. (JP 5-0)

mobile defense

A type of defensive operation that concentrates on the destruction or defeat of the enemy through a decisive attack by a striking force. (ADP 3-90)

mounted movement

The movement of troops and equipment by combat and tactical vehicles. (FM 3-90)

movement

The positioning of combat power to establish the conditions for maneuver. (ADP 3-90)

movement formation

An ordered arrangement of forces for a specific purpose and describes the general configuration of a unit on the ground. (ADP 3-90)

movement to contact

(Army) A type of offensive operation designed to establish or regain contact to develop the situation. (FM 3-90)

multidomain operations

The combined arms employment of joint and Army capabilities to create and exploit relative advantages to achieve objectives, defeat enemy forces, and consolidate gains on behalf of joint force commanders. (FM 3-0)

named area of interest

A geospatial area or systems node or link against which information that will satisfy a specific information requirement can be collected, usually to capture indications of enemy and adversary courses of action. (JP 2-0)

networked munitions

Remotely controlled, interconnected, weapon systems designed to provide rapidly emplaced ground-based countermobility and protection capability through scalable application of lethal and nonlethal means. (JP 3-15)

neutralize

A tactical mission task in which a unit renders the enemy incapable of interfering with an operation. (FM 3-90)

nontactical movement

A movement in which troops and vehicles are arranged to expedite their movement and conserve time and energy when no enemy ground interference is anticipated. (FM 3-90)

objective area

A geographical area, defined by competent authority, within which is located an objective to be captured or reached by the military forces. (JP 3-06)

objective rally point

An easily identifiable point where all elements of the infiltrating unit assemble and prepare to attack the objective. (ADP 3-90)

obscuration

The employment of materials into the environment that degrade optical and/or electro-optical capabilities within select portions of the electromagnetic spectrum in order to deny acquisition by or deceive an enemy or adversary. (ATP 3-11.50)

observation post

A position from which observations are made or fires are directed and adjusted. (FM 3-90)

occupy

A tactical mission task in which a unit moves into an area to control it without enemy opposition. (FM 3-90)

offensive operation

An operation to defeat or destroy enemy forces and gain control of terrain, resources, and population centers. (ADP 3-0)

operation

A sequence of tactical actions with a common purpose or unifying theme. (JP 1, Vol 1)

operational environment

The aggregate of the conditions, circumstances, and influences that affect the employment of capabilities and bear on the decisions of the commander. (JP 3-0)

operational framework

A cognitive tool used to assist commanders and staffs in clearly visualizing and describing the application of combat power in time, space, purpose, and resources in the concept of operations. (ADP 1-01)

operations in depth

The simultaneous application of combat power throughout an area of operations. (ADP 3-90)

operations process

The major command and control activities performed during operations: planning, preparing, executing, and continuously assessing the operation. (ADP 5-0)

operations security

A capability that identifies and controls critical information, indicators of friendly force actions attendant to military operations, and incorporates countermeasures to reduce the risk of an adversary exploiting vulnerabilities. (JP 3-13.3)

overwatch

A task that positions an element to support the movement of another element with immediate fire. (ATP 3-21.10)

passage lane

A lane through an obstacle that provides safe passage for a passing force. (FM 3-90)

passage of lines

An operation in which a force moves forward or rearward through another force's combat positions with the intention of moving into or out of contact with the enemy. (JP 3-18)

passage point

A designated place where passing units pass through the stationary unit. (FM 3-90)

***patrol**

A detachment sent out by a larger unit to conduct a specific mission that operates semi-independently and returns to the main body upon completion of mission.

penetration

A form of maneuver in which a force attacks on a narrow front. (FM 3-90)

physical dimension

The material characteristics and capabilities, both natural and manufactured, within an operational environment. (FM 3-0)

physical security

That part of security concerned with physical measures designed to safeguard personnel; to prevent unauthorized access to equipment, installations, material, and documents; and to safeguard them against espionage, sabotage, damage, and theft. (JP 3-0)

point of breach

The location at an obstacle where the creation of a lane is being attempted. (ATP 3-90.4)

point of departure

The point where the unit crosses the line of departure and begins moving along a direction of attack. (ADP 3-90)

point of penetration

(Army) Point of penetration is the location, identified on the ground, where the commanders concentrate their efforts to seize a foothold on the far side objective. (ATP 3-90.4)

primary position

The position that covers the enemy's most likely avenue of approach into the assigned area. (FM 3-90)

probable line of deployment

A phase line that designates the location where the commander intends to deploy the unit into assault formation before beginning the assault. (ADP 3-90)

pursuit

A type of offensive operation to catch or cut off a disorganized hostile force attempting to escape, with the aim of destroying it. (FM 3-90)

quartering party

A group dispatched to a new assigned area in advance of the main body. (FM 3-90)

quick reaction force

A commander designated force to respond to threat attacks or emergencies. (FM 3-90)

raid

(Army) A variation of attack to temporarily seize an objective with a planned withdrawal. (FM 3-90)

rally point

An easily identifiable point on the ground at which units can reassemble and reorganize if they become dispersed. (FM 3-90)

rear boundary

A boundary that delineates the rearward limits of a unit's assigned area. (FM 3-90)

rearward passage of lines

Occurs when a unit passes through another unit's positions while moving away from the enemy. (ADP 3-90)

reconnaissance

A mission undertaken to obtain information about the activities and resources of an enemy or adversary, or to secure data concerning the meteorological, hydrographic, geographic, or other characteristics of a particular area, by visual observation or other detection methods. (JP 2-0)

reconnaissance by fire

A technique in which a unit fires on a suspected enemy position. (FM 3-90)

reconnaissance objective

The most important result desired from that specific reconnaissance effort. (FM 3-90)

reduce

A tactical mission task in which a unit destroys an encircled or bypassed enemy force. (FM 3-90)

reduction

The creation of lanes through a minefield or obstacle to enable passage of the attacking ground force. (JP 3-15)

reduction area

A number of adjacent points of breach that are under the control of the breaching commander. (ATP 3-90.4)

rehearsal

A session in which the commander and staff or unit practices expected actions to improve performance during execution. (ADP 5-0)

release point

A designated place on a route where elements are released from centralized control. (FM 3-90)

relief in place

An operation in which, by direction of higher authority, all or part of a unit is replaced in an area by the incoming unit and the responsibilities of the replaced elements for the mission and the assigned zone of operations are transferred to the incoming unit (JP 3-07.3)

reorganization

All measures taken by the commander to maintain unit combat effectiveness or return it to a specified level of combat capability. (ATP 3-94.4)

reserved obstacle

(Army) Obstacles of any type, for which the commander restricts execution authority. (ATP 3-90.8)

retain

A tactical mission task in which a unit prevents enemy occupation or use of terrain. (FM 3-90)

retirement

When a force out of contact moves away from the enemy. (ADP 3-90)

retrograde

(Army) A type of defensive operation that involves organized movement away from the enemy. (ADP 3-90)

retrograde movement

Any movement to the rear or away from the enemy. (FM 3-90)

risk management

The process to identify, assess, and mitigate risks and make decisions that balance risk cost with mission benefits. (JP 3-0)

route

The prescribed course to be traveled from a point of origin to a destination. (FM 3-90)

route reconnaissance

A form of reconnaissance operation to obtain detailed information of a specified route and all terrain from which the enemy could influence movement along that route. (FM 3-90)

scheme of fires

(Joint) The detailed, logical sequence of targets and fire support events to find and engage targets to support the commander's objectives. (JP 3-09)

screen

A type of security operation that primarily provides early warning to the protected force. (ADP 3-90)

search and attack

A variation of a movement to contact where a friendly force conducts coordinated attacks to defeat a distributed enemy force. (FM 3-90)

sector

An operational area assigned to a unit in the defense that has rear and lateral boundaries and interlocking fires. (FM 3-0)

sector of fire

That area assigned to a unit or weapon system in which it will engage the enemy according to the established engagement priorities. (FM 3-90)

secure

A tactical mission task in which a unit prevents the enemy from damaging or destroying a force, facility, or geographical location. (FM 3-90)

security

Measures taken by a military unit, activity, or installation to protect itself against all acts designed to, or which may, impair its effectiveness. (JP 3-10)

security area

That area occupied by a unit's security elements and includes the areas of influence of those security elements. (ADP 3-90)

security objective

The most important entity to protect during that specific security effort. (FM 3-90)

security operations

Those operations performed by commanders to provide early and accurate warning of enemy operations, to provide the forces being protected with time and maneuver space within which to react to the enemy, and to develop the situation to allow commanders to effectively use their protected forces. (ADP 3-90)

seize

(Army) A tactical mission task in which a unit takes possession of a designated area by using overwhelming force. (FM 3-90)

sequential relief in place

Occurs when each element within the relieved unit is relieved in successions, from right to left or left to right, depending on how it is deployed. (ADP 3-90).

simultaneous relief in place

Occurs when all elements are relieved at the same time. (ADP 3-90)

single envelopment

A variation of envelopment where a force attacks along one flank of an enemy force. (FM 3-90)

situational obstacle

An obstacle that a unit plans and possibly prepares prior to starting an operation, but does not execute unless specific criteria are met. (ATP 3-90.8)

spoiling attack

A variation of an attack employed against an enemy preparing for an attack. (FM 3-90)

staggered relief in place

Occurs when a commander relieves each element in a sequence determined by the tactical situation, not its geographical orientation. (ADP 3-90)

start point

A designated place on a route where elements fall under the control of a designated march commander. (FM 3-90)

stay-behind operation

An operation in which a unit remains in position to conduct a specified mission while the remainder of the force withdraws or retires from an area. (FM 3-90)

striking force

A dedicated counterattack force in a mobile defense constituted with the bulk of available combat power. (ADP 3-90)

strong point

A heavily fortified battle position tied to a natural or reinforcing obstacle to create an anchor for the defense or to deny the enemy decisive or key terrain. (ADP 3-90)

subsequent position

A position that a unit expects to move to during the course of battle. (FM 3-90)

supplementary position

A defensive position located within a unit's assigned area that provides the best sectors of fire and defensive terrain along an avenue of approach that is not the primary avenue where the enemy is expected to attack. (FM 3-90)

support by fire

A tactical mission task in which a unit engages the enemy by direct fire in support of another maneuvering force. (FM 3-90)

support by fire position

The general position from which a unit performs the tactical mission task of support by fire. (ADP 3-90)

supporting effort

A designated subordinate unit with a mission that supports the success of the main effort. (ADP 3-0)

suppress

A tactical mission task in which a unit temporarily degrades a force or weapon system from accomplishing its mission. (FM 3-90)

suppression

(Joint) Temporary or transient degradation by an opposing force of the performance of a weapons system below the level needed to fulfill its mission objectives. (JP 3-01) (Army) In the context of the computed effects of field artillery fires, renders a target ineffective for a short period of time producing suppression of enemy air defenses at least 3-percent casualties or materiel damage. (FM 3-09)

surveillance

The systematic observation of aerospace, cyberspace, surface or subsurface areas, places, persons, or things by visual, aural, electronic, photographic, or other means. (JP 3-0)

survivability

(Army, Marine Corps) A quality or capability of military forces which permits them to avoid or withstand hostile actions or environmental conditions while retaining the ability to fulfill their primary mission. (ATP 3-37.34)

survivability operations

Those protection activities that alter the physical environment by providing or improving cover, camouflage, and concealment. (ATP 3-37.34)

sustainment

(Army) The provision of logistics, financial management, personnel services, and health service support necessary to maintain operations until successful mission completion. (ADP 4-0)

tactical assembly area

An area that is generally out of the reach of light artillery and the location where units make final preparations (precombat checks and inspections) and rest, prior to moving to the line of departure. (JP 3-35)

tactical mission task

The specific activity a unit performs while executing a tactical operation or form of maneuver. (FM 3-90)

tactical movement

A movement in which troops and vehicles are arranged to protect combat forces during movement when a threat of enemy interference is possible. (FM 3-90)

tactical road march

A rapid movement used to relocate units within an assigned area to prepare for combat operations. (FM 3-90)

tactics

(Army) The employment, ordered arrangement, and directed actions of forces in relation to each other. (ADP 3-90)

tailgate medical support

An economy of force device employed primarily to retain maximum mobility during movement halts or to avoid the time and effort required to set up a formal, operational treatment facility (for example, during rapid advance and retrograde operations). (FM 4-02)

target reference point

A predetermined point of reference, normally a permanent structure or terrain feature that can be used when describing a target location. (JP 3-09.3)

tempo

The relative speed and rhythm of military operations over time with respect to the enemy. (ADP 3-0)

terminal attack control

The authority to control the maneuver of and grant weapons release clearance to attacking aircraft. (JP 3-09.3)

terrain management

The process of allocating terrain by specifying locations for units and activities to deconflict activities that might interfere with each other. (FM 3-90)

traveling

A movement technique used when speed is necessary and contact with enemy forces is not likely. (FM 3-90)

traveling overwatch

A movement technique used when contact with enemy forces is possible. (FM 3-90)

trigger line

A phase line located on identifiable terrain used to initiate and mass fires into an engagement area at a predetermined range. (FM 3-90)

troop leading procedures

A dynamic process used by small-unit leaders to analyze a mission, develop a plan, and prepare for an operation. (ADP 5-0)

troop movement

The movement of Soldiers and units from one place to another by any available means. (FM 3-90)

turn

An obstacle effect that integrates fire planning and obstacle effort to divert an enemy formation from one avenue of approach to an adjacent avenue of approach or into an engagement area. (FM 3-90)

turning movement

(Army) A form of maneuver in which the attacking force seeks to avoid the enemy's principal defensive positions by attacking to the rear of their current positions forcing them to move or divert forces to meet the threat. (FM 3-90)

vertical envelopment

A variation of envelopment where air-dropped or airlanded troops attack an enemy forces rear, flank, or both. (FM 3-90)

warfighting function

A group of tasks and systems united by a common purpose that commanders use to accomplish missions and training objectives. (ADP 3-0)

Table D-7. Effects of the M141 bunker defeat munition on field fortifications or bunkers

<table>
<tr><th>Aimpoint</th><th colspan="2">Effect When Munition Is Fired at Aimpoint</th><th>Recommended Firing Technique</th></tr>
<tr><td>Bunkers</td><td colspan="2">Rounds fired into firing ports or apertures can destroy standard earth and timber bunkers, and hasty urban fighting positions (example, vehicles, and metal dumpsters). Rounds will detonate inside the rear of the position, causing major structural damage. Damage to enemy equipment may be minor unless it is hit directly. The round will cause injury or death to occupants.</td><td>Coordinate fire: Fire shoulder-launched munitions at and through firing ports.</td></tr>
<tr><td>Buildings</td><td>Windows/ Doorways</td><td>Rounds fired through windows and doorways can destroy the contents of the building. Destruction may not be contained within a single room. Rounds and debris from the round and materiel may pass through into other sections of the building, causing collateral damage. Damage to enemy equipment may be minor unless it is hit directly. The round will cause injury or death to occupants.</td><td>Coordinate fire: Fire an M141 bunker defeat munitions (BDMs) at the center of the visible part of a window or door.</td></tr>
<tr><td>Buildings (continued)</td><td>Walls</td><td>Rounds fired at walls will penetrate double- reinforced concrete walls up to 8 inches thick, and triple-brick structures. The initial blast will open a hole in the wall but may or may not completely penetrate the building.</td><td>Coordinate fire: Fire one or more M141 BDMs at the center of the desired location for the opening.
Fire a second round through the opening to destroy targets within the structure.
Note. It takes more than one round to create a Soldier-size hole. Use pair or volley fire, placing the rounds about 12 to 18 inches apart.</td></tr>
<tr><td>Underground Opening</td><td colspan="2">Rounds fired through underground openings can collapse the opening or destroy the contents within it. Destruction may not be contained within the opening. Rounds and debris may pass through into other sections of the opening, causing more damage. Damage to enemy equipment may be minor unless it is hit directly. The round will cause injury or death to occupants at the front entrance, and others farther into the opening may be incapacitated or die from the concussion, heat, and debris caused by the explosion.</td><td>Coordinate fire: Fire one or more M141 BDMs.</td></tr>
</table>

Javelin Employment Considerations in Urban Terrain

D-31. Urban employment considerations for Javelin include engagement distance, thermal crossover, backblast, weapon penetration, and breaching structural walls. Details follow:

- Engagement distance. The Javelin missile has a minimum engagement distance (150 meters in the top attack mode and 65 meters in the direct attack mode), which limits its use in built-up areas.
- Crossover. Sometimes the Javelin seeker will not be able to distinguish between the background and target because the two have the same temperature (crossover).
- Time. When a gunner comes across a target of opportunity, the gunner may not be able to take advantage of it. The cooldown time of the Javelin's night vision sight is 2.5 to 3.5 minutes. Javelin seeker cooldown takes about 10 seconds. Once the battery coolant unit is activated, the gunner has a maximum of 4 minutes to engage the target before the battery coolant unit is depleted.
- Backblast. The soft launch capability of the Javelin enables the gunner to fire from inside buildings because there is little overpressure or flying debris.
- Weapon penetration. The dual-charge Javelin warhead penetrates typical urban targets. The direct attack mode is selected when engaging targets in a building. Enemy positions or bunkers in the open closer than 150 meters are engaged using the direct attack mode. Positions in the open farther than 150 meters are engaged using either the top or direct attack mode, depending on the situation.
- Breaching structural walls. The Javelin is not effective when breaching structural walls. AT guided missiles are not designed to breach structural walls. All Javelin is designed to produce a small hole, penetrate armor, and deliver the explosive charge. Breaching calls for the creation of a large hole. Javelins are better used against armored vehicles or the destruction of enemy-fortified fighting positions.

M3 Multi-Role, Antiarmor, Antipersonnel Weapon System Ammunition Safety Precautions

D-32. Leaders and Soldiers must be familiar with the safety precautions during training and combat operations for the M3 MAAWS. The two basic safety rules to keep in mind when operating the M3 is to always wear double hearing protection and to always ensure the backblast area is clear. Firing teams will be observant of all safe separation and surface safety distances. Failure to comply with this warning could result in injury or death to personnel. Table D-8 details the ammunition safe separation and fuze arming distances.

Table D-8. Ammunition safe separation and fuze arming distances

	HEAT 551C RS	*HE 441D RS*	*HEDP 502 RS*	*ASM 509*
Minimum Fuze Arming Distance	72 Feet 22 Meters	151 Feet 46 Meters	56 to 66 Feet 17 to 20 Meters	39 Feet 12 Meters
*Safe Distance of Gunner and Assistant Gunner	279 Feet 85 Meters	820 Feet 250 Meters	600 Feet 183 Meters	No Fragmentation Hazard at 39 Feet or 12 Meters
**Safe Distance of All Other Personnel	279 Feet 85 Meters	984 Feet 300 Meters	1,083 Feet 330 Meters	591 Feet 180 Meters

Remarks

*Leaders must train teams not to engage targets closer than safe separation distance because fragmentation may cause injury or death.

**Teams shall use extreme caution when firing within these distances of adjacent personnel and friendly troops.

Measure distances from intended target or point of detonation.

Legend: ASM–Antistructure Munition; HE–High Explosive; HEAT–High Explosive Antitank; HEDP–High Explosive Dual-Purpose; RS–Reduced Sensitivity

---

Note. The reduced sensitivity (known as RS) of the bursting charge; for example, withstands outer heat better than the Octol charge in the standard round. See operator's manual (TM 9-1015-262-10) for additional information.

D-33. Due to excessive noise and blast over pressurization that occurs when firing the M3, the gunner and assistant gunner may only fire a limited number of rounds during training and combat. Table D-9 on page D-18 details the ammunition firing restrictions (rounds) by ammunition type and firing position. The table provides the maximum amount of rounds the gunner and assistant gunner can fire in a 24-hour period. Each type of ammunition and firing position have a point value assigned. The gunner and assistant gunner cannot exceed a total of 6 points within a 24-hour period regardless of the firing position or ammunition type. Leaders must use the ammunition type and firing position information to determine the maximum number of rounds that can be fired. For example, if the Soldier fired 3 rounds of HEDP 502 RS from the standing, kneeling, and prone positions (1 round from each position), the total point value would be 5. The Soldier may now only fire an additional round from the standing or kneeling position to ensure the 6-point rule is not exceeded.

> ***Note.*** Ammunition firing restrictions for ammunition type and firing position apply to both gunner and assistant gunner over a 24-hour period. Failure to comply with this warning could result in injury or death to personnel. Seek immediate medical attention if injury occurs. Double hearing protection is required. Failure to wear double hearing protection when firing the M3 will cause permanent hearing loss. Failure to comply with this warning could result in injury or death to personnel. Seek immediate medical attention if injury occurs.

Table D-9. Ammunition firing restrictions (rounds)

Round Type	Firing Position (Firing from Fighting Position or Enclosure is Prohibited)							
	Standing		Kneeling		Sitting		Prone	
	Rounds	Points	Rounds	Points	Rounds	Points	Rounds	Points
TP 552	1	1	1	1	1	3	1	1
HEAT 551C RS	1	1	1	1.5	1	1	1	1
HE 441D RS	1	1.5	1	1	1	3	1	3
HEDP 502 RS	1	1	1	1	1	2	1	3
ADM 401 (All Variants)	1	1	1	1	1	2	1	1.5
Smoke 469C	1	1.5	1	1	1	1.5	1	3
Illuminator 545C	1	1	1	1	1	1	0	N/A
TPT 141	1	1	1	1	1	1	1	1
ASM 509	1	1	1	1	1	1	1	1
SCA 553B	44	1	27	1	27	1	10	1

Legend: ADM–Area Deterrent Munition; ASM–Antistructure Munition; HE–High Explosive; HEAT–High Explosive Anti-Tank; HEDP–High Explosive Dual-Purpose; N/A–Not Applicable; RS–Reduced Sensitivity; SCA–Subcaliber Adapter; TP–Target Practice; TPT–Target Practice Tracer

Exploiting Armored Vehicle Weaknesses

D-34. Because they are designed mainly for offense against other armored vehicles (see figure D-10), armored vehicles usually have their heaviest armor in front. All vehicles are vulnerable to repeated hits on their flanks and rear, though the flank offers the largest possible target. (See TC 3-20.31-4.) Shooters always should aim center of mass to increase the probability of a hit. The older the vehicle model, the less protection it has against SLM and Javelins. Newer versions of older vehicle models may use bolt-on (appliqué) armor to improve their survivability. Reactive armor usually covers the forward-facing portions and sides of the vehicle and can defeat shaped-charge weapons such as the SLM. When reactive armor detonates, it disperses metal fragments to 200 meters. SLM cause only a small entry hole in an armored vehicle target, though some fragmentation or spall may occur.

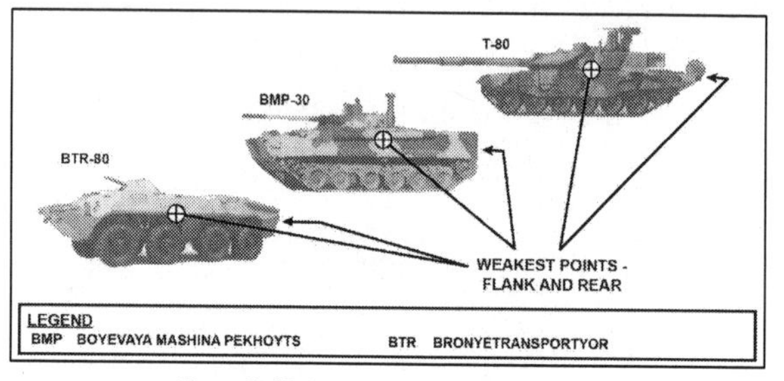

Figure D-10. Armored vehicle weak points

D-35. Natural or man-made obstacles can be used to force the armored vehicle to slow, stop, or change direction. This pause enables the shooter to achieve a first-round hit. If the gunner does not achieve a catastrophic kill on the first round, the gunner or another shooter must be ready to engage the target vehicle immediately with another round.

D-36. The gray area in figure D-11 shows the most favorable direction of attack when the turret is facing to the front. The white area shows the vehicle's principal direction of fire (known as PDF) and observation when the turret is facing to the front. Volley fires can degrade the additional protection appliqué and reactive armors provide to the target vehicle greatly.

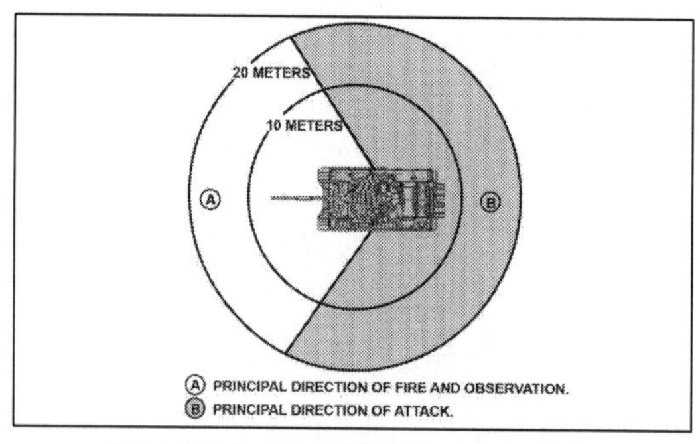

Figure D-11. Limited visibility of armored vehicles

D-37. Armored vehicle kills are classified according to the level of damage achieved. (See table D-10.)

Table D-10. Armored vehicle kills

Type of Kill	Part of Vehicle Damaged or Destroyed	Capability After Kill
Mobility Kill	Suspension (track, wheels, or road wheels) or power train (engine or transmission) has been damaged.	Vehicle cannot move, but it can still return fire.
Firepower Kill	Main armament has been disabled.	Vehicle still can move, so it can get away.
Catastrophic Kill	Ammunition or fuel storage section has been hit by more than one round.	Vehicle completely destroyed.

ANTIARMOR ROLE (SHOULDER-LAUNCHED MUNITIONS)

D-38. When Soldiers employ the M136 AT4, M136A1 AT4-CS, and M72-series LAW to defeat threat armored vehicles, it requires Soldiers to engage threat vehicles using single or paired shots. Gunners require positions allowing engagement against the flank or rear of the target vehicles. Gunners prepare standard range cards (Individual Task 071-317-0000 [*Prepare an Antiarmor Range Card*] outlines the process in preparing an antiarmor range card) for the SLM's primary position. If time permits, gunners also prepare standard range cards for the SLM's alternate and supplementary positions. They must seek covered and concealed positions from where targets can be engaged. The M136A1 AT4-CS is the only SLM that can be fired safely from within an enclosure because of its countermass propulsion system. TM 3-23.25 advises firing the M136 AT4 and M141 BDM from an enclosure under combat conditions only when no other tactical option exists due to the risk of both auditory and nonauditory injury.

Shoulder-Launched Munitions Warhead Effects on Armor

D-39. SLM warheads have excellent armor penetration ability and lethal after-armor effects (especially the M136 AT4, M136A1 AT4-CS, and M72A7). The extremely destructive shaped-charge explosives can penetrate more than 14 inches (35.6 centimeters) of rolled homogeneous armor. Types of warhead armor effects follow and are illustrated in figure D-12:

- Impact. The nosecone crushes: the impact sensor activates the fuze.
- Ignition. The fuze element activates the electric detonator. The booster detonates, initiating the main charge.
- Penetration. The main charge fires and forces the warhead body liner into a directional gas jet-penetrating armor plate.
- Spalling (after-armor effects). The projectile fragments and incendiary effects produce blinding light and highly destructive results.

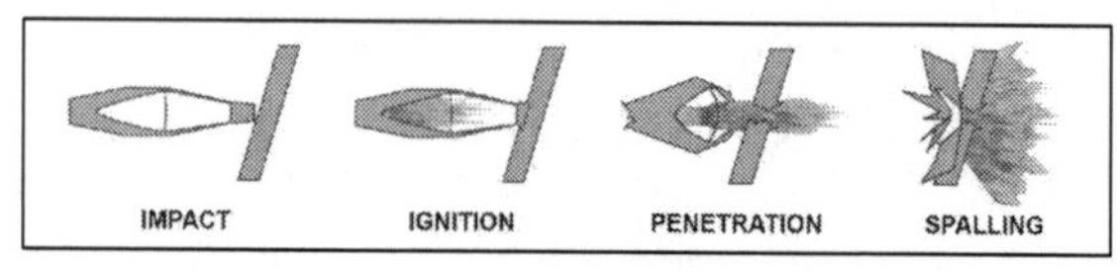

Figure D-12. Effects of shoulder-launched munition warheads on armor targets

Employment Against of Other Vehicles

D-40. The M72-series LAW proves more effective against light-armored vehicles. The M136 AT4 proves more effective against armored vehicles. Nonarmored vehicles such as trucks, cars, and boats are considered soft targets. Firing along their length offers the greatest chance of a kill, because this type of shot is most likely to hit their engine block or fuel tank. Effects of different munitions on vehicle types are listed in table D-11.

Table D-11. Effects of different munitions on vehicle types

<table>
<tr><td colspan="3"><u>Heavy-Armored Vehicle</u>
The older the vehicle model, the less protection it has against shoulder-launched munitions. Newer versions may use bolt-on (appliqué) armor to improve their survivability. Some vehicles are equipped with reactive armor, which consists of metal plates and plastic explosives.</td></tr>
<tr><th>Munition</th><th>Effects</th><th>Remarks</th></tr>
<tr><td>M141 BDM</td><td>Can cause mobility kills by disabling the vehicle's suspension system.</td><td>The M141 BDM should be a last resort when engaging armored vehicles.</td></tr>
<tr><td>M136-series</td><td>Causes only a small entry hole, though some fragmentation or spalling may occur.</td><td rowspan="2">Reactive armor usually covers the front and sides of the vehicle, and can defeat shaped-charge weapons; however, the munitions can restrict the vehicle's mobility and may destroy the vehicle if the round hits a vulnerable spot, such as the engine compartment area.</td></tr>
<tr><td>M72-series</td><td>Causes only a small entry hole, though some fragmentation or spalling may occur.</td></tr>
<tr><td colspan="3">Legend: BDM–bunker defeat munition</td></tr>
</table>

Table D-11. Effects of different munitions on vehicle types (continued)

Munition	Effects	Remarks
Light-Armored Vehicle All current shoulder-launched munitions are capable of destroying most light-armored vehicles, if the round hits a vulnerable spot, such as the engine compartment area, or fuel tank. Unit leaders should provide squad and platoon supporting fires when engaging light-armored troop carriers. Any Infantry troops who survive the initial assault may dismount and return fire.		
Munition	**Effects**	**Remarks**
M141 BDM	Can cause a catastrophic kill, if the round hits a vulnerable spot, such as the engine compartment area or fuel tank.	
M136-series	Can cause a catastrophic kill, if the round hits a vulnerable spot, such as the engine compartment area or fuel tank.	
M72-series	Can cause a catastrophic kill, if the round hits a vulnerable spot, such as the engine compartment area or fuel tank.	
Nonarmored Vehicles Nonarmored vehicles, such as trucks and cars, are considered soft targets. Firing along their length (flank) offers the greatest chance of a kill, because this type of shot is most likely to hit their engine block or fuel tank. Front and rear angles offer a much smaller target, reducing the chance of a first time hit.		
Munition	**Effects**	**Remarks**
M141 BDM	Causes a catastrophic kill.	
M136-series	May penetrate but will pass through the body with limited damage unless the rocket hits a vital part of the engine.	When engaging enemy-used privately owned vehicles with M136- or the M72-series munitions, do not fire at the main body. Instead, fire at the engine compartment area.
M72-series	May penetrate but will pass through the body with limited damage unless the rocket hits a vital part of the engine.	
Legend: BDM–bunker defeat munition		

Methods of Employment

D-41. The four employment methods for SLM include single, sequence, pair, and volley firing. The leader evaluates the situation on the ground to determine which of these methods to use. Regardless of whether they are used singly or in combination, communications are needed as well. The methods of employment are rehearsed in accordance with the unit SOP.

Single Firing

D-42. A single Soldier with one SLM may engage an armored vehicle, but this is not the preferred method of employment. Several SLM normally are required to kill an armored vehicle. A single gunner firing one round must hit a vital part of the target in order to do damage. (See figure D-13.) A single shooter can engage targets out to 225 meters with the LAW, or 300 meters with the M136 AT4 (when the actual range is known).

Figure D-13. Shoulder-launched munitions single firing

Sequence Firing

D-43. A single shooter is employed, equipped with two or more SLMs prepared for firing, to engage a target. After engaging with the first round and observing the impact, the shooter adjusts the point of aim. The shooter then engages with another round until the target is destroy or all rounds are expended. (See figure D-14.)

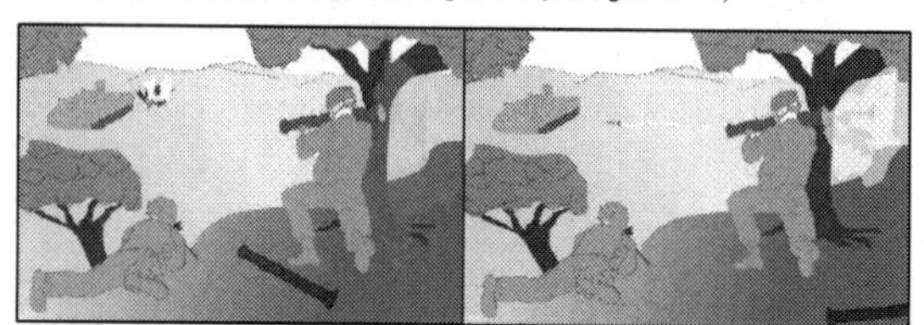

Figure D-14. Shoulder-launched munitions sequence firing

Pair Firing

D-44. Two or more shooters, equipped with two or more SLMs prepared for firing, are employed to engage a single target. Before firing, the first shooter informs the others of the estimated speed and distance to the target. If the impact of the first shooter's round proves to be correct, the other shooters engage the target until it is destroyed. If the impact of the first shooter's round proves to be incorrect, the second shooter informs the

others shooter of the estimate and engages the target. This continues until the target is destroyed or all rounds are expended. (See figure D-15).

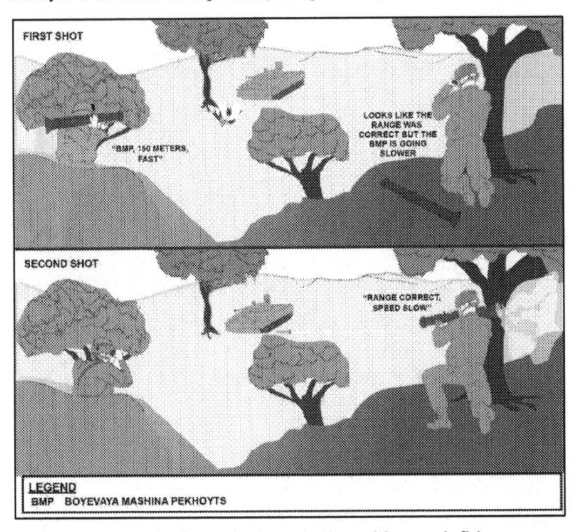

Figure D-15. Shoulder-launched munitions pair firing

Volley Firing

D-45. Volley firing involves two or more firers employed against a single target (when the range is known) at the same time on a prearranged signal (such as a command, whistle, mine, or TRP). Volley firing can be the most effective means of employment, as it places the most possible rounds on one target at one time, increasing the possibility of a kill. (See figure D-16.)

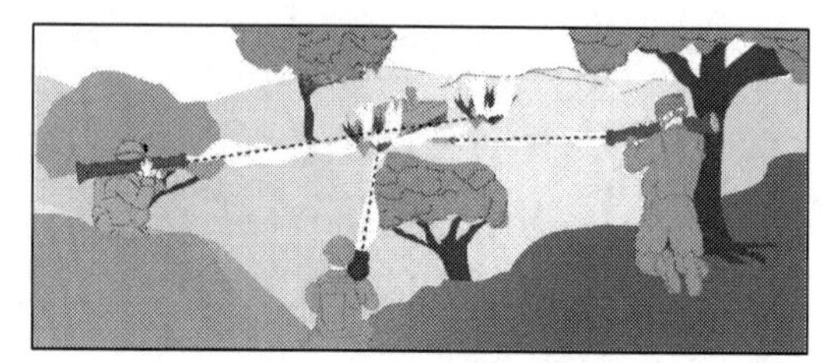

Figure D-16. Shoulder-launched munitions volley firing

ANTIARMOR ROLE (JAVELIN-CLOSE COMBAT MISSILE SYSTEM)

D-46. The platoon conducts an antiarmor ambush (see paragraphs 7-176 to 7-182 for additional information) to destroy small groups of armored vehicles, force the enemy to move slowly and cautiously, or force the enemy into a choke point. The Javelin's 2,000-meter range allows flexibility in choosing ambush positions. Leaders must understand with the Javelin's slow rate of fire, while the Javelin gunners attach the CLU to new rounds, other weapon systems must be prepared to engage the vehicle(s). In addition to fires into the kill zones (see figure 7-20 on page 7-53), the Javelin can be employed in a security role to guard high-speed avenues of approach, to slow or stop enemy reinforcements, or to destroy vehicles attempting to flee the kill zone.

D-47. In the offense, the Javelin CCMS contributes by providing long-range fires to destroy enemy armor and by protecting the platoon from armored counterattacks. In the absence of armored targets, the Javelin can engage enemy fortifications and hovering helicopters. Javelins are normally used in a support by fire or an attack by fire role during the offense. The primary consideration for such employment is the availability of appropriate fields of fire and armored threat. Javelin teams can protect flanks against armored threats and can provide overwatch for the platoon's movement. In the offense, once the objective is secured, leaders must rapidly position Javelin CCMS and other available SLM to defend against enemy counterattack.

D-48. During the defense, the platoon leader considers the enemy armor threat, then positions antiarmor weapons accordingly to cover armor avenues of approach. The leader considers the fields of fire, tracking time, and minimum engagement distance of each weapon and selects a primary position and sector of fire for each antiarmor weapon. The leader then picks alternate and supplementary positions for each system. Each position should allow flank fire and have cover and concealment. The leader integrates the CLU into the platoon's limited visibility security and observation plan.

D-49. The platoon leader selects Javelin fighting positions and assigns their sectors of fire. Considering the fundamentals of antiarmor employment will improve the team's survivability greatly in the defense. The weapons squad leader and the Javelin team

coordinate with adjacent units to ensure security. The Javelin's maximum effective engagement range (direct attack and top attack modes): 2,000 meters for the M98A1 and 2,500 meters for the M98A2 makes it difficult for the enemy to engage the team with direct fire, which forces the enemy to deploy earlier than intended. The gunner prepares a standard range card (TC 3-22.37 outlines the process for preparing the standard range card for the Javelin) for the system's primary position. If time permits, the gunner also prepares standard range cards for the system's alternate and supplementary positions. (See table D-12 for Javelin team duties and responsibilities.)

Note. The duties and responsibilities stated in table D-12 also applied when the M3 MAAWS is used in place of the Javelin.

Table D-12. Javelin team duties and responsibilities

TASKS TO BE PERFORMED	*Weapons Squad Leader*	*JG*	*AH*
Integrate Javelins into the platoon tactical plan:			
• Select general weapons positions.	X		
• Assign sectors of fires.	X		
• Coordinate mutual support.	X	X	
• Coordinate with adjacent units.	X	X	X
Positions (primary, alternate, and supplementary) and routes between positions.	X	X	
Supervise continual preparation and improvement of positions.	X	X	
Coordinate security of the Javelin teams.	X	X	
Confirm or make adjustments.	X	X	
Supervise preparation of DA Form 5517 (*Standard Range Card*).	X	X	
Control movement of gunners between positions.	X	X	
Issue fire commands to gunners.	X	X	
Coordinate resupply and collection of extra rounds carried in platoon.	X	X	
Identify enemy avenues of approach.	X	X	
Prepare fighting position (primary, alternate, supplementary).		X	X
Prepare standard range card.		X	X
Designate target reference point.	X	X	
Pre-stock rounds.		X	X
Prepare round for firing.		X	X
React to fire commands.		X	X
Engage targets.		X	
Legend: AH–ammunition handler; DA–Department of Army; JG–Javelin gunner			

D-50. Reserve forces armed with the Javelin (and to a lesser degree SLMs) may be employed to assist counterattacks to regain essential positions. They are used to block enemy penetrations, to meet unexpected enemy thrusts, and to provide support by fire

to endangered friendly units during disengagements and withdrawals. In the event defensive positions are in danger of being overrun by enemy armored vehicles, they may be used against armored vehicles and lightly armored vehicles posing an immediate threat. The maximum range provides the platoon leader with greater flexibility in positioning each round and provides a means of achieving overlapping sectors of fire for increased survivability.

ANTIPERSONNEL AND ANTIARMOR ROLE (M3 MULTI-ROLE, ANTIARMOR, ANTIPERSONNEL WEAPON SYSTEM)

D-51. The primary role of the M3 MAAWS gunner is to engage a target with a well-aimed shot. Consistently hitting a target with precision is a complex interaction of factors immediately before, during, and after firing the round. The interactions include maintaining postural steadiness, establishing and maintaining the proper aim on the target, stabilizing the weapon while pressing the trigger, and adjusting for environmental and battlefield conditions. (See TC 3-22.84 for additional information.)

Shot Process

D-52. The shot process is the basic outline of an individual engagement sequence all firers consider during an engagement, regardless of the weapon employed. The shot process formulates all decisions, calculations, and actions that lead to taking the shot. Soldiers may interrupt the shot process at any point before the trigger sear disengages and fires the weapon should the situation change. The shot process has three distinct phases: pre-shot, shot, and post-shot.

D-53. Soldiers must understand and correctly apply the shot process to achieve consistent, accurate, well-aimed shots. The sequence of the shot process does not change; however, the application of each element varies based on the conditions of the engagement. Every shot that the Soldier fires has a complete shot process. The shot process allows the Soldier to focus on one cognitive task at a time. The Soldier must maintain the ability to mentally organize the shot process's tasks and actions into a disciplined mental checklist, and focus their attention on activities, which produce the desired outcome—a well-aimed shot.

Types of Ammunition

D-54. Ammunition for use in the M3 MAAWS is described as a cartridge. A cartridge is an assembly consisting of a cartridge case, a percussion cap, a quantity of propellant, a projectile, and a base plate. The following terminology describes the general components (see figure D-17 on page D-28) of an M3 MAAWS round:

- Cartridge case. The cartridge case is made of light alloy that holds the other components of the cartridge.
- Propellant. The propellant (or powder) provides the energy to propel the projectile through the barrel and downrange towards a target through combustion.

- Percussion cap. The percussion cap is a small explosive charge that provides an ignition source for the propellant.
- Projectile. The projectile is the component that travels to the target.
- Base plate. The rear portion of the cartridge. Upon firing, the case base and seal assembly provide rear obturation.

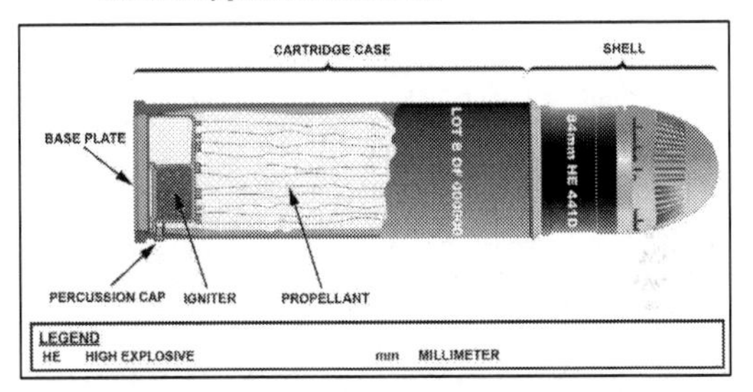

Figure D-17. General components of a cartridge

84-millimeter High Explosive 441D Reduced Sensitivity Cartridge

D-55. The 84-mm HE 441D RS round is intended for use against troops in the open, behind cover, or in trenches as well as soft-skinned transport vehicles, and similar types of targets. The cartridge is subject to firing limitations. See table D-8 on page D-17 and table D-9 on page D-18 or the operator's manual (TM 9-1015-262-10) for information on firing considerations and the maximum allowable quantity of cartridges that may be fired per Soldier per 24-hour period. Additional firing considerations include:

- Do not engage targets closer than 250 meters (820 feet).
- Use extreme caution when firing within 300 meters (984 feet) of friendly troops.
- Teams shall be trained not to engage targets closer than minimum engagement distance because fragmentation may cause injury or death.
- Teams shall use extreme caution when firing within these distances of adjacent personnel or friendly troops.
- Distances are measured from intended target or point of detonation.

84-millimeter Area Deterrent Munition 401 (Variants) Cartridge

D-56. The 84-mm area deterrent munition 401 is intended for area protection; used as an antipersonnel munition for close in protection in tight conditions of jungle or urban warfare. The area deterrent munition 401 round will spread approximately 1,100 flechettes against the target area. At a distance of 100 meters and a target of (h x w) 2 x 7 meters divided into 14 square meters, at least 10 of the squares will be fully penetrated.

The 84-mm, area deterrent munition 401B round contains approximately 300 ball bearings with an effective range of 200 meters. Both cartridges are subject to firing limitations. See table D-8 on page D-17 and table D-9 on page D-18 or the operator's manual (TM 9-1015-262-10) for information on the maximum allowable quantity of cartridges that may be fired per Soldier per 24-hour period. Additional firing considerations include:

- Do not engage targets closer than 150 meters (492 feet).
- Use extreme caution when firing within 330 meters (1,083 feet) of friendly troops.
- Distances are measured from intended target or point of detonation.

84-millimeter High Explosive Dual Purpose 502 Reduced Sensitivity Cartridge

D-57. The 84-mm HEDP 502 RS is a high explosive (HE), dual purpose round for use against armored personnel carriers and field fortifications. The round has an impact (I) mode and a delay (D) mode. The impact mode provides HEAT capability against armored personnel carriers and reinforced concrete walls. The delay mode provides delay burst capability to achieve maximum lethality against armored vehicles and delay action for bursting inside field fortifications and buildings. The cartridge is subject to firing limitations. See table D-8 on page D-17 and table D-9 on page D-18 or the operator's manual (TM 9-1015-262-10) for information on firing considerations and the maximum allowable quantity of cartridges that may be fired per Soldier per 24-hour period. Additional firing considerations include:

- Do not engage targets closer than 183 meters (600 feet).
- Use extreme caution when firing with 330 meters (1,083 feet) of friendly troops.
- Teams shall be trained not to engage targets closer than safe separation distance because fragmentation may cause injury or death.
- Teams shall use extreme caution when firing within these distances of adjacent personnel and friendly troops.
- Distances are measured from intended target or point of detonation.

84-millimeter High Explosive Antitank 551 Reduced Sensitivity Cartridge

D-58. The 84-mm HEAT 551 RS round is intended for use against all types of armored fighting vehicles with armor less than 16 inches thick including those fitted with protective devices such as skirting plates, grids, and other devices. It is also effective against concrete bunkers, landing craft, and similar hard targets. In addition to its great penetrating power, the fragments of the projectile body have a high lethal effect on personnel near the target. The design of the fuze system makes it possible to fire the projectile through brush and scrub without the projectile initiating. The cartridge is subject to firing limitations. See table D-8 on page D-17 and table D-9 on page D-18 or the operator's manual (TM 9-1015-262-10) for information on firing considerations and the maximum allowable quantity of cartridges that may be fired per Soldier per 24-hour period. Additional firing considerations include:

- Do not engage targets closer than 85 meters (279 feet).

- Use extreme caution when firing within 85 meters (984 feet) of friendly troops.
- Teams shall be trained not to engage targets closer than minimum engagement distance because fragmentation may cause injury or death.
- Teams shall use extreme caution when firing within these distances of adjacent personnel or friendly troops.
- Distances are measured from intended target or point of detonation.

84-millimeter Antistructure Munition 509 Cartridge

D-59. The 84-mm antistructure munition 509 is an antistructure munition that provides a defeat capability against masonry walls, parapets made of bricks and light concrete, and light armored vehicles. The fuze system has an impact (I) mode and a delay (D) mode. The impact mode defeats masonry walls and light, armored vehicles. The delay mode defeats enemy inside structures and field fortifications as well as any enemy behind masonry walls when they are close to the impact point. The cartridge is subject to firing limitations. See table D-8 on page D-17 and table D-9 on page D-18 or the operator's manual (TM 9-1015-262-10) for information on firing considerations and the maximum allowable quantity of cartridges that may be fired per Soldier per 24-hour period. Additional firing considerations include:

- Do not engage targets closer than 12 meters (39 feet).
- Use extreme caution when firing with 180 meters (591 feet) of friendly troops.
- Teams shall be trained not to engage targets closer than safe separation distance because fragmentation may cause injury of death.
- Teams shall use extreme caution when firing within these distances of adjacent personnel and friendly troops.
- Distances are measured from intended target or point of detonation.

84-millimeter Illuminator 545C Cartridge

D-60. The 84-mm illuminator 545C is designed to meet the requirement for a very quick illumination of target areas, offering facilities for all types of direct firing weapons and guided antitank weapons to engage armored fighting vehicles including support and other weapons. The illuminator 545C round is also intended to facilitate for the subunits of a battalion to supply their own illumination of a battlefield, even continuously, when required. The cartridge is subject to firing limitations. See table D-8 on page D-17 and table D-9 on page D-18 or the operator's manual (TM 9-1015-262-10) for information on firing considerations and the maximum allowable quantity of cartridges that may be fired per Soldier per 24-hour period. Teams must train not to engage targets closer than 150 meters when firing illumination rounds in impact mode. Firing the illuminator 545C from the prone position is limited to a horizontal firing elevation (0 degrees). Failure to comply with this warning could result in injury or death to personnel. Seek immediate medical attention if injury occurs.

84-millimeter Multi-target 756 Cartridge

D-61. The 84-mm multi-target 756 is intended mainly for combat in urban areas and is therefore, a wall piercing round. The projectile is provided with a tandem warhead, which gives good effect in different types of targets as 8 inches reinforced concrete walls, 12 inches triple brick walls and earth and timber bunker. The tandem warhead consists of precursor and follow-through-charge. When hitting the wall, the precursor detonates and creates a hole in the wall. The follow-through charge passes through the hole, detonating behind the wall distributing steel fragments and creating a blast effect. The cartridge is subject to firing limitations. See the operator's manual (TM 9-1015-262-10) for information on firing considerations and the maximum allowable quantity of cartridges that may be fired per Soldier per 24-hour period. Additional firing considerations include:

- Do not engage targets closer than 77 meters (252 feet).
- Use extreme caution when firing within 115 meters (377 feet) of friendly troops.
- Teams shall be trained not to engage targets closer than safe separation distance because fragmentation may cause injury of death.
- Teams shall use extreme caution when firing within these distances of adjacent personnel/friendly troops.
- Distances are measured from intended target or point of detonation.

84-millimeter Smoke 469C Cartridge

D-62. The 84-mm smoke 469C is intended for tactical use on the battlefield to obscure direct-fire weapons such as supporting tanks, self-propelled artillery, armored fighting vehicles, machine guns, and so forth. On impact, a smoke screen with a 10- to 15-meter width and with good screening effect is instantly obtained. Using this round enables the subunit to lay a smoke screen rapidly when required. Use this round in the combat situations, to include blinding, launch round directly at target; screening, between enemy and friendly position; and marking, show position of target to artillery or CAS. The cartridge is subject to firing limitations. See table D-8 on page D-17 and table D-9 on page D-18 or the operator's manual (TM 9-1015-262-10) for information on firing considerations and the maximum allowable quantity of cartridges that may be fired per Soldier per 24-hour period.

Appendix E
Battle Drills

Infantry rifle platoons and squads undergo extensive training to conduct combat operations in all operational environments. In preparation for these operations, battle drills are conducted to mitigate the risk of making contact with the enemy before maneuvering. *Battle drills* are rehearsed and well understood actions made in response to common battlefield occurrences (ADP 3-90).

OVERVIEW

E-1. Battle drills are tasks (individual and collective) designed to teach a Soldier or small unit to react and survive in common combat situations. A platoon, squad, team, or crew performs these actions when initiated by a predetermined cue (verbal or visual). Battle drills are performed instinctively; they require minimal leader direction and have little to no notice. When initiated they are considered vital to the success of the combat operation or critical to preserving life (see FM 7-0). A drill has the following advantages:

- It is based on unit missions and the specific tasks, conditions, standards, and performance measures required to support mission proficiency.
- It builds from simple to complex but focuses on the basics.
- It links how-to-train and how-to-fight at the small-unit level.
- It provides an agenda for continuous coaching and analyzing.
- It develops leaders and builds teamwork and cohesion under stress.
- It enhances the chance for individual and unit survival on the battlefield.

SELECTED BATTLE DRILLS

E-2. This section identifies essential battle drills that an Infantry rifle platoon and squad must train on to ensure success. (See the Army Training Network or the Central Army Registry websites to view selected battle drills.) They include:

- Battle Drill 1: React to Direct Fire Contact While Dismounted – Platoon (07-PLT-D9501).
- Battle Drill 1A: React to Direct Fire Contact While Dismounted – Squad (07-SQD-D9501).
- Battle Drill 2: Conduct a Platoon Assault (07-PLT-D9514).
- Battle Drill 2A: Conduct a Squad Assault (07-SQD-D9515).
- Battle Drill 3: Break Contact – Platoon (07-PLT-D9505).
- Battle Drill 3A: Break Contact – Squad (07-SQD-D9505).

- Battle Drill 4: React to Ambush (Dismounted) – Platoon (07-PLT-D9502).
- Battle Drill 4A: React to Ambush (Dismounted) – Squad (07-SQD-D9502).
- Battle Drill 5: Knock Out a Bunker – Squad (07-SQD-D9406).
- Battle Drill 6: Enter and Clear a Room – Squad (07-SQD-D9509).
- Battle Drill 7: Enter a Trench to Secure a Foothold – Platoon (07-PLT-D9510).
- Battle Drill 8: Breach of a Mined Wire Obstacle – Platoon (07-PLT-D9412).
- Battle Drill 9: React to Indirect Fire While Dismounted – Platoon (07-PLT-D9504).
- Battle Drill 9A: React to Indirect Fire While Dismounted – Squad (07-SQD-D9504).
- Battle Drill 10: React to Aircraft While Dismounted – Platoon (07-PLT-D8015).
- Battle Drill 11: Establish Security at the Halt – Platoon (07-PLT-D9508).
- Battle Drill 12: Conduct a Hasty Attack – Platoon (07-PLT-D9511).
- Battle Drill 13: Dismount a Vehicle Under Direct Fire – Squad (07-SQD-D9506).
- Battle Drill 14: React to Nuclear Attack – Platoon (07-PLT-D9483).
- Battle Drill 15: React to a Chemical Agent Attack – Platoon (03-PLT-D0071).

Glossary

The glossary lists acronyms and terms with Army or joint definitions. Where Army and joint definitions differ, (Army) precedes the definition. Terms for which ATP 3-21.8 is the proponent publication (the authority) are marked with an asterisk (*). The proponent publication for other terms is listed in parentheses after the definition.

SECTION I – ACRONYMS AND ABBREVIATION

AO	area of operations
AA	assembly area
ADP	Army doctrine publication
AFTTP	Air Force tactics, techniques, and procedures
APL	assistant patrol leader
ASCOPE	areas, structures, capabilities, organizations, people, and events
AT	antitank
ATP	Army techniques publication
ATTP	Army tactics, techniques, and procedures
BDM	bunker defeat munition
BEWL	biometric-enabled watchlist
BHL	battle handover line
BMP	Boyevaya Mashina Pekhotys
BRDM	Boyevaya Razvedyvatelnaya Dozornaya Mashina
BSA	brigade support area
BTR	bronyetransportyor
CAS	close air support
CASEVAC	casualty evacuation
CBRN	chemical, biological, radiological, and nuclear
CCIR	commander's critical information requirement
CCMS	close combat missile system

CCP	casualty collection point
CGTTP	Coast Guard tactics, techniques, and procedures
CLS	combat lifesaver
CLU	command launch unit
COA	course of action
CS	confined space
DA	Department of the Army
DA Pam	Department of Army pamphlet
DD	Department of Defense
DLIC	detachment left in contact
DOD	Department of Defense
DODD	Department of Defense Directive
DRAW-D	defend, reinforce, attack, withdraw, or delay
EMCON	emission control
EP	electromagnetic protection
EW	electromagnetic warfare
FAC (A)	forward air controller (airborne)
FIST	fire support team
FLOT	forward line of own troops
FM	field manual
FO	forward observer
FPF	final protective fire
FPL	final protective line
FRAGORD	fragmentary order
GOTWA	going, others, time, what, actions
GPS	Global Positioning System
GTA	graphic training aid
GTAO	graphic terrain analysis overlay
HE	high explosive
HEAT	high explosive antitank
HEDP	high explosive dual purpose
HOPE-LW	higher, operational, planning, enemy, light/weather
IBCT	Infantry brigade combat team
IPOE	intelligence preparation of the operational environment